STEAM TABLES

STEAM TABLES

Thermodynamic Properties of Water
Including
Vapor, Liquid, and Solid Phases

(English Units)

JOSEPH H. KEENAN, B.S., LL.D.

Professor of Mechanical Engineering Emeritus
Senior Lecturer
Massachusetts Institute of Technology

FREDERICK G. KEYES, B.S. Sc.D., Ph.D.

Professor of Physical Chemistry Emeritus
Lecturer in Chemistry
Massachusetts Institute of Technology

PHILIP G. HILL, B.Sc., Sc.D.

Professor of Mechanical Engineering
Queen's University, Kingston, Ontario

JOAN G. MOORE, B.S.

Member of Research Staff
Massachusetts Institute of Technology

A WILEY-INTERSCIENCE PUBLICATION

JOHN WILEY & SONS
New York • Chichester • Brisbane • Toronto • Singapore

Library of Congress Catalog Card Number: 68 – 54568

7 8 9 10

Printed in the United States of America

ISBN 471 46501 1

Preface

The tables by Keenan and Keyes of the properties of water, published in 1936, were based on critical correlation and formulation of all the experimental data available at that time. Included were the Keyes and Smith measurements of specific volume of the liquid; the Keyes, Smith, and Gerry measurements of specific volume of vapor; the Osborne, Stimson, Ginnings, and Fiock observations of saturation states; the enthalpy measurements of Callendar and Egerton and of Havlicek and Miskowsky, and the Gordon calculation from spectrographic data of the heat capacity at zero pressure. For specific volumes from 10 cm³/g to infinity the result was the p-v-T relation given by Equation 13 of the Keenan and Keyes tables and the formulation of the spectroscopic observations of specific heat capacity given by Gordon and expressed by Equation 15 in Keenan and Keyes. These two, taken together, were equivalent to a fundamental equation of the form $\zeta = \zeta(\mathrm{p},\ T)$, where ζ denotes specific Gibbs free energy, p pressure, and T temperature on the Kelvin scale. This fundamental equation can be stated explicitly and completely. It was completed before, and its results made available to, the Third International Conference on the Properties of Steam in 1934.

Values for the compressed liquid region in the Keenan and Keyes tables were taken from a semialgebraic, semigraphic study by Keenan in 1931. These values were also made available to the Third International Conference. Values for the narrow region of states between the critical temperature on the liquid side and a specific volume of 10 cm³/g were the subject of a subsequent algebraic-graphic study based largely on available enthalpy measurements.

Research activity in the U.S.S.R. early in the decade of the 1950's stimulated an international effort to extend experimental knowledge of the properties of water to higher pressures and temperatures than before. The results of this effort make possible a reliable reformulation of the properties of steam which not only extends the range of dependable knowledge but also permits refinement of knowledge of regions previously investigated. Moreover, the modern computer provides a correlation tool that is far more powerful than anything available 30 years ago.

With a background of formulation techniques, discussed by F. G. Keyes from time to time, an extensive examination and correlation of all available experi-

mental data on liquid and vapor water was undertaken at the Massachusetts
Institute of Technology about four years ago. Although the International Skeleton
Tables of 1963 were then available, it was felt that an independent study was
required in order to take full advantage of the new experimental information and
computational tools. Consequently the present tables are in no sense a formulation of the International Skeleton Tables. In general they are in good accord with
and their values fall within the tolerances of the International Skeleton Tables of
1963. The exceptions to this statement and their implications are discussed in
the appendix.

The result of this study was a fundamental equation of the form $\psi = \psi(T, v)$,
where ψ denotes specific Helmholtz free energy and v specific volume. This
equation represents a continuity of single-phase states which cover both liquid and
vapor regions to about 1400°C in temperature and 1000 bars in pressure and provides saturation states that agree in regard to Gibbs free energy. The agreement
with experimental observations, which is discussed in detail in the appendix, is
generally within the uncertainty in the observations. This fundamental equation
is a new formulation. It was used to generate all tabulated values of thermodynamic properties corresponding to liquid and vapor states.

Because the principal objective was a table of thermodynamic properties, no
similar independent study of transport properties, namely, viscosity and thermal
conductivity, was undertaken. A substantial amount of new experimental observations of these properties has appeared since publication of the Keenan and
Keyes tables. Much of it is of high quality. A committee of the Sixth International Conference on the Properties of Steam has studied all the available mate-
rial and has arrived at reasonable compromises where differences between
various sets of observations have seemed irreconcilable. Moreover, the International Skeleton Tables are, in this instance, sufficiently detailed for practical
purposes. They were therefore adopted without modification.

The present tables extend the range of the Keenan and Keyes tables to 2400°F
in temperature and to 15,000 psi in pressure. The changes made in values of
properties for states given in the Keenan and Keyes tables are generally small.
They are discussed in the appendix.

Extension of the range of temperature makes desirable a change from lines of
constant pressure to columns of constant pressure. Internal energies are given
for all states in view of their importance in the wider variety of problems encoun-
tered in practice and in teaching. The engineering importance of metastable
states is recognized by including a substantial band of them on both the liquid
side and the vapor side of the two-phase region. A more convenient table of properties for compressed and metastable liquid is provided, along with a separate
and detailed table for states near the critical point.

The fundamental equation itself, as distinguished from the tables, will doubtless prove useful in the design of machinery and the analysis of engineering data.
The coefficients of the equation and its extension into expressions for the usual
thermodynamic properties are exhibited in the appendix.

In order to ensure accuracy, the typesetting of the table of thermodynamic
properties has been done by computer; that is, the numbers were generated from
the fundamental equation on an IBM 360, converted to Photon control codes,

and punched onto paper tape which was fed directly into a Photon typesetter to obtain the tables.

It is a pleasure to acknowledge indebtedness to many colleagues. The experimenters whose data were drawn upon in the development of the fundamental equation devoted untold time, skill, and ingenuity to providing a body of data that is unique in quality and extent. Thanks are gratefully extended to them in general. Thanks are due in particular, however, to those colleagues who have responded generously to requests for information; for example, G. S. Kell and E. Whalley transmitted preliminary data and E. J. LeFevre offered several helpful suggestions. Dr. Oscar Bridgeman of the U.S. National Bureau of Standards sent reports of his work before publication, and J. Kestin and E. A. Bruges reported frequently on their own and other work on transport properties. At one stage in the development of the fundamental equation some assistance was obtained from the Ford Foundation.

Thanks are due in greatest measure, however, to the Massachusetts Institute of Technology for support of three of the authors as members of its faculty and for the use of its extensive computer center. Glenn Peterson wrote the necessary routines for the IBM 360 computer that made the computer typesetting of the tables possible. The M.I.T. Instrumentation Laboratory provided the facilities for producing the paper tapes.

Joseph H. Keenan
Frederick G. Keyes
Philip G. Hill
Joan G. Moore

Cambridge, Massachusetts
January 1969

Contents

STEAM TABLES

Symbols Used in Tables

c_{po} specific heat capacity at constant pressure for zero pressure, Btu* per lb degree Fahr.

c_{vo} specific heat capacity at constant volume for zero pressure, Btu per lb degree Fahr.

h specific enthalpy, Btu per lb.

h_o specific enthalpy at zero pressure, Btu per lb.

k isentropic exponent,$-(\partial \log p/\partial \log v)_s$.

p   pressure, lbf per in².

p_r relative pressure, pressure of semiperfect vapor at zero entropy multiplied by 10^{-7}, lbf per in².

s specific entropy, Btu per lb degree Rankine.

s_1 specific entropy of semiperfect vapor at 1 lbf per in², Btu per lb degree Rankine.

t thermodynamic temperature, degrees Fahrenheit.

T thermodynamic temperature, degrees Rankine.

u specific internal energy, Btu per lb.

u_o specific internal energy at zero pressure, Btu per lb.

v specific volume, ft³ per lb.

v_r relative specific volume, specific volume of semiperfect vapor at zero entropy multiplied by 10^7, ft³ per lb.

ζ_1 specific Gibbs free energy of semiperfect vapor at 1 lbf per in², Btu per lb.

ψ_1 specific Helmholtz free energy of semiperfect vapor at 1 lbf per in², Btu per lb.

*Units are defined and compared in Table 10.

SUBSCRIPTS

f refers to a property of liquid in equilibrium with vapor
g refers to a property of vapor in equilibrium with liquid
i refers to a property of solid in equilibrium with vapor
fg refers to a change by evaporation
ig refers to a change by sublimation

Table 1. Saturation: Temperatures

Temp Fahr.	Press. Lbf. Sq.In.	Specific Volume		Internal Energy			Enthalpy			Entropy		
		Sat. Liquid	Sat. Vapor	Sat. Liquid	Evap.	Sat. Vapor	Sat. Liquid	Evap.	Sat. Vapor	Sat. Liquid	Evap.	Sat. Vapor
t	p	v_f	v_g	u_f	u_{fg}	u_g	h_f	h_{fg}	h_g	s_f	s_{fg}	s_g
32	.08859	.016022	3305.	-.01	1021.2	1021.2	-.01	1075.4	1075.4	-.00003	2.1870	2.1870
32.018	.08866	.016022	3302.	.00	1021.2	1021.2	.01	1075.4	1075.4	.00000	2.1869	2.1869
33	.09223	.016022	3181.	.99	1020.6	1021.5	.99	1074.8	1075.8	.00201	2.1815	2.1835
34	.09601	.016021	3062.	1.99	1019.9	1021.9	1.99	1074.3	1076.3	.00404	2.1759	2.1799
35	.09992	.016021	2948.	2.99	1019.2	1022.2	3.00	1073.7	1076.7	.00607	2.1704	2.1764
36	.10397	.016021	2839.	4.00	1018.5	1022.5	4.00	1073.1	1077.1	.00810	2.1648	2.1729
37	.10816	.016021	2734.	5.00	1017.9	1022.9	5.00	1072.6	1077.6	.01012	2.1594	2.1695
38	.11250	.016020	2634.	6.01	1017.2	1023.2	6.01	1072.0	1078.0	.01215	2.1539	2.1660
39	.11700	.016020	2538.	7.01	1016.5	1023.5	7.01	1071.5	1078.5	.01416	2.1484	2.1626
40	.12166	.016020	2445.	8.02	1015.8	1023.9	8.02	1070.9	1078.9	.01617	2.1430	2.1592
41	.12648	.016020	2357.	9.02	1015.2	1024.2	9.02	1070.3	1079.3	.01818	2.1376	2.1558
42	.13146	.016021	2272.	10.02	1014.5	1024.5	10.03	1069.8	1079.8	.02018	2.1322	2.1524
43	.13662	.016021	2190.	11.03	1013.8	1024.8	11.03	1069.2	1080.2	.02218	2.1268	2.1490
44	.14196	.016021	2112.	12.03	1013.1	1025.2	12.03	1068.6	1080.7	.02418	2.1215	2.1457
45	.14748	.016021	2037.	13.04	1012.5	1025.5	13.04	1068.1	1081.1	.02618	2.1162	2.1423
46	.15319	.016022	1965.	14.04	1011.8	1025.8	14.04	1067.5	1081.5	.02815	2.1109	2.1390
47	.15909	.016022	1896.	15.04	1011.1	1026.2	15.05	1066.9	1082.0	.03014	2.1056	2.1357
48	.16520	.016023	1829.	16.05	1010.4	1026.5	16.05	1066.4	1082.4	.03212	2.1003	2.1324
49	.17151	.016024	1766.	17.05	1009.8	1026.8	17.05	1065.8	1082.9	.03410	2.0951	2.1292
50	.17803	.016024	1704.2	18.06	1009.1	1027.2	18.06	1065.2	1083.3	.03607	2.0899	2.1259
51	.18477	.016025	1645.2	19.06	1008.4	1027.5	19.06	1064.7	1083.7	.03804	2.0847	2.1227
52	.19173	.016026	1588.6	20.06	1007.8	1027.8	20.07	1064.1	1084.2	.04000	2.0795	2.1195
53	.19892	.016027	1534.1	21.06	1007.1	1028.1	21.07	1063.5	1084.6	.04195	2.0743	2.1163
54	.20635	.016028	1481.7	22.07	1006.4	1028.5	22.07	1063.0	1085.1	.04391	2.0692	2.1131
55	.2140	.016029	1431.4	23.07	1005.7	1028.8	23.07	1062.4	1085.5	.04586	2.0641	2.1099
56	.2219	.016030	1382.9	24.08	1005.1	1029.1	24.08	1061.9	1085.9	.04781	2.0590	2.1068
57	.2301	.016031	1336.3	25.08	1004.4	1029.5	25.08	1061.3	1086.4	.04975	2.0539	2.1036
58	.2386	.016032	1291.5	26.08	1003.7	1029.8	26.08	1060.7	1086.8	.05169	2.0488	2.1005
59	.2473	.016033	1248.4	27.08	1003.0	1030.1	27.08	1060.2	1087.2	.05362	2.0438	2.0974
60	.2563	.016035	1206.9	28.08	1002.4	1030.4	28.08	1059.6	1087.7	.05555	2.0388	2.0943
61	.2655	.016036	1166.9	29.08	1001.7	1030.8	29.09	1059.0	1088.1	.05747	2.0338	2.0912
62	.2751	.016037	1128.5	30.09	1001.0	1031.1	30.09	1058.5	1088.6	.05940	2.0288	2.0882
63	.2850	.016039	1091.4	31.09	1000.3	1031.4	31.09	1057.9	1089.0	.06131	2.0238	2.0851
64	.2952	.016040	1055.8	32.09	999.7	1031.8	32.09	1057.3	1089.4	.06323	2.0189	2.0821
65	.3057	.016042	1021.5	33.09	999.0	1032.1	33.09	1056.8	1089.9	.06514	2.0140	2.0791
66	.3165	.016044	988.4	34.09	998.3	1032.4	34.09	1056.2	1090.3	.06704	2.0091	2.0761
67	.3276	.016045	956.5	35.09	997.7	1032.7	35.09	1055.6	1090.7	.06895	2.0042	2.0731
68	.3391	.016047	925.8	36.09	997.0	1033.1	36.09	1055.1	1091.2	.07084	1.9993	2.0701
69	.3510	.016049	896.2	37.09	996.3	1033.4	37.09	1054.5	1091.6	.07273	1.9944	2.0672
70	.3632	.016051	867.7	38.09	995.6	1033.7	38.09	1054.0	1092.0	.07463	1.9896	2.0642
71	.3758	.016053	840.2	39.09	995.0	1034.1	39.09	1053.4	1092.5	.07651	1.9848	2.0613
72	.3887	.016055	813.7	40.09	994.3	1034.4	40.09	1052.8	1092.9	.07839	1.9800	2.0584
73	.4021	.016057	788.1	41.09	993.6	1034.7	41.09	1052.3	1093.4	.08028	1.9752	2.0555
74	.4158	.016059	763.5	42.09	992.9	1035.0	42.09	1051.7	1093.8	.08215	1.9705	2.0526
75	.4300	.016061	739.7	43.09	992.3	1035.4	43.09	1051.1	1094.2	.08402	1.9657	2.0497
76	.4446	.016063	716.8	44.09	991.6	1035.7	44.09	1050.6	1094.7	.08589	1.9610	2.0469
77	.4596	.016065	694.6	45.09	990.9	1036.0	45.09	1050.0	1095.1	.08775	1.9563	2.0440
78	.4750	.016068	673.3	46.09	990.3	1036.3	46.09	1049.4	1095.5	.08961	1.9516	2.0412
79	.4909	.016070	652.7	47.09	989.6	1036.7	47.09	1048.9	1096.0	.09146	1.9469	2.0384
80	.5073	.016073	632.8	48.08	988.9	1037.0	48.09	1048.3	1096.4	.09332	1.9423	2.0356
81	.5241	.016075	613.5	49.08	988.2	1037.3	49.09	1047.7	1096.8	.09517	1.9376	2.0328
82	.5414	.016077	595.0	50.08	987.6	1037.6	50.08	1047.2	1097.3	.09701	1.9330	2.0300
83	.5593	.016080	577.1	51.08	986.9	1038.0	51.08	1046.6	1097.7	.09885	1.9284	2.0273
84	.5776	.016083	559.8	52.08	986.2	1038.3	52.08	1046.0	1098.1	.10069	1.9238	2.0245
85	.5964	.016085	543.1	53.08	985.5	1038.6	53.08	1045.5	1098.6	.10252	1.9193	2.0218
86	.6158	.016088	527.0	54.08	984.9	1038.9	54.08	1044.9	1099.0	.10436	1.9147	2.0190
87	.6357	.016091	511.4	55.07	984.2	1039.3	55.08	1044.4	1099.4	.10618	1.9102	2.0163
88	.6562	.016094	496.3	56.07	983.5	1039.6	56.07	1043.8	1099.9	.10801	1.9056	2.0136
89	.6772	.016096	481.7	57.07	982.9	1039.9	57.07	1043.2	1100.3	.10983	1.9011	2.0110

Temp Fahr.	Press. Lbf. Sq.In.	Specific Volume		Internal Energy			Enthalpy			Entropy		
		Sat. Liquid	Sat. Vapor	Sat. Liquid	Evap.	Sat. Vapor	Sat. Liquid	Evap.	Sat. Vapor	Sat. Liquid	Evap.	Sat. Vapor
t	p	v_f	v_g	u_f	u_{fg}	u_g	h_f	h_{fg}	h_g	s_f	s_{fg}	s_g
90	.6988	.016099	467.7	58.07	982.2	1040.2	58.07	1042.7	1100.7	.11165	1.8966	2.0083
91	.7211	.016102	454.0	59.06	981.5	1040.6	59.07	1042.1	1101.2	.11346	1.8922	2.0056
92	.7439	.016105	440.9	60.06	980.8	1040.9	60.06	1041.5	1101.6	.11527	1.8877	2.0030
93	.7674	.016108	428.2	61.06	980.2	1041.2	61.06	1041.0	1102.0	.11708	1.8833	2.0003
94	.7914	.016111	415.9	62.06	979.5	1041.5	62.06	1040.4	1102.4	.11888	1.8788	1.9977
95	.8162	.016114	404.0	63.06	978.8	1041.9	63.06	1039.8	1102.9	.12068	1.8744	1.9951
96	.8416	.016117	392.4	64.05	978.1	1042.2	64.06	1039.2	1103.3	.12248	1.8700	1.9925
97	.8677	.016121	381.3	65.05	977.5	1042.5	65.05	1038.7	1103.7	.12427	1.8657	1.9899
98	.8945	.016124	370.5	66.05	976.8	1042.8	66.05	1038.1	1104.2	.12606	1.8613	1.9874
99	.9220	.016127	360.1	67.05	976.1	1043.2	67.05	1037.5	1104.6	.12785	1.8569	1.9848
100	.9503	.016130	350.0	68.04	975.4	1043.5	68.05	1037.0	1105.0	.12963	1.8526	1.9822
101	.9792	.016134	340.2	69.04	974.8	1043.8	69.04	1036.4	1105.5	.13141	1.8483	1.9797
102	1.0090	.016137	330.8	70.04	974.1	1044.1	70.04	1035.8	1105.9	.13319	1.8440	1.9772
103	1.0395	.016141	321.6	71.04	973.4	1044.4	71.04	1035.3	1106.3	.13496	1.8397	1.9747
104	1.0708	.016144	312.8	72.03	972.7	1044.8	72.04	1034.7	1106.7	.13674	1.8354	1.9722
105	1.1029	.016148	304.2	73.03	972.1	1045.1	73.03	1034.1	1107.2	.13850	1.8312	1.9697
106	1.1359	.016151	295.9	74.03	971.4	1045.4	74.03	1033.6	1107.6	.14027	1.8269	1.9672
107	1.1697	.016155	287.8	75.03	970.7	1045.7	75.03	1033.0	1108.0	.14203	1.8227	1.9647
108	1.2044	.016158	280.0	76.02	970.0	1046.0	76.03	1032.4	1108.4	.14379	1.8185	1.9623
109	1.2399	.016162	272.4	77.02	969.3	1046.4	77.02	1031.9	1108.9	.14554	1.8143	1.9598
110	1.2763	.016166	265.1	78.02	968.7	1046.7	78.02	1031.3	1109.3	.14730	1.8101	1.9574
111	1.3137	.016169	258.0	79.01	968.0	1047.0	79.02	1030.7	1109.7	.14905	1.8059	1.9549
112	1.3520	.016173	251.1	80.01	967.3	1047.3	80.02	1030.1	1110.2	.15079	1.8017	1.9525
113	1.3913	.016177	244.4	81.01	966.6	1047.6	81.02	1029.6	1110.6	.15254	1.7976	1.9501
114	1.4315	.016181	238.0	82.01	966.0	1048.0	82.01	1029.0	1111.0	.15427	1.7935	1.9477
115	1.4727	.016185	231.7	83.00	965.3	1048.3	83.01	1028.4	1111.4	.15601	1.7894	1.9454
116	1.5150	.016189	225.6	84.00	964.6	1048.6	84.01	1027.8	1111.9	.15775	1.7852	1.9430
117	1.5583	.016193	219.7	85.00	963.9	1048.9	85.01	1027.3	1112.3	.15948	1.7812	1.9406
118	1.6026	.016197	214.0	86.00	963.2	1049.2	86.00	1026.7	1112.7	.16121	1.7771	1.9383
119	1.6480	.016201	208.4	86.99	962.6	1049.6	87.00	1026.1	1113.1	.16293	1.7730	1.9359
120	1.6945	.016205	203.0	87.99	961.9	1049.9	88.00	1025.5	1113.5	.16465	1.7690	1.9336
121	1.7422	.016209	197.8	88.99	961.2	1050.2	89.00	1025.0	1114.0	.16637	1.7649	1.9313
122	1.7910	.016213	192.7	89.99	960.5	1050.5	89.99	1024.4	1114.4	.16809	1.7609	1.9290
123	1.8409	.016217	187.8	90.99	959.8	1050.8	90.99	1023.8	1114.8	.16980	1.7569	1.9267
124	1.8921	.016221	183.1	91.98	959.2	1051.1	91.99	1023.2	1115.2	.17152	1.7529	1.9244
125	1.9444	.016226	178.41	92.98	958.5	1051.5	92.99	1022.7	1115.7	.17322	1.7489	1.9221
126	1.9980	.016230	173.91	93.98	957.8	1051.8	93.99	1022.1	1116.1	.17493	1.7449	1.9199
127	2.0529	.016234	169.54	94.98	957.1	1052.1	94.98	1021.5	1116.5	.17663	1.7410	1.9176
128	2.1090	.016239	165.30	95.97	956.4	1052.4	95.98	1020.9	1116.9	.17833	1.7370	1.9154
129	2.1664	.016243	161.18	96.97	955.7	1052.7	96.98	1020.4	1117.3	.18003	1.7331	1.9131
130	2.225	.016247	157.17	97.97	955.1	1053.0	97.98	1019.8	1117.8	.18172	1.7292	1.9109
131	2.285	.016252	153.28	98.97	954.4	1053.3	98.98	1019.2	1118.2	.18341	1.7253	1.9087
132	2.347	.016256	149.51	99.97	953.7	1053.7	99.98	1018.6	1118.6	.18510	1.7214	1.9065
133	2.410	.016261	145.84	100.97	953.0	1054.0	100.97	1018.0	1119.0	.18679	1.7175	1.9043
134	2.474	.016265	142.27	101.96	952.3	1054.3	101.97	1017.5	1119.4	.18847	1.7136	1.9021
135	2.540	.016270	138.81	102.96	951.6	1054.6	102.97	1016.9	1119.8	.19015	1.7098	1.8999
136	2.607	.016275	135.44	103.96	951.0	1054.9	103.97	1016.3	1120.3	.19183	1.7059	1.8978
137	2.676	.016279	132.17	104.96	950.3	1055.2	104.97	1015.7	1120.7	.19350	1.7021	1.8956
138	2.746	.016284	128.98	105.96	949.6	1055.5	105.97	1015.1	1121.1	.19518	1.6983	1.8934
139	2.818	.016289	125.89	106.96	948.9	1055.9	106.97	1014.5	1121.5	.19685	1.6945	1.8913
140	2.892	.016293	122.88	107.95	948.2	1056.2	107.96	1014.0	1121.9	.19851	1.6907	1.8892
141	2.967	.016298	119.96	108.95	947.5	1056.5	108.96	1013.4	1122.3	.20017	1.6869	1.8871
142	3.044	.016303	117.12	109.95	946.8	1056.8	109.96	1012.8	1122.8	.20184	1.6831	1.8849
143	3.122	.016308	114.35	110.95	946.1	1057.1	110.96	1012.2	1123.2	.20350	1.6793	1.8828
144	3.203	.016313	111.66	111.95	945.5	1057.4	111.96	1011.6	1123.6	.20515	1.6756	1.8807

Table 1.   Saturation: Temperatures

Temp Fahr.	Press. Lbf. Sq.In.	Specific Volume		Internal Energy			Enthalpy			Entropy		
		Sat. Liquid	Sat. Vapor	Sat. Liquid	Evap.	Sat. Vapor	Sat. Liquid	Evap.	Sat. Vapor	Sat. Liquid	Evap.	Sat. Vapor
t	p	v_f	v_g	u_f	u_{fg}	u_g	h_f	h_{fg}	h_g	s_f	s_{fg}	s_g
145	3.285	.016318	109.04	112.95	944.8	1057.7	112.96	1011.0	1124.0	.20681	1.6718	1.8787
146	3.368	.016323	106.50	113.95	944.1	1058.0	113.96	1010.5	1124.4	.20846	1.6681	1.8766
147	3.454	.016328	104.02	114.95	943.4	1058.3	114.96	1009.9	1124.8	.21010	1.6644	1.8745
148	3.541	.016333	101.61	115.95	942.7	1058.6	115.96	1009.3	1125.2	.21175	1.6607	1.8725
149	3.630	.016338	99.27	116.95	942.0	1059.0	116.96	1008.7	1125.6	.21339	1.6570	1.8704
150	3.722	.016343	96.99	117.95	941.3	1059.3	117.96	1008.1	1126.1	.21503	1.6533	1.8684
151	3.815	.016348	94.76	118.94	940.6	1059.6	118.96	1007.5	1126.5	.21667	1.6497	1.8663
152	3.910	.016353	92.60	119.94	939.9	1059.9	119.96	1006.9	1126.9	.21831	1.6460	1.8643
153	4.007	.016358	90.50	120.94	939.2	1060.2	120.96	1006.3	1127.3	.21994	1.6423	1.8623
154	4.106	.016363	88.45	121.94	938.6	1060.5	121.96	1005.7	1127.7	.22157	1.6387	1.8603
155	4.207	.016369	86.45	122.94	937.9	1060.8	122.96	1005.1	1128.1	.22320	1.6351	1.8583
156	4.310	.016374	84.51	123.94	937.2	1061.1	123.96	1004.6	1128.5	.22482	1.6315	1.8563
157	4.416	.016379	82.62	124.94	936.5	1061.4	124.96	1004.0	1128.9	.22645	1.6279	1.8543
158	4.523	.016384	80.77	125.94	935.8	1061.7	125.96	1003.4	1129.3	.22807	1.6243	1.8523
159	4.633	.016390	78.98	126.94	935.1	1062.0	126.96	1002.8	1129.7	.22969	1.6207	1.8504
160	4.745	.016395	77.23	127.94	934.4	1062.3	127.96	1002.2	1130.1	.23130	1.6171	1.8484
161	4.859	.016400	75.53	128.94	933.7	1062.6	128.96	1001.6	1130.5	.23291	1.6135	1.8465
162	4.976	.016406	73.87	129.94	933.0	1062.9	129.96	1001.0	1131.0	.23452	1.6100	1.8445
163	5.095	.016411	72.25	130.94	932.3	1063.2	130.96	1000.4	1131.4	.23613	1.6064	1.8426
164	5.216	.016417	70.67	131.95	931.6	1063.5	131.96	999.8	1131.8	.23774	1.6029	1.8407
165	5.340	.016422	69.14	132.95	930.9	1063.8	132.96	999.2	1132.2	.23934	1.5994	1.8387
166	5.466	.016428	67.64	133.95	930.2	1064.2	133.96	998.6	1132.6	.24094	1.5959	1.8368
167	5.595	.016434	66.18	134.95	929.5	1064.5	134.97	998.0	1133.0	.24254	1.5924	1.8349
168	5.726	.016439	64.76	135.95	928.8	1064.8	135.97	997.4	1133.4	.24414	1.5889	1.8330
169	5.860	.016445	63.37	136.95	928.1	1065.1	136.97	996.8	1133.8	.24573	1.5854	1.8311
170	5.996	.016450	62.02	137.95	927.4	1065.4	137.97	996.2	1134.2	.24732	1.5819	1.8293
171	6.136	.016456	60.70	138.95	926.7	1065.7	138.97	995.6	1134.6	.24891	1.5785	1.8274
172	6.277	.016462	59.42	139.95	926.0	1066.0	139.97	995.0	1135.0	.25050	1.5750	1.8255
173	6.422	.016468	58.16	140.96	925.3	1066.3	140.97	994.4	1135.4	.25208	1.5716	1.8237
174	6.569	.016473	56.94	141.96	924.6	1066.6	141.98	993.8	1135.8	.25366	1.5682	1.8218
175	6.720	.016479	55.75	142.96	923.9	1066.9	142.98	993.2	1136.2	.25524	1.5647	1.8200
176	6.873	.016485	54.58	143.96	923.2	1067.2	143.98	992.6	1136.6	.25682	1.5613	1.8181
177	7.029	.016491	53.45	144.96	922.5	1067.5	144.98	992.0	1137.0	.25839	1.5579	1.8163
178	7.188	.016497	52.34	145.96	921.8	1067.8	145.99	991.4	1137.4	.25997	1.5545	1.8145
179	7.350	.016503	51.26	146.97	921.1	1068.1	146.99	990.8	1137.8	.26154	1.5512	1.8127
180	7.515	.016509	50.20	147.97	920.4	1068.3	147.99	990.2	1138.2	.26311	1.5478	1.8109
181	7.683	.016515	49.17	148.97	919.7	1068.6	149.00	989.6	1138.6	.26467	1.5444	1.8091
182	7.854	.016521	48.17	149.97	919.0	1068.9	150.00	989.0	1139.0	.26623	1.5411	1.8073
183	8.029	.016527	47.19	150.98	918.3	1069.2	151.00	988.3	1139.3	.26780	1.5377	1.8055
184	8.206	.016533	46.23	151.98	917.6	1069.5	152.01	987.7	1139.7	.26936	1.5344	1.8037
185	8.387	.016539	45.30	152.98	916.8	1069.8	153.01	987.1	1140.1	.27091	1.5311	1.8020
186	8.572	.016545	44.38	153.98	916.1	1070.1	154.01	986.5	1140.5	.27247	1.5277	1.8002
187	8.759	.016551	43.49	154.99	915.4	1070.4	155.02	985.9	1140.9	.27402	1.5244	1.7985
188	8.951	.016558	42.62	155.99	914.7	1070.7	156.02	985.3	1141.3	.27557	1.5211	1.7967
189	9.145	.016564	41.78	156.99	914.0	1071.0	157.02	984.7	1141.7	.27712	1.5178	1.7950
190	9.343	.016570	40.95	158.00	913.3	1071.3	158.03	984.1	1142.1	.27866	1.5146	1.7932
191	9.545	.016576	40.14	159.00	912.6	1071.6	159.03	983.4	1142.5	.28021	1.5113	1.7915
192	9.750	.016583	39.34	160.00	911.9	1071.9	160.04	982.8	1142.9	.28175	1.5080	1.7898
193	9.959	.016589	38.57	161.01	911.2	1072.2	161.04	982.2	1143.3	.28329	1.5048	1.7881
194	10.172	.016596	37.82	162.01	910.4	1072.5	162.04	981.6	1143.6	.28482	1.5015	1.7863
195	10.388	.016602	37.08	163.02	909.7	1072.7	163.05	981.0	1144.0	.28636	1.4983	1.7846
196	10.609	.016608	36.36	164.02	909.0	1073.0	164.05	980.4	1144.4	.28789	1.4951	1.7830
197	10.833	.016615	35.65	165.02	908.3	1073.3	165.06	979.7	1144.8	.28942	1.4918	1.7813
198	11.061	.016621	34.96	166.03	907.6	1073.6	166.06	979.1	1145.2	.29095	1.4886	1.7796
199	11.293	.016628	34.29	167.03	906.9	1073.9	167.07	978.5	1145.6	.29248	1.4854	1.7779

Temp Fahr.	Press. Lbf. Sq.In.	Specific Volume		Internal Energy			Enthalpy			Entropy		
		Sat. Liquid	Sat. Vapor	Sat. Liquid	Evap.	Sat. Vapor	Sat. Liquid	Evap.	Sat. Vapor	Sat. Liquid	Evap.	Sat. Vapor
t	p	v_f	v_g	u_f	u_{fg}	u_g	h_f	h_{fg}	h_g	s_f	s_{fg}	s_g
200	11.529	.016634	33.63	168.04	906.2	1074.2	168.07	977.9	1145.9	.29400	1.4822	1.7762
202	12.014	.016648	32.36	170.05	904.7	1074.8	170.09	976.6	1146.7	.29704	1.4759	1.7729
204	12.515	.016661	31.15	172.06	903.3	1075.3	172.10	975.4	1147.5	.30008	1.4695	1.7696
206	13.034	.016674	29.99	174.07	901.8	1075.9	174.11	974.1	1148.2	.30310	1.4632	1.7664
208	13.570	.016688	28.88	176.08	900.4	1076.5	176.13	972.9	1149.0	.30612	1.4570	1.7631
210	14.125	.016702	27.82	178.10	898.9	1077.0	178.14	971.6	1149.7	.30913	1.4508	1.7599
212	14.698	.016716	26.80	180.11	897.5	1077.6	180.16	970.3	1150.5	.31213	1.4446	1.7567
214	15.291	.016729	25.83	182.12	896.0	1078.2	182.17	969.1	1151.2	.31513	1.4384	1.7535
216	15.903	.016743	24.90	184.14	894.6	1078.7	184.18	967.8	1152.0	.31811	1.4322	1.7504
218	16.535	.016758	24.00	186.15	893.1	1079.3	186.20	966.5	1152.7	.32109	1.4261	1.7472
220	17.188	.016772	23.15	188.17	891.7	1079.8	188.22	965.3	1153.5	.32406	1.4201	1.7441
222	17.861	.016786	22.33	190.18	890.2	1080.4	190.24	964.0	1154.2	.32702	1.4140	1.7410
224	18.557	.016801	21.55	192.20	888.7	1080.9	192.26	962.7	1154.9	.32998	1.4080	1.7380
226	19.275	.016816	20.80	194.22	887.3	1081.5	194.28	961.4	1155.7	.33292	1.4020	1.7349
228	20.015	.016830	20.08	196.24	885.8	1082.0	196.30	960.1	1156.4	.33586	1.3961	1.7319
230	20.78	.016845	19.386	198.26	884.3	1082.6	198.32	958.8	1157.1	.33880	1.3901	1.7289
232	21.57	.016860	18.723	200.28	882.9	1083.1	200.34	957.5	1157.9	.34172	1.3842	1.7260
234	22.38	.016875	18.087	202.30	881.4	1083.7	202.37	956.2	1158.6	.34464	1.3784	1.7230
236	23.22	.016891	17.476	204.32	879.9	1084.2	204.39	954.9	1159.3	.34755	1.3725	1.7201
238	24.08	.016906	16.890	206.34	878.4	1084.7	206.42	953.6	1160.0	.35045	1.3667	1.7172
240	24.97	.016922	16.327	208.36	876.9	1085.3	208.44	952.3	1160.7	.35335	1.3609	1.7143
242	25.88	.016937	15.786	210.39	875.4	1085.8	210.47	950.9	1161.4	.35624	1.3552	1.7114
244	26.82	.016953	15.267	212.41	873.9	1086.3	212.49	949.6	1162.1	.35912	1.3494	1.7085
246	27.79	.016969	14.767	214.44	872.4	1086.9	214.52	948.3	1162.8	.36199	1.3437	1.7057
248	28.79	.016985	14.287	216.46	870.9	1087.4	216.55	947.0	1163.5	.36486	1.3380	1.7029
250	29.82	.017001	13.826	218.49	869.4	1087.9	218.59	945.6	1164.2	.36772	1.3324	1.7001
252	30.88	.017017	13.382	220.52	867.9	1088.4	220.62	944.3	1164.9	.37058	1.3267	1.6973
254	31.97	.017034	12.955	222.55	866.4	1088.9	222.65	942.9	1165.6	.37342	1.3211	1.6945
256	33.09	.017050	12.544	224.58	864.9	1089.4	224.68	941.6	1166.2	.37626	1.3155	1.6918
258	34.24	.017067	12.149	226.61	863.3	1090.0	226.72	940.2	1166.9	.37910	1.3100	1.6891
260	35.42	.017084	11.768	228.64	861.8	1090.5	228.76	938.8	1167.6	.38193	1.3044	1.6864
262	36.64	.017101	11.402	230.68	860.3	1091.0	230.79	937.5	1168.3	.38475	1.2989	1.6837
264	37.89	.017118	11.049	232.71	858.8	1091.5	232.83	936.1	1168.9	.38756	1.2935	1.6810
266	39.17	.017135	10.709	234.75	857.2	1092.0	234.87	934.7	1169.6	.39037	1.2880	1.6784
268	40.49	.017152	10.382	236.78	855.7	1092.5	236.91	933.3	1170.2	.39317	1.2825	1.6757
270	41.85	.017170	10.066	238.82	854.1	1093.0	238.95	932.0	1170.9	.39597	1.2771	1.6731
272	43.24	.017187	9.762	240.86	852.6	1093.4	241.00	930.6	1171.6	.39876	1.2717	1.6705
274	44.67	.017205	9.469	242.90	851.0	1093.9	243.04	929.2	1172.2	.40154	1.2664	1.6679
276	46.13	.017223	9.186	244.94	849.5	1094.4	245.08	927.8	1172.8	.40432	1.2610	1.6653
278	47.64	.017241	8.913	246.98	847.9	1094.9	247.13	926.3	1173.5	.40709	1.2557	1.6628
280	49.18	.017259	8.650	249.02	846.3	1095.4	249.18	924.9	1174.1	.40986	1.2504	1.6602
282	50.77	.017277	8.397	251.07	844.8	1095.8	251.23	923.5	1174.7	.41262	1.2451	1.6577
284	52.40	.017296	8.152	253.11	843.2	1096.3	253.28	922.1	1175.4	.41537	1.2398	1.6552
286	54.07	.017314	7.915	255.16	841.6	1096.8	255.33	920.6	1176.0	.41812	1.2346	1.6527
288	55.78	.017333	7.687	257.21	840.0	1097.3	257.38	919.2	1176.6	.42086	1.2293	1.6502
290	57.53	.017352	7.467	259.25	838.5	1097.7	259.44	917.8	1177.2	.42360	1.2241	1.6477
292	59.33	.017371	7.254	261.31	836.9	1098.2	261.50	916.3	1177.8	.42633	1.2189	1.6453
294	61.17	.017390	7.048	263.36	835.3	1098.6	263.55	914.9	1178.4	.42906	1.2138	1.6428
296	63.06	.017409	6.849	265.41	833.7	1099.1	265.61	913.4	1179.0	.43178	1.2086	1.6404
298	65.00	.017429	6.657	267.46	832.1	1099.5	267.67	911.9	1179.6	.43449	1.2035	1.6380
300	66.98	.017448	6.472	269.52	830.5	1100.0	269.73	910.4	1180.2	.43720	1.1984	1.6356
302	69.01	.017468	6.292	271.57	828.8	1100.4	271.79	909.0	1180.8	.43990	1.1933	1.6332
304	71.09	.017488	6.119	273.63	827.2	1100.8	273.86	907.5	1181.3	.44260	1.1882	1.6308
306	73.22	.017508	5.951	275.69	825.6	1101.3	275.93	906.0	1181.9	.44530	1.1832	1.6285
308	75.40	.017528	5.789	277.75	824.0	1101.7	278.00	904.5	1182.5	.44799	1.1781	1.6261

Table 1. Saturation: Temperatures

Temp Fahr.	Press. Lbf. Sq.In.	Specific Volume		Internal Energy			Enthalpy			Entropy		
		Sat. Liquid	Sat. Vapor	Sat. Liquid	Evap.	Sat. Vapor	Sat. Liquid	Evap.	Sat. Vapor	Sat. Liquid	Evap.	Sat. Vapor
t	p	v_f	v_g	u_f	u_{fg}	u_g	h_f	h_{fg}	h_g	s_f	s_{fg}	s_g
310	77.64	.017548	5.632	279.81	822.3	1102.1	280.06	903.0	1183.0	.45067	1.1731	1.6238
312	79.92	.017569	5.480	281.87	820.7	1102.6	282.13	901.5	1183.6	.45334	1.1681	1.6214
314	82.26	.017589	5.333	283.94	819.0	1103.0	284.21	899.9	1184.1	.45602	1.1631	1.6191
316	84.65	.017610	5.190	286.01	817.4	1103.4	286.28	898.4	1184.7	.45868	1.1581	1.6168
318	87.10	.017631	5.052	288.07	815.7	1103.8	288.36	896.9	1185.2	.46135	1.1532	1.6145
320	89.60	.017652	4.919	290.14	814.1	1104.2	290.43	895.3	1185.8	.46400	1.1483	1.6123
322	92.16	.017673	4.790	292.21	812.4	1104.6	292.51	893.8	1186.3	.46666	1.1433	1.6100
324	94.78	.017695	4.665	294.28	810.7	1105.0	294.59	892.2	1186.8	.46930	1.1384	1.6077
326	97.46	.017716	4.543	296.36	809.0	1105.4	296.67	890.7	1187.3	.47194	1.1335	1.6055
328	100.20	.017738	4.426	298.43	807.4	1105.8	298.76	889.1	1187.9	.47458	1.1287	1.6033
330	103.00	.017760	4.312	300.51	805.7	1106.2	300.84	887.5	1188.4	.47722	1.1238	1.6010
332	105.86	.017782	4.201	302.58	804.0	1106.6	302.93	885.9	1188.9	.47984	1.1190	1.5988
334	108.78	.017804	4.094	304.66	802.3	1106.9	305.02	884.3	1189.4	.48247	1.1141	1.5966
336	111.76	.017826	3.991	306.74	800.6	1107.3	307.11	882.7	1189.9	.48509	1.1093	1.5944
338	114.82	.017849	3.890	308.83	798.9	1107.7	309.21	881.1	1190.3	.48770	1.1045	1.5922
340	117.93	.017872	3.792	310.91	797.1	1108.0	311.30	879.5	1190.8	.49031	1.0997	1.5901
342	121.11	.017894	3.698	313.00	795.4	1108.4	313.39	877.9	1191.3	.49291	1.0950	1.5879
344	124.36	.017917	3.606	315.08	793.7	1108.8	315.49	876.3	1191.7	.49552	1.0902	1.5857
346	127.68	.017941	3.517	317.17	791.9	1109.1	317.59	874.6	1192.2	.49811	1.0855	1.5836
348	131.07	.017964	3.430	319.26	790.2	1109.5	319.70	873.0	1192.7	.50071	1.0807	1.5814
350	134.53	.017988	3.346	321.35	788.4	1109.8	321.80	871.3	1193.1	.50329	1.0760	1.5793
352	138.06	.018011	3.265	323.45	786.7	1110.1	323.91	869.6	1193.5	.50588	1.0713	1.5772
354	141.66	.018035	3.185	325.54	784.9	1110.5	326.02	868.0	1194.0	.50846	1.0666	1.5751
356	145.34	.018059	3.109	327.64	783.2	1110.8	328.13	866.3	1194.4	.51103	1.0619	1.5730
358	149.09	.018083	3.034	329.74	781.4	1111.1	330.24	864.6	1194.8	.51360	1.0573	1.5709
360	152.92	.018108	2.961	331.84	779.6	1111.4	332.35	862.9	1195.2	.51617	1.0526	1.5688
362	156.82	.018133	2.891	333.94	777.8	1111.8	334.47	861.2	1195.6	.51873	1.0480	1.5667
364	160.80	.018157	2.822	336.05	776.0	1112.1	336.59	859.5	1196.0	.52129	1.0433	1.5646
366	164.87	.018182	2.756	338.16	774.2	1112.4	338.71	857.7	1196.4	.52385	1.0387	1.5626
368	169.01	.018208	2.691	340.26	772.4	1112.7	340.83	856.0	1196.8	.52640	1.0341	1.5605
370	173.23	.018233	2.628	342.37	770.6	1112.9	342.96	854.2	1197.2	.52894	1.0295	1.5585
372	177.53	.018258	2.567	344.49	768.7	1113.2	345.08	852.5	1197.6	.53149	1.0249	1.5564
374	181.92	.018284	2.508	346.60	766.9	1113.5	347.21	850.7	1197.9	.53403	1.0204	1.5544
376	186.39	.018310	2.450	348.72	765.1	1113.8	349.35	848.9	1198.3	.53656	1.0158	1.5524
378	190.95	.018336	2.394	350.83	763.2	1114.1	351.48	847.2	1198.6	.53910	1.0112	1.5503
380	195.60	.018363	2.339	352.95	761.4	1114.3	353.62	845.4	1199.0	.54163	1.0067	1.5483
382	200.33	.018389	2.286	355.07	759.5	1114.6	355.76	843.6	1199.3	.54415	1.0021	1.5463
384	205.15	.018416	2.234	357.20	757.6	1114.8	357.90	841.7	1199.6	.54667	.9976	1.5443
386	210.06	.018443	2.183	359.32	755.8	1115.1	360.04	839.9	1200.0	.54919	.9931	1.5423
388	215.06	.018470	2.134	361.45	753.9	1115.3	362.19	838.1	1200.3	.55171	.9886	1.5403
390	220.2	.018498	2.087	363.58	752.0	1115.6	364.34	836.2	1200.6	.55422	.9841	1.5383
392	225.3	.018525	2.040	365.72	750.1	1115.8	366.49	834.4	1200.9	.55672	.9796	1.5363
394	230.6	.018553	1.995	367.85	748.2	1116.0	368.64	832.5	1201.1	.55923	.9751	1.5344
396	236.0	.018581	1.951	369.99	746.2	1116.2	370.80	830.6	1201.4	.56173	.9706	1.5324
398	241.5	.018609	1.908	372.13	744.3	1116.4	372.96	828.7	1201.7	.56423	.9662	1.5304
400	247.1	.018638	1.8661	374.27	742.4	1116.6	375.12	826.8	1202.0	.56672	.9617	1.5284
405	261.4	.018710	1.7663	379.63	737.5	1117.1	380.53	822.0	1202.6	.57295	.9506	1.5236
410	276.5	.018784	1.6726	385.01	732.6	1117.6	385.97	817.2	1203.1	.57916	.9395	1.5187
415	292.1	.018859	1.5848	390.40	727.6	1118.0	391.42	812.2	1203.6	.58534	.9285	1.5139
420	308.5	.018936	1.5024	395.81	722.5	1118.3	396.89	807.2	1204.1	.59152	.9175	1.5091
425	325.6	.019014	1.4249	401.24	717.4	1118.6	402.38	802.1	1204.5	.59767	.9066	1.5043
430	343.3	.019094	1.3521	406.68	712.2	1118.9	407.89	796.9	1204.8	.60381	.8957	1.4995
435	361.9	.019176	1.2836	412.14	707.0	1119.1	413.42	791.7	1205.1	.60994	.8848	1.4948
440	381.2	.019260	1.2192	417.62	701.7	1119.3	418.98	786.3	1205.3	.61605	.8740	1.4900
445	401.2	.019345	1.1584	423.12	696.3	1119.5	424.55	780.9	1205.5	.62215	.8631	1.4853

Temp Fahr.	Press. Lbf. Sq.In.	Specific Volume		Internal Energy			Enthalpy			Entropy		
		Sat. Liquid	Sat. Vapor	Sat. Liquid	Evap.	Sat. Vapor	Sat. Liquid	Evap.	Sat. Vapor	Sat. Liquid	Evap.	Sat. Vapor
t	p	v_f	v_g	u_f	u_{fg}	u_g	h_f	h_{fg}	h_g	s_f	s_{fg}	s_g
450	422.1	.019433	1.1011	428.6	690.9	1119.5	430.2	775.4	1205.6	.6282	.8523	1.4806
455	443.8	.019522	1.0471	434.2	685.4	1119.6	435.8	769.8	1205.6	.6343	.8415	1.4759
460	466.3	.019614	.9961	439.7	679.8	1119.6	441.4	764.1	1205.5	.6404	.8308	1.4712
465	489.8	.019707	.9480	445.3	674.2	1119.5	447.1	758.3	1205.4	.6465	.8200	1.4665
470	514.1	.019803	.9025	450.9	668.4	1119.4	452.8	752.4	1205.2	.6525	.8093	1.4618
475	539.3	.019901	.8594	456.6	662.6	1119.2	458.5	746.4	1204.9	.6586	.7985	1.4571
480	565.5	.020002	.8187	462.2	656.7	1118.9	464.3	740.3	1204.6	.6646	.7878	1.4524
485	592.6	.020106	.7801	467.9	650.7	1118.6	470.1	734.1	1204.2	.6707	.7770	1.4477
490	620.7	.020211	.7436	473.6	644.7	1118.3	475.9	727.8	1203.7	.6767	.7663	1.4430
495	649.8	.020320	.7090	479.3	638.5	1117.9	481.8	721.3	1203.1	.6827	.7555	1.4383
500	680.0	.02043	.6761	485.1	632.3	1117.4	487.7	714.8	1202.5	.6888	.7448	1.4335
505	711.2	.02055	.6449	490.9	625.9	1116.8	493.6	708.1	1201.7	.6948	.7340	1.4288
510	743.5	.02067	.6153	496.8	619.5	1116.2	499.6	701.3	1200.9	.7009	.7232	1.4240
515	776.9	.02079	.5872	502.6	612.9	1115.5	505.6	694.3	1200.0	.7069	.7123	1.4193
520	811.4	.02091	.5605	508.5	606.2	1114.8	511.7	687.3	1198.9	.7130	.7015	1.4145
525	847.1	.02104	.5350	514.5	599.5	1113.9	517.8	680.0	1197.8	.7191	.6906	1.4097
530	884.0	.02117	.5108	520.5	592.6	1113.0	523.9	672.7	1196.6	.7252	.6796	1.4048
535	922.1	.02131	.4878	526.5	585.5	1112.0	530.1	665.1	1195.3	.7313	.6687	1.3999
540	961.5	.02145	.4658	532.6	578.4	1111.0	536.4	657.5	1193.8	.7374	.6576	1.3950
545	1002.1	.02160	.4449	538.7	571.1	1109.8	542.7	649.6	1192.3	.7435	.6466	1.3901
550	1044.0	.02175	.4249	544.9	563.7	1108.6	549.1	641.6	1190.6	.7497	.6354	1.3851
555	1087.2	.02191	.4059	551.1	556.1	1107.2	555.5	633.4	1188.9	.7558	.6242	1.3800
560	1131.8	.02207	.3877	557.4	548.4	1105.8	562.0	625.0	1187.0	.7620	.6129	1.3749
565	1177.8	.02224	.3703	563.7	540.5	1104.2	568.5	616.4	1184.9	.7683	.6015	1.3698
570	1225.1	.02241	.3537	570.1	532.5	1102.6	575.2	607.6	1182.8	.7745	.5901	1.3646
575	1274.0	.02259	.3378	576.5	524.3	1100.8	581.9	598.6	1180.4	.7808	.5785	1.3593
580	1324.3	02278	.3225	583.1	515.9	1098.9	588.6	589.3	1178.0	.7872	.5668	1.3540
585	1376.1	.02298	.3079	589.7	507.3	1096.9	595.5	579.8	1175.3	.7935	.5550	1.3486
590	1429.5	.02319	.2940	596.3	498.4	1094.8	602.5	570.1	1172.5	.8000	.5431	1.3430
595	1484.5	.02340	.2805	603.1	489.4	1092.5	609.5	560.0	1169.5	.8064	.5310	1.3374
600	1541.0	.02363	.2677	609.9	480.1	1090.0	616.7	549.7	1166.4	.8130	.5187	1.3317
605	1599.3	.02386	.2553	616.8	470.6	1087.4	623.9	539.1	1163.0	.8196	.5063	1.3259
610	1659.2	.02411	.2434	623.9	460.8	1084.6	631.3	528.1	1159.4	.8262	.4937	1.3199
615	1720.9	.02437	.2320	631.0	450.6	1081.7	638.8	516.8	1155.5	.8329	.4808	1.3138
620	1784.4	.02465	.2209	638.3	440.2	1078.5	646.4	505.0	1151.4	.8398	.4677	1.3075
625	1849.7	.02494	.2103	645.7	429.4	1075.1	654.2	492.9	1147.0	.8467	.4544	1.3010
630	1916.9	.02525	.2000	653.2	418.2	1071.4	662.1	480.2	1142.4	.8537	.4407	1.2944
635	1986.0	.02558	.1901	660.9	406.6	1067.5	670.3	467.1	1137.3	.8608	.4267	1.2875
640	2057.1	.02593	.1805	668.7	394.5	1063.2	678.6	453.4	1131.9	.8681	.4122	1.2803
645	2130.2	.02632	.1711	676.7	381.9	1058.6	687.1	439.0	1126.1	.8755	.3974	1.2729
650	2205.	.02673	.16206	685.0	368.7	1053.7	695.9	423.9	1119.8	.8831	.3820	1.2651
655	2283.	.02718	.15323	693.5	354.8	1048.2	705.0	408.0	1113.0	.8909	.3660	1.2569
660	2362.	.02767	.14459	702.3	340.0	1042.3	714.4	391.1	1105.5	.8990	.3493	1.2483
665	2444.	.02821	.13613	711.4	324.4	1035.7	724.1	373.2	1097.3	.9073	.3318	1.2391
670	2529.	.02882	.12779	720.9	307.5	1028.5	734.4	353.9	1088.3	.9160	.3132	1.2293
675	2616.	.02951	.11952	731.0	289.3	1020.3	745.3	332.9	1078.2	.9252	.2934	1.2186
680	2705.	.03032	.11127	741.7	269.3	1011.0	756.9	309.8	1066.7	.9350	.2718	1.2068
685	2797.	.03128	.10291	753.4	246.8	1000.2	769.6	283.9	1053.4	.9456	.2480	1.1936
690	2892.	.03248	.09428	766.4	220.8	987.2	783.8	253.9	1037.7	.9575	.2208	1.1784
695	2990.	.03410	.08506	781.6	189.2	970.8	800.5	217.4	1017.9	.9715	.1882	1.1598
700	3090.	.03666	.07438	801.7	145.9	947.7	822.7	167.5	990.2	.9902	.1444	1.1346
702	3131.	.03837	.06911	813.1	121.1	934.2	835.4	138.9	974.3	1.0008	.1196	1.1204
704	3173.	.04139	.06214	830.6	83.3	913.9	854.9	95.5	950.4	1.0173	.0821	1.0994
705	3194.	.04507	.05652	848.9	46.3	895.2	875.5	53.0	928.6	1.0349	.0455	1.0805
705.44	3204.	.05053	.05053	872.6	0	872.6	902.5	0	902.5	1.0580	0	1.0580

Table 2.   Saturation: Pressures

Press. Lbf. Sq.In.	Temp. Fahr.	Specific Volume		Internal Energy			Enthalpy			Entropy		
		Sat. Liquid	Sat. Vapor	Sat. Liquid	Evap.	Sat. Vapor	Sat. Liquid	Evap.	Sat. Vapor	Sat. Liquid	Evap.	Sat. Vapor
p	t	v_f	v_g	u_f	u_{fg}	u_g	h_f	h_{fg}	h_g	s_f	s_{fg}	s_g
.08866	32.02	.016022	3302.	.00	1021.2	1021.2	.01	1075.4	1075.4	.00000	2.1869	2.1869
.090	32.39	.016022	3256.	.38	1021.0	1021.3	.38	1075.2	1075.6	.00077	2.1849	2.1856
.095	33.74	.016021	3093.	1.73	1020.1	1021.8	1.73	1074.4	1076.1	.00350	2.1774	2.1809
.10	35.02	.016021	2946.	3.02	1019.2	1022.2	3.02	1073.7	1076.7	.00612	2.1702	2.1764
.11	37.43	.016020	2691.	5.43	1017.6	1023.0	5.43	1072.3	1077.8	.01098	2.1570	2.1680
.12	39.65	.016020	2477.	7.66	1016.1	1023.7	7.67	1071.1	1078.8	.01547	2.1449	2.1604
.13	41.71	.016021	2296.	9.73	1014.7	1024.4	9.73	1069.9	1079.7	.01960	2.1338	2.1534
.14	43.64	.016021	2140.	11.67	1013.4	1025.1	11.67	1068.8	1080.5	.02346	2.1234	2.1469
.15	45.45	.016022	2005.	13.49	1012.2	1025.7	13.49	1067.8	1081.3	.02706	2.1138	2.1409
.16	47.15	.016022	1886.	15.20	1011.0	1026.2	15.20	1066.9	1082.0	.03044	2.1048	2.1352
.17	48.76	.016023	1780.	16.82	1009.9	1026.7	16.82	1065.9	1082.8	.03363	2.0963	2.1299
.18	50.30	.016024	1687.	18.35	1008.9	1027.3	18.36	1065.1	1083.4	.03665	2.0883	2.1250
.19	51.75	.016026	1602.	19.81	1007.9	1027.7	19.82	1064.3	1084.1	.03951	2.0808	2.1203
.20	53.15	.016027	1526.3	21.22	1007.0	1028.2	21.22	1063.5	1084.7	.04225	2.0736	2.1158
.21	54.48	.016028	1457.3	22.55	1006.1	1028.6	22.55	1062.7	1085.3	.04485	2.0667	2.1116
.22	55.76	.016029	1394.5	23.83	1005.2	1029.1	23.83	1062.0	1085.8	.04734	2.0602	2.1075
.23	56.99	.016031	1337.0	25.06	1004.4	1029.5	25.06	1061.3	1086.4	.04972	2.0540	2.1037
.24	58.17	.016032	1284.2	26.24	1003.6	1029.8	26.25	1060.6	1086.9	.05200	2.0480	2.1000
.25	59.31	.016034	1235.5	27.39	1002.8	1030.2	27.39	1060.0	1087.4	.05421	2.0422	2.0965
.26	60.41	.016035	1190.5	28.49	1002.1	1030.6	28.49	1059.4	1087.9	.05633	2.0367	2.0931
.27	61.47	.016037	1148.7	29.56	1001.4	1030.9	29.56	1058.8	1088.3	.05838	2.0314	2.0898
.28	62.50	.016038	1109.8	30.58	1000.7	1031.3	30.58	1058.2	1088.8	.06035	2.0263	2.0867
.29	63.50	.016040	1073.6	31.58	1000.0	1031.6	31.59	1057.6	1089.2	.06226	2.0214	2.0836
.30	64.46	.016041	1039.7	32.55	999.4	1031.9	32.56	1057.1	1089.6	.06411	2.0166	2.0807
.32	66.32	.016044	978.1	34.41	998.1	1032.5	34.41	1056.0	1090.4	.06765	2.0075	2.0751
.34	68.07	.016047	923.6	36.16	996.9	1033.1	36.16	1055.0	1091.2	.07098	1.9989	2.0699
.36	69.74	.016050	875.0	37.83	995.8	1033.6	37.84	1054.1	1091.9	.07414	1.9909	2.0650
.38	71.33	.016053	831.4	39.42	994.7	1034.2	39.42	1053.2	1092.6	.07713	1.9832	2.0604
.40	72.84	.016056	792.0	40.94	993.7	1034.7	40.94	1052.3	1093.3	.07998	1.9760	2.0559
.42	74.30	.016059	756.3	42.39	992.7	1035.1	42.39	1051.5	1093.9	.08270	1.9691	2.0518
.44	75.69	.016063	723.8	43.78	991.8	1035.6	43.78	1050.7	1094.5	.08531	1.9625	2.0478
.46	77.03	.016066	694.0	45.12	990.9	1036.0	45.12	1050.0	1095.1	.08780	1.9562	2.0440
.48	78.32	.016069	666.7	46.40	990.0	1036.4	46.40	1049.3	1095.7	.09019	1.9501	2.0403
.50	79.56	.016071	641.5	47.64	989.2	1036.9	47.65	1048.6	1096.2	.09250	1.9443	2.0368
.60	85.19	.016086	540.0	53.26	985.4	1038.7	53.27	1045.4	1098.6	.10287	1.9184	2.0213
.70	90.05	.016099	466.9	58.12	982.1	1040.3	58.12	1042.6	1100.7	.11174	1.8964	2.0081
.80	94.35	.016112	411.7	62.41	979.2	1041.7	62.41	1040.2	1102.6	.11951	1.8773	1.9968
.90	98.20	.016124	368.4	66.25	976.6	1042.9	66.25	1038.0	1104.3	.12642	1.8604	1.9868
1.0	101.70	.016136	333.6	69.74	974.3	1044.0	69.74	1036.0	1105.8	.13266	1.8453	1.9779
1.1	104.91	.016147	304.9	72.94	972.1	1045.1	72.94	1034.2	1107.1	.13834	1.8315	1.9699
1.2	107.88	.016158	280.9	75.90	970.1	1046.0	75.90	1032.5	1108.4	.14357	1.8190	1.9626
1.3	110.64	.016168	260.6	78.65	968.2	1046.9	78.66	1030.9	1109.6	.14841	1.8074	1.9558
1.4	113.22	.016178	243.0	81.23	966.5	1047.7	81.23	1029.4	1110.7	.15292	1.7967	1.9496
1.5	115.65	.016187	227.7	83.65	964.8	1048.5	83.65	1028.0	1111.7	.15714	1.7867	1.9438
1.6	117.94	.016196	214.3	85.94	963.3	1049.2	85.94	1026.7	1112.7	.16111	1.7773	1.9384
1.7	120.12	.016205	202.4	88.11	961.8	1049.9	88.11	1025.5	1113.6	.16485	1.7685	1.9334
1.8	122.18	.016214	191.8	90.17	960.4	1050.6	90.18	1024.3	1114.5	.16840	1.7602	1.9286
1.9	124.15	.016222	182.3	92.14	959.1	1051.2	92.14	1023.2	1115.3	.17178	1.7523	1.9241
2.0	126.04	.016230	173.75	94.02	957.8	1051.8	94.02	1022.1	1116.1	.17499	1.7448	1.9198
2.2	129.57	.016246	158.86	97.55	955.4	1052.9	97.55	1020.0	1117.6	.18100	1.7309	1.9119
2.4	132.85	.016260	146.40	100.81	953.1	1053.9	100.82	1018.1	1118.9	.18653	1.7181	1.9046
2.6	135.89	.016274	135.79	103.85	951.0	1054.9	103.86	1016.4	1120.2	.19165	1.7063	1.8980
2.8	138.75	.016288	126.67	106.70	949.1	1055.8	106.71	1014.7	1121.4	.19642	1.6954	1.8918
3.0	141.43	.016300	118.72	109.38	947.2	1056.6	109.39	1013.1	1122.5	.20089	1.6852	1.8861
3.2	143.97	.016313	111.75	111.92	945.5	1057.4	111.93	1011.6	1123.6	.20510	1.6757	1.8808
3.4	146.37	.016325	105.57	114.32	943.8	1058.1	114.33	1010.2	1124.6	.20907	1.6667	1.8758
3.6	148.66	.016336	100.05	116.61	942.2	1058.9	116.62	1008.9	1125.5	.21284	1.6583	1.8711
3.8	150.84	.016347	95.11	118.79	940.7	1059.5	118.80	1007.6	1126.4	.21642	1.6502	1.8666

(8)

		Specific Volume		Internal Energy			Enthalpy			Entropy		
Press. Lbf. Sq.In.	Temp. Fahr.	Sat. Liquid	Sat. Vapor	Sat. Liquid	Evap.	Sat. Vapor	Sat. Liquid	Evap.	Sat. Vapor	Sat. Liquid	Evap.	Sat. Vapor
p	t	v_f	v_g	u_f	u_{fg}	u_g	h_f	h_{fg}	h_g	s_f	s_{fg}	s_g
4.0	152.93	.016358	90.64	120.88	939.3	1060.2	120.89	1006.4	1127.3	.21983	1.6426	1.8624
4.2	154.93	.016368	86.59	122.88	937.9	1060.8	122.89	1005.2	1128.1	.22309	1.6353	1.8584
4.4	156.85	.016378	82.89	124.80	936.6	1061.4	124.81	1004.1	1128.9	.22621	1.6284	1.8546
4.6	158.70	.016388	79.51	126.65	935.3	1061.9	126.66	1003.0	1129.6	.22920	1.6217	1.8509
4.8	160.49	.016398	76.40	128.43	934.1	1062.5	128.45	1001.9	1130.3	.23208	1.6154	1.8475
5.0	162.21	.016407	73.53	130.15	932.9	1063.0	130.17	1000.9	1131.0	.23486	1.6093	1.8441
5.5	166.27	.016429	67.25	134.21	930.0	1064.2	134.23	998.4	1132.7	.24137	1.5949	1.8363
6.0	170.03	.016451	61.98	137.98	927.4	1065.4	138.00	996.2	1134.2	.24736	1.5819	1.8292
6.5	173.53	.016471	57.51	141.49	924.9	1066.4	141.51	994.1	1135.6	.25292	1.5698	1.8227
7.0	176.82	.016490	53.65	144.78	922.6	1067.4	144.80	992.1	1136.9	.25811	1.5585	1.8167
7.5	179.91	.016508	50.30	147.88	920.4	1068.3	147.90	990.2	1138.1	.26297	1.5481	1.8110
8.0	182.84	.016526	47.35	150.81	918.4	1069.2	150.84	988.4	1139.3	.26754	1.5383	1.8058
8.5	185.61	.016543	44.74	153.60	916.4	1070.0	153.62	986.7	1140.4	.27187	1.5290	1.8009
9.0	188.26	.016559	42.41	156.25	914.5	1070.8	156.27	985.1	1141.4	.27596	1.5203	1.7963
9.5	190.78	.016575	40.31	158.78	912.7	1071.5	158.81	983.6	1142.4	.27986	1.5120	1.7919
10	193.19	.016590	38.42	161.20	911.0	1072.2	161.23	982.1	1143.3	.28358	1.5041	1.7877
11	197.73	.016620	35.15	165.76	907.8	1073.5	165.80	979.3	1145.1	.29055	1.4895	1.7800
12	201.94	.016647	32.40	169.99	904.8	1074.7	170.03	976.7	1146.7	.29696	1.4761	1.7730
13	205.87	.016674	30.06	173.94	901.9	1075.9	173.98	974.2	1148.2	.30291	1.4637	1.7666
14	209.55	.016699	28.05	177.65	899.3	1076.9	177.69	971.9	1149.6	.30846	1.4521	1.7606
14.696	211.99	.016715	26.80	180.10	897.5	1077.6	180.15	970.4	1150.5	.31212	1.4446	1.7567
15	213.03	.016723	26.29	181.14	896.8	1077.9	181.19	969.7	1150.9	.31367	1.4414	1.7551
16	216.31	.016746	24.75	184.45	894.4	1078.8	184.50	967.6	1152.1	.31858	1.4313	1.7499
17	219.43	.016768	23.39	187.59	892.1	1079.7	187.65	965.6	1153.3	.32322	1.4218	1.7450
18	222.40	.016789	22.17	190.59	889.9	1080.5	190.64	963.7	1154.4	.32762	1.4128	1.7404
19	225.24	.016810	21.08	193.45	887.8	1081.3	193.51	961.9	1155.4	.33181	1.4043	1.7361
20	227.96	.016830	20.09	196.19	885.8	1082.0	196.26	960.1	1156.4	.33580	1.3962	1.7320
21	230.57	.016849	19.20	198.83	883.9	1082.7	198.89	958.4	1157.3	.33963	1.3885	1.7281
22	233.08	.016868	18.38	201.36	882.1	1083.4	201.43	956.8	1158.2	.34329	1.3811	1.7244
23	235.49	.016887	17.63	203.80	880.3	1084.1	203.88	955.2	1159.1	.34681	1.3740	1.7208
24	237.82	.016905	16.94	206.16	878.5	1084.7	206.23	953.7	1159.9	.35019	1.3672	1.7174
25	240.08	.016922	16.306	208.44	876.9	1085.3	208.52	952.2	1160.7	.35345	1.3607	1.7142
26	242.26	.016939	15.719	210.65	875.2	1085.9	210.73	950.8	1161.5	.35660	1.3544	1.7110
27	244.37	.016956	15.173	212.78	873.6	1086.4	212.87	949.4	1162.2	.35965	1.3484	1.7080
28	246.42	.016972	14.665	214.86	872.1	1087.0	214.95	948.0	1163.0	.36259	1.3425	1.7051
29	248.41	.016988	14.192	216.88	870.6	1087.5	216.97	946.7	1163.6	.36545	1.3369	1.7023
30	250.34	.017004	13.748	218.84	869.2	1088.0	218.93	945.4	1164.3	.36821	1.3314	1.6996
31	252.23	.017019	13.333	220.75	867.7	1088.5	220.85	944.1	1165.0	.37090	1.3261	1.6970
32	254.06	.017034	12.942	222.61	866.3	1089.0	222.71	942.9	1165.6	.37351	1.3210	1.6945
33	255.85	.017049	12.575	224.43	865.0	1089.4	224.53	941.7	1166.2	.37605	1.3160	1.6920
34	257.59	.017064	12.228	226.20	863.7	1089.9	226.31	940.5	1166.8	.37852	1.3111	1.6896
35	259.30	.017078	11.900	227.93	862.4	1090.3	228.04	939.3	1167.4	.38093	1.3064	1.6873
36	260.96	.017092	11.591	229.62	861.1	1090.7	229.73	938.2	1167.9	.38328	1.3018	1.6851
37	262.59	.017106	11.297	231.27	859.8	1091.1	231.39	937.1	1168.5	.38558	1.2973	1.6829
38	264.18	.017119	11.018	232.89	858.6	1091.5	233.01	936.0	1169.0	.38782	1.2930	1.6808
39	265.74	.017133	10.753	234.48	857.4	1091.9	234.60	934.9	1169.5	.39000	1.2887	1.6787
40	267.26	.017146	10.501	236.03	856.2	1092.3	236.16	933.8	1170.0	.39214	1.2845	1.6767
41	268.76	.017159	10.261	237.56	855.1	1092.6	237.69	932.8	1170.5	.39424	1.2805	1.6747
42	270.22	.017172	10.032	239.05	854.0	1093.0	239.18	931.8	1171.0	.39628	1.2765	1.6728
43	271.66	.017184	9.813	240.51	852.8	1093.4	240.65	930.8	1171.4	.39829	1.2726	1.6709
44	273.07	.017197	9.603	241.95	851.8	1093.7	242.09	929.8	1171.9	.40026	1.2688	1.6691
45	274.46	.017209	9.403	243.37	850.7	1094.0	243.51	928.8	1172.3	.40218	1.2651	1.6673
46	275.82	.017221	9.211	244.75	849.6	1094.4	244.90	927.9	1172.8	.40407	1.2615	1.6656
47	277.16	.017233	9.027	246.12	848.6	1094.7	246.27	926.9	1173.2	.40593	1.2579	1.6638
48	278.47	.017245	8.851	247.46	847.5	1095.0	247.61	926.0	1173.6	.40775	1.2544	1.6622
49	279.76	.017257	8.681	248.78	846.5	1095.3	248.94	925.1	1174.0	.40953	1.2510	1.6605
50	281.03	.017269	8.518	250.08	845.5	1095.6	250.24	924.2	1174.4	.41129	1.2476	1.6589
51	282.28	.017280	8.361	251.36	844.6	1095.9	251.52	923.3	1174.8	.41301	1.2443	1.6573
52	283.52	.017291	8.210	252.62	843.6	1096.2	252.78	922.4	1175.2	.41471	1.2411	1.6558
53	284.73	.017303	8.065	253.86	842.6	1096.5	254.03	921.6	1175.6	.41638	1.2379	1.6543
54	285.92	.017314	7.924	255.08	841.7	1096.8	255.25	920.7	1176.0	.41802	1.2348	1.6528

Table 2. Saturation: Pressures

Press. Lbf. Sq.In.	Temp. Fahr.	Specific Volume		Internal Energy			Enthalpy			Entropy		
		Sat. Liquid	Sat. Vapor	Sat. Liquid	Evap.	Sat. Vapor	Sat. Liquid	Evap.	Sat. Vapor	Sat. Liquid	Evap.	Sat. Vapor
p	t	v_f	v_g	u_f	u_{fg}	u_g	h_f	h_{fg}	h_g	s_f	s_{fg}	s_g
55	287.10	.017325	7.789	256.28	840.8	1097.0	256.46	919.9	1176.3	.41963	1.2317	1.6513
56	288.26	.017336	7.658	257.47	839.8	1097.3	257.65	919.0	1176.7	.42122	1.2287	1.6499
57	289.40	.017346	7.532	258.64	838.9	1097.6	258.82	918.2	1177.0	.42278	1.2257	1.6485
58	290.53	.017357	7.410	259.79	838.0	1097.8	259.98	917.4	1177.4	.42432	1.2228	1.6471
59	291.64	.017368	7.292	260.93	837.2	1098.1	261.12	916.6	1177.7	.42584	1.2199	1.6457
60	292.73	.017378	7.177	262.06	836.3	1098.3	262.25	915.8	1178.0	.42733	1.2170	1.6444
61	293.81	.017388	7.067	263.17	835.4	1098.6	263.36	915.0	1178.4	.42881	1.2143	1.6431
62	294.88	.017399	6.960	264.26	834.6	1098.8	264.46	914.2	1178.7	.43026	1.2115	1.6418
63	295.93	.017409	6.856	265.34	833.7	1099.1	265.55	913.4	1179.0	.43169	1.2088	1.6405
64	296.98	.017419	6.755	266.41	832.9	1099.3	266.61	912.7	1179.3	.43310	1.2061	1.6392
65	298.00	.017429	6.657	267.46	832.1	1099.5	267.67	911.9	1179.6	.43450	1.2035	1.6380
66	299.02	.017439	6.562	268.51	831.2	1099.8	268.72	911.2	1179.9	.43587	1.2009	1.6368
67	300.02	.017449	6.470	269.54	830.4	1100.0	269.75	910.4	1180.2	.43723	1.1983	1.6356
68	301.01	.017458	6.380	270.56	829.6	1100.2	270.77	909.7	1180.5	.43857	1.1958	1.6344
69	301.99	.017468	6.293	271.56	828.8	1100.4	271.78	909.0	1180.8	.43989	1.1933	1.6332
70	302.96	.017478	6.209	272.56	828.1	1100.6	272.79	908.3	1181.0	.44120	1.1909	1.6321
71	303.91	.017487	6.126	273.54	827.3	1100.8	273.77	907.5	1181.3	.44249	1.1884	1.6309
72	304.86	.017496	6.046	274.52	826.5	1101.0	274.75	906.8	1181.6	.44376	1.1860	1.6298
73	305.79	.017506	5.968	275.48	825.8	1101.2	275.72	906.1	1181.9	.44502	1.1837	1.6287
74	306.72	.017515	5.892	276.43	825.0	1101.4	276.67	905.4	1182.1	.44626	1.1814	1.6276
75	307.63	.017524	5.818	277.37	824.3	1101.6	277.61	904.8	1182.4	.44749	1.1790	1.6265
76	308.54	.017534	5.746	278.31	823.5	1101.8	278.55	904.1	1182.6	.44871	1.1768	1.6255
77	309.44	.017543	5.675	279.23	822.8	1102.0	279.48	903.4	1182.9	.44991	1.1745	1.6244
78	310.32	.017552	5.607	280.14	822.1	1102.2	280.40	902.7	1183.1	.45110	1.1723	1.6234
79	311.20	.017561	5.540	281.05	821.3	1102.4	281.30	902.1	1183.4	.45227	1.1701	1.6224
80	312.07	.017570	5.474	281.95	820.6	1102.6	282.21	901.4	1183.6	.45344	1.1679	1.6214
81	312.93	.017578	5.411	282.83	819.9	1102.8	283.10	900.8	1183.9	.45459	1.1658	1.6204
82	313.78	.017587	5.348	283.71	819.2	1102.9	283.98	900.1	1184.1	.45572	1.1637	1.6194
83	314.62	.017596	5.288	284.58	818.5	1103.1	284.85	899.5	1184.3	.45685	1.1616	1.6184
84	315.46	.017605	5.228	285.45	817.8	1103.3	285.72	898.8	1184.5	.45796	1.1595	1.6175
85	316.29	.017613	5.170	286.30	817.1	1103.5	286.58	898.2	1184.8	.45907	1.1574	1.6165
86	317.11	.017622	5.113	287.15	816.5	1103.6	287.43	897.6	1185.0	.46016	1.1554	1.6156
87	317.92	.017630	5.058	287.99	815.8	1103.8	288.27	896.9	1185.2	.46124	1.1534	1.6146
88	318.72	.017639	5.004	288.82	815.1	1104.0	289.11	896.3	1185.4	.46231	1.1514	1.6137
89	319.52	.017647	4.950	289.65	814.5	1104.1	289.94	895.7	1185.6	.46337	1.1494	1.6128
90	320.31	.017655	4.898	290.46	813.8	1104.3	290.76	895.1	1185.9	.46442	1.1475	1.6119
91	321.10	.017664	4.848	291.28	813.2	1104.4	291.57	894.5	1186.1	.46546	1.1456	1.6110
92	321.87	.017672	4.798	292.08	812.5	1104.6	292.38	893.9	1186.3	.46649	1.1436	1.6101
93	322.64	.017680	4.749	292.88	811.9	1104.7	293.18	893.3	1186.5	.46751	1.1418	1.6093
94	323.41	.017688	4.701	293.67	811.2	1104.9	293.98	892.7	1186.7	.46852	1.1399	1.6084
95	324.16	.017696	4.654	294.45	810.6	1105.0	294.76	892.1	1186.9	.46952	1.1380	1.6076
96	324.91	.017705	4.609	295.23	810.0	1105.2	295.55	891.5	1187.1	.47051	1.1362	1.6067
97	325.66	.017713	4.564	296.00	809.3	1105.3	296.32	890.9	1187.3	.47149	1.1344	1.6059
98	326.40	.017721	4.520	296.77	808.7	1105.5	297.09	890.4	1187.4	.47247	1.1326	1.6050
99	327.13	.017729	4.476	297.53	808.1	1105.6	297.85	889.8	1187.6	.47343	1.1308	1.6042
100	327.86	.017736	4.434	298.28	807.5	1105.8	298.61	889.2	1187.8	.47439	1.1290	1.6034
101	328.58	.017744	4.392	299.03	806.9	1105.9	299.36	888.6	1188.0	.47534	1.1273	1.6026
102	329.29	.017752	4.352	299.77	806.3	1106.0	300.11	888.1	1188.2	.47629	1.1255	1.6018
103	330.00	.017760	4.312	300.51	805.7	1106.2	300.85	887.5	1188.4	.47722	1.1238	1.6010
104	330.71	.017768	4.272	301.24	805.1	1106.3	301.58	887.0	1188.5	.47814	1.1221	1.6002
105	331.41	.017775	4.234	301.97	804.5	1106.5	302.31	886.4	1188.7	.47906	1.1204	1.5995
106	332.10	.017783	4.196	302.69	803.9	1106.6	303.04	885.9	1188.9	.47997	1.1187	1.5987
107	332.79	.017791	4.159	303.40	803.3	1106.7	303.75	885.3	1189.1	.48088	1.1171	1.5979
108	333.47	.017798	4.122	304.11	802.7	1106.8	304.47	884.8	1189.2	.48177	1.1154	1.5972
109	334.15	.017806	4.087	304.82	802.2	1107.0	305.18	884.2	1189.4	.48266	1.1138	1.5964
110	334.82	.017813	4.051	305.52	801.6	1107.1	305.88	883.7	1189.6	.48355	1.1122	1.5957
111	335.49	.017821	4.017	306.22	801.0	1107.2	306.58	883.1	1189.7	.48442	1.1106	1.5950
112	336.16	.017828	3.983	306.91	800.4	1107.3	307.28	882.6	1189.9	.48529	1.1090	1.5942
113	336.81	.017836	3.949	307.59	799.9	1107.5	307.97	882.1	1190.0	.48615	1.1074	1.5935
114	337.47	.017843	3.916	308.27	799.3	1107.6	308.65	881.6	1190.2	.48701	1.1058	1.5928

		Specific Volume		Internal Energy			Enthalpy			Entropy		
Press. Lbf. Sq.In.	Temp. Fahr.	Sat. Liquid	Sat. Vapor	Sat. Liquid	Evap.	Sat. Vapor	Sat. Liquid	Evap.	Sat. Vapor	Sat. Liquid	Evap.	Sat. Vapor
p	t	v_f	v_g	u_f	u_{fg}	u_g	h_f	h_{fg}	h_g	s_f	s_{fg}	s_g
115	338.12	.017850	3.884	308.95	798.8	1107.7	309.33	881.0	1190.4	.48786	1.1042	1.5921
116	338.77	.017858	3.852	309.62	798.2	1107.8	310.01	880.5	1190.5	.48870	1.1027	1.5914
117	339.41	.017865	3.821	310.29	797.6	1107.9	310.68	880.0	1190.7	.48954	1.1012	1.5907
118	340.04	.017872	3.790	310.96	797.1	1108.1	311.34	879.5	1190.8	.49037	1.0996	1.5900
119	340.68	.017879	3.760	311.62	796.6	1108.2	312.01	879.0	1191.0	.49119	1.0981	1.5893
120	341.30	.017886	3.730	312.27	796.0	1108.3	312.67	878.5	1191.1	.49201	1.0966	1.5886
121	341.93	.017894	3.701	312.92	795.5	1108.4	313.32	877.9	1191.3	.49282	1.0951	1.5880
122	342.55	.017901	3.672	313.57	794.9	1108.5	313.97	877.4	1191.4	.49363	1.0937	1.5873
123	343.17	.017908	3.644	314.21	794.4	1108.6	314.62	876.9	1191.6	.49443	1.0922	1.5866
124	343.78	.017915	3.616	314.85	793.9	1108.7	315.26	876.4	1191.7	.49523	1.0907	1.5860
125	344.39	.017922	3.588	315.49	793.3	1108.8	315.90	875.9	1191.8	.49602	1.0893	1.5853
126	344.99	.017929	3.561	316.12	792.8	1108.9	316.54	875.4	1192.0	.49680	1.0879	1.5847
127	345.59	.017936	3.535	316.75	792.3	1109.0	317.17	874.9	1192.1	.49758	1.0864	1.5840
128	346.19	.017943	3.508	317.37	791.8	1109.1	317.79	874.5	1192.2	.49836	1.0850	1.5834
129	346.78	.017950	3.482	317.99	791.3	1109.3	318.42	874.0	1192.4	.49913	1.0836	1.5828
130	347.37	.017957	3.457	318.61	790.7	1109.4	319.04	873.5	1192.5	.49989	1.0822	1.5821
131	347.96	.017964	3.432	319.22	790.2	1109.5	319.65	873.0	1192.6	.50065	1.0808	1.5815
132	348.54	.017970	3.407	319.83	789.7	1109.6	320.27	872.5	1192.8	.50141	1.0795	1.5809
133	349.12	.017977	3.383	320.43	789.2	1109.7	320.88	872.0	1192.9	.50216	1.0781	1.5803
134	349.70	.017984	3.359	321.04	788.7	1109.8	321.48	871.6	1193.0	.50290	1.0767	1.5796
135	350.27	.017991	3.335	321.64	788.2	1109.8	322.08	871.1	1193.2	.50364	1.0754	1.5790
136	350.84	.017997	3.312	322.23	787.7	1109.9	322.68	870.6	1193.3	.50438	1.0741	1.5784
137	351.40	.018004	3.289	322.82	787.2	1110.0	323.28	870.1	1193.4	.50511	1.0727	1.5778
138	351.97	.018011	3.266	323.41	786.7	1110.1	323.87	869.7	1193.5	.50583	1.0714	1.5772
139	352.53	.018018	3.244	324.00	786.2	1110.2	324.46	869.2	1193.7	.50655	1.0701	1.5766
140	353.08	.018024	3.221	324.58	785.7	1110.3	325.05	868.7	1193.8	.50727	1.0688	1.5761
141	353.63	.018031	3.200	325.16	785.3	1110.4	325.63	868.3	1193.9	.50799	1.0675	1.5755
142	354.18	.018037	3.178	325.74	784.8	1110.5	326.21	867.8	1194.0	.50870	1.0662	1.5749
143	354.73	.018044	3.157	326.31	784.3	1110.6	326.79	867.3	1194.1	.50940	1.0649	1.5743
144	355.28	.018051	3.136	326.88	783.8	1110.7	327.36	866.9	1194.2	.51010	1.0636	1.5737
145	355.82	.018057	3.115	327.45	783.3	1110.8	327.93	866.4	1194.4	.51079	1.0624	1.5732
146	356.35	.018064	3.095	328.01	782.8	1110.9	328.50	866.0	1194.5	.51149	1.0611	1.5726
147	356.89	.018070	3.075	328.58	782.4	1110.9	329.07	865.5	1194.6	.51218	1.0599	1.5720
148	357.42	.018076	3.055	329.13	781.9	1111.0	329.63	865.1	1194.7	.51286	1.0586	1.5715
149	357.95	.018083	3.036	329.69	781.4	1111.1	330.19	864.6	1194.8	.51354	1.0574	1.5709
150	358.48	.018089	3.016	330.24	781.0	1111.2	330.75	864.2	1194.9	.51422	1.0562	1.5704
152	359.52	.018102	2.978	331.34	780.0	1111.4	331.85	863.3	1195.1	.51556	1.0537	1.5693
154	360.56	.018115	2.941	332.43	779.1	1111.5	332.94	862.4	1195.3	.51689	1.0513	1.5682
156	361.58	.018127	2.905	333.50	778.2	1111.7	334.03	861.5	1195.6	.51820	1.0490	1.5671
158	362.59	.018140	2.870	334.57	777.3	1111.8	335.10	860.7	1195.8	.51949	1.0466	1.5661
160	363.60	.018152	2.836	335.63	776.4	1112.0	336.16	859.8	1196.0	.52078	1.0443	1.5651
162	364.59	.018165	2.802	336.67	775.5	1112.1	337.22	858.9	1196.2	.52205	1.0420	1.5640
164	365.58	.018177	2.770	337.71	774.6	1112.3	338.26	858.1	1196.4	.52331	1.0397	1.5630
166	366.55	.018189	2.738	338.74	773.7	1112.4	339.29	857.2	1196.5	.52455	1.0375	1.5620
168	367.52	.018201	2.707	339.75	772.8	1112.6	340.32	856.4	1196.7	.52578	1.0352	1.5610
170	368.47	.018214	2.676	340.76	772.0	1112.7	341.33	855.6	1196.9	.52700	1.0330	1.5600
172	369.42	.018226	2.646	341.76	771.1	1112.9	342.34	854.7	1197.1	.52821	1.0308	1.5591
174	370.36	.018238	2.617	342.76	770.2	1113.0	343.34	853.9	1197.3	.52940	1.0287	1.5581
176	371.29	.018249	2.589	343.74	769.4	1113.1	344.33	853.1	1197.4	.53059	1.0266	1.5571
178	372.21	.018261	2.561	344.71	768.6	1113.3	345.31	852.3	1197.6	.53176	1.0244	1.5562
180	373.13	.018273	2.533	345.68	767.7	1113.4	346.29	851.5	1197.8	.53292	1.0223	1.5553
182	374.04	.018285	2.507	346.64	766.9	1113.5	347.25	850.7	1197.9	.53407	1.0203	1.5543
184	374.93	.018296	2.480	347.59	766.1	1113.8	348.21	849.9	1198.1	.53521	1.0182	1.5534
186	375.83	.018308	2.455	348.53	765.2	1113.8	349.16	849.1	1198.3	.53634	1.0162	1.5525
188	376.71	.018319	2.430	349.47	764.4	1113.9	350.10	848.3	1198.4	.53746	1.0142	1.5516
190	377.59	.018331	2.405	350.39	763.6	1114.0	351.04	847.5	1198.6	.53857	1.0122	1.5507
192	378.45	.018342	2.381	351.32	762.8	1114.1	351.97	846.7	1198.7	.53967	1.0102	1.5499
194	379.32	.018354	2.357	352.23	762.0	1114.2	352.89	846.0	1198.9	.54076	1.0082	1.5490
196	380.17	.018365	2.334	353.14	761.2	1114.3	353.80	845.2	1199.0	.54184	1.0063	1.5481
198	381.02	.018376	2.312	354.04	760.4	1114.5	354.71	844.4	1199.1	.54291	1.0044	1.5473

Table 2.   Saturation: Pressures

Press. Lbf. Sq.In. p	Temp. Fahr. t	Specific Volume		Internal Energy			Enthalpy			Entropy		
		Sat. Liquid v_f	Sat. Vapor v_g	Sat. Liquid u_f	Evap. u_{fg}	Sat. Vapor u_g	Sat. Liquid h_f	Evap. h_{fg}	Sat. Vapor h_g	Sat. Liquid s_f	Evap. s_{fg}	Sat. Vapor s_g
200	381.86	.018387	2.289	354.9	759.6	1114.6	355.6	843.7	1199.3	.5440	1.0025	1.5464
205	383.94	.018415	2.235	357.1	757.7	1114.8	357.8	841.8	1199.6	.5466	.9978	1.5444
210	385.97	.018443	2.184	359.3	755.8	1115.1	360.0	839.9	1200.0	.5492	.9932	1.5423
215	387.97	.018470	2.135	361.4	753.9	1115.3	362.2	838.1	1200.3	.5517	.9887	1.5403
220	389.94	.018497	2.088	363.5	752.0	1115.5	364.3	836.3	1200.6	.5541	.9842	1.5384
225	391.87	.018523	2.043	365.6	750.2	1115.8	366.3	834.5	1200.8	.5566	.9799	1.5365
230	393.76	.018550	2.000	367.6	748.4	1116.0	368.4	832.7	1201.1	.5589	.9757	1.5346
235	395.63	.018576	1.959	369.6	746.6	1116.2	370.4	831.0	1201.4	.5613	.9715	1.5327
240	397.46	.018602	1.919	371.5	744.8	1116.4	372.4	829.3	1201.6	.5636	.9674	1.5309
245	399.26	.018627	1.881	373.5	743.1	1116.6	374.3	827.5	1201.9	.5658	.9634	1.5292
250	401.04	.018653	1.8448	375.4	741.4	1116.7	376.2	825.8	1202.1	.5680	.9594	1.5274
255	402.79	.018678	1.8097	377.3	739.7	1116.9	378.1	824.2	1202.3	.5702	.9555	1.5257
260	404.51	.018703	1.7758	379.1	738.0	1117.1	380.0	822.5	1202.5	.5723	.9517	1.5240
265	406.20	.018728	1.7432	380.9	736.3	1117.2	381.8	820.9	1202.7	.5744	.9479	1.5224
270	407.87	.018752	1.7117	382.7	734.7	1117.4	383.7	819.2	1202.9	.5765	.9442	1.5208
275	409.52	.018777	1.6813	384.5	733.0	1117.5	385.4	817.6	1203.1	.5786	.9406	1.5192
280	411.15	.018801	1.6520	386.2	731.4	1117.7	387.2	816.0	1203.3	.5806	.9370	1.5176
285	412.75	.018825	1.6237	388.0	729.8	1117.8	389.0	814.5	1203.4	.5826	.9335	1.5160
290	414.33	.018849	1.5963	389.7	728.2	1117.9	390.7	812.9	1203.6	.5845	.9300	1.5145
295	415.89	.018873	1.5698	391.4	726.7	1118.0	392.4	811.3	1203.7	.5864	.9266	1.5130
300	417.43	.018896	1.5442	393.0	725.1	1118.2	394.1	809.8	1203.9	.5883	.9232	1.5115
310	420.45	.018943	1.4953	396.3	722.1	1118.4	397.4	806.8	1204.1	.5921	.9166	1.5086
320	423.39	.018989	1.4493	399.5	719.1	1118.6	400.6	803.8	1204.4	.5957	.9101	1.5058
330	426.27	.019034	1.4061	402.6	716.1	1118.7	403.8	800.8	1204.6	.5992	.9038	1.5031
340	429.07	.019079	1.3653	405.7	713.2	1118.9	406.9	797.9	1204.8	.6027	.8977	1.5004
350	431.82	.019124	1.3267	408.7	710.3	1119.0	409.9	795.0	1204.9	.6060	.8917	1.4978
360	434.50	.019168	1.2903	411.6	707.5	1119.1	412.9	792.2	1205.1	.6093	.8859	1.4952
370	437.13	.019212	1.2557	414.5	704.8	1119.2	415.8	789.4	1205.2	.6125	.8802	1.4927
380	439.70	.019255	1.2229	417.3	702.0	1119.3	418.6	786.7	1205.3	.6157	.8746	1.4903
390	442.23	.019298	1.1916	420.1	699.3	1119.4	421.5	783.9	1205.4	.6188	.8691	1.4879
400	444.70	.019340	1.1620	422.8	696.7	1119.5	424.2	781.2	1205.5	.6218	.8638	1.4856
410	447.12	.019382	1.1337	425.5	694.0	1119.5	426.9	778.6	1205.5	.6247	.8585	1.4833
420	449.50	.019424	1.1067	428.1	691.4	1119.5	429.6	776.0	1205.5	.6276	.8534	1.4810
430	451.84	.019465	1.0809	430.7	688.9	1119.6	432.2	773.3	1205.6	.6305	.8484	1.4788
440	454.14	.019507	1.0562	433.2	686.4	1119.6	434.8	770.8	1205.6	.6333	.8434	1.4767
450	456.39	.019547	1.0326	435.7	683.9	1119.6	437.4	768.2	1205.6	.6360	.8385	1.4746
460	458.61	.019588	1.0100	438.2	681.4	1119.6	439.9	765.7	1205.5	.6387	.8338	1.4725
470	460.79	.019628	.9883	440.6	678.9	1119.6	442.3	763.2	1205.5	.6414	.8291	1.4704
480	462.94	.019668	.9675	443.0	676.5	1119.5	444.8	760.7	1205.5	.6440	.8244	1.4684
490	465.05	.019708	.9475	445.4	674.1	1119.5	447.2	758.2	1205.4	.6465	.8199	1.4664
500	467.13	.019748	.9283	447.7	671.7	1119.4	449.5	755.8	1205.3	.6490	.8154	1.4645
510	469.17	.019787	.9098	450.0	669.4	1119.4	451.9	753.4	1205.2	.6515	.8110	1.4626
520	471.19	.019826	.8920	452.3	667.1	1119.3	454.2	751.0	1205.2	.6540	.8067	1.4607
530	473.18	.019865	.8748	454.5	664.8	1119.3	456.4	748.6	1205.1	.6564	.8024	1.4588
540	475.14	.019904	.8583	456.7	662.5	1119.2	458.7	746.2	1204.9	.6587	.7982	1.4570
550	477.07	.019943	.8423	458.9	660.2	1119.1	460.9	743.9	1204.8	.6611	.7941	1.4551
560	478.97	.019981	.8269	461.0	657.9	1119.0	463.1	741.6	1204.7	.6634	.7900	1.4534
570	480.85	.020020	.8120	463.2	655.7	1118.9	465.3	739.3	1204.5	.6656	.7860	1.4516
580	482.70	.020058	.7976	465.3	653.5	1118.8	467.4	737.0	1204.4	.6679	.7820	1.4499
590	484.53	.020096	.7837	467.4	651.3	1118.7	469.6	734.7	1204.2	.6701	.7780	1.4481
600	486.33	.02013	.7702	469.4	649.1	1118.6	471.7	732.4	1204.1	.6723	.7742	1.4464
620	489.88	.02021	.7445	473.5	644.8	1118.3	475.8	727.9	1203.7	.6766	.7665	1.4431
640	493.33	.02028	.7203	477.4	640.6	1118.0	479.8	723.5	1203.3	.6807	.7591	1.4398
660	496.71	.02036	.6975	481.3	636.4	1117.7	483.8	719.1	1202.9	.6848	.7519	1.4367
680	500.00	.02043	.6761	485.1	632.3	1117.4	487.7	714.8	1202.5	.6888	.7447	1.4335
700	503.23	.02051	.6558	488.9	628.2	1117.0	491.5	710.5	1202.0	.6927	.7378	1.4305
720	506.38	.02058	.6366	492.5	624.1	1116.7	495.3	706.2	1201.5	.6965	.7310	1.4275
740	509.47	.02065	.6184	496.1	620.2	1116.3	499.0	702.0	1201.0	.7002	.7243	1.4246
760	512.49	.02073	.6011	499.7	616.2	1115.9	502.6	697.8	1200.4	.7039	.7178	1.4217
780	515.46	.02080	.5847	503.2	612.3	1115.5	506.2	693.7	1199.9	.7075	.7113	1.4188

Press. Lbf. Sq.In.	Temp. Fahr.	Specific Volume		Internal Energy			Enthalpy			Entropy		
		Sat. Liquid	Sat. Vapor	Sat. Liquid	Evap.	Sat. Vapor	Sat. Liquid	Evap.	Sat. Vapor	Sat. Liquid	Evap.	Sat. Vapor
p	t	v_f	v_g	u_f	u_{fg}	u_g	h_f	h_{fg}	h_g	s_f	s_{fg}	s_g
800	518.36	.02087	.5691	506.6	608.4	1115.0	509.7	689.6	1199.3	.7110	.7050	1.4160
820	521.21	.02094	.5542	510.0	604.6	1114.6	513.2	685.5	1198.7	.7145	.6988	1.4133
840	524.01	.02101	.5400	513.3	600.8	1114.1	516.6	681.5	1198.0	.7179	.6927	1.4106
860	526.76	.02109	.5264	516.6	597.0	1113.6	519.9	677.5	1197.4	.7212	.6867	1.4080
880	529.46	.02116	.5134	519.8	593.3	1113.1	523.3	673.5	1196.7	.7245	.6808	1.4053
900	532.12	.02123	.5009	523.0	589.6	1112.6	526.6	669.5	1196.0	.7277	.6750	1.4027
920	534.73	.02130	.4890	526.2	585.9	1112.1	529.8	665.6	1195.4	.7309	.6693	1.4002
940	537.29	.02137	.4776	529.3	582.3	1111.6	533.0	661.6	1194.6	.7341	.6636	1.3977
960	539.82	.02145	.4666	532.4	578.7	1111.0	536.2	657.7	1193.9	.7372	.6580	1.3952
980	542.30	.02152	.4561	535.4	575.1	1110.4	539.3	653.9	1193.2	.7402	.6525	1.3927
1000	544.75	.02159	.4459	538.4	571.5	1109.9	542.4	650.0	1192.4	.7432	.6471	1.3903
1050	550.71	.02177	.4222	545.7	562.6	1108.4	550.0	640.4	1190.4	.7505	.6338	1.3844
1100	556.45	.02195	.4005	552.9	553.9	1106.8	557.4	631.0	1188.3	.7576	.6209	1.3786
1150	562.00	.02214	.3806	559.9	545.3	1105.2	564.6	621.6	1186.2	.7645	.6084	1.3729
1200	567.37	.02232	.3623	566.7	536.8	1103.5	571.7	612.3	1183.9	.7712	.5961	1.3673
1250	572.56	.02250	.3454	573.4	528.3	1101.7	578.6	603.0	1181.6	.7778	.5841	1.3619
1300	577.60	.02269	.3297	579.9	519.9	1099.8	585.4	593.8	1179.2	.7841	.5724	1.3565
1350	582.50	.02288	.3152	586.3	511.6	1097.9	592.1	584.6	1176.7	.7903	.5609	1.3513
1400	587.25	.02307	.3016	592.7	503.3	1096.0	598.6	575.5	1174.1	.7964	.5497	1.3461
1450	591.88	.02326	.2888	598.8	495.1	1093.9	605.1	566.3	1171.4	.8024	.5385	1.3409
1500	596.39	.02346	.2769	605.0	486.9	1091.8	611.5	557.2	1168.7	.8082	.5276	1.3359
1550	600.78	.02366	.2657	611.0	478.7	1089.6	617.8	548.1	1165.9	.8140	.5168	1.3308
1600	605.06	.02386	.2552	616.9	470.5	1087.4	624.0	538.9	1162.9	.8196	.5062	1.3258
1650	609.24	.02407	.2452	622.8	462.3	1085.1	630.2	529.8	1159.9	.8252	.4956	1.3208
1700	613.32	.02428	.2358	628.6	454.1	1082.7	636.2	520.6	1156.9	.8307	.4852	1.3159
1750	617.31	.02450	.2268	634.4	445.9	1080.2	642.3	511.4	1153.7	.8361	.4748	1.3109
1800	621.21	.02472	.2183	640.0	437.6	1077.7	648.3	502.1	1150.4	.8414	.4645	1.3060
1850	625.02	.02494	.2102	645.7	429.4	1075.1	654.2	492.8	1147.0	.8467	.4543	1.3010
1900	628.76	.02517	.2025	651.3	421.0	1072.3	660.1	483.4	1143.5	.8519	.4441	1.2961
1950	632.41	.02541	.1952	656.9	412.7	1069.5	666.0	473.9	1140.0	.8571	.4340	1.2911
2000	636.00	.02565	.18813	662.4	404.2	1066.6	671.9	464.4	1136.3	.8623	.4238	1.2861
2050	639.51	.02590	.18139	667.9	395.7	1063.7	677.7	454.7	1132.5	.8674	.4137	1.2811
2100	642.95	.02616	.17491	673.4	387.1	1060.6	683.6	445.0	1128.5	.8725	.4035	1.2760
2150	646.33	.02642	.16869	678.9	378.4	1057.3	689.4	435.1	1124.5	.8775	.3933	1.2709
2200	649.64	.02670	.16270	684.4	369.6	1054.0	695.3	425.0	1120.3	.8826	.3831	1.2657
2250	652.90	.02698	.15692	689.9	360.7	1050.6	701.1	414.8	1115.9	.8876	.3728	1.2604
2300	656.09	.02728	.15133	695.4	351.6	1047.0	707.0	404.4	1111.4	.8927	.3624	1.2551
2350	659.23	.02759	.14592	700.9	342.4	1043.3	712.9	393.8	1106.7	.8977	.3520	1.2497
2400	662.31	.02791	.14067	706.4	332.9	1039.3	718.8	383.0	1101.8	.9028	.3413	1.2441
2450	665.33	.02825	.13556	712.0	323.3	1035.3	724.8	371.9	1096.7	.9079	.3306	1.2385
2500	668.31	.02860	.13059	717.7	313.4	1031.0	730.9	360.5	1091.4	.9131	.3196	1.2327
2550	671.24	.02898	.12574	723.4	303.2	1026.5	737.0	348.8	1085.9	.9183	.3084	1.2267
2600	674.11	.02938	.12099	729.2	292.7	1021.8	743.3	336.7	1080.0	.9236	.2970	1.2205
2650	676.94	.02981	.11632	735.1	281.8	1016.8	749.7	324.2	1073.9	.9289	.2852	1.2141
2700	679.73	.03027	.11172	741.1	270.4	1011.5	756.2	311.1	1067.4	.9345	.2730	1.2075
2750	682.46	.03077	.10717	747.3	258.6	1005.9	763.0	297.4	1060.4	.9401	.2604	1.2005
2800	685.16	.03131	.10264	753.8	246.0	999.8	770.0	283.0	1053.0	.9460	.2472	1.1932
2850	687.81	.03192	.09811	760.5	232.7	993.2	777.3	267.6	1045.0	.9521	.2332	1.1853
2900	690.42	.03260	.09353	767.6	218.4	986.0	785.1	251.1	1036.2	.9586	.2183	1.1769
2950	692.99	.03338	.08887	775.1	202.8	978.0	793.4	233.1	1026.5	.9656	.2022	1.1678
3000	695.52	.03431	.08404	783.4	185.4	968.8	802.5	213.0	1015.5	.9732	.1843	1.1575
3050	698.02	.03546	.07891	792.9	165.3	958.1	812.9	189.8	1002.6	.9819	.1639	1.1458
3100	700.47	.03701	.07322	804.2	140.7	944.8	825.4	161.5	986.8	.9924	.1392	1.1316
3150	702.89	.03944	.06632	819.7	106.8	926.4	842.7	122.4	965.1	1.0070	.1053	1.1123
3200	705.27	.04805	.05444	862.1	25.6	887.7	890.6	29.3	919.9	1.0478	.0252	1.0730
3203.6	705.44	.05053	.05053	872.6	0	872.6	902.5	0	902.5	1.0580	0	1.0580

Table 3. Vapor

Water Vapor at Low Pressures $\left(\dfrac{pv}{RT} = 1,\ R = 0.110216\ \dfrac{\text{Btu.}}{\text{Lb. }^\circ R} \right)$

t	T	pv	u_o	h_o	s_1	ψ_1	ζ_1	p_r	v_r	c_{po}	c_{vo}	k
32	491.7	292.9	1021.3	1075.5	1.9200	77.3	131.5	3.668	79.84	.4439	.3336	1.3304
40	499.7	297.6	1023.9	1079.0	1.9271	61.0	116.1	3.915	76.03	.4440	.3338	1.3302
50	509.7	303.6	1027.3	1083.5	1.9359	40.6	96.8	4.240	71.60	.4443	.3340	1.3300
60	519.7	309.6	1030.6	1087.9	1.9446	20.1	77.4	4.585	67.51	.4445	.3343	1.3297
70	529.7	315.5	1034.0	1092.3	1.9530	- .5	57.9	4.952	63.71	.4448	.3346	1.3294
80	539.7	321.5	1037.3	1096.8	1.9614	- 21.2	38.3	5.340	60.20	.4452	.3349	1.3291
90	549.7	327.4	1040.7	1101.3	1.9695	- 41.9	18.7	5.751	56.93	.4455	.3353	1.3288
100	559.7	333.4	1044.0	1105.7	1.9776	- 62.8	- 1.1	6.186	53.89	.4459	.3356	1.3284
110	569.7	339.3	1047.4	1110.2	1.9855	- 83.7	- 20.9	6.646	51.06	.4463	.3360	1.3280
120	579.7	345.3	1050.7	1114.6	1.9932	- 104.7	- 40.8	7.131	48.42	.4467	.3364	1.3276
130	589.7	351.3	1054.1	1119.1	2.0009	- 125.8	- 60.8	7.643	45.95	.4471	.3369	1.3272
140	599.7	357.2	1057.5	1123.6	2.0084	- 146.9	- 80.8	8.183	43.65	.4476	.3373	1.3268
150	609.7	363.2	1060.8	1128.1	2.0158	- 168.1	- 100.9	8.752	41.50	.4480	.3378	1.3263
160	619.7	369.1	1064.2	1132.5	2.0231	- 189.4	- 121.1	9.351	39.48	.4485	.3383	1.3259
170	629.7	375.1	1067.6	1137.0	2.0303	- 210.8	- 141.4	9.980	37.58	.4490	.3388	1.3254
180	639.7	381.0	1071.0	1141.5	2.0374	- 232.2	- 161.7	10.642	35.80	.4495	.3393	1.3249
190	649.7	387.0	1074.4	1146.0	2.0443	- 253.7	- 182.1	11.338	34.13	.4501	.3398	1.3244
200	659.7	393.0	1077.8	1150.5	2.0512	- 275.3	- 202.6	12.068	32.56	.4506	.3404	1.3239
210	669.7	398.9	1081.2	1155.0	2.0580	- 297.0	- 223.2	12.834	31.08	.4512	.3409	1.3233
220	679.7	404.9	1084.6	1159.5	2.0647	- 318.7	- 243.8	13.637	29.69	.4517	.3415	1.3228
230	689.7	410.8	1088.0	1164.1	2.0713	- 340.5	- 264.5	14.479	28.37	.4523	.3421	1.3222
240	699.7	416.8	1091.5	1168.6	2.0778	- 362.3	- 285.2	15.360	27.13	.4529	.3427	1.3217
250	709.7	422.7	1094.9	1173.1	2.0843	- 384.2	- 306.0	16.283	25.96	.4535	.3433	1.3211
260	719.7	428.7	1098.3	1177.7	2.0906	- 406.2	- 326.9	17.249	24.85	.4541	.3439	1.3205
270	729.7	434.6	1101.8	1182.2	2.0969	- 428.3	- 347.8	18.258	23.81	.4548	.3445	1.3199
280	739.7	440.6	1105.2	1186.8	2.1031	- 450.4	- 368.8	19.314	22.81	.4554	.3452	1.3194
290	749.7	446.6	1108.7	1191.3	2.1092	- 472.5	- 389.9	20.416	21.87	.4560	.3458	1.3188
300	759.7	452.5	1112.1	1195.9	2.1152	- 494.7	- 411.0	21.57	20.98	.4567	.3465	1.3182
320	779.7	464.4	1119.1	1205.0	2.1271	- 539.4	- 453.4	24.02	19.33	.4580	.3478	1.3169
340	799.7	476.3	1126.0	1214.2	2.1387	- 584.2	- 496.1	26.69	17.84	.4594	.3492	1.3157
360	819.7	488.3	1133.0	1223.4	2.1501	- 629.3	- 539.0	29.59	16.50	.4608	.3506	1.3144
380	839.7	500.2	1140.1	1232.6	2.1612	- 674.6	- 582.1	32.74	15.28	.4622	.3520	1.3131
400	859.7	512.1	1147.1	1241.9	2.1721	- 720.2	- 625.4	36.14	14.170	.4637	.3535	1.3119
420	879.7	524.0	1154.2	1251.2	2.1828	- 765.9	- 669.0	39.81	13.161	.4652	.3549	1.3106
440	899.7	535.9	1161.3	1260.5	2.1933	- 811.9	- 712.7	43.78	12.240	.4667	.3564	1.3092
460	919.7	547.8	1168.5	1269.8	2.2036	- 858.1	- 756.7	48.06	11.398	.4682	.3580	1.3079
480	939.7	559.7	1175.6	1279.2	2.2136	- 904.5	- 800.9	52.67	10.627	.4698	.3595	1.3066
500	959.7	571.7	1182.8	1288.6	2.2236	- 951.0	- 845.3	57.62	9.920	.4713	.3611	1.3053
550	1009.7	601.4	1201.0	1312.3	2.2476	- 1068.3	- 957.0	71.66	8.392	.4753	.3651	1.3019
600	1059.7	631.2	1219.4	1336.2	2.2707	- 1186.8	- 1070.0	88.35	7.145	.4794	.3692	1.2986
650	1109.7	661.0	1237.9	1360.2	2.2929	- 1306.4	- 1184.1	108.06	6.117	.4836	.3734	1.2952
700	1159.7	690.8	1256.7	1384.5	2.3143	- 1427.1	- 1299.3	131.22	5.264	.4879	.3776	1.2919
750	1209.7	720.6	1275.7	1409.0	2.3350	- 1548.8	- 1415.5	158.31	4.552	.4922	.3820	1.2886
800	1259.7	750.4	1294.9	1433.7	2.3550	- 1671.6	- 1532.8	189.84	3.953	.4966	.3864	1.2853
850	1309.7	780.1	1314.3	1458.7	2.3744	- 1795.4	- 1651.0	226.40	3.446	.5010	.3908	1.2821
900	1359.7	809.9	1334.0	1483.9	2.3933	- 1920.1	- 1770.2	268.64	3.015	.5055	.3953	1.2789
950	1409.7	839.7	1353.9	1509.2	2.4116	- 2045.7	- 1890.3	317.26	2.647	.5101	.3998	1.2757
1000	1459.7	869.5	1374.0	1534.9	2.4294	- 2172.2	- 2011.3	373.1	2.331	.5146	.4044	1.2726
1100	1559.7	929.1	1414.9	1586.8	2.4639	- 2427.9	- 2256.0	509.7	1.823	.5239	.4136	1.2665
1200	1659.7	988.6	1456.7	1639.6	2.4967	- 2687.0	- 2504.1	686.6	1.440	.5332	.4230	1.2606
1300	1759.7	1048.2	1499.5	1693.4	2.5282	- 2949.3	- 2755.3	913.5	1.147	.5425	.4323	1.2550
1400	1859.7	1107.8	1543.1	1748.1	2.5584	- 3214.6	- 3009.6	1201.8	.922	.5519	.4416	1.2496
1500	1959.7	1167.3	1587.8	1803.8	2.5875	- 3483.0	- 3266.9	1565.6	.746	.5612	.4509	1.2444
1600	2060.	1226.9	1633.3	1860.4	2.6157	- 3754.2	- 3527.1	2021.	.6070	.5704	.4601	1.2396
1800	2260.	1346.0	1727.2	1976.2	2.6694	- 4304.8	- 4055.7	3289.	.4092	.5882	.4780	1.2306
2000	2460.	1465.2	1824.5	2095.6	2.7200	- 4865.8	- 4594.7	5205.	.2815	.6050	.4947	1.2228
2200	2660.	1584.3	1925.0	2218.1	2.7679	- 5436.7	- 5143.5	8038.	.1971	.6201	.5099	1.2162
2400	2860.	1703.4	2028.3	2343.5	2.8133	- 6016.9	- 5701.6	12 138.	.1403	.6329	.5227	1.2109

	.2 (53.15)				.4 (72.84)				.6 (85.19)			p (t Sat.)
v	u	h	s	v	u	h	s	v	u	h	s	t
1526.3	1028.2	1084.7	2.1158	792.0	1034.7	1093.3	2.0559	540.0	1038.7	1098.6	2.0213	Sat.
1463.1	*1021.1*	*1075.3*	*2.0971*	*730.9*	*1021.0*	*1075.1*	*2.0204*	*486.9*	*1020.9*	*1074.9*	*1.9755*	32
1487.0	*1023.8*	*1078.8*	*2.1043*	*742.9*	*1023.7*	*1078.7*	*2.0276*	*494.9*	*1023.5*	*1078.5*	*1.9826*	40
1516.9	*1027.1*	*1083.3*	*2.1131*	*757.9*	*1027.0*	*1083.1*	*2.0364*	*504.9*	*1026.9*	*1082.9*	*1.9914*	50
1546.7	1030.5	1087.7	2.1217	*772.8*	*1030.4*	*1087.6*	*2.0451*	*514.9*	*1030.2*	*1087.4*	*2.0001*	60
1576.6	1033.8	1092.2	2.1302	*787.8*	*1033.7*	*1092.0*	*2.0536*	*524.9*	*1033.6*	*1091.9*	*2.0086*	70
1606.4	1037.2	1096.6	2.1385	802.7	1037.1	1096.5	2.0619	*534.8*	*1036.9*	*1096.3*	*2.0170*	80
1636.3	1040.5	1101.1	2.1467	817.7	1040.4	1100.9	2.0701	544.8	1040.3	1100.8	2.0252	90
1666.1	1043.9	1105.6	2.1548	832.6	1043.8	1105.4	2.0782	554.8	1043.7	1105.3	2.0333	100
1695.9	1047.3	1110.0	2.1627	847.6	1047.2	1109.9	2.0861	564.8	1047.1	1109.8	2.0412	110
1725.8	1050.6	1114.5	2.1705	862.5	1050.5	1114.4	2.0939	574.8	1050.4	1114.2	2.0490	120
1755.6	1054.0	1119.0	2.1781	877.5	1053.9	1118.9	2.1016	584.7	1053.8	1118.7	2.0567	130
1785.4	1057.4	1123.5	2.1857	892.4	1057.3	1123.3	2.1091	594.7	1057.2	1123.2	2.0643	140
1815.2	1060.8	1127.9	2.1931	907.3	1060.7	1127.8	2.1165	604.7	1060.6	1127.7	2.0717	150
1845.1	1064.1	1132.4	2.2004	922.3	1064.1	1132.3	2.1239	614.6	1064.0	1132.2	2.0790	160
1874.9	1067.5	1136.9	2.2076	937.2	1067.5	1136.8	2.1311	624.6	1067.4	1136.7	2.0862	170
1904.7	1070.9	1141.4	2.2147	952.1	1070.9	1141.3	2.1382	634.6	1070.8	1141.3	2.0934	180
1934.5	1074.3	1145.9	2.2217	967.0	1074.3	1145.8	2.1451	644.5	1074.2	1145.8	2.1004	190
1964.3	1077.7	1150.4	2.2285	981.9	1077.7	1150.4	2.1520	654.5	1077.6	1150.3	2.1073	200
1994.1	1081.2	1155.0	2.2353	996.9	1081.1	1154.9	2.1588	664.4	1081.0	1154.8	2.1141	210
2023.9	1084.6	1159.5	2.2420	1011.8	1084.5	1159.4	2.1655	674.4	1084.5	1159.3	2.1208	220
2053.7	1088.0	1164.0	2.2486	1026.7	1087.9	1163.9	2.1722	684.3	1087.9	1163.9	2.1274	230
2083.5	1091.4	1168.5	2.2552	1041.6	1091.4	1168.5	2.1787	694.3	1091.3	1168.4	2.1339	240
2113.	1094.8	1173.1	2.2616	1056.5	1094.8	1173.0	2.1851	704.2	1094.8	1172.9	2.1404	250
2143.	1098.3	1177.6	2.2680	1071.4	1098.2	1177.5	2.1915	714.2	1098.2	1177.5	2.1467	260
2173.	1101.7	1182.1	2.2742	1086.3	1101.7	1182.1	2.1978	724.1	1101.6	1182.0	2.1530	270
2203.	1105.2	1186.7	2.2804	1101.2	1105.1	1186.7	2.2040	734.1	1105.1	1186.6	2.1592	280
2233.	1108.6	1191.3	2.2865	1116.1	1108.6	1191.2	2.2101	744.0	1108.6	1191.2	2.1654	290
2262.	1112.1	1195.8	2.2926	1131.0	1112.1	1195.8	2.2161	753.9	1112.0	1195.7	2.1714	300
2322.	1119.0	1205.0	2.3045	1160.8	1119.0	1204.9	2.2280	773.8	1119.0	1204.9	2.1833	320
2382.	1126.0	1214.2	2.3161	1190.6	1126.0	1214.1	2.2397	793.7	1126.0	1214.1	2.1949	340
2441.	1133.0	1223.4	2.3275	1220.4	1133.0	1223.3	2.2510	813.6	1133.0	1223.3	2.2063	360
2501.	1140.0	1232.6	2.3386	1250.2	1140.0	1232.6	2.2622	833.4	1140.0	1232.5	2.2175	380
2560.	1147.1	1241.9	2.3495	1280.0	1147.1	1241.8	2.2731	853.3	1147.1	1241.8	2.2284	400
2620.	1154.2	1251.1	2.3602	1309.8	1154.2	1251.1	2.2838	873.2	1154.2	1251.1	2.2391	420
2679.	1161.3	1260.5	2.3707	1339.6	1161.3	1260.4	2.2942	893.0	1161.3	1260.4	2.2495	440
2739.	1168.5	1269.8	2.3810	1369.4	1168.4	1269.8	2.3045	912.9	1168.4	1269.8	2.2598	460
2799.	1175.6	1279.2	2.3910	1399.2	1175.6	1279.2	2.3146	932.8	1175.6	1279.2	2.2699	480
2858.	1182.8	1288.6	2.4010	1429.0	1182.8	1288.6	2.3245	952.6	1182.8	1288.6	2.2798	500
3007.	1201.0	1312.3	2.4250	1503.5	1201.0	1312.3	2.3486	1002.3	1201.0	1312.2	2.3039	550
3156.	1219.3	1336.2	2.4481	1578.0	1219.3	1336.1	2.3717	1051.9	1219.3	1336.1	2.3269	600
3305.	1237.9	1360.2	2.4703	1652.4	1237.9	1360.2	2.3939	1101.6	1237.9	1360.2	2.3492	650
3454.	1256.7	1384.5	2.4917	1726.9	1256.7	1384.5	2.4153	1151.2	1256.7	1384.5	2.3706	700
3603.	1275.7	1409.0	2.5124	1801.4	1275.7	1409.0	2.4359	1200.9	1275.7	1409.0	2.3912	750
3752.	1294.9	1433.7	2.5324	1875.8	1294.9	1433.7	2.4560	1250.5	1294.9	1433.7	2.4113	800
3901.	1314.3	1458.7	2.5518	1950.3	1314.3	1458.7	2.4754	1300.2	1314.3	1458.7	2.4307	850
4050.	1334.0	1483.8	2.5707	2024.8	1334.0	1483.8	2.4942	1349.8	1334.0	1483.8	2.4495	900
4198.	1353.8	1509.2	2.5890	2099.2	1353.8	1509.2	2.5126	1399.5	1353.8	1509.2	2.4679	950
4347.	1374.0	1534.9	2.6069	2174.	1373.9	1534.8	2.5304	1449.1	1373.9	1534.8	2.4857	1000
4645.	1414.9	1586.8	2.6413	2323.	1414.9	1586.8	2.5648	1548.4	1414.8	1586.8	2.5202	1100
4943.	1456.7	1639.6	2.6741	2472.	1456.7	1639.6	2.5977	1647.7	1456.7	1639.6	2.5530	1200
5241.	1499.4	1693.4	2.7056	2620.	1499.4	1693.4	2.6292	1747.0	1499.4	1693.4	2.5845	1300
5539.	1543.1	1748.1	2.7358	2769.	1543.1	1748.1	2.6594	1846.3	1543.1	1748.1	2.6147	1400
5837.	1587.8	1803.8	2.7650	2918.	1587.8	1803.8	2.6885	1945.5	1587.8	1803.8	2.6439	1500
6134.	1633.3	1860.4	2.7931	3067.	1633.3	1860.4	2.7167	2045.	1633.3	1860.4	2.6720	1600
6730.	1727.2	1976.2	2.8468	3365.	1727.2	1976.2	2.7704	2243.	1727.2	1976.2	2.7257	1800
7326.	1824.5	2095.6	2.8974	3663.	1824.5	2095.6	2.8210	2442.	1824.5	2095.6	2.7763	2000
7922.	1924.9	2218.1	2.9453	3961.	1924.9	2218.1	2.8689	2641.	1924.9	2218.1	2.8242	2200
8517.	2028.2	2343.5	2.9907	4259.	2028.2	2343.5	2.9143	2839.	2028.2	2343.5	2.8696	2400

Table 3. Vapor

p (t Sat.)	.8 (94.35)				1.0 (101.70)				1.5 (115.65)			
t	v	u	h	s	v	u	h	s	v	u	h	s
Sat.	411.7	1041.7	1102.6	1.9968	333.6	1044.0	1105.8	1.9779	227.7	1048.5	1111.7	1.9438
50	*378.4*	*1026.7*	*1082.8*	*1.9595*	*302.5*	*1026.6*	*1082.6*	*1.9346*	*201.3*	*1026.3*	*1082.1*	*1.8893*
60	*385.9*	*1030.1*	*1087.2*	*1.9681*	*308.5*	*1030.0*	*1087.0*	*1.9433*	*205.3*	*1029.6*	*1086.6*	*1.8980*
70	*393.4*	*1033.4*	*1091.7*	*1.9767*	*314.5*	*1033.3*	*1091.5*	*1.9518*	*209.3*	*1033.0*	*1091.1*	*1.9065*
80	*400.9*	*1036.8*	*1096.2*	*1.9850*	*320.5*	*1036.7*	*1096.0*	*1.9602*	*213.4*	*1036.4*	*1095.6*	*1.9149*
90	*408.4*	*1040.2*	*1100.6*	*1.9933*	*326.5*	*1040.1*	*1100.5*	*1.9685*	*217.4*	*1039.8*	*1100.1*	*1.9232*
100	415.9	1043.6	1105.1	2.0014	*332.6*	*1043.5*	*1105.0*	*1.9766*	*221.4*	*1043.2*	*1104.6*	*1.9314*
110	423.4	1046.9	1109.6	2.0093	338.6	1046.8	1109.5	1.9845	*225.5*	*1046.6*	*1109.1*	*1.9394*
120	430.9	1050.3	1114.1	2.0171	344.6	1050.2	1114.0	1.9924	229.5	1050.0	1113.7	1.9472
130	438.4	1053.7	1118.6	2.0248	350.6	1053.6	1118.5	2.0001	233.5	1053.4	1118.2	1.9550
140	445.9	1057.1	1123.1	2.0324	356.6	1057.0	1123.0	2.0077	237.5	1056.8	1122.7	1.9626
150	453.4	1060.5	1127.6	2.0399	362.6	1060.4	1127.5	2.0151	241.5	1060.2	1127.3	1.9701
160	460.8	1063.9	1132.1	2.0472	368.6	1063.8	1132.0	2.0225	245.5	1063.6	1131.8	1.9775
170	468.3	1067.3	1136.6	2.0544	374.6	1067.2	1136.6	2.0297	249.5	1067.1	1136.3	1.9847
180	475.8	1070.7	1141.2	2.0615	380.5	1070.7	1141.1	2.0368	253.5	1070.5	1140.9	1.9919
190	483.3	1074.1	1145.7	2.0685	386.5	1074.1	1145.6	2.0438	257.5	1073.9	1145.4	1.9989
200	490.7	1077.6	1150.2	2.0755	392.5	1077.5	1150.1	2.0508	261.5	1077.3	1149.9	2.0058
210	498.2	1081.0	1154.7	2.0823	398.5	1080.9	1154.7	2.0576	265.5	1080.8	1154.5	2.0127
220	505.7	1084.4	1159.3	2.0890	404.5	1084.3	1159.2	2.0643	269.5	1084.2	1159.0	2.0194
230	513.2	1087.8	1163.8	2.0956	410.4	1087.8	1163.7	2.0709	273.5	1087.7	1163.6	2.0261
240	520.6	1091.3	1168.3	2.1021	416.4	1091.2	1168.3	2.0775	277.5	1091.1	1168.1	2.0326
250	528.1	1094.7	1172.9	2.1086	422.4	1094.7	1172.8	2.0839	281.5	1094.6	1172.7	2.0391
260	535.5	1098.2	1177.4	2.1150	428.4	1098.1	1177.4	2.0903	285.5	1098.0	1177.3	2.0455
270	543.0	1101.6	1182.0	2.1212	434.3	1101.6	1181.9	2.0966	289.5	1101.5	1181.8	2.0518
280	550.5	1105.1	1186.6	2.1275	440.3	1105.0	1186.5	2.1028	293.4	1104.9	1186.4	2.0580
290	557.9	1108.5	1191.1	2.1336	446.3	1108.5	1191.1	2.1089	297.4	1108.4	1191.0	2.0641
300	565.4	1112.0	1195.7	2.1397	452.3	1112.0	1195.7	2.1150	301.4	1111.9	1195.5	2.0702
320	580.3	1119.0	1204.9	2.1516	464.2	1118.9	1204.8	2.1269	309.4	1118.9	1204.7	2.0821
340	595.2	1125.9	1214.1	2.1632	476.1	1125.9	1214.0	2.1386	317.3	1125.8	1213.9	2.0938
360	610.1	1132.9	1223.3	2.1746	488.1	1132.9	1223.2	2.1500	325.3	1132.9	1223.2	2.1052
380	625.0	1140.0	1232.5	2.1857	500.0	1140.0	1232.5	2.1611	333.3	1139.9	1232.4	2.1163
400	639.9	1147.0	1241.8	2.1966	511.9	1147.0	1241.8	2.1720	341.2	1147.0	1241.7	2.1273
420	654.8	1154.1	1251.1	2.2073	523.8	1154.1	1251.1	2.1827	349.2	1154.1	1251.0	2.1380
440	669.7	1161.3	1260.4	2.2178	535.8	1161.2	1260.4	2.1932	357.1	1161.2	1260.3	2.1484
460	684.6	1168.4	1269.8	2.2281	547.7	1168.4	1269.7	2.2035	365.1	1168.3	1269.7	2.1587
480	699.5	1175.6	1279.1	2.2382	559.6	1175.6	1279.1	2.2136	373.0	1175.5	1279.1	2.1688
500	714.4	1182.8	1288.6	2.2481	571.5	1182.8	1288.5	2.2235	381.0	1182.7	1288.5	2.1788
550	751.7	1201.0	1312.2	2.2721	601.3	1200.9	1312.2	2.2475	400.8	1200.9	1312.2	2.2028
600	788.9	1219.3	1336.1	2.2952	631.1	1219.3	1336.1	2.2706	420.7	1219.3	1336.1	2.2259
650	826.2	1237.9	1360.2	2.3174	660.9	1237.9	1360.2	2.2928	440.6	1237.9	1360.2	2.2481
700	863.4	1256.7	1384.5	2.3388	690.7	1256.7	1384.5	2.3142	460.5	1256.6	1384.4	2.2695
750	900.7	1275.7	1409.0	2.3595	720.5	1275.7	1409.0	2.3349	480.3	1275.6	1409.0	2.2902
800	937.9	1294.9	1433.7	2.3796	750.3	1294.9	1433.7	2.3550	500.2	1294.8	1433.7	2.3102
850	975.1	1314.3	1458.7	2.3990	780.1	1314.3	1458.6	2.3744	520.0	1314.3	1458.6	2.3297
900	1012.4	1334.0	1483.8	2.4178	809.9	1333.9	1483.8	2.3932	539.9	1333.9	1483.8	2.3485
950	1049.6	1353.8	1509.2	2.4362	839.7	1353.8	1509.2	2.4116	559.8	1353.8	1509.2	2.3669
1000	1086.8	1373.9	1534.8	2.4540	869.5	1373.9	1534.8	2.4294	579.6	1373.9	1534.8	2.3847
1100	1161.3	1414.8	1586.8	2.4884	929.0	1414.8	1586.8	2.4638	619.3	1414.8	1586.7	2.4191
1200	1235.8	1456.7	1639.6	2.5213	988.6	1456.7	1639.6	2.4967	659.1	1456.7	1639.6	2.4520
1300	1310.2	1499.4	1693.4	2.5527	1048.2	1499.4	1693.4	2.5282	698.8	1499.4	1693.4	2.4835
1400	1384.7	1543.1	1748.1	2.5830	1107.7	1543.1	1748.1	2.5584	738.5	1543.1	1748.1	2.5137
1500	1459.1	1587.8	1803.8	2.6121	1167.3	1587.8	1803.8	2.5875	778.2	1587.8	1803.8	2.5428
1600	1533.6	1633.3	1860.4	2.6403	1226.9	1633.3	1860.4	2.6157	817.9	1633.3	1860.4	2.5710
1800	1682.5	1727.2	1976.2	2.6940	1346.0	1727.2	1976.2	2.6694	897.3	1727.1	1976.2	2.6247
2000	1831.5	1824.5	2095.6	2.7446	1465.2	1824.4	2095.6	2.7200	976.8	1824.4	2095.6	2.6753
2200	1980.4	1924.9	2218.1	2.7925	1584.3	1924.9	2218.1	2.7679	1056.2	1924.9	2218.1	2.7232
2400	2129.3	2028.2	2343.5	2.8379	1703.4	2028.2	2343.5	2.8133	1135.6	2028.2	2343.5	2.7686

	2.0 (126.04)				2.5 (134.40)				3.0 (141.43)			p (t Sat.)
v	u	h	s	v	u	h	s	v	u	h	s	t
173.75	1051.8	1116.1	1.9198	140.89	1054.4	1119.6	1.9012	118.72	1056.6	1122.5	1.8861	Sat.
150.65	*1025.9*	*1081.7*	*1.8569*	*120.29*	*1025.6*	*1081.3*	*1.8316*	*100.05*	*1025.3*	*1080.8*	*1.8109*	**50**
153.70	*1029.3*	*1086.2*	*1.8656*	*122.75*	*1029.0*	*1085.8*	*1.8404*	*102.11*	*1028.6*	*1085.3*	*1.8196*	**60**
156.75	*1032.7*	*1090.7*	*1.8742*	*125.20*	*1032.4*	*1090.3*	*1.8490*	*104.16*	*1032.0*	*1089.9*	*1.8283*	**70**
159.79	*1036.1*	*1095.2*	*1.8827*	*127.64*	*1035.8*	*1094.8*	*1.8575*	*106.21*	*1035.4*	*1094.4*	*1.8368*	**80**
162.83	*1039.5*	*1099.7*	*1.8910*	*130.09*	*1039.2*	*1099.4*	*1.8658*	*108.26*	*1038.9*	*1099.0*	*1.8452*	**90**
165.87	*1042.9*	*1104.3*	*1.8991*	*132.53*	*1042.6*	*1103.9*	*1.8740*	*110.30*	*1042.3*	*1103.5*	*1.8534*	**100**
168.90	*1046.3*	*1108.8*	*1.9072*	*134.96*	*1046.0*	*1108.5*	*1.8821*	*112.34*	*1045.8*	*1108.1*	*1.8615*	**110**
171.92	*1049.7*	*1113.3*	*1.9151*	*137.39*	*1049.5*	*1113.0*	*1.8900*	*114.37*	*1049.2*	*1112.7*	*1.8695*	**120**
174.95	1053.1	1117.9	1.9229	*139.82*	*1052.9*	*1117.6*	*1.8978*	116.40	1052.7	1117.3	1.8773	**130**
177.97	1056.6	1122.4	1.9305	142.25	1056.3	1122.2	1.9055	*118.43*	*1056.1*	*1121.9*	*1.8850*	**140**
180.99	1060.0	1127.0	1.9380	144.67	1059.8	1126.7	1.9131	120.46	1059.6	1126.4	1.8926	**150**
184.00	1063.4	1131.5	1.9454	147.09	1063.2	1131.3	1.9205	122.48	1063.0	1131.0	1.9001	**160**
187.01	1066.9	1136.1	1.9527	149.51	1066.7	1135.8	1.9278	124.50	1066.5	1135.6	1.9074	**170**
190.02	1070.3	1140.6	1.9599	151.92	1070.1	1140.4	1.9350	126.52	1070.0	1140.2	1.9146	**180**
193.03	1073.7	1145.2	1.9669	154.33	1073.6	1145.0	1.9421	128.53	1073.4	1144.8	1.9217	**190**
196.04	1077.2	1149.7	1.9739	156.74	1077.0	1149.5	1.9491	130.54	1076.9	1149.3	1.9287	**200**
199.04	1080.6	1154.3	1.9807	159.15	1080.5	1154.1	1.9559	132.55	1080.3	1153.9	1.9356	**210**
202.04	1084.1	1158.9	1.9875	161.55	1083.9	1158.7	1.9627	134.56	1083.8	1158.5	1.9424	**220**
205.04	1087.5	1163.4	1.9942	163.96	1087.4	1163.3	1.9694	136.57	1087.3	1163.1	1.9491	**230**
208.04	1091.0	1168.0	2.0007	166.36	1090.9	1167.8	1.9760	138.57	1090.7	1167.7	1.9557	**240**
211.0	1094.4	1172.5	2.0072	168.76	1094.3	1172.4	1.9825	140.58	1094.2	1172.3	1.9622	**250**
214.0	1097.9	1177.1	2.0136	171.16	1097.8	1177.0	1.9889	142.58	1097.7	1176.8	1.9686	**260**
217.0	1101.4	1181.7	2.0199	173.56	1101.3	1181.6	1.9952	144.58	1101.2	1181.4	1.9750	**270**
220.0	1104.8	1186.3	2.0262	175.95	1104.7	1186.1	2.0014	146.58	1104.7	1186.0	1.9812	**280**
223.0	1108.3	1190.9	2.0323	178.35	1108.2	1190.7	2.0076	148.58	1108.1	1190.6	1.9874	**290**
226.0	1111.8	1195.4	2.0384	180.74	1111.7	1195.3	2.0137	150.58	1111.6	1195.2	1.9935	**300**
232.0	1118.8	1204.6	2.0503	185.53	1118.7	1204.5	2.0256	154.57	1118.6	1204.4	2.0054	**320**
238.0	1125.8	1213.8	2.0620	190.32	1125.7	1213.8	2.0373	158.56	1125.6	1213.7	2.0171	**340**
243.9	1132.8	1223.1	2.0734	195.10	1132.7	1223.0	2.0487	162.55	1132.7	1222.9	2.0286	**360**
249.9	1139.8	1232.3	2.0846	199.88	1139.8	1232.3	2.0599	166.53	1139.7	1232.2	2.0397	**380**
255.9	1146.9	1241.6	2.0955	204.7	1146.9	1241.6	2.0708	170.52	1146.8	1241.5	2.0507	**400**
261.8	1154.0	1250.9	2.1062	209.4	1154.0	1250.9	2.0815	174.50	1153.9	1250.8	2.0614	**420**
267.8	1161.2	1260.3	2.1167	214.2	1161.1	1260.2	2.0920	178.49	1161.1	1260.2	2.0719	**440**
273.8	1168.3	1269.6	2.1270	219.0	1168.3	1269.6	2.1023	182.47	1168.2	1269.5	2.0822	**460**
279.7	1175.5	1279.0	2.1371	223.8	1175.5	1279.0	2.1124	186.45	1175.4	1278.9	2.0923	**480**
285.7	1182.7	1288.4	2.1470	228.5	1182.7	1288.4	2.1224	190.43	1182.6	1288.4	2.1022	**500**
300.6	1200.9	1312.1	2.1711	240.5	1200.9	1312.1	2.1464	200.37	1200.8	1312.1	2.1263	**550**
315.5	1219.3	1336.0	2.1942	252.4	1219.2	1336.0	2.1695	210.31	1219.2	1336.0	2.1494	**600**
330.4	1237.8	1360.1	2.2164	264.3	1237.8	1360.1	2.1918	220.25	1237.8	1360.1	2.1716	**650**
345.3	1256.6	1384.4	2.2378	276.2	1256.6	1384.4	2.2132	230.19	1256.6	1384.4	2.1931	**700**
360.2	1275.6	1408.9	2.2585	288.2	1275.6	1408.9	2.2339	240.1	1275.6	1408.9	2.2138	**750**
375.1	1294.8	1433.7	2.2785	300.1	1294.8	1433.6	2.2539	250.1	1294.8	1433.6	2.2338	**800**
390.0	1314.3	1458.6	2.2979	312.0	1314.3	1458.6	2.2733	260.0	1314.2	1458.6	2.2532	**850**
404.9	1333.9	1483.8	2.3168	323.9	1333.9	1483.8	2.2922	269.9	1333.9	1483.7	2.2721	**900**
419.8	1353.8	1509.2	2.3352	335.8	1353.8	1509.2	2.3105	279.9	1353.8	1509.1	2.2904	**950**
434.7	1373.9	1534.8	2.3530	347.8	1373.9	1534.8	2.3284	289.8	1373.9	1534.8	2.3083	**1000**
464.5	1414.8	1586.7	2.3874	371.6	1414.8	1586.7	2.3628	309.7	1414.8	1586.7	2.3427	**1100**
494.3	1456.7	1639.6	2.4203	395.4	1456.7	1639.6	2.3957	329.5	1456.6	1639.6	2.3756	**1200**
524.1	1499.4	1693.4	2.4517	419.3	1499.4	1693.4	2.4271	349.4	1499.4	1693.4	2.4070	**1300**
553.9	1543.1	1748.1	2.4820	443.1	1543.1	1748.1	2.4574	369.2	1543.1	1748.1	2.4373	**1400**
583.7	1587.8	1803.8	2.5111	466.9	1587.8	1803.8	2.4865	389.1	1587.8	1803.8	2.4664	**1500**
613.4	1633.3	1860.3	2.5393	490.7	1633.3	1860.3	2.5147	409.0	1633.3	1860.3	2.4946	**1600**
673.0	1727.1	1976.2	2.5930	538.4	1727.1	1976.2	2.5684	448.7	1727.1	1976.2	2.5483	**1800**
732.6	1824.4	2095.6	2.6436	586.1	1824.4	2095.6	2.6190	488.4	1824.4	2095.6	2.5989	**2000**
792.2	1924.9	2218.1	2.6915	633.7	1924.9	2218.1	2.6669	528.1	1924.9	2218.1	2.6468	**2200**
851.7	2028.2	2343.5	2.7369	681.4	2028.2	2343.5	2.7123	567.8	2028.2	2343.5	2.6922	**2400**

Table 3. Vapor

p (t Sat.)	3.5 (147.53)				4.0 (152.93)				4.5 (157.79)			
t	v	u	h	s	v	u	h	s	v	u	h	s
Sat.	102.73	1058.5	1125.0	1.8734	90.64	1060.2	1127.3	1.8624	81.16	1061.7	1129.2	1.8527
100	*94.42*	*1042.0*	*1103.2*	*1.8359*	*82.51*	*1041.7*	*1102.8*	*1.8207*	*73.25*	*1041.5*	*1102.5*	*1.8072*
110	*96.18*	*1045.5*	*1107.8*	*1.8441*	*84.06*	*1045.2*	*1107.4*	*1.8289*	*74.63*	*1044.9*	*1107.1*	*1.8154*
120	*97.93*	*1049.0*	*1112.4*	*1.8521*	*85.60*	*1048.7*	*1112.1*	*1.8369*	*76.00*	*1048.4*	*1111.7*	*1.8235*
130	*99.68*	*1052.4*	*1117.0*	*1.8599*	*87.13*	*1052.2*	*1116.7*	*1.8448*	*77.37*	*1051.9*	*1116.4*	*1.8314*
140	*101.42*	*1055.9*	*1121.6*	*1.8677*	*88.66*	*1055.7*	*1121.3*	*1.8526*	*78.74*	*1055.4*	*1121.0*	*1.8392*
150	103.16	1059.4	1126.2	1.8753	*90.19*	*1059.1*	*1125.9*	*1.8602*	*80.10*	*1058.9*	*1125.6*	*1.8469*
160	104.90	1062.8	1130.8	1.8828	91.72	1062.6	1130.5	1.8677	81.46	1062.4	1130.3	1.8544
170	106.64	1066.3	1135.4	1.8901	93.24	1066.1	1135.1	1.8751	82.82	1065.9	1134.9	1.8618
180	108.37	1069.8	1140.0	1.8974	94.76	1069.6	1139.7	1.8824	84.18	1069.4	1139.5	1.8691
190	110.10	1073.2	1144.6	1.9045	96.28	1073.1	1144.3	1.8895	85.53	1072.9	1144.1	1.8763
200	111.83	1076.7	1149.1	1.9115	97.80	1076.6	1149.0	1.8965	86.88	1076.4	1148.8	1.8833
210	113.56	1080.2	1153.7	1.9184	99.31	1080.0	1153.6	1.9035	88.23	1079.9	1153.4	1.8903
220	115.28	1083.7	1158.3	1.9252	100.82	1083.5	1158.2	1.9103	89.57	1083.4	1158.0	1.8971
230	117.00	1087.1	1162.9	1.9319	102.33	1087.0	1162.8	1.9170	90.92	1086.9	1162.6	1.9038
240	118.73	1090.6	1167.5	1.9385	103.84	1090.5	1167.4	1.9236	92.26	1090.4	1167.2	1.9105
250	120.45	1094.1	1172.1	1.9450	105.35	1094.0	1172.0	1.9302	93.60	1093.9	1171.8	1.9170
260	122.16	1097.6	1176.7	1.9515	106.85	1097.5	1176.6	1.9366	94.94	1097.4	1176.4	1.9235
270	123.88	1101.1	1181.3	1.9578	108.36	1101.0	1181.2	1.9430	96.28	1100.9	1181.0	1.9298
280	125.60	1104.6	1185.9	1.9641	109.86	1104.5	1185.8	1.9492	97.62	1104.4	1185.7	1.9361
290	127.31	1108.1	1190.5	1.9703	111.36	1108.0	1190.4	1.9554	98.96	1107.9	1190.3	1.9423
300	129.03	1111.5	1195.1	1.9764	112.87	1111.5	1195.0	1.9615	100.30	1111.4	1194.9	1.9484
320	132.45	1118.5	1204.3	1.9884	115.87	1118.5	1204.2	1.9735	102.97	1118.4	1204.1	1.9605
340	135.88	1125.6	1213.6	2.0001	118.86	1125.5	1213.5	1.9852	105.63	1125.4	1213.4	1.9722
360	139.30	1132.6	1222.8	2.0115	121.86	1132.6	1222.8	1.9967	108.30	1132.5	1222.7	1.9836
380	142.72	1139.7	1232.1	2.0227	124.85	1139.6	1232.0	2.0079	110.96	1139.6	1232.0	1.9948
400	146.13	1146.8	1241.4	2.0336	127.85	1146.7	1241.3	2.0188	113.62	1146.7	1241.3	2.0058
420	149.55	1153.9	1250.7	2.0443	130.84	1153.8	1250.7	2.0296	116.28	1153.8	1250.6	2.0165
440	152.97	1161.0	1260.1	2.0549	133.83	1161.0	1260.0	2.0401	118.94	1160.9	1260.0	2.0271
460	156.38	1168.2	1269.5	2.0652	136.81	1168.1	1269.4	2.0504	121.60	1168.1	1269.4	2.0374
480	159.79	1175.4	1278.9	2.0753	139.80	1175.3	1278.8	2.0605	124.25	1175.3	1278.8	2.0475
500	163.20	1182.6	1288.3	2.0852	142.79	1182.6	1288.3	2.0705	126.91	1182.5	1288.2	2.0574
520	166.61	1189.9	1297.8	2.0950	145.77	1189.8	1297.7	2.0802	129.56	1189.8	1297.7	2.0672
540	170.03	1197.1	1307.3	2.1046	148.76	1197.1	1307.2	2.0898	132.22	1197.1	1307.2	2.0768
560	173.44	1204.5	1316.8	2.1140	151.74	1204.4	1316.8	2.0993	134.87	1204.4	1316.7	2.0862
580	176.85	1211.8	1326.3	2.1233	154.73	1211.8	1326.3	2.1085	137.52	1211.8	1326.3	2.0955
600	180.25	1219.2	1335.9	2.1324	157.71	1219.2	1335.9	2.1177	140.18	1219.1	1335.9	2.1047
650	188.78	1237.8	1360.0	2.1546	165.17	1237.7	1360.0	2.1399	146.81	1237.7	1360.0	2.1269
700	197.30	1256.6	1384.3	2.1761	172.62	1256.5	1384.3	2.1613	153.44	1256.5	1384.3	2.1483
750	205.81	1275.6	1408.9	2.1968	180.08	1275.5	1408.8	2.1820	160.06	1275.5	1408.8	2.1690
800	214.33	1294.8	1433.6	2.2168	187.53	1294.8	1433.6	2.2021	166.69	1294.8	1433.6	2.1891
850	222.8	1314.2	1458.6	2.2362	194.98	1314.2	1458.5	2.2215	173.31	1314.2	1458.5	2.2085
900	231.4	1333.9	1483.7	2.2551	202.43	1333.9	1483.7	2.2404	179.94	1333.9	1483.7	2.2274
950	239.9	1353.8	1509.1	2.2734	209.89	1353.8	1509.1	2.2587	186.56	1353.7	1509.1	2.2457
1000	248.4	1373.9	1534.8	2.2913	217.34	1373.9	1534.7	2.2766	193.18	1373.9	1534.7	2.2636
1050	256.9	1394.2	1560.6	2.3087	224.79	1394.2	1560.6	2.2940	199.81	1394.2	1560.6	2.2810
1100	265.4	1414.8	1586.7	2.3257	232.2	1414.8	1586.7	2.3110	206.4	1414.8	1586.7	2.2980
1200	282.4	1456.6	1639.6	2.3586	247.1	1456.6	1639.6	2.3438	219.7	1456.6	1639.5	2.3309
1300	299.5	1499.4	1693.4	2.3900	262.0	1499.4	1693.4	2.3753	232.9	1499.4	1693.3	2.3623
1400	316.5	1543.1	1748.1	2.4203	276.9	1543.1	1748.1	2.4056	246.2	1543.1	1748.1	2.3926
1500	333.5	1587.7	1803.7	2.4494	291.8	1587.7	1803.7	2.4347	259.4	1587.7	1803.7	2.4217
1600	350.5	1633.3	1860.3	2.4776	306.7	1633.3	1860.3	2.4629	272.6	1633.3	1860.3	2.4499
1800	384.6	1727.1	1976.2	2.5313	336.5	1727.1	1976.2	2.5166	299.1	1727.1	1976.2	2.5036
2000	418.6	1824.4	2095.6	2.5819	366.3	1824.4	2095.6	2.5672	325.6	1824.4	2095.6	2.5542
2200	452.7	1924.9	2218.1	2.6298	396.1	1924.9	2218.1	2.6150	352.1	1924.9	2218.1	2.6021
2400	486.7	2028.2	2343.4	2.6752	425.9	2028.2	2343.4	2.6605	378.5	2028.2	2343.4	2.6475

	5.0 (162.21)				5.5 (166.27)				6.0 (170.03)			p (t Sat.)
v	u	h	s	v	u	h	s	v	u	h	s	t
73.53	1063.0	1131.0	1.8441	67.25	1064.2	1132.7	1.8363	61.98	1065.4	1134.2	1.8292	Sat.
65.84	*1041.2*	*1102.1*	*1.7951*	*59.78*	*1040.9*	*1101.7*	*1.7840*	*54.73*	*1040.6*	*1101.4*	*1.7739*	**100**
67.09	*1044.7*	*1106.7*	*1.8033*	*60.92*	*1044.4*	*1106.4*	*1.7923*	*55.77*	*1044.1*	*1106.0*	*1.7822*	**110**
68.33	*1048.2*	*1111.4*	*1.8114*	*62.05*	*1047.9*	*1111.1*	*1.8005*	*56.82*	*1047.7*	*1110.7*	*1.7904*	**120**
69.57	*1051.7*	*1116.1*	*1.8194*	*63.18*	*1051.4*	*1115.8*	*1.8085*	*57.86*	*1051.2*	*1115.4*	*1.7985*	**130**
70.80	*1055.2*	*1120.7*	*1.8272*	*64.31*	*1055.0*	*1120.4*	*1.8163*	*58.89*	*1054.7*	*1120.1*	*1.8063*	**140**
72.03	*1058.7*	*1125.4*	*1.8349*	*65.43*	*1058.5*	*1125.1*	*1.8240*	*59.93*	*1058.3*	*1124.8*	*1.8141*	**150**
73.26	*1062.2*	*1130.0*	*1.8425*	*66.55*	*1062.0*	*1129.8*	*1.8316*	*60.96*	*1061.8*	*1129.5*	*1.8217*	**160**
74.49	1065.7	1134.7	1.8499	67.67	1065.5	1134.4	1.8391	*61.98*	*1065.4*	*1134.2*	*1.8292*	**170**
75.71	1069.2	1139.3	1.8572	68.78	1069.1	1139.1	1.8464	63.01	1068.9	1138.8	1.8366	**180**
76.93	1072.7	1143.9	1.8644	69.89	1072.6	1143.7	1.8536	64.03	1072.4	1143.5	1.8438	**190**
78.15	1076.3	1148.6	1.8715	71.00	1076.1	1148.4	1.8607	65.05	1075.9	1148.2	1.8509	**200**
79.36	1079.8	1153.2	1.8784	72.11	1079.6	1153.0	1.8677	66.07	1079.5	1152.8	1.8579	**210**
80.58	1083.3	1157.8	1.8853	73.22	1083.1	1157.6	1.8746	67.08	1083.0	1157.5	1.8648	**220**
81.79	1086.8	1162.4	1.8920	74.32	1086.6	1162.3	1.8813	68.10	1086.5	1162.1	1.8716	**230**
83.00	1090.3	1167.1	1.8987	75.42	1090.1	1166.9	1.8880	69.11	1090.0	1166.7	1.8782	**240**
84.21	1093.8	1171.7	1.9052	76.52	1093.6	1171.5	1.8946	70.12	1093.5	1171.4	1.8848	**250**
85.42	1097.3	1176.3	1.9117	77.62	1097.2	1176.2	1.9011	71.13	1097.1	1176.0	1.8913	**260**
86.62	1100.8	1180.9	1.9181	78.72	1100.7	1180.8	1.9074	72.14	1100.6	1180.7	1.8977	**270**
87.83	1104.3	1185.5	1.9244	79.82	1104.2	1185.4	1.9137	73.14	1104.1	1185.3	1.9040	**280**
89.04	1107.8	1190.2	1.9306	80.92	1107.7	1190.0	1.9200	74.15	1107.6	1189.9	1.9103	**290**
90.24	1111.3	1194.8	1.9367	82.01	1111.2	1194.7	1.9261	75.16	1111.1	1194.6	1.9164	**300**
92.64	1118.3	1204.0	1.9487	84.20	1118.2	1203.9	1.9381	77.16	1118.2	1203.8	1.9285	**320**
95.05	1125.4	1213.3	1.9605	86.39	1125.3	1213.2	1.9499	79.17	1125.2	1213.1	1.9402	**340**
97.45	1132.4	1222.6	1.9719	88.57	1132.4	1222.5	1.9614	81.17	1132.3	1222.4	1.9517	**360**
99.85	1139.5	1231.9	1.9832	90.75	1139.5	1231.8	1.9726	83.17	1139.4	1231.7	1.9629	**380**
102.24	1146.6	1241.2	1.9941	92.93	1146.6	1241.1	1.9836	85.17	1146.5	1241.1	1.9739	**400**
104.64	1153.7	1250.6	2.0049	95.11	1153.7	1250.5	1.9943	87.17	1153.6	1250.4	1.9847	**420**
107.03	1160.9	1259.9	2.0154	97.29	1160.9	1259.9	2.0048	89.17	1160.8	1259.8	1.9952	**440**
109.42	1168.1	1269.3	2.0257	99.46	1168.0	1269.3	2.0152	91.16	1168.0	1269.2	2.0055	**460**
111.81	1175.3	1278.7	2.0358	101.64	1175.2	1278.7	2.0253	93.16	1175.2	1278.6	2.0157	**480**
114.20	1182.5	1288.2	2.0458	103.81	1182.5	1288.1	2.0353	95.15	1182.4	1288.1	2.0256	**500**
116.59	1189.8	1297.6	2.0556	105.98	1189.7	1297.6	2.0450	97.14	1189.7	1297.6	2.0354	**520**
118.98	1197.1	1307.1	2.0652	108.16	1197.0	1307.1	2.0546	99.13	1197.0	1307.1	2.0450	**540**
121.37	1204.4	1316.7	2.0746	110.33	1204.3	1316.6	2.0641	101.13	1204.3	1316.6	2.0544	**560**
123.76	1211.7	1326.2	2.0839	112.50	1211.7	1326.2	2.0734	103.12	1211.7	1326.2	2.0637	**580**
126.15	1219.1	1335.8	2.0930	114.67	1219.1	1335.8	2.0825	105.11	1219.1	1335.8	2.0729	**600**
132.12	1237.7	1359.9	2.1153	120.10	1237.7	1359.9	2.1047	110.08	1237.7	1359.9	2.0951	**650**
138.08	1256.5	1384.2	2.1367	125.52	1256.5	1384.2	2.1262	115.06	1256.5	1384.2	2.1166	**700**
144.05	1275.5	1408.8	2.1574	130.95	1275.5	1408.8	2.1469	120.03	1275.5	1408.7	2.1373	**750**
150.01	1294.7	1433.5	2.1775	136.37	1294.7	1433.5	2.1669	125.00	1294.7	1433.5	2.1573	**800**
155.98	1314.2	1458.5	2.1969	141.79	1314.2	1458.5	2.1864	129.97	1314.2	1458.5	2.1768	**850**
161.94	1333.8	1483.7	2.2158	147.21	1333.8	1483.7	2.2052	134.94	1333.8	1483.6	2.1956	**900**
167.90	1353.7	1509.1	2.2341	152.63	1353.7	1509.1	2.2236	139.91	1353.7	1509.1	2.2140	**950**
173.86	1373.9	1534.7	2.2520	158.05	1373.8	1534.7	2.2415	144.88	1373.8	1534.7	2.2319	**1000**
179.82	1394.2	1560.6	2.2694	163.47	1394.2	1560.6	2.2589	149.85	1394.2	1560.6	2.2493	**1050**
185.78	1414.8	1586.7	2.2864	168.89	1414.8	1586.7	2.2759	154.81	1414.8	1586.6	2.2663	**1100**
197.70	1456.6	1639.5	2.3192	179.73	1456.6	1639.5	2.3087	164.75	1456.6	1639.5	2.2991	**1200**
209.62	1499.4	1693.3	2.3507	190.56	1499.4	1693.3	2.3402	174.68	1499.4	1693.3	2.3306	**1300**
221.54	1543.1	1748.1	2.3810	201.39	1543.1	1748.1	2.3705	184.61	1543.1	1748.1	2.3609	**1400**
233.45	1587.7	1803.7	2.4101	212.23	1587.7	1803.7	2.3996	194.54	1587.7	1803.7	2.3900	**1500**
245.4	1633.3	1860.3	2.4383	223.1	1633.3	1860.3	2.4278	204.5	1633.3	1860.3	2.4182	**1600**
269.2	1727.1	1976.2	2.4920	244.7	1727.1	1976.2	2.4815	224.3	1727.1	1976.2	2.4719	**1800**
293.0	1824.4	2095.5	2.5426	266.4	1824.4	2095.5	2.5320	244.2	1824.4	2095.5	2.5225	**2000**
316.9	1924.9	2218.1	2.5904	288.1	1924.9	2218.1	2.5799	264.0	1924.9	2218.1	2.5703	**2200**
340.7	2028.2	2343.4	2.6359	309.7	2028.2	2343.4	2.6254	283.9	2028.2	2343.4	2.6158	**2400**

Table 3. Vapor

p (t Sat.)	6.5 (173.53)				7.0 (176.82)				7.5 (179.91)			
t	v	u	h	s	v	u	h	s	v	u	h	s
Sat.	57.51	1066.4	1135.6	1.8227	53.65	1067.4	1136.9	1.8167	50.30	1068.3	1138.1	1.8110
100	*50.45*	*1040.3*	*1101.0*	*1.7646*	*46.79*	*1040.0*	*1100.6*	*1.7559*	*43.61*	*1039.7*	*1100.2*	*1.7478*
110	*51.42*	*1043.8*	*1105.7*	*1.7729*	*47.69*	*1043.6*	*1105.3*	*1.7643*	*44.46*	*1043.3*	*1105.0*	*1.7562*
120	*52.39*	*1047.4*	*1110.4*	*1.7811*	*48.59*	*1047.1*	*1110.1*	*1.7725*	*45.30*	*1046.9*	*1109.8*	*1.7645*
130	*53.35*	*1051.0*	*1115.1*	*1.7892*	*49.49*	*1050.7*	*1114.8*	*1.7806*	*46.15*	*1050.5*	*1114.5*	*1.7726*
140	*54.31*	*1054.5*	*1119.8*	*1.7971*	*50.39*	*1054.3*	*1119.6*	*1.7886*	*46.98*	*1054.1*	*1119.3*	*1.7806*
150	*55.27*	*1058.1*	*1124.5*	*1.8049*	*51.28*	*1057.9*	*1124.3*	*1.7964*	*47.82*	*1057.6*	*1124.0*	*1.7884*
160	*56.22*	*1061.6*	*1129.2*	*1.8126*	*52.17*	*1061.4*	*1129.0*	*1.8041*	*48.65*	*1061.2*	*1128.7*	*1.7961*
170	*57.17*	*1065.2*	*1133.9*	*1.8201*	*53.05*	*1065.0*	*1133.7*	*1.8116*	*49.48*	*1064.8*	*1133.5*	*1.8037*
180	58.12	1068.7	1138.6	1.8274	53.93	1068.5	1138.4	1.8190	50.30	1068.4	1138.2	1.8111
190	59.07	1072.2	1143.3	1.8347	54.81	1072.1	1143.1	1.8263	51.13	1071.9	1142.9	1.8184
200	60.01	1075.8	1148.0	1.8418	55.69	1075.6	1147.8	1.8334	51.95	1075.5	1147.6	1.8256
210	60.95	1079.3	1152.6	1.8489	56.57	1079.2	1152.4	1.8405	52.77	1079.0	1152.3	1.8326
220	61.89	1082.8	1157.3	1.8558	57.44	1082.7	1157.1	1.8474	53.58	1082.6	1156.9	1.8396
230	62.83	1086.4	1161.9	1.8626	58.31	1086.2	1161.8	1.8542	54.40	1086.1	1161.6	1.8464
240	63.76	1089.9	1166.6	1.8693	59.18	1089.8	1166.4	1.8609	55.21	1089.7	1166.3	1.8531
250	64.70	1093.4	1171.2	1.8758	60.05	1093.3	1171.1	1.8675	56.03	1093.2	1170.9	1.8597
260	65.63	1096.9	1175.9	1.8823	60.92	1096.8	1175.7	1.8740	56.84	1096.7	1175.6	1.8663
270	66.56	1100.5	1180.5	1.8888	61.79	1100.4	1180.4	1.8804	57.65	1100.3	1180.3	1.8727
280	67.49	1104.0	1185.2	1.8951	62.65	1103.9	1185.1	1.8868	58.46	1103.8	1184.9	1.8790
290	68.42	1107.5	1189.8	1.9013	63.52	1107.4	1189.7	1.8930	59.26	1107.3	1189.6	1.8853
300	69.35	1111.0	1194.5	1.9075	64.38	1111.0	1194.3	1.8992	60.07	1110.9	1194.2	1.8915
320	71.21	1118.1	1203.7	1.9195	66.11	1118.0	1203.6	1.9113	61.68	1117.9	1203.6	1.9036
340	73.06	1125.2	1213.0	1.9313	67.83	1125.1	1213.0	1.9230	63.29	1125.0	1212.9	1.9154
360	74.91	1132.2	1222.4	1.9428	69.55	1132.2	1222.3	1.9346	64.90	1132.1	1222.2	1.9269
380	76.76	1139.3	1231.7	1.9540	71.26	1139.3	1231.6	1.9458	66.50	1139.2	1231.5	1.9381
400	78.61	1146.5	1241.0	1.9650	72.98	1146.4	1240.9	1.9568	68.10	1146.4	1240.9	1.9491
420	80.45	1153.6	1250.4	1.9758	74.69	1153.6	1250.3	1.9676	69.70	1153.5	1250.2	1.9599
440	82.30	1160.8	1259.8	1.9863	76.41	1160.7	1259.7	1.9781	71.30	1160.7	1259.6	1.9705
460	84.14	1168.0	1269.2	1.9967	78.12	1167.9	1269.1	1.9885	72.90	1167.9	1269.0	1.9808
480	85.98	1175.2	1278.6	2.0068	79.83	1175.1	1278.5	1.9986	74.50	1175.1	1278.5	1.9910
500	87.82	1182.4	1288.0	2.0168	81.54	1182.4	1288.0	2.0086	76.09	1182.3	1287.9	2.0009
520	89.66	1189.7	1297.5	2.0265	83.25	1189.6	1297.5	2.0183	77.69	1189.6	1297.4	2.0107
540	91.50	1197.0	1307.0	2.0361	84.96	1196.9	1307.0	2.0280	79.29	1196.9	1306.9	2.0203
560	93.34	1204.3	1316.6	2.0456	86.66	1204.3	1316.5	2.0374	80.88	1204.2	1316.5	2.0298
580	95.18	1211.6	1326.1	2.0549	88.37	1211.6	1326.1	2.0467	82.47	1211.6	1326.1	2.0391
600	97.02	1219.0	1335.7	2.0640	90.08	1219.0	1335.7	2.0558	84.07	1219.0	1335.7	2.0482
650	101.61	1237.6	1359.9	2.0863	94.35	1237.6	1359.8	2.0781	88.05	1237.6	1359.8	2.0705
700	106.20	1256.4	1384.2	2.1077	98.61	1256.4	1384.2	2.0995	92.03	1256.4	1384.1	2.0919
750	110.79	1275.5	1408.7	2.1284	102.87	1275.4	1408.7	2.1203	96.01	1275.4	1408.7	2.1126
800	115.38	1294.7	1433.5	2.1485	107.14	1294.7	1433.5	2.1403	99.99	1294.7	1433.4	2.1327
850	119.97	1314.1	1458.4	2.1679	111.40	1314.1	1458.4	2.1598	103.97	1314.1	1458.4	2.1521
900	124.56	1333.8	1483.6	2.1868	115.66	1333.8	1483.6	2.1786	107.94	1333.8	1483.6	2.1710
950	129.14	1353.7	1509.0	2.2052	119.92	1353.7	1509.0	2.1970	111.92	1353.7	1509.0	2.1894
1000	133.73	1373.8	1534.7	2.2230	124.18	1373.8	1534.7	2.2148	115.89	1373.8	1534.6	2.2072
1050	138.32	1394.2	1560.5	2.2404	128.43	1394.2	1560.5	2.2323	119.87	1394.2	1560.5	2.2247
1100	142.90	1414.7	1586.6	2.2574	132.69	1414.7	1586.6	2.2493	123.84	1414.7	1586.6	2.2417
1200	152.07	1456.6	1639.5	2.2903	141.21	1456.6	1639.5	2.2821	131.79	1456.6	1639.5	2.2745
1300	161.24	1499.4	1693.3	2.3218	149.72	1499.4	1693.3	2.3136	139.74	1499.4	1693.3	2.3060
1400	170.41	1543.1	1748.0	2.3520	158.23	1543.1	1748.0	2.3439	147.68	1543.1	1748.0	2.3363
1500	179.57	1587.7	1803.7	2.3812	166.75	1587.7	1803.7	2.3730	155.63	1587.7	1803.7	2.3654
1600	188.74	1633.3	1860.3	2.4093	175.26	1633.3	1860.3	2.4012	163.57	1633.3	1860.3	2.3936
1800	207.07	1727.1	1976.2	2.4630	192.28	1727.1	1976.2	2.4549	179.46	1727.1	1976.2	2.4473
2000	225.41	1824.4	2095.5	2.5136	209.31	1824.4	2095.5	2.5055	195.35	1824.4	2095.5	2.4979
2200	243.74	1924.9	2218.1	2.5615	226.33	1924.9	2218.1	2.5534	211.24	1924.9	2218.1	2.5457
2400	262.07	2028.2	2343.4	2.6070	243.35	2028.2	2343.4	2.5988	227.13	2028.2	2343.4	2.5912

8.0 (182.84)				8.5 (185.61)				9.0 (188.26)				p (t Sat.)
v	u	h	s	v	u	h	s	v	u	h	s	t
47.35	1069.2	1139.3	1.8058	44.74	1070.0	1140.4	1.8009	42.41	1070.8	1141.4	1.7963	Sat.
40.83	*1039.4*	*1099.9*	*1.7401*	*38.38*	*1039.1*	*1099.5*	*1.7329*	*36.20*	*1038.8*	*1099.1*	*1.7261*	**100**
41.63	*1043.0*	*1104.6*	*1.7486*	*39.13*	*1042.7*	*1104.3*	*1.7414*	*36.91*	*1042.5*	*1103.9*	*1.7346*	**110**
42.42	*1046.6*	*1109.4*	*1.7569*	*39.88*	*1046.4*	*1109.1*	*1.7498*	*37.63*	*1046.1*	*1108.8*	*1.7430*	**120**
43.22	*1050.2*	*1114.2*	*1.7651*	*40.63*	*1050.0*	*1113.9*	*1.7580*	*38.34*	*1049.7*	*1113.6*	*1.7512*	**130**
44.01	*1053.8*	*1119.0*	*1.7731*	*41.38*	*1053.6*	*1118.7*	*1.7660*	*39.04*	*1053.4*	*1118.4*	*1.7593*	**140**
44.79	*1057.4*	*1123.7*	*1.7810*	*42.12*	*1057.2*	*1123.4*	*1.7739*	*39.74*	*1057.0*	*1123.2*	*1.7672*	**150**
45.57	*1061.0*	*1128.5*	*1.7887*	*42.86*	*1060.8*	*1128.2*	*1.7817*	*40.44*	*1060.6*	*1128.0*	*1.7750*	**160**
46.35	*1064.6*	*1133.2*	*1.7963*	*43.59*	*1064.4*	*1133.0*	*1.7893*	*41.14*	*1064.2*	*1132.7*	*1.7827*	**170**
47.13	*1068.2*	*1137.9*	*1.8037*	*44.32*	*1068.0*	*1137.7*	*1.7967*	*41.83*	*1067.8*	*1137.5*	*1.7902*	**180**
47.90	1071.7	1142.7	1.8110	45.06	1071.6	1142.4	1.8041	42.53	1071.4	1142.2	1.7975	**190**
48.67	1075.3	1147.4	1.8182	45.78	1075.2	1147.2	1.8113	43.21	1075.0	1147.0	1.8048	**200**
49.44	1078.9	1152.1	1.8253	46.51	1078.7	1151.9	1.8184	43.90	1078.6	1151.7	1.8119	**210**
50.21	1082.4	1156.8	1.8323	47.23	1082.3	1156.6	1.8254	44.59	1082.2	1156.4	1.8189	**220**
50.98	1086.0	1161.4	1.8391	47.96	1085.9	1161.3	1.8322	45.27	1085.7	1161.1	1.8257	**230**
51.74	1089.5	1166.1	1.8458	48.68	1089.4	1166.0	1.8390	45.95	1089.3	1165.8	1.8325	**240**
52.50	1093.1	1170.8	1.8525	49.39	1093.0	1170.7	1.8456	46.63	1092.8	1170.5	1.8392	**250**
53.26	1096.6	1175.5	1.8590	50.11	1096.5	1175.3	1.8522	47.31	1096.4	1175.2	1.8457	**260**
54.03	1100.2	1180.1	1.8654	50.83	1100.1	1180.0	1.8586	47.99	1100.0	1179.9	1.8522	**270**
54.78	1103.7	1184.8	1.8718	51.54	1103.6	1184.7	1.8650	48.66	1103.5	1184.6	1.8586	**280**
55.54	1107.2	1189.5	1.8781	52.26	1107.2	1189.4	1.8713	49.34	1107.1	1189.2	1.8648	**290**
56.30	1110.8	1194.1	1.8842	52.97	1110.7	1194.0	1.8774	50.01	1110.6	1193.9	1.8710	**300**
57.81	1117.9	1203.5	1.8963	54.40	1117.8	1203.4	1.8896	51.36	1117.7	1203.3	1.8832	**320**
59.32	1125.0	1212.8	1.9082	55.82	1124.9	1212.7	1.9014	52.70	1124.8	1212.6	1.8950	**340**
60.83	1132.1	1222.1	1.9197	57.24	1132.0	1222.0	1.9129	54.05	1131.9	1221.9	1.9066	**360**
62.33	1139.2	1231.5	1.9309	58.65	1139.1	1231.4	1.9242	55.38	1139.1	1231.3	1.9178	**380**
63.83	1146.3	1240.8	1.9420	60.07	1146.3	1240.7	1.9352	56.72	1146.2	1240.7	1.9289	**400**
65.34	1153.5	1250.2	1.9527	61.48	1153.4	1250.1	1.9460	58.06	1153.4	1250.1	1.9397	**420**
66.84	1160.6	1259.6	1.9633	62.90	1160.6	1259.5	1.9566	59.39	1160.5	1259.5	1.9502	**440**
68.34	1167.8	1269.0	1.9737	64.31	1167.8	1268.9	1.9669	60.73	1167.8	1268.9	1.9606	**460**
69.83	1175.1	1278.4	1.9838	65.72	1175.0	1278.4	1.9771	62.06	1175.0	1278.3	1.9707	**480**
71.33	1182.3	1287.9	1.9938	67.13	1182.3	1287.9	1.9871	63.39	1182.2	1287.8	1.9807	**500**
72.83	1189.6	1297.4	2.0036	68.54	1189.5	1297.3	1.9968	64.72	1189.5	1297.3	1.9905	**520**
74.32	1196.9	1306.9	2.0132	69.94	1196.8	1306.9	2.0065	66.05	1196.8	1306.8	2.0001	**540**
75.82	1204.2	1316.4	2.0226	71.35	1204.2	1316.4	2.0159	67.38	1204.1	1316.4	2.0096	**560**
77.31	1211.6	1326.0	2.0319	72.76	1211.5	1326.0	2.0252	68.71	1211.5	1325.9	2.0189	**580**
78.81	1219.0	1335.6	2.0411	74.17	1218.9	1335.6	2.0344	70.04	1218.9	1335.6	2.0280	**600**
82.54	1237.6	1359.8	2.0633	77.68	1237.5	1359.7	2.0566	73.36	1237.5	1359.7	2.0503	**650**
86.28	1256.4	1384.1	2.0848	81.20	1256.4	1384.1	2.0781	76.68	1256.3	1384.1	2.0718	**700**
90.01	1275.4	1408.7	2.1055	84.71	1275.4	1408.6	2.0988	80.00	1275.4	1408.6	2.0925	**750**
93.74	1294.6	1433.4	2.1256	88.22	1294.6	1433.4	2.1189	83.32	1294.6	1433.4	2.1126	**800**
97.47	1314.1	1458.4	2.1450	91.73	1314.1	1458.4	2.1383	86.63	1314.1	1458.4	2.1320	**850**
101.19	1333.8	1483.6	2.1639	95.24	1333.8	1483.6	2.1572	89.95	1333.7	1483.5	2.1509	**900**
104.92	1353.7	1509.0	2.1822	98.75	1353.7	1509.0	2.1756	93.26	1353.6	1509.0	2.1692	**950**
108.65	1373.8	1534.6	2.2001	102.26	1373.8	1534.6	2.1934	96.57	1373.8	1534.6	2.1871	**1000**
112.38	1394.1	1560.5	2.2175	105.76	1394.1	1560.5	2.2109	99.89	1394.1	1560.5	2.2045	**1050**
116.10	1414.7	1586.6	2.2345	109.27	1414.7	1586.6	2.2279	103.20	1414.7	1586.6	2.2216	**1100**
123.55	1456.6	1639.5	2.2674	116.28	1456.6	1639.5	2.2607	109.82	1456.6	1639.5	2.2544	**1200**
131.00	1499.3	1693.3	2.2989	123.30	1499.3	1693.3	2.2922	116.45	1499.3	1693.3	2.2859	**1300**
138.45	1543.1	1748.0	2.3291	130.31	1543.1	1748.0	2.3225	123.07	1543.0	1748.0	2.3161	**1400**
145.90	1587.7	1803.7	2.3583	137.32	1587.7	1803.7	2.3516	129.69	1587.7	1803.7	2.3453	**1500**
153.35	1633.3	1860.3	2.3865	144.33	1633.3	1860.3	2.3798	136.31	1633.3	1860.3	2.3735	**1600**
168.25	1727.1	1976.2	2.4401	158.35	1727.1	1976.2	2.4335	149.55	1727.1	1976.2	2.4272	**1800**
183.14	1824.4	2095.5	2.4907	172.37	1824.4	2095.5	2.4841	162.79	1824.4	2095.5	2.4778	**2000**
198.04	1924.9	2218.1	2.5386	186.39	1924.9	2218.1	2.5319	176.03	1924.9	2218.1	2.5256	**2200**
212.93	2028.2	2343.4	2.5841	200.41	2028.2	2343.4	2.5774	189.27	2028.2	2343.4	2.5711	**2400**

Table 3. Vapor

p (t Sat.)	9.5 (190.78)				10 (193.19)				11 (197.73)			
t	v	u	h	s	v	u	h	s	v	u	h	s
Sat.	40.31	1071.5	1142.4	1.7919	38.42	1072.2	1143.3	1.7877	35.15	1073.5	1145.1	1.7800
150	*37.62*	*1056.8*	*1122.9*	*1.7609*	*35.71*	*1056.5*	*1122.6*	*1.7549*	*32.40*	*1056.1*	*1122.1*	*1.7437*
160	*38.28*	*1060.4*	*1127.7*	*1.7687*	*36.34*	*1060.2*	*1127.4*	*1.7627*	*32.98*	*1059.8*	*1126.9*	*1.7516*
170	*38.95*	*1064.0*	*1132.5*	*1.7764*	*36.97*	*1063.8*	*1132.2*	*1.7704*	*33.56*	*1063.4*	*1131.8*	*1.7593*
180	*39.61*	*1067.6*	*1137.3*	*1.7839*	*37.60*	*1067.5*	*1137.0*	*1.7780*	*34.13*	*1067.1*	*1136.6*	*1.7669*
190	*40.26*	*1071.2*	*1142.0*	*1.7913*	*38.22*	*1071.1*	*1141.8*	*1.7854*	*34.71*	*1070.7*	*1141.4*	*1.7744*
200	40.92	1074.8	1146.8	1.7986	38.85	1074.7	1146.6	1.7927	35.27	1074.4	1146.2	1.7817
210	41.57	1078.4	1151.5	1.8057	39.47	1078.3	1151.3	1.7998	35.84	1078.0	1150.9	1.7889
220	42.22	1082.0	1156.2	1.8127	40.09	1081.9	1156.1	1.8068	36.41	1081.6	1155.7	1.7959
230	42.87	1085.6	1161.0	1.8196	40.70	1085.5	1160.8	1.8137	36.97	1085.2	1160.5	1.8029
240	43.51	1089.2	1165.7	1.8264	41.32	1089.0	1165.5	1.8205	37.53	1088.8	1165.2	1.8097
250	44.16	1092.7	1170.4	1.8330	41.93	1092.6	1170.2	1.8272	38.09	1092.4	1169.9	1.8164
260	44.80	1096.3	1175.1	1.8396	42.55	1096.2	1174.9	1.8338	38.65	1096.0	1174.6	1.8230
270	45.45	1099.9	1179.7	1.8461	43.16	1099.8	1179.6	1.8403	39.21	1099.6	1179.4	1.8295
280	46.09	1103.4	1184.4	1.8525	43.77	1103.3	1184.3	1.8467	39.76	1103.1	1184.1	1.8359
290	46.73	1107.0	1189.1	1.8588	44.38	1106.9	1189.0	1.8530	40.32	1106.7	1188.8	1.8422
300	47.37	1110.5	1193.8	1.8650	44.99	1110.4	1193.7	1.8592	40.87	1110.3	1193.5	1.8485
310	48.01	1114.1	1198.5	1.8711	45.59	1114.0	1198.4	1.8653	41.43	1113.8	1198.2	1.8546
320	48.64	1117.6	1203.2	1.8771	46.20	1117.6	1203.1	1.8714	41.98	1117.4	1202.9	1.8607
330	49.28	1121.2	1207.8	1.8831	46.81	1121.1	1207.7	1.8773	42.53	1121.0	1207.5	1.8666
340	49.92	1124.8	1212.5	1.8890	47.41	1124.7	1212.4	1.8832	43.08	1124.5	1212.2	1.8725
350	50.55	1128.3	1217.2	1.8948	48.02	1128.2	1217.1	1.8890	43.63	1128.1	1216.9	1.8784
360	51.19	1131.9	1221.9	1.9005	48.62	1131.8	1221.8	1.8948	44.18	1131.7	1221.6	1.8841
370	51.83	1135.4	1226.5	1.9062	49.22	1135.4	1226.5	1.9005	44.73	1135.3	1226.3	1.8898
380	52.46	1139.0	1231.2	1.9118	49.83	1139.0	1231.2	1.9061	45.28	1138.8	1231.0	1.8954
390	53.09	1142.6	1235.9	1.9174	50.43	1142.5	1235.8	1.9116	45.83	1142.4	1235.7	1.9010
400	53.73	1146.2	1240.6	1.9228	51.03	1146.1	1240.5	1.9171	46.38	1146.0	1240.4	1.9065
420	54.99	1153.3	1250.0	1.9336	52.24	1153.3	1249.9	1.9279	47.47	1153.2	1249.8	1.9173
440	56.26	1160.5	1259.4	1.9442	53.44	1160.5	1259.3	1.9385	48.57	1160.4	1259.2	1.9279
460	57.52	1167.7	1268.8	1.9546	54.64	1167.7	1268.8	1.9489	49.66	1167.6	1268.7	1.9383
480	58.79	1174.9	1278.3	1.9647	55.84	1174.9	1278.2	1.9590	50.75	1174.8	1278.1	1.9485
500	60.05	1182.2	1287.8	1.9747	57.04	1182.2	1287.7	1.9690	51.84	1182.1	1287.6	1.9585
520	61.31	1189.5	1297.3	1.9845	58.24	1189.4	1297.2	1.9788	52.93	1189.4	1297.1	1.9683
540	62.57	1196.8	1306.8	1.9941	59.44	1196.8	1306.7	1.9885	54.02	1196.7	1306.7	1.9779
560	63.83	1204.1	1316.3	2.0036	60.63	1204.1	1316.3	1.9979	55.11	1204.0	1316.2	1.9874
580	65.09	1211.5	1325.9	2.0129	61.83	1211.5	1325.9	2.0072	56.20	1211.4	1325.8	1.9967
600	66.35	1218.9	1335.5	2.0221	63.03	1218.9	1335.5	2.0164	57.29	1218.8	1335.4	2.0058
620	67.61	1226.3	1345.2	2.0311	64.22	1226.3	1345.1	2.0254	58.38	1226.2	1345.1	2.0148
640	68.87	1233.8	1354.8	2.0399	65.42	1233.7	1354.8	2.0343	59.46	1233.7	1354.7	2.0237
660	70.13	1241.3	1364.5	2.0487	66.62	1241.2	1364.5	2.0430	60.55	1241.2	1364.4	2.0325
680	71.38	1248.8	1374.3	2.0573	67.81	1248.8	1374.2	2.0516	61.64	1248.7	1374.2	2.0411
700	72.64	1256.3	1384.0	2.0658	69.01	1256.3	1384.0	2.0601	62.73	1256.3	1384.0	2.0496
720	73.90	1263.9	1393.8	2.0742	70.20	1263.9	1393.8	2.0685	63.81	1263.9	1393.8	2.0580
740	75.16	1271.5	1403.7	2.0824	71.40	1271.5	1403.6	2.0768	64.90	1271.5	1403.6	2.0662
760	76.41	1279.2	1413.5	2.0906	72.59	1279.2	1413.5	2.0849	65.98	1279.1	1413.5	2.0744
780	77.67	1286.9	1423.4	2.0986	73.78	1286.9	1423.4	2.0930	67.07	1286.8	1423.4	2.0824
800	78.93	1294.6	1433.4	2.1066	74.98	1294.6	1433.3	2.1009	68.16	1294.6	1433.3	2.0904
850	82.07	1314.1	1458.3	2.1260	77.96	1314.0	1458.3	2.1204	70.87	1314.0	1458.3	2.1098
900	85.21	1333.7	1483.5	2.1449	80.95	1333.7	1483.5	2.1393	73.58	1333.7	1483.5	2.1287
950	88.35	1353.6	1509.0	2.1633	83.93	1353.6	1508.9	2.1576	76.30	1353.6	1508.9	2.1471
1000	91.49	1373.8	1534.6	2.1812	86.91	1373.8	1534.6	2.1755	79.01	1373.7	1534.6	2.1650
1100	97.77	1414.7	1586.6	2.2156	92.88	1414.7	1586.6	2.2099	84.43	1414.7	1586.5	2.1994
1200	104.04	1456.5	1639.4	2.2484	98.84	1456.5	1639.4	2.2428	89.85	1456.5	1639.4	2.2323
1300	110.32	1499.3	1693.3	2.2799	104.80	1499.3	1693.3	2.2743	95.27	1499.3	1693.2	2.2638
1400	116.59	1543.0	1748.0	2.3102	110.76	1543.0	1748.0	2.3045	100.69	1543.0	1748.0	2.2940
1500	122.86	1587.7	1803.7	2.3393	116.72	1587.7	1803.7	2.3337	106.11	1587.7	1803.7	2.3232
1600	129.14	1633.3	1860.3	2.3675	122.68	1633.2	1860.3	2.3618	111.52	1633.2	1860.2	2.3513
1800	141.68	1727.1	1976.2	2.4212	134.60	1727.1	1976.2	2.4155	122.36	1727.1	1976.1	2.4050
2000	154.22	1824.4	2095.5	2.4718	146.51	1824.4	2095.5	2.4661	133.19	1824.4	2095.5	2.4556
2200	166.77	1924.9	2218.1	2.5197	158.43	1924.9	2218.1	2.5140	144.03	1924.9	2218.0	2.5035
2400	179.31	2028.2	2343.4	2.5651	170.35	2028.2	2343.4	2.5595	154.86	2028.2	2343.4	2.5490

	12 (201.94)				13 (205.87)				14 (209.55)			p (t Sat.)
v	u	h	s	v	u	h	s	v	u	h	s	t
32.40	1074.7	1146.7	1.7730	30.06	1075.9	1148.2	1.7666	28.05	1076.9	1149.6	1.7606	Sat.
29.65	1055.7	1121.5	1.7334	27.32	1055.2	1120.9	1.7238	25.33	1054.8	1120.4	1.7149	150
30.19	1059.4	1126.4	1.7413	27.82	1058.9	1125.9	1.7318	25.79	1058.5	1125.3	1.7230	160
30.72	1063.0	1131.3	1.7491	28.31	1062.7	1130.8	1.7397	26.25	1062.3	1130.3	1.7309	170
31.25	1066.7	1136.1	1.7567	28.80	1066.4	1135.6	1.7473	26.71	1066.0	1135.2	1.7386	180
31.77	1070.4	1140.9	1.7642	29.29	1070.0	1140.5	1.7549	27.16	1069.7	1140.1	1.7462	190
32.30	1074.0	1145.8	1.7716	29.78	1073.7	1145.4	1.7623	27.62	1073.4	1144.9	1.7536	200
32.82	1077.7	1150.6	1.7788	30.26	1077.4	1150.2	1.7695	28.07	1077.1	1149.8	1.7609	210
33.34	1081.3	1155.3	1.7859	30.74	1081.0	1155.0	1.7767	28.52	1080.7	1154.6	1.7681	220
33.86	1084.9	1160.1	1.7929	31.22	1084.7	1159.8	1.7837	28.96	1084.4	1159.4	1.7751	230
34.37	1088.5	1164.9	1.7997	31.70	1088.3	1164.6	1.7906	29.41	1088.1	1164.2	1.7820	240
34.89	1092.2	1169.6	1.8065	32.18	1091.9	1169.3	1.7973	29.85	1091.7	1169.0	1.7888	250
35.40	1095.8	1174.4	1.8131	32.65	1095.5	1174.1	1.8040	30.30	1095.3	1173.8	1.7955	260
35.91	1099.3	1179.1	1.8196	33.13	1099.1	1178.8	1.8105	30.74	1098.9	1178.6	1.8021	270
36.42	1102.9	1183.8	1.8261	33.60	1102.7	1183.6	1.8170	31.18	1102.5	1183.3	1.8085	280
36.93	1106.5	1188.5	1.8324	34.07	1106.3	1188.3	1.8233	31.62	1106.2	1188.1	1.8149	290
37.44	1110.1	1193.2	1.8386	34.54	1109.9	1193.0	1.8296	32.06	1109.8	1192.8	1.8212	300
37.95	1113.7	1198.0	1.8448	35.01	1113.5	1197.7	1.8358	32.49	1113.4	1197.5	1.8274	310
38.46	1117.3	1202.7	1.8509	35.48	1117.1	1202.5	1.8418	32.93	1116.9	1202.3	1.8335	320
38.97	1120.8	1207.4	1.8569	35.95	1120.7	1207.2	1.8479	33.37	1120.5	1207.0	1.8395	330
39.47	1124.4	1212.1	1.8628	36.42	1124.3	1211.9	1.8538	33.80	1124.1	1211.7	1.8454	340
39.98	1128.0	1216.8	1.8686	36.89	1127.9	1216.6	1.8596	34.24	1127.7	1216.4	1.8513	350
40.48	1131.6	1221.5	1.8744	37.35	1131.4	1221.3	1.8654	34.67	1131.3	1221.1	1.8571	360
40.99	1135.1	1226.2	1.8801	37.82	1135.0	1226.0	1.8711	35.10	1134.9	1225.8	1.8628	370
41.49	1138.7	1230.9	1.8857	38.28	1138.6	1230.7	1.8768	35.54	1138.5	1230.6	1.8685	380
41.99	1142.3	1235.6	1.8913	38.75	1142.2	1235.4	1.8823	35.97	1142.1	1235.3	1.8740	390
42.50	1145.9	1240.3	1.8968	39.21	1145.8	1240.1	1.8878	36.40	1145.7	1240.0	1.8796	400
43.50	1153.1	1249.7	1.9076	40.14	1153.0	1249.6	1.8987	37.26	1152.9	1249.4	1.8904	420
44.51	1160.3	1259.1	1.9182	41.07	1160.2	1259.0	1.9093	38.13	1160.1	1258.9	1.9010	440
45.51	1167.5	1268.6	1.9286	42.00	1167.4	1268.5	1.9197	38.99	1167.4	1268.4	1.9114	460
46.51	1174.8	1278.0	1.9388	42.92	1174.7	1277.9	1.9299	39.85	1174.6	1277.8	1.9216	480
47.51	1182.0	1287.5	1.9488	43.85	1182.0	1287.4	1.9399	40.71	1181.9	1287.3	1.9317	500
48.51	1189.3	1297.0	1.9586	44.77	1189.3	1297.0	1.9497	41.56	1189.2	1296.9	1.9415	520
49.51	1196.6	1306.6	1.9682	45.69	1196.6	1306.5	1.9594	42.42	1196.5	1306.4	1.9511	540
50.51	1204.0	1316.1	1.9777	46.62	1203.9	1316.1	1.9688	43.28	1203.9	1316.0	1.9606	560
51.51	1211.4	1325.7	1.9870	47.54	1211.3	1325.7	1.9782	44.14	1211.2	1325.6	1.9699	580
52.51	1218.8	1335.4	1.9962	48.46	1218.7	1335.3	1.9873	44.99	1218.7	1335.2	1.9791	600
53.50	1226.2	1345.0	2.0052	49.38	1226.1	1344.9	1.9963	45.85	1226.1	1344.9	1.9881	620
54.50	1233.7	1354.7	2.0141	50.30	1233.6	1354.6	2.0052	46.70	1233.6	1354.6	1.9970	640
55.50	1241.1	1364.4	2.0228	51.22	1241.1	1364.3	2.0140	47.56	1241.1	1364.3	2.0058	660
56.50	1248.7	1374.1	2.0315	52.14	1248.6	1374.1	2.0226	48.41	1248.6	1374.0	2.0144	680
57.49	1256.2	1383.9	2.0400	53.06	1256.2	1383.8	2.0311	49.27	1256.2	1383.8	2.0229	700
58.49	1263.8	1393.7	2.0483	53.98	1263.8	1393.7	2.0395	50.12	1263.7	1393.6	2.0313	720
59.49	1271.4	1403.5	2.0566	54.90	1271.4	1403.5	2.0478	50.98	1271.4	1403.4	2.0396	740
60.48	1279.1	1413.4	2.0648	55.82	1279.1	1413.4	2.0559	51.83	1279.0	1413.3	2.0477	760
61.48	1286.8	1423.3	2.0728	56.74	1286.8	1423.3	2.0640	52.69	1286.7	1423.2	2.0558	780
62.47	1294.5	1433.2	2.0808	57.66	1294.5	1433.2	2.0719	53.54	1294.5	1433.2	2.0637	800
64.96	1314.0	1458.2	2.1002	59.96	1314.0	1458.2	2.0914	55.67	1313.9	1458.2	2.0832	850
67.45	1333.7	1483.4	2.1191	62.26	1333.6	1483.4	2.1103	57.81	1333.6	1483.4	2.1021	900
69.93	1353.6	1508.9	2.1375	64.55	1353.6	1508.8	2.1286	59.94	1353.5	1508.8	2.1205	950
72.42	1373.7	1534.5	2.1554	66.85	1373.7	1534.5	2.1465	62.07	1373.7	1534.5	2.1383	1000
77.39	1414.7	1586.5	2.1898	71.44	1414.6	1586.5	2.1810	66.33	1414.6	1586.5	2.1728	1100
82.36	1456.5	1639.4	2.2227	76.02	1456.5	1639.4	2.2138	70.59	1456.5	1639.4	2.2057	1200
87.33	1499.3	1693.2	2.2542	80.61	1499.3	1693.2	2.2453	74.85	1499.3	1693.2	2.2372	1300
92.30	1543.0	1748.0	2.2844	85.20	1543.0	1748.0	2.2756	79.11	1543.0	1747.9	2.2674	1400
97.26	1587.7	1803.6	2.3136	89.78	1587.7	1803.6	2.3047	83.37	1587.6	1803.6	2.2966	1500
102.23	1633.2	1860.2	2.3417	94.36	1633.2	1860.2	2.3329	87.62	1633.2	1860.2	2.3247	1600
112.16	1727.1	1976.1	2.3954	103.53	1727.1	1976.1	2.3866	96.14	1727.1	1976.1	2.3784	1800
122.09	1824.4	2095.5	2.4460	112.70	1824.4	2095.5	2.4372	104.65	1824.4	2095.5	2.4290	2000
132.02	1924.9	2218.0	2.4939	121.87	1924.9	2218.0	2.4851	113.16	1924.9	2218.0	2.4769	2200
141.95	2028.2	2343.4	2.5394	131.04	2028.2	2343.4	2.5305	121.68	2028.2	2343.4	2.5224	2400

Table 3. Vapor

p (t Sat.)	14.696 (211.99)				15 (213.03)				16 (216.31)			
t	v	u	h	s	v	u	h	s	v	u	h	s
Sat.	26.80	1077.6	1150.5	1.7567	26.29	1077.9	1150.9	1.7551	24.75	1078.8	1152.1	1.7499
150	*24.10*	*1054.5*	*1120.0*	*1.7090*	*23.59*	*1054.3*	*1119.8*	*1.7065*	*22.08*	*1053.9*	*1119.2*	*1.6987*
160	*24.54*	*1058.2*	*1125.0*	*1.7171*	*24.03*	*1058.1*	*1124.8*	*1.7147*	*22.49*	*1057.7*	*1124.3*	*1.7069*
170	*24.98*	*1062.0*	*1129.9*	*1.7251*	*24.46*	*1061.9*	*1129.8*	*1.7226*	*22.90*	*1061.5*	*1129.3*	*1.7149*
180	*25.42*	*1065.7*	*1134.9*	*1.7328*	*24.89*	*1065.6*	*1134.7*	*1.7304*	*23.30*	*1065.3*	*1134.2*	*1.7227*
190	*25.85*	*1069.5*	*1139.8*	*1.7405*	*25.32*	*1069.4*	*1139.6*	*1.7380*	*23.71*	*1069.0*	*1139.2*	*1.7304*
200	*26.29*	*1073.2*	*1144.7*	*1.7479*	*25.74*	*1073.1*	*1144.5*	*1.7455*	*24.11*	*1072.8*	*1144.1*	*1.7379*
210	*26.72*	*1076.9*	*1149.5*	*1.7553*	*26.17*	*1076.8*	*1149.4*	*1.7529*	*24.50*	*1076.5*	*1149.0*	*1.7453*
220	27.15	1080.6	1154.4	1.7624	26.59	1080.5	1154.3	1.7601	24.90	1080.2	1153.9	1.7525
230	27.57	1084.2	1159.2	1.7695	27.01	1084.1	1159.1	1.7671	25.29	1083.9	1158.8	1.7596
240	28.00	1087.9	1164.0	1.7764	27.42	1087.8	1163.9	1.7741	25.69	1087.6	1163.6	1.7666
250	28.42	1091.5	1168.8	1.7832	27.84	1091.5	1168.7	1.7809	26.08	1091.2	1168.4	1.7734
260	28.85	1095.2	1173.6	1.7899	28.25	1095.1	1173.5	1.7876	26.47	1094.9	1173.2	1.7802
270	29.27	1098.8	1178.4	1.7965	28.67	1098.7	1178.3	1.7942	26.86	1098.5	1178.0	1.7868
280	29.69	1102.4	1183.1	1.8030	29.08	1102.4	1183.1	1.8007	27.24	1102.2	1182.8	1.7933
290	30.11	1106.0	1187.9	1.8094	29.49	1106.0	1187.8	1.8071	27.63	1105.8	1187.6	1.7997
300	30.52	1109.6	1192.6	1.8157	29.90	1109.6	1192.6	1.8134	28.01	1109.4	1192.4	1.8060
310	30.94	1113.2	1197.4	1.8219	30.31	1113.2	1197.3	1.8196	28.40	1113.0	1197.1	1.8122
320	31.36	1116.8	1202.1	1.8280	30.72	1116.8	1202.1	1.8257	28.78	1116.6	1201.9	1.8184
330	31.77	1120.4	1206.8	1.8340	31.13	1120.4	1206.8	1.8317	29.17	1120.2	1206.6	1.8244
340	32.19	1124.0	1211.6	1.8400	31.53	1124.0	1211.5	1.8377	29.55	1123.9	1211.3	1.8304
350	32.60	1127.6	1216.3	1.8458	31.94	1127.6	1216.2	1.8435	29.93	1127.5	1216.1	1.8363
360	33.02	1131.2	1221.0	1.8516	32.34	1131.2	1221.0	1.8493	30.31	1131.1	1220.8	1.8421
370	33.43	1134.8	1225.7	1.8574	32.75	1134.8	1225.7	1.8551	30.69	1134.7	1225.5	1.8478
380	33.84	1138.4	1230.5	1.8630	33.15	1138.4	1230.4	1.8607	31.07	1138.3	1230.3	1.8535
390	34.26	1142.0	1235.2	1.8686	33.56	1142.0	1235.1	1.8663	31.45	1141.9	1235.0	1.8591
400	34.67	1145.6	1239.9	1.8741	33.96	1145.6	1239.9	1.8718	31.83	1145.5	1239.7	1.8646
420	35.49	1152.8	1249.3	1.8850	34.77	1152.8	1249.3	1.8827	32.59	1152.7	1249.2	1.8755
440	36.31	1160.1	1258.8	1.8956	35.57	1160.0	1258.8	1.8933	33.34	1159.9	1258.7	1.8861
460	37.13	1167.3	1268.3	1.9060	36.38	1167.3	1268.2	1.9038	34.10	1167.2	1268.1	1.8965
480	37.95	1174.6	1277.8	1.9162	37.18	1174.5	1277.7	1.9140	34.85	1174.5	1277.6	1.9068
500	38.77	1181.8	1287.3	1.9263	37.98	1181.8	1287.3	1.9240	35.60	1181.7	1287.2	1.9168
520	39.59	1189.1	1296.8	1.9361	38.79	1189.1	1296.8	1.9338	36.35	1189.1	1296.7	1.9266
540	40.41	1196.5	1306.4	1.9457	39.59	1196.5	1306.3	1.9435	37.11	1196.4	1306.3	1.9363
560	41.22	1203.8	1315.9	1.9552	40.39	1203.8	1315.9	1.9529	37.86	1203.8	1315.8	1.9458
580	42.04	1211.2	1325.5	1.9645	41.19	1211.2	1325.5	1.9623	38.61	1211.1	1325.4	1.9551
600	42.86	1218.6	1335.2	1.9737	41.99	1218.6	1335.1	1.9714	39.36	1218.6	1335.1	1.9643
620	43.67	1226.1	1344.8	1.9827	42.79	1226.0	1344.8	1.9805	40.11	1226.0	1344.7	1.9733
640	44.49	1233.5	1354.5	1.9916	43.58	1233.5	1354.5	1.9894	40.86	1233.5	1354.4	1.9822
660	45.30	1241.0	1364.2	2.0004	44.38	1241.0	1364.2	1.9981	41.60	1241.0	1364.2	1.9910
680	46.12	1248.6	1374.0	2.0090	45.18	1248.5	1374.0	2.0068	42.35	1248.5	1373.9	1.9996
700	46.93	1256.1	1383.8	2.0175	45.98	1256.1	1383.7	2.0153	43.10	1256.1	1383.7	2.0081
720	47.75	1263.7	1393.6	2.0259	46.78	1263.7	1393.6	2.0237	43.85	1263.7	1393.5	2.0165
740	48.56	1271.4	1403.4	2.0342	47.57	1271.3	1403.4	2.0319	44.60	1271.3	1403.3	2.0248
760	49.37	1279.0	1413.3	2.0424	48.37	1279.0	1413.3	2.0401	45.34	1279.0	1413.2	2.0329
780	50.19	1286.7	1423.2	2.0504	49.17	1286.7	1423.2	2.0481	46.09	1286.7	1423.1	2.0410
800	51.00	1294.4	1433.1	2.0584	49.97	1294.4	1433.1	2.0561	46.84	1294.4	1433.1	2.0490
850	53.03	1313.9	1458.1	2.0778	51.96	1313.9	1458.1	2.0756	48.71	1313.9	1458.1	2.0684
900	55.07	1333.6	1483.4	2.0967	53.95	1333.6	1483.3	2.0945	50.57	1333.6	1483.3	2.0873
950	57.10	1353.5	1508.8	2.1151	55.94	1353.5	1508.8	2.1128	52.44	1353.5	1508.8	2.1057
1000	59.13	1373.7	1534.5	2.1330	57.93	1373.6	1534.4	2.1307	54.31	1373.6	1534.4	2.1236
1100	63.19	1414.6	1586.4	2.1674	61.91	1414.6	1586.4	2.1652	58.04	1414.6	1586.4	2.1581
1200	67.25	1456.5	1639.3	2.2003	65.88	1456.5	1639.3	2.1981	61.76	1456.5	1639.3	2.1909
1300	71.30	1499.3	1693.2	2.2318	69.86	1499.3	1693.2	2.2295	65.49	1499.2	1693.2	2.2224
1400	75.36	1543.0	1747.9	2.2621	73.83	1543.0	1747.9	2.2598	69.22	1543.0	1747.9	2.2527
1500	79.42	1587.6	1803.6	2.2912	77.81	1587.6	1803.6	2.2890	72.94	1587.6	1803.6	2.2818
1600	83.47	1633.2	1860.2	2.3194	81.78	1633.2	1860.2	2.3171	76.67	1633.2	1860.2	2.3100
1800	91.58	1727.0	1976.1	2.3731	89.73	1727.0	1976.1	2.3708	84.12	1727.0	1976.1	2.3637
2000	99.69	1824.4	2095.5	2.4237	97.67	1824.3	2095.5	2.4214	91.57	1824.3	2095.5	2.4143
2200	107.80	1924.8	2218.0	2.4716	105.62	1924.8	2218.0	2.4693	99.02	1924.8	2218.0	2.4622
2400	115.91	2028.1	2343.4	2.5170	113.56	2028.1	2343.4	2.5148	106.47	2028.1	2343.4	2.5076

	17 (219.43)				18 (222.40)				19 (225.24)			p (t Sat.)
v	u	h	s	v	u	h	s	v	u	h	s	t
23.39	1079.7	1153.3	1.7450	22.17	1080.5	1154.4	1.7404	21.08	1081.3	1155.4	1.7361	Sat.
20.74	*1053.4*	*1118.7*	*1.6912*	*19.553*	*1053.0*	*1118.1*	*1.6842*	*18.490*	*1052.5*	*1117.5*	*1.6775*	**150**
21.13	*1057.3*	*1123.7*	*1.6995*	*19.923*	*1056.8*	*1123.2*	*1.6925*	*18.843*	*1056.4*	*1122.7*	*1.6858*	**160**
21.52	*1061.1*	*1128.8*	*1.7075*	*20.291*	*1060.7*	*1128.3*	*1.7006*	*19.192*	*1060.3*	*1127.8*	*1.6940*	**170**
21.90	*1064.9*	*1133.8*	*1.7154*	*20.655*	*1064.5*	*1133.3*	*1.7086*	*19.539*	*1064.1*	*1132.8*	*1.7020*	**180**
22.28	*1068.7*	*1138.8*	*1.7232*	*21.016*	*1068.3*	*1138.3*	*1.7163*	*19.884*	*1068.0*	*1137.9*	*1.7098*	**190**
22.66	*1072.4*	*1143.7*	*1.7307*	*21.38*	*1072.1*	*1143.3*	*1.7239*	*20.23*	*1071.8*	*1142.9*	*1.7175*	**200**
23.04	*1076.2*	*1148.6*	*1.7381*	*21.73*	*1075.9*	*1148.3*	*1.7314*	*20.57*	*1075.6*	*1147.9*	*1.7250*	**210**
23.41	1079.9	1153.5	1.7454	*22.09*	*1079.6*	*1153.2*	*1.7387*	*20.90*	*1079.3*	*1152.8*	*1.7323*	**220**
23.78	1083.6	1158.4	1.7525	22.44	1083.3	1158.1	1.7459	21.24	1083.1	1157.7	1.7395	**230**
24.15	1087.3	1163.3	1.7595	22.79	1087.1	1163.0	1.7529	21.57	1086.8	1162.6	1.7466	**240**
24.52	1091.0	1168.1	1.7664	23.14	1090.7	1167.8	1.7598	21.90	1090.5	1167.5	1.7535	**250**
24.89	1094.7	1173.0	1.7732	23.49	1094.4	1172.7	1.7666	22.24	1094.2	1172.4	1.7603	**260**
25.26	1098.3	1177.8	1.7798	23.84	1098.1	1177.5	1.7732	22.57	1097.9	1177.2	1.7670	**270**
25.62	1102.0	1182.6	1.7863	24.18	1101.8	1182.3	1.7798	22.89	1101.6	1182.1	1.7735	**280**
25.99	1105.6	1187.4	1.7928	24.53	1105.4	1187.1	1.7862	23.22	1105.2	1186.9	1.7800	**290**
26.35	1109.2	1192.1	1.7991	24.87	1109.1	1191.9	1.7926	23.55	1108.9	1191.7	1.7864	**300**
26.71	1112.9	1196.9	1.8053	25.21	1112.7	1196.7	1.7988	23.87	1112.5	1196.5	1.7926	**310**
27.07	1116.5	1201.7	1.8115	25.56	1116.3	1201.5	1.8050	24.20	1116.2	1201.3	1.7988	**320**
27.44	1120.1	1206.4	1.8175	25.90	1120.0	1206.2	1.8110	24.52	1119.8	1206.0	1.8049	**330**
27.80	1123.7	1211.2	1.8235	26.24	1123.6	1211.0	1.8170	24.85	1123.4	1210.8	1.8109	**340**
28.16	1127.3	1215.9	1.8294	26.58	1127.2	1215.7	1.8229	25.17	1127.1	1215.6	1.8168	**350**
28.51	1130.9	1220.6	1.8352	26.92	1130.8	1220.5	1.8288	25.49	1130.7	1220.3	1.8227	**360**
28.87	1134.5	1225.4	1.8410	27.26	1134.4	1225.2	1.8345	25.81	1134.3	1225.1	1.8284	**370**
29.23	1138.2	1230.1	1.8466	27.60	1138.0	1230.0	1.8402	26.13	1137.9	1229.8	1.8341	**380**
29.59	1141.8	1234.9	1.8523	27.93	1141.7	1234.7	1.8458	26.45	1141.6	1234.6	1.8397	**390**
29.95	1145.4	1239.6	1.8578	28.27	1145.3	1239.5	1.8514	26.77	1145.2	1239.3	1.8453	**400**
30.66	1152.6	1249.1	1.8687	28.95	1152.5	1248.9	1.8623	27.41	1152.4	1248.8	1.8562	**420**
31.37	1159.9	1258.5	1.8793	29.62	1159.8	1258.4	1.8729	28.05	1159.7	1258.3	1.8669	**440**
32.08	1167.1	1268.0	1.8898	30.29	1167.0	1267.9	1.8834	28.69	1167.0	1267.8	1.8773	**460**
32.79	1174.4	1277.5	1.9000	30.96	1174.3	1277.4	1.8936	29.33	1174.2	1277.3	1.8876	**480**
33.50	1181.7	1287.1	1.9100	31.63	1181.6	1287.0	1.9037	29.96	1181.5	1286.9	1.8976	**500**
34.21	1189.0	1296.6	1.9199	32.30	1188.9	1296.5	1.9135	30.59	1188.9	1296.4	1.9075	**520**
34.92	1196.3	1306.2	1.9295	32.97	1196.3	1306.1	1.9232	31.23	1196.2	1306.0	1.9172	**540**
35.62	1203.7	1315.8	1.9390	33.64	1203.6	1315.7	1.9327	31.86	1203.6	1315.6	1.9267	**560**
36.33	1211.1	1325.4	1.9484	34.31	1211.0	1325.3	1.9420	32.49	1211.0	1325.2	1.9360	**580**
37.04	1218.5	1335.0	1.9576	34.97	1218.5	1334.9	1.9512	33.13	1218.4	1334.9	1.9452	**600**
37.74	1226.0	1344.7	1.9666	35.64	1225.9	1344.6	1.9602	33.76	1225.9	1344.6	1.9542	**620**
38.45	1233.4	1354.4	1.9755	36.31	1233.4	1354.3	1.9692	34.39	1233.3	1354.3	1.9631	**640**
39.15	1240.9	1364.1	1.9843	36.97	1240.9	1364.0	1.9779	35.02	1240.8	1364.0	1.9719	**660**
39.86	1248.5	1373.9	1.9929	37.64	1248.4	1373.8	1.9866	35.65	1248.4	1373.7	1.9806	**680**
40.56	1256.0	1383.6	2.0014	38.30	1256.0	1383.6	1.9951	36.28	1256.0	1383.5	1.9891	**700**
41.27	1263.6	1393.5	2.0098	38.97	1263.6	1393.4	2.0035	36.91	1263.6	1393.4	1.9975	**720**
41.97	1271.3	1403.3	2.0181	39.63	1271.2	1403.3	2.0117	37.54	1271.2	1403.2	2.0058	**740**
42.67	1278.9	1413.2	2.0262	40.30	1278.9	1413.1	2.0199	38.18	1278.9	1413.1	2.0139	**760**
43.38	1286.6	1423.1	2.0343	40.96	1286.6	1423.1	2.0280	38.81	1286.6	1423.0	2.0220	**780**
44.08	1294.4	1433.0	2.0423	41.63	1294.3	1433.0	2.0359	39.43	1294.3	1433.0	2.0299	**800**
45.84	1313.8	1458.1	2.0617	43.29	1313.8	1458.0	2.0554	41.01	1313.8	1458.0	2.0494	**850**
47.60	1333.5	1483.3	2.0806	44.95	1333.5	1483.2	2.0743	42.58	1333.5	1483.2	2.0683	**900**
49.35	1353.5	1508.7	2.0990	46.61	1353.4	1508.7	2.0927	44.15	1353.4	1508.7	2.0867	**950**
51.11	1373.6	1534.4	2.1169	48.27	1373.6	1534.4	2.1106	45.73	1373.6	1534.3	2.1046	**1000**
54.62	1414.6	1586.4	2.1514	51.58	1414.5	1586.4	2.1450	48.87	1414.5	1586.3	2.1391	**1100**
58.13	1456.4	1639.3	2.1842	54.90	1456.4	1639.3	2.1779	52.01	1456.4	1639.3	2.1720	**1200**
61.64	1499.2	1693.1	2.2157	58.21	1499.2	1693.1	2.2094	55.15	1499.2	1693.1	2.2035	**1300**
65.15	1543.0	1747.9	2.2460	61.52	1542.9	1747.9	2.2397	58.29	1542.9	1747.9	2.2337	**1400**
68.65	1587.6	1803.6	2.2752	64.84	1587.6	1803.6	2.2689	61.42	1587.6	1803.6	2.2629	**1500**
72.16	1633.2	1860.2	2.3033	68.15	1633.2	1860.2	2.2970	64.56	1633.2	1860.2	2.2911	**1600**
79.17	1727.0	1976.1	2.3570	74.77	1727.0	1976.1	2.3507	70.84	1727.0	1976.1	2.3448	**1800**
86.18	1824.3	2095.5	2.4076	81.39	1824.3	2095.4	2.4013	77.11	1824.3	2095.4	2.3954	**2000**
93.19	1924.8	2218.0	2.4555	88.02	1924.8	2218.0	2.4492	83.38	1924.8	2218.0	2.4433	**2200**
100.20	2028.1	2343.4	2.5010	94.64	2028.1	2343.3	2.4947	89.66	2028.1	2343.3	2.4887	**2400**

Table 3. Vapor

p (t Sat.)	20 (227.96)				22 (233.08)				24 (237.82)			
t	v	u	h	s	v	u	h	s	v	u	h	s
Sat.	20.09	1082.0	1156.4	1.7320	18.378	1083.4	1158.2	1.7244	16.941	1084.7	1159.9	1.7174
150	*17.532*	*1052.0*	*1116.9*	*1.6710*	*15.878*	*1051.1*	*1115.8*	*1.6590*	*14.498*	*1050.2*	*1114.6*	*1.6479*
160	*17.870*	*1056.0*	*1122.1*	*1.6795*	*16.188*	*1055.1*	*1121.0*	*1.6675*	*14.787*	*1054.2*	*1119.9*	*1.6565*
170	*18.204*	*1059.9*	*1127.2*	*1.6877*	*16.496*	*1059.1*	*1126.2*	*1.6759*	*15.073*	*1058.3*	*1125.2*	*1.6650*
180	*18.535*	*1063.8*	*1132.4*	*1.6957*	*16.801*	*1063.0*	*1131.4*	*1.6840*	*15.355*	*1062.2*	*1130.4*	*1.6732*
190	*18.864*	*1067.6*	*1137.4*	*1.7036*	*17.103*	*1066.9*	*1136.5*	*1.6920*	*15.635*	*1066.2*	*1135.6*	*1.6813*
200	*19.191*	*1071.4*	*1142.5*	*1.7113*	*17.402*	*1070.8*	*1141.6*	*1.6998*	*15.912*	*1070.1*	*1140.8*	*1.6892*
210	*19.515*	*1075.2*	*1147.5*	*1.7188*	*17.700*	*1074.6*	*1146.7*	*1.7074*	*16.187*	*1074.0*	*1145.9*	*1.6969*
220	*19.837*	*1079.0*	*1152.4*	*1.7262*	*17.995*	*1078.5*	*1151.7*	*1.7149*	*16.460*	*1077.9*	*1151.0*	*1.7044*
230	20.157	1082.8	1157.4	1.7335	*18.288*	*1082.3*	*1156.7*	*1.7222*	*16.730*	*1081.7*	*1156.0*	*1.7118*
240	20.475	1086.5	1162.3	1.7405	18.579	1086.0	1161.7	1.7293	16.999	1085.5	1161.0	1.7190
250	20.79	1090.3	1167.2	1.7475	18.869	1089.8	1166.6	1.7363	17.267	1089.3	1166.0	1.7260
260	21.11	1094.0	1172.1	1.7543	19.158	1093.5	1171.5	1.7432	17.533	1093.1	1171.0	1.7330
270	21.42	1097.7	1177.0	1.7610	19.445	1097.3	1176.4	1.7499	17.798	1096.8	1175.9	1.7398
280	21.73	1101.4	1181.8	1.7676	19.731	1101.0	1181.3	1.7566	18.061	1100.6	1180.8	1.7464
290	22.05	1105.0	1186.6	1.7741	20.016	1104.7	1186.2	1.7631	18.324	1104.3	1185.7	1.7530
300	22.36	1108.7	1191.5	1.7805	20.30	1108.4	1191.0	1.7695	18.585	1108.0	1190.5	1.7595
310	22.67	1112.4	1196.3	1.7868	20.58	1112.0	1195.8	1.7758	18.845	1111.7	1195.4	1.7658
320	22.98	1116.0	1201.0	1.7930	20.86	1115.7	1200.6	1.7820	19.105	1115.4	1200.2	1.7721
330	23.28	1119.7	1205.8	1.7991	21.15	1119.4	1205.4	1.7882	19.364	1119.1	1205.1	1.7782
340	23.59	1123.3	1210.6	1.8051	21.43	1123.0	1210.2	1.7942	19.622	1122.7	1209.9	1.7843
350	23.90	1126.9	1215.4	1.8110	21.71	1126.7	1215.0	1.8002	19.879	1126.4	1214.7	1.7902
360	24.21	1130.6	1220.1	1.8168	21.99	1130.3	1219.8	1.8060	20.136	1130.1	1219.5	1.7961
370	24.51	1134.2	1224.9	1.8226	22.27	1133.9	1224.6	1.8118	20.393	1133.7	1224.3	1.8019
380	24.82	1137.8	1229.7	1.8283	22.54	1137.6	1229.4	1.8175	20.649	1137.4	1229.1	1.8077
390	25.12	1141.4	1234.4	1.8340	22.82	1141.2	1234.1	1.8232	20.904	1141.0	1233.8	1.8133
400	25.43	1145.1	1239.2	1.8395	23.10	1144.9	1238.9	1.8288	21.16	1144.7	1238.6	1.8189
420	26.03	1152.3	1248.7	1.8504	23.65	1152.1	1248.4	1.8397	21.67	1151.9	1248.2	1.8299
440	26.64	1159.6	1258.2	1.8611	24.21	1159.4	1258.0	1.8504	22.18	1159.2	1257.7	1.8407
460	27.25	1166.9	1267.7	1.8716	24.76	1166.7	1267.5	1.8609	22.68	1166.5	1267.3	1.8512
480	27.85	1174.2	1277.2	1.8819	25.31	1174.0	1277.0	1.8712	23.19	1173.9	1276.8	1.8614
500	28.46	1181.5	1286.8	1.8919	25.86	1181.3	1286.6	1.8813	23.69	1181.2	1286.4	1.8715
520	29.06	1188.8	1296.3	1.9018	26.41	1188.7	1296.2	1.8911	24.20	1188.5	1296.0	1.8814
540	29.66	1196.1	1305.9	1.9114	26.95	1196.0	1305.8	1.9008	24.70	1195.9	1305.6	1.8911
560	30.26	1203.5	1315.5	1.9210	27.50	1203.4	1315.4	1.9103	25.20	1203.3	1315.2	1.9006
580	30.87	1210.9	1325.2	1.9303	28.05	1210.8	1325.0	1.9197	25.70	1210.7	1324.9	1.9100
600	31.47	1218.4	1334.8	1.9395	28.60	1218.3	1334.7	1.9289	26.21	1218.2	1334.5	1.9192
620	32.07	1225.8	1344.5	1.9485	29.14	1225.7	1344.4	1.9380	26.71	1225.6	1344.2	1.9283
640	32.67	1233.3	1354.2	1.9575	29.69	1233.2	1354.1	1.9469	27.21	1233.1	1353.9	1.9372
660	33.27	1240.8	1363.9	1.9662	30.24	1240.7	1363.8	1.9556	27.71	1240.6	1363.7	1.9460
680	33.87	1248.3	1373.7	1.9749	30.78	1248.3	1373.6	1.9643	28.21	1248.2	1373.5	1.9546
700	34.47	1255.9	1383.5	1.9834	31.33	1255.8	1383.4	1.9728	28.71	1255.8	1383.3	1.9632
720	35.07	1263.5	1393.3	1.9918	31.87	1263.5	1393.2	1.9812	29.21	1263.4	1393.1	1.9716
740	35.66	1271.2	1403.2	2.0001	32.42	1271.1	1403.1	1.9895	29.71	1271.0	1403.0	1.9799
760	36.26	1278.8	1413.0	2.0082	32.96	1278.8	1413.0	1.9977	30.21	1278.7	1412.9	1.9880
780	36.86	1286.5	1423.0	2.0163	33.51	1286.5	1422.9	2.0058	30.71	1286.4	1422.8	1.9961
800	37.46	1294.3	1432.9	2.0243	34.05	1294.2	1432.8	2.0137	31.21	1294.2	1432.7	2.0041
850	38.96	1313.8	1457.9	2.0438	35.41	1313.7	1457.9	2.0332	32.45	1313.7	1457.8	2.0236
900	40.45	1333.5	1483.2	2.0627	36.77	1333.4	1483.1	2.0521	33.70	1333.4	1483.0	2.0425
950	41.94	1353.4	1508.6	2.0810	38.13	1353.4	1508.6	2.0705	34.95	1353.3	1508.5	2.0609
1000	43.44	1373.5	1534.3	2.0989	39.48	1373.5	1534.2	2.0884	36.19	1373.5	1534.2	2.0788
1100	46.42	1414.5	1586.3	2.1334	42.20	1414.5	1586.3	2.1229	38.68	1414.4	1586.2	2.1133
1200	49.41	1456.4	1639.2	2.1663	44.91	1456.4	1639.2	2.1558	41.17	1456.3	1639.2	2.1462
1300	52.39	1499.2	1693.1	2.1978	47.62	1499.2	1693.1	2.1873	43.65	1499.1	1693.0	2.1777
1400	55.37	1542.9	1747.9	2.2281	50.34	1542.9	1747.8	2.2175	46.14	1542.9	1747.8	2.2079
1500	58.35	1587.6	1803.5	2.2572	53.05	1587.6	1803.5	2.2467	48.62	1587.5	1803.5	2.2371
1600	61.33	1633.2	1860.1	2.2854	55.76	1633.1	1860.1	2.2749	51.11	1633.1	1860.1	2.2653
1800	67.29	1727.0	1976.1	2.3391	61.18	1727.0	1976.0	2.3286	56.08	1727.0	1976.0	2.3190
2000	73.25	1824.3	2095.4	2.3897	66.59	1824.3	2095.4	2.3792	61.04	1824.3	2095.4	2.3696
2200	79.21	1924.8	2218.0	2.4376	72.01	1924.8	2218.0	2.4271	66.01	1924.8	2218.0	2.4175
2400	85.17	2028.1	2343.3	2.4830	77.43	2028.1	2343.3	2.4725	70.98	2028.1	2343.3	2.4629

v	u	h	s	v	u	h	s	v	u	h	s	t
15.719	1085.9	1161.5	1.7110	14.665	1087.0	1163.0	1.7051	13.748	1088.0	1164.3	1.6996	**Sat.**
13.330	*1049.2*	*1113.4*	*1.6375*	*12.329*	*1048.3*	*1112.2*	*1.6277*	*11.460*	*1047.3*	*1111.0*	*1.6185*	**150**
13.601	*1053.4*	*1118.8*	*1.6463*	*12.583*	*1052.5*	*1117.7*	*1.6366*	*11.701*	*1051.6*	*1116.5*	*1.6275*	**160**
13.867	*1057.4*	*1124.1*	*1.6548*	*12.834*	*1056.6*	*1123.1*	*1.6453*	*11.938*	*1055.8*	*1122.0*	*1.6364*	**170**
14.131	*1061.5*	*1129.5*	*1.6632*	*13.082*	*1060.7*	*1128.5*	*1.6538*	*12.172*	*1059.9*	*1127.5*	*1.6449*	**180**
14.392	*1065.5*	*1134.7*	*1.6714*	*13.327*	*1064.7*	*1133.8*	*1.6621*	*12.403*	*1064.0*	*1132.9*	*1.6533*	**190**
14.650	*1069.4*	*1139.9*	*1.6793*	*13.569*	*1068.8*	*1139.1*	*1.6701*	*12.631*	*1068.1*	*1138.2*	*1.6615*	**200**
14.906	*1073.4*	*1145.1*	*1.6871*	*13.809*	*1072.7*	*1144.3*	*1.6780*	*12.857*	*1072.1*	*1143.5*	*1.6694*	**210**
15.160	*1077.3*	*1150.2*	*1.6947*	*14.046*	*1076.7*	*1149.5*	*1.6856*	*13.081*	*1076.1*	*1148.7*	*1.6771*	**220**
15.412	*1081.2*	*1155.3*	*1.7021*	*14.282*	*1080.6*	*1154.6*	*1.6931*	*13.303*	*1080.0*	*1153.9*	*1.6847*	**230**
15.662	*1085.0*	*1160.4*	*1.7094*	*14.516*	*1084.5*	*1159.7*	*1.7005*	*13.523*	*1084.0*	*1159.0*	*1.6921*	**240**
15.911	1088.8	1165.4	1.7165	14.748	1088.3	1164.8	1.7077	*13.741*	*1087.9*	*1164.1*	*1.6994*	**250**
16.158	1092.6	1170.4	1.7235	14.979	1092.2	1169.8	1.7147	13.958	1091.7	1169.2	1.7064	**260**
16.404	1096.4	1175.3	1.7304	15.209	1096.0	1174.8	1.7216	14.173	1095.6	1174.2	1.7134	**270**
16.648	1100.2	1180.3	1.7371	15.437	1099.8	1179.8	1.7284	14.387	1099.4	1179.2	1.7202	**280**
16.892	1103.9	1185.2	1.7437	15.664	1103.5	1184.7	1.7350	14.600	1103.2	1184.2	1.7269	**290**
17.134	1107.6	1190.1	1.7502	15.890	1107.3	1189.6	1.7415	14.812	1106.9	1189.2	1.7334	**300**
17.375	1111.4	1195.0	1.7565	16.115	1111.0	1194.5	1.7479	15.023	1110.7	1194.1	1.7399	**310**
17.616	1115.1	1199.8	1.7628	16.339	1114.8	1199.4	1.7542	15.233	1114.4	1199.0	1.7462	**320**
17.856	1118.8	1204.7	1.7690	16.563	1118.5	1204.3	1.7604	15.442	1118.2	1203.9	1.7525	**330**
18.095	1122.4	1209.5	1.7751	16.786	1122.2	1209.1	1.7666	15.651	1121.9	1208.8	1.7586	**340**
18.333	1126.1	1214.3	1.7811	17.008	1125.9	1214.0	1.7726	15.859	1125.6	1213.6	1.7646	**350**
18.571	1129.8	1219.1	1.7870	17.229	1129.5	1218.8	1.7785	16.067	1129.3	1218.5	1.7706	**360**
18.808	1133.5	1224.0	1.7928	17.450	1133.2	1223.6	1.7844	16.273	1133.0	1223.3	1.7765	**370**
19.045	1137.1	1228.8	1.7986	17.671	1136.9	1228.5	1.7901	16.480	1136.7	1228.1	1.7822	**380**
19.282	1140.8	1233.6	1.8043	17.891	1140.6	1233.3	1.7958	16.686	1140.3	1233.0	1.7880	**390**
19.518	1144.4	1238.3	1.8099	18.110	1144.2	1238.1	1.8014	16.891	1144.0	1237.8	1.7936	**400**
19.988	1151.8	1247.9	1.8209	18.549	1151.6	1247.7	1.8125	17.301	1151.4	1247.4	1.8047	**420**
20.458	1159.1	1257.5	1.8316	18.985	1158.9	1257.3	1.8233	17.709	1158.7	1257.0	1.8155	**440**
20.926	1166.4	1267.1	1.8422	19.421	1166.2	1266.8	1.8338	18.116	1166.1	1266.6	1.8260	**460**
21.393	1173.7	1276.6	1.8525	19.856	1173.6	1276.4	1.8441	18.523	1173.4	1276.2	1.8364	**480**
21.86	1181.1	1286.2	1.8625	20.29	1180.9	1286.0	1.8542	18.928	1180.8	1285.9	1.8465	**500**
22.33	1188.4	1295.8	1.8724	20.72	1188.3	1295.7	1.8641	19.333	1188.2	1295.5	1.8564	**520**
22.79	1195.8	1305.4	1.8822	21.15	1195.7	1305.3	1.8739	19.737	1195.5	1305.1	1.8661	**540**
23.26	1203.2	1315.1	1.8917	21.59	1203.1	1314.9	1.8834	20.140	1203.0	1314.8	1.8757	**560**
23.72	1210.6	1324.7	1.9011	22.02	1210.5	1324.6	1.8928	20.543	1210.4	1324.4	1.8851	**580**
24.18	1218.0	1334.4	1.9103	22.45	1217.9	1334.3	1.9020	20.95	1217.8	1334.1	1.8943	**600**
24.65	1225.5	1344.1	1.9194	22.88	1225.4	1344.0	1.9111	21.35	1225.3	1343.8	1.9034	**620**
25.11	1233.0	1353.8	1.9283	23.31	1232.9	1353.7	1.9200	21.75	1232.8	1353.6	1.9123	**640**
25.57	1240.5	1363.6	1.9371	23.74	1240.5	1363.5	1.9288	22.15	1240.4	1363.3	1.9211	**660**
26.03	1248.1	1373.4	1.9457	24.17	1248.0	1373.2	1.9375	22.55	1247.9	1373.1	1.9298	**680**
26.50	1255.7	1383.2	1.9543	24.60	1255.6	1383.1	1.9460	22.95	1255.5	1383.0	1.9384	**700**
26.96	1263.3	1393.0	1.9627	25.03	1263.2	1392.9	1.9544	23.35	1263.2	1392.8	1.9468	**720**
27.42	1271.0	1402.9	1.9710	25.46	1270.9	1402.8	1.9627	23.75	1270.8	1402.7	1.9551	**740**
27.88	1278.6	1412.8	1.9792	25.88	1278.6	1412.7	1.9709	24.15	1278.5	1412.6	1.9633	**760**
28.34	1286.3	1422.7	1.9872	26.31	1286.3	1422.6	1.9790	24.55	1286.2	1422.5	1.9714	**780**
28.80	1294.1	1432.7	1.9952	26.74	1294.0	1432.6	1.9870	24.95	1294.0	1432.5	1.9793	**800**
29.95	1313.6	1457.7	2.0147	27.81	1313.5	1457.6	2.0065	25.95	1313.5	1457.6	1.9988	**850**
31.10	1333.3	1483.0	2.0336	28.88	1333.3	1482.9	2.0254	26.95	1333.2	1482.8	2.0178	**900**
32.25	1353.3	1508.4	2.0520	29.95	1353.2	1508.4	2.0438	27.95	1353.2	1508.3	2.0362	**950**
33.40	1373.4	1534.1	2.0699	31.02	1373.4	1534.1	2.0617	28.95	1373.3	1534.0	2.0541	**1000**
35.70	1414.4	1586.2	2.1044	33.15	1414.4	1586.1	2.0962	30.94	1414.3	1586.1	2.0886	**1100**
38.00	1456.3	1639.1	2.1373	35.28	1456.3	1639.1	2.1291	32.93	1456.2	1639.1	2.1215	**1200**
40.29	1499.1	1693.0	2.1688	37.41	1499.1	1693.0	2.1606	34.92	1499.1	1692.9	2.1530	**1300**
42.59	1542.9	1747.8	2.1991	39.55	1542.8	1747.7	2.1909	36.91	1542.8	1747.7	2.1833	**1400**
44.88	1587.5	1803.5	2.2283	41.68	1587.5	1803.4	2.2201	38.90	1587.5	1803.4	2.2125	**1500**
47.18	1633.1	1860.1	2.2565	43.81	1633.1	1860.1	2.2483	40.88	1633.1	1860.0	2.2407	**1600**
51.76	1727.0	1976.0	2.3102	48.06	1726.9	1976.0	2.3020	44.86	1726.9	1976.0	2.2944	**1800**
56.35	1824.3	2095.4	2.3608	52.32	1824.3	2095.4	2.3526	48.83	1824.2	2095.3	2.3450	**2000**
60.93	1924.8	2217.9	2.4087	56.58	1924.8	2217.9	2.4005	52.81	1924.7	2217.9	2.3929	**2200**
65.52	2028.1	2343.3	2.4541	60.84	2028.1	2343.3	2.4459	56.78	2028.0	2343.3	2.4383	**2400**

Table 3. Vapor

p (t Sat.)	32 (254.06)				34 (257.59)				36 (260.96)			
t	v	u	h	s	v	u	h	s	v	u	h	s
Sat.	12.942	1089.0	1165.6	1.6945	12.228	1089.9	1166.8	1.6896	11.591	1090.7	1167.9	1.6851
200	*11.810*	*1067.4*	*1137.3*	*1.6533*	*11.086*	*1066.7*	*1136.4*	*1.6455*	*10.442*	*1066.0*	*1135.6*	*1.6381*
210	*12.024*	*1071.5*	*1142.7*	*1.6613*	*11.289*	*1070.8*	*1141.8*	*1.6536*	*10.635*	*1070.2*	*1141.0*	*1.6463*
220	*12.236*	*1075.5*	*1147.9*	*1.6691*	*11.490*	*1074.9*	*1147.2*	*1.6616*	*10.827*	*1074.3*	*1146.4*	*1.6543*
230	*12.445*	*1079.5*	*1153.2*	*1.6768*	*11.689*	*1078.9*	*1152.5*	*1.6693*	*11.016*	*1078.4*	*1151.7*	*1.6621*
240	*12.653*	*1083.4*	*1158.4*	*1.6842*	*11.886*	*1082.9*	*1157.7*	*1.6768*	*11.203*	*1082.4*	*1157.0*	*1.6697*
250	*12.859*	*1087.4*	*1163.5*	*1.6915*	*12.081*	*1086.9*	*1162.9*	*1.6842*	*11.389*	*1086.4*	*1162.2*	*1.6771*
260	13.063	1091.3	1168.6	1.6987	12.274	1090.8	1168.0	1.6913	*11.573*	*1090.3*	*1167.4*	*1.6844*
270	13.267	1095.1	1173.7	1.7057	12.467	1094.7	1173.1	1.6984	11.755	1094.2	1172.6	1.6915
280	13.468	1099.0	1178.7	1.7125	12.658	1098.6	1178.2	1.7053	11.937	1098.1	1177.7	1.6984
290	13.669	1102.8	1183.7	1.7193	12.847	1102.4	1183.2	1.7121	12.117	1102.0	1182.7	1.7052
300	13.869	1106.6	1188.7	1.7258	13.036	1106.2	1188.2	1.7187	12.296	1105.8	1187.8	1.7119
310	14.067	1110.4	1193.7	1.7323	13.224	1110.0	1193.2	1.7252	12.474	1109.7	1192.8	1.7184
320	14.265	1114.1	1198.6	1.7387	13.411	1113.8	1198.2	1.7316	12.651	1113.5	1197.7	1.7249
330	14.462	1117.9	1203.5	1.7450	13.597	1117.6	1203.1	1.7379	12.828	1117.2	1202.7	1.7312
340	14.658	1121.6	1208.4	1.7511	13.782	1121.3	1208.0	1.7441	13.004	1121.0	1207.6	1.7374
350	14.854	1125.3	1213.3	1.7572	13.967	1125.0	1212.9	1.7502	13.179	1124.8	1212.6	1.7435
360	15.049	1129.0	1218.1	1.7632	14.151	1128.8	1217.8	1.7562	13.353	1128.5	1217.5	1.7495
370	15.243	1132.7	1223.0	1.7690	14.335	1132.5	1222.7	1.7621	13.527	1132.2	1222.4	1.7555
380	15.437	1136.4	1227.8	1.7749	14.518	1136.2	1227.5	1.7679	13.700	1136.0	1227.2	1.7613
390	15.631	1140.1	1232.7	1.7806	14.700	1139.9	1232.4	1.7736	13.873	1139.7	1232.1	1.7671
400	15.824	1143.8	1237.5	1.7862	14.882	1143.6	1237.2	1.7793	14.045	1143.4	1237.0	1.7728
420	16.209	1151.2	1247.2	1.7973	15.246	1151.0	1246.9	1.7904	14.389	1150.8	1246.6	1.7839
440	16.592	1158.5	1256.8	1.8081	15.607	1158.4	1256.6	1.8013	14.731	1158.2	1256.3	1.7948
460	16.975	1165.9	1266.4	1.8187	15.968	1165.7	1266.2	1.8119	15.072	1165.6	1266.0	1.8054
480	17.356	1173.3	1276.0	1.8291	16.327	1173.1	1275.8	1.8222	15.413	1173.0	1275.6	1.8158
500	17.737	1180.6	1285.7	1.8392	16.686	1180.5	1285.5	1.8324	15.752	1180.4	1285.3	1.8259
520	18.117	1188.0	1295.3	1.8492	17.044	1187.9	1295.1	1.8423	16.091	1187.8	1294.9	1.8359
540	18.496	1195.4	1304.9	1.8589	17.401	1195.3	1304.8	1.8521	16.428	1195.2	1304.6	1.8457
560	18.875	1202.8	1314.6	1.8685	17.758	1202.7	1314.5	1.8617	16.766	1202.6	1314.3	1.8553
580	19.253	1210.3	1324.3	1.8779	18.114	1210.2	1324.1	1.8711	17.102	1210.1	1324.0	1.8647
600	19.631	1217.7	1334.0	1.8871	18.470	1217.6	1333.9	1.8803	17.439	1217.5	1333.7	1.8739
620	20.008	1225.2	1343.7	1.8962	18.826	1225.1	1343.6	1.8894	17.775	1225.0	1343.5	1.8830
640	20.385	1232.7	1353.5	1.9052	19.181	1232.7	1353.3	1.8984	18.110	1232.6	1353.2	1.8920
660	20.761	1240.3	1363.2	1.9140	19.535	1240.2	1363.1	1.9072	18.446	1240.1	1363.0	1.9008
680	21.138	1247.9	1373.0	1.9226	19.890	1247.8	1372.9	1.9159	18.781	1247.7	1372.8	1.9095
700	21.51	1255.5	1382.8	1.9312	20.24	1255.4	1382.7	1.9244	19.115	1255.3	1382.6	1.9181
720	21.89	1263.1	1392.7	1.9396	20.60	1263.0	1392.6	1.9329	19.450	1262.9	1392.5	1.9265
740	22.27	1270.7	1402.6	1.9479	20.95	1270.7	1402.5	1.9412	19.784	1270.6	1402.4	1.9348
760	22.64	1278.4	1412.5	1.9561	21.31	1278.4	1412.4	1.9494	20.118	1278.3	1412.3	1.9430
780	23.02	1286.2	1422.4	1.9642	21.66	1286.1	1422.4	1.9575	20.452	1286.0	1422.3	1.9511
800	23.39	1293.9	1432.4	1.9722	22.01	1293.8	1432.3	1.9654	20.79	1293.8	1432.2	1.9591
850	24.33	1313.4	1457.5	1.9917	22.89	1313.4	1457.4	1.9850	21.62	1313.3	1457.3	1.9786
900	25.26	1333.2	1482.8	2.0106	23.78	1333.1	1482.7	2.0039	22.45	1333.1	1482.6	1.9976
950	26.20	1353.1	1508.3	2.0290	24.66	1353.1	1508.2	2.0223	23.28	1353.0	1508.1	2.0160
1000	27.13	1373.3	1534.0	2.0470	25.54	1373.3	1533.9	2.0403	24.12	1373.2	1533.9	2.0339
1100	29.00	1414.3	1586.0	2.0815	27.29	1414.3	1586.0	2.0748	25.78	1414.2	1586.0	2.0684
1200	30.87	1456.2	1639.0	2.1144	29.05	1456.2	1639.0	2.1077	27.44	1456.2	1638.9	2.1014
1300	32.74	1499.0	1692.9	2.1459	30.81	1499.0	1692.9	2.1392	29.10	1499.0	1692.8	2.1329
1400	34.60	1542.8	1747.7	2.1762	32.56	1542.8	1747.7	2.1695	30.75	1542.8	1747.6	2.1632
1500	36.46	1587.5	1803.4	2.2054	34.32	1587.4	1803.4	2.1987	32.41	1587.4	1803.3	2.1924
1600	38.33	1633.1	1860.0	2.2335	36.07	1633.0	1860.0	2.2268	34.07	1633.0	1860.0	2.2205
1800	42.06	1726.9	1975.9	2.2872	39.58	1726.9	1975.9	2.2806	37.38	1726.9	1975.9	2.2743
2000	45.78	1824.2	2095.3	2.3379	43.09	1824.2	2095.3	2.3312	40.69	1824.2	2095.3	2.3249
2200	49.51	1924.7	2217.9	2.3858	46.60	1924.7	2217.9	2.3791	44.01	1924.7	2217.9	2.3728
2400	53.23	2028.0	2343.3	2.4312	50.10	2028.0	2343.2	2.4245	47.32	2028.0	2343.2	2.4182

	38 (264.18)				**40** (267.26)				**42** (270.22)			**p** (t Sat.)
v	u	h	s	v	u	h	s	v	u	h	s	t
11.018	1091.5	1169.0	1.6808	10.501	1092.3	1170.0	1.6767	10.032	1093.0	1171.0	1.6728	**Sat.**
9.865	*1065.3*	*1134.7*	*1.6311*	*9.346*	*1064.6*	*1133.8*	*1.6243*	*8.876*	*1063.9*	*1132.8*	*1.6178*	**200**
10.050	*1069.5*	*1140.2*	*1.6394*	*9.523*	*1068.8*	*1139.3*	*1.6327*	*9.047*	*1068.2*	*1138.5*	*1.6263*	**210**
10.233	*1073.7*	*1145.6*	*1.6475*	*9.699*	*1073.0*	*1144.8*	*1.6409*	*9.215*	*1072.4*	*1144.0*	*1.6346*	**220**
10.414	*1077.8*	*1151.0*	*1.6553*	*9.872*	*1077.2*	*1150.3*	*1.6488*	*9.381*	*1076.6*	*1149.5*	*1.6426*	**230**
10.593	*1081.8*	*1156.3*	*1.6630*	*10.043*	*1081.3*	*1155.6*	*1.6565*	*9.545*	*1080.8*	*1155.0*	*1.6504*	**240**
10.770	*1085.9*	*1161.6*	*1.6705*	*10.212*	*1085.4*	*1161.0*	*1.6641*	*9.708*	*1084.9*	*1160.3*	*1.6580*	**250**
10.945	*1089.9*	*1166.8*	*1.6778*	*10.380*	*1089.4*	*1166.2*	*1.6714*	*9.869*	*1088.9*	*1165.6*	*1.6654*	**260**
11.119	1093.8	1172.0	1.6849	10.546	1093.4	1171.4	1.6786	*10.028*	*1092.9*	*1170.9*	*1.6726*	**270**
11.292	1097.7	1177.1	1.6919	10.711	1097.3	1176.6	1.6857	10.186	1096.9	1176.1	1.6797	**280**
11.463	1101.6	1182.2	1.6987	10.875	1101.2	1181.7	1.6926	10.343	1100.8	1181.2	1.6867	**290**
11.634	1105.5	1187.3	1.7055	11.038	1105.1	1186.8	1.6993	10.498	1104.7	1186.3	1.6934	**300**
11.803	1109.3	1192.3	1.7120	11.200	1109.0	1191.9	1.7059	10.653	1108.6	1191.4	1.7001	**310**
11.972	1113.1	1197.3	1.7185	11.360	1112.8	1196.9	1.7124	10.807	1112.5	1196.5	1.7066	**320**
12.140	1116.9	1202.3	1.7248	11.520	1116.6	1201.9	1.7188	10.960	1116.3	1201.5	1.7130	**330**
12.307	1120.7	1207.3	1.7311	11.680	1120.4	1206.9	1.7251	11.112	1120.1	1206.5	1.7193	**340**
12.473	1124.5	1212.2	1.7372	11.838	1124.2	1211.8	1.7312	11.264	1123.9	1211.5	1.7255	**350**
12.639	1128.2	1217.1	1.7433	11.996	1128.0	1216.8	1.7373	11.414	1127.7	1216.4	1.7316	**360**
12.804	1132.0	1222.0	1.7492	12.153	1131.7	1221.7	1.7432	11.565	1131.5	1221.4	1.7376	**370**
12.969	1135.7	1226.9	1.7551	12.310	1135.5	1226.6	1.7491	11.715	1135.3	1226.3	1.7435	**380**
13.133	1139.5	1231.8	1.7608	12.467	1139.2	1231.5	1.7549	11.864	1139.0	1231.2	1.7493	**390**
13.297	1143.2	1236.7	1.7665	12.623	1143.0	1236.4	1.7606	12.013	1142.7	1236.1	1.7550	**400**
13.623	1150.6	1246.4	1.7777	12.933	1150.4	1246.1	1.7718	12.309	1150.2	1245.9	1.7662	**420**
13.948	1158.0	1256.1	1.7886	13.243	1157.8	1255.8	1.7828	12.604	1157.6	1255.6	1.7772	**440**
14.272	1165.4	1265.8	1.7993	13.551	1165.2	1265.5	1.7934	12.898	1165.1	1265.3	1.7879	**460**
14.594	1172.8	1275.4	1.8097	13.858	1172.7	1275.2	1.8038	13.191	1172.5	1275.0	1.7983	**480**
14.916	1180.2	1285.1	1.8198	14.164	1180.1	1284.9	1.8140	13.483	1179.9	1284.7	1.8085	**500**
15.237	1187.6	1294.8	1.8298	14.469	1187.5	1294.6	1.8240	13.775	1187.4	1294.4	1.8185	**520**
15.558	1195.1	1304.5	1.8396	14.774	1194.9	1304.3	1.8338	14.065	1194.8	1304.1	1.8283	**540**
15.878	1202.5	1314.1	1.8492	15.078	1202.4	1314.0	1.8434	14.355	1202.3	1313.8	1.8379	**560**
16.197	1210.0	1323.9	1.8586	15.382	1209.9	1323.7	1.8529	14.645	1209.7	1323.6	1.8474	**580**
16.516	1217.4	1333.6	1.8679	15.685	1217.3	1333.4	1.8621	14.934	1217.2	1333.3	1.8567	**600**
16.834	1224.9	1343.3	1.8770	15.988	1224.8	1343.2	1.8713	15.223	1224.8	1343.1	1.8658	**620**
17.153	1232.5	1353.1	1.8860	16.291	1232.4	1353.0	1.8802	15.511	1232.3	1352.8	1.8748	**640**
17.470	1240.0	1362.9	1.8948	16.593	1239.9	1362.8	1.8891	15.799	1239.9	1362.6	1.8836	**660**
17.788	1247.6	1372.7	1.9035	16.895	1247.5	1372.6	1.8977	16.087	1247.4	1372.5	1.8923	**680**
18.105	1255.2	1382.5	1.9120	17.196	1255.1	1382.4	1.9063	16.374	1255.1	1382.3	1.9009	**700**
18.422	1262.9	1392.4	1.9205	17.498	1262.8	1392.3	1.9147	16.661	1262.7	1392.2	1.9093	**720**
18.739	1270.5	1402.3	1.9288	17.799	1270.5	1402.2	1.9231	16.948	1270.4	1402.1	1.9176	**740**
19.056	1278.2	1412.2	1.9370	18.100	1278.2	1412.1	1.9313	17.235	1278.1	1412.0	1.9258	**760**
19.372	1286.0	1422.2	1.9451	18.401	1285.9	1422.1	1.9394	17.521	1285.8	1422.0	1.9340	**780**
19.688	1293.7	1432.2	1.9531	18.701	1293.7	1432.1	1.9474	17.808	1293.6	1432.0	1.9419	**800**
20.478	1313.3	1457.3	1.9726	19.452	1313.2	1457.2	1.9669	18.523	1313.2	1457.1	1.9615	**850**
21.268	1333.0	1482.6	1.9916	20.202	1333.0	1482.5	1.9859	19.238	1332.9	1482.4	1.9805	**900**
22.056	1353.0	1508.1	2.0100	20.951	1352.9	1508.0	2.0043	19.951	1352.9	1508.0	1.9989	**950**
22.844	1373.2	1533.8	2.0279	21.700	1373.1	1533.8	2.0223	20.665	1373.1	1533.7	2.0168	**1000**
24.42	1414.2	1585.9	2.0625	23.20	1414.2	1585.9	2.0568	22.09	1414.1	1585.8	2.0514	**1100**
25.99	1456.1	1638.9	2.0954	24.69	1456.1	1638.9	2.0897	23.51	1456.1	1638.8	2.0843	**1200**
27.56	1499.0	1692.8	2.1269	26.18	1498.9	1692.8	2.1212	24.94	1498.9	1692.7	2.1159	**1300**
29.13	1542.7	1747.6	2.1572	27.68	1542.7	1747.6	2.1515	26.36	1542.7	1747.5	2.1461	**1400**
30.70	1587.4	1803.3	2.1864	29.17	1587.4	1803.3	2.1807	27.78	1587.4	1803.3	2.1753	**1500**
32.27	1633.0	1859.9	2.2146	30.66	1633.0	1859.9	2.2089	29.20	1633.0	1859.9	2.2035	**1600**
35.41	1726.9	1975.9	2.2683	33.64	1726.9	1975.9	2.2626	32.04	1726.8	1975.9	2.2572	**1800**
38.55	1824.2	2095.3	2.3189	36.62	1824.2	2095.3	2.3132	34.88	1824.2	2095.3	2.3079	**2000**
41.69	1924.7	2217.8	2.3668	39.61	1924.7	2217.8	2.3611	37.72	1924.7	2217.8	2.3558	**2200**
44.83	2028.0	2343.2	2.4122	42.59	2028.0	2343.2	2.4066	40.56	2028.0	2343.2	2.4012	**2400**

Table 3. Vapor

p (t Sat.)	44 (273.07)				46 (275.82)				48 (278.47)			
t	v	u	h	s	v	u	h	s	v	u	h	s
Sat.	9.603	1093.7	1171.9	1.6691	9.211	1094.4	1172.8	1.6656	8.851	1095.0	1173.6	1.6622
200	8.448	1063.1	1131.9	1.6116	8.058	1062.4	1131.0	1.6055	7.699	1061.7	1130.1	1.5997
210	8.613	1067.5	1137.6	1.6202	8.217	1066.8	1136.8	1.6142	7.853	1066.1	1135.9	1.6085
220	8.775	1071.8	1143.2	1.6285	8.373	1071.2	1142.4	1.6226	8.005	1070.5	1141.6	1.6170
230	8.935	1076.0	1148.8	1.6366	8.527	1075.4	1148.0	1.6308	8.154	1074.8	1147.3	1.6252
240	9.093	1080.2	1154.3	1.6444	8.680	1079.7	1153.5	1.6387	8.301	1079.1	1152.8	1.6332
250	9.249	1084.3	1159.7	1.6521	8.830	1083.8	1159.0	1.6465	8.446	1083.3	1158.3	1.6410
260	9.404	1088.4	1165.0	1.6596	8.979	1087.9	1164.4	1.6540	8.590	1087.5	1163.8	1.6486
270	9.557	1092.5	1170.3	1.6669	9.126	1092.0	1169.7	1.6614	8.731	1091.6	1169.1	1.6560
280	9.708	1096.5	1175.5	1.6740	9.272	1096.1	1175.0	1.6685	8.872	1095.6	1174.4	1.6633
290	9.859	1100.4	1180.7	1.6810	9.417	1100.0	1180.2	1.6756	9.011	1099.6	1179.7	1.6703
300	10.008	1104.4	1185.9	1.6878	9.560	1104.0	1185.4	1.6824	9.150	1103.6	1184.9	1.6772
310	10.156	1108.3	1191.0	1.6945	9.703	1107.9	1190.5	1.6891	9.287	1107.6	1190.1	1.6840
320	10.304	1112.2	1196.1	1.7011	9.844	1111.8	1195.6	1.6957	9.423	1111.5	1195.2	1.6906
330	10.450	1116.0	1201.1	1.7075	9.985	1115.7	1200.7	1.7022	9.558	1115.4	1200.3	1.6971
340	10.596	1119.8	1206.1	1.7138	10.125	1119.6	1205.7	1.7085	9.693	1119.3	1205.4	1.7035
350	10.741	1123.7	1211.1	1.7200	10.264	1123.4	1210.8	1.7148	9.827	1123.1	1210.4	1.7097
360	10.886	1127.5	1216.1	1.7261	10.403	1127.2	1215.7	1.7209	9.960	1126.9	1215.4	1.7159
370	11.030	1131.2	1221.0	1.7321	10.541	1131.0	1220.7	1.7269	10.093	1130.7	1220.4	1.7219
380	11.173	1135.0	1226.0	1.7381	10.678	1134.8	1225.7	1.7329	10.225	1134.5	1225.4	1.7279
390	11.316	1138.8	1230.9	1.7439	10.816	1138.5	1230.6	1.7387	10.357	1138.3	1230.3	1.7337
400	11.458	1142.5	1235.8	1.7496	10.952	1142.3	1235.5	1.7445	10.488	1142.1	1235.3	1.7395
420	11.742	1150.0	1245.6	1.7609	11.224	1149.8	1245.4	1.7558	10.749	1149.6	1245.1	1.7508
440	12.024	1157.5	1255.4	1.7718	11.495	1157.3	1255.1	1.7667	11.009	1157.1	1254.9	1.7619
460	12.305	1164.9	1265.1	1.7825	11.764	1164.7	1264.9	1.7775	11.268	1164.6	1264.7	1.7726
480	12.585	1172.3	1274.8	1.7930	12.032	1172.2	1274.6	1.7879	11.525	1172.0	1274.4	1.7831
500	12.865	1179.8	1284.5	1.8032	12.300	1179.6	1284.3	1.7982	11.782	1179.5	1284.2	1.7933
520	13.143	1187.2	1294.2	1.8132	12.566	1187.1	1294.1	1.8082	12.038	1187.0	1293.9	1.8034
540	13.421	1194.7	1304.0	1.8231	12.832	1194.6	1303.8	1.8180	12.293	1194.4	1303.6	1.8132
560	13.698	1202.2	1313.7	1.8327	13.098	1202.0	1313.5	1.8277	12.547	1201.9	1313.4	1.8229
580	13.975	1209.6	1323.4	1.8421	13.363	1209.5	1323.3	1.8371	12.801	1209.4	1323.1	1.8323
600	14.251	1217.1	1333.2	1.8514	13.627	1217.0	1333.0	1.8464	13.055	1216.9	1332.9	1.8417
620	14.526	1224.7	1342.9	1.8606	13.891	1224.6	1342.8	1.8556	13.308	1224.5	1342.7	1.8508
640	14.802	1232.2	1352.7	1.8696	14.155	1232.1	1352.6	1.8646	13.561	1232.0	1352.5	1.8598
660	15.077	1239.8	1362.5	1.8784	14.418	1239.7	1362.4	1.8734	13.814	1239.6	1362.3	1.8686
680	15.352	1247.4	1372.4	1.8871	14.681	1247.3	1372.3	1.8821	14.066	1247.2	1372.1	1.8774
700	15.626	1255.0	1382.2	1.8957	14.944	1254.9	1382.1	1.8907	14.318	1254.8	1382.0	1.8859
720	15.901	1262.6	1392.1	1.9041	15.206	1262.6	1392.0	1.8992	14.570	1262.5	1391.9	1.8944
740	16.175	1270.3	1402.0	1.9124	15.469	1270.2	1401.9	1.9075	14.821	1270.2	1401.8	1.9027
760	16.449	1278.0	1412.0	1.9207	15.731	1278.0	1411.9	1.9157	15.073	1277.9	1411.8	1.9110
780	16.722	1285.8	1421.9	1.9288	15.993	1285.7	1421.8	1.9238	15.324	1285.6	1421.7	1.9191
800	16.996	1293.5	1431.9	1.9368	16.254	1293.5	1431.8	1.9318	15.575	1293.4	1431.7	1.9271
850	17.679	1313.1	1457.0	1.9563	16.908	1313.0	1457.0	1.9514	16.201	1313.0	1456.9	1.9467
900	18.361	1332.9	1482.4	1.9753	17.561	1332.8	1482.3	1.9704	16.827	1332.8	1482.2	1.9656
950	19.043	1352.8	1507.9	1.9937	18.213	1352.8	1507.8	1.9888	17.452	1352.8	1507.8	1.9841
1000	19.724	1373.0	1533.6	2.0117	18.865	1373.0	1533.6	2.0068	18.077	1373.0	1533.5	2.0020
1100	21.08	1414.1	1585.8	2.0462	20.17	1414.1	1585.7	2.0413	19.325	1414.0	1585.7	2.0366
1200	22.44	1456.0	1638.8	2.0792	21.47	1456.0	1638.7	2.0743	20.571	1456.0	1638.7	2.0695
1300	23.80	1498.9	1692.7	2.1107	22.77	1498.9	1692.7	2.1058	21.816	1498.8	1692.6	2.1011
1400	25.16	1542.7	1747.5	2.1410	24.06	1542.6	1747.5	2.1361	23.061	1542.6	1747.5	2.1314
1500	26.52	1587.3	1803.2	2.1702	25.36	1587.3	1803.2	2.1653	24.305	1587.3	1803.2	2.1606
1600	27.87	1632.9	1859.9	2.1984	26.66	1632.9	1859.9	2.1935	25.55	1632.9	1859.8	2.1888
1800	30.58	1726.8	1975.8	2.2521	29.25	1726.8	1975.8	2.2472	28.03	1726.8	1975.8	2.2425
2000	33.29	1824.1	2095.2	2.3027	31.85	1824.1	2095.2	2.2978	30.52	1824.1	2095.2	2.2931
2200	36.01	1924.6	2217.8	2.3506	34.44	1924.6	2217.8	2.3457	33.00	1924.6	2217.8	2.3410
2400	38.72	2027.9	2343.2	2.3961	37.03	2027.9	2343.2	2.3912	35.49	2027.9	2343.1	2.3865

	50 (281.03)				52 (283.52)				54 (285.92)			p (t Sat.)
v	u	h	s	v	u	h	s	v	u	h	s	t
8.518	1095.6	1174.4	1.6589	8.210	1096.2	1175.2	1.6558	7.924	1096.8	1176.0	1.6528	Sat.
7.370	1060.9	1129.1	1.5940	7.065	1060.2	1128.2	1.5886	6.783	1059.4	1127.2	1.5832	200
7.519	1065.4	1135.0	1.6029	7.210	1064.7	1134.1	1.5975	6.924	1064.0	1133.2	1.5923	210
7.665	1069.9	1140.8	1.6115	7.352	1069.2	1140.0	1.6062	7.062	1068.6	1139.1	1.6011	220
7.810	1074.2	1146.5	1.6198	7.492	1073.6	1145.7	1.6146	7.198	1073.0	1145.0	1.6096	230
7.952	1078.6	1152.1	1.6279	7.630	1078.0	1151.4	1.6228	7.332	1077.4	1150.7	1.6178	240
8.092	1082.8	1157.7	1.6358	7.766	1082.3	1157.0	1.6307	7.464	1081.7	1156.3	1.6258	250
8.231	1087.0	1163.1	1.6434	7.900	1086.5	1162.5	1.6384	7.594	1086.0	1161.9	1.6336	260
8.368	1091.1	1168.5	1.6509	8.033	1090.7	1167.9	1.6459	7.722	1090.2	1167.4	1.6411	270
8.504	1095.2	1173.9	1.6582	8.164	1094.8	1173.3	1.6533	7.850	1094.3	1172.8	1.6485	280
8.639	1099.2	1179.2	1.6653	8.294	1098.8	1178.7	1.6604	7.975	1098.4	1178.1	1.6557	290
8.772	1103.2	1184.4	1.6722	8.423	1102.9	1183.9	1.6674	8.100	1102.5	1183.4	1.6627	300
8.904	1107.2	1189.6	1.6790	8.551	1106.9	1189.1	1.6742	8.224	1106.5	1188.7	1.6696	310
9.036	1111.2	1194.8	1.6857	8.678	1110.8	1194.3	1.6809	8.346	1110.5	1193.9	1.6763	320
9.166	1115.1	1199.9	1.6922	8.804	1114.8	1199.5	1.6875	8.468	1114.4	1199.1	1.6829	330
9.296	1119.0	1205.0	1.6986	8.929	1118.7	1204.6	1.6939	8.589	1118.4	1204.2	1.6893	340
9.425	1122.8	1210.0	1.7049	9.053	1122.5	1209.7	1.7002	8.709	1122.3	1209.3	1.6957	350
9.553	1126.7	1215.1	1.7110	9.177	1126.4	1214.7	1.7064	8.829	1126.1	1214.4	1.7019	360
9.681	1130.5	1220.1	1.7171	9.300	1130.2	1219.7	1.7125	8.948	1130.0	1219.4	1.7080	370
9.808	1134.3	1225.0	1.7231	9.423	1134.1	1224.7	1.7185	9.067	1133.8	1224.4	1.7140	380
9.935	1138.1	1230.0	1.7290	9.545	1137.9	1229.7	1.7244	9.184	1137.6	1229.4	1.7199	390
10.061	1141.9	1235.0	1.7348	9.667	1141.7	1234.7	1.7302	9.302	1141.4	1234.4	1.7258	400
10.312	1149.4	1244.8	1.7461	9.909	1149.2	1244.6	1.7416	9.536	1149.0	1244.3	1.7372	420
10.562	1156.9	1254.6	1.7572	10.150	1156.7	1254.4	1.7526	9.768	1156.6	1254.2	1.7483	440
10.811	1164.4	1264.4	1.7679	10.389	1164.2	1264.2	1.7634	9.999	1164.1	1264.0	1.7591	460
11.059	1171.9	1274.2	1.7784	10.628	1171.7	1274.0	1.7739	10.229	1171.6	1273.8	1.7696	480
11.305	1179.4	1284.0	1.7887	10.865	1179.2	1283.8	1.7842	10.458	1179.1	1283.6	1.7799	500
11.551	1186.8	1293.7	1.7988	11.102	1186.7	1293.5	1.7943	10.687	1186.6	1293.4	1.7900	520
11.796	1194.3	1303.5	1.8086	11.338	1194.2	1303.3	1.8042	10.914	1194.1	1303.1	1.7999	540
12.041	1201.8	1313.2	1.8183	11.574	1201.7	1313.1	1.8138	11.141	1201.6	1312.9	1.8095	560
12.285	1209.3	1323.0	1.8277	11.809	1209.2	1322.8	1.8233	11.368	1209.1	1322.7	1.8190	580
12.529	1216.8	1332.8	1.8371	12.043	1216.7	1332.6	1.8326	11.594	1216.6	1332.5	1.8284	600
12.772	1224.4	1342.5	1.8462	12.278	1224.3	1342.4	1.8418	11.820	1224.2	1342.3	1.8375	620
13.015	1231.9	1352.4	1.8552	12.511	1231.8	1352.2	1.8508	12.045	1231.7	1352.1	1.8466	640
13.258	1239.5	1362.2	1.8641	12.745	1239.4	1362.1	1.8597	12.270	1239.3	1361.9	1.8554	660
13.500	1247.1	1372.0	1.8728	12.978	1247.0	1371.9	1.8684	12.494	1247.0	1371.8	1.8642	680
13.742	1254.8	1381.9	1.8814	13.211	1254.7	1381.8	1.8770	12.719	1254.6	1381.7	1.8728	700
13.984	1262.4	1391.8	1.8898	13.444	1262.3	1391.7	1.8854	12.943	1262.3	1391.6	1.8812	720
14.226	1270.1	1401.7	1.8982	13.676	1270.0	1401.6	1.8938	13.167	1270.0	1401.5	1.8896	740
14.467	1277.8	1411.7	1.9064	13.908	1277.8	1411.6	1.9020	13.391	1277.7	1411.5	1.8978	760
14.708	1285.6	1421.7	1.9145	14.140	1285.5	1421.6	1.9101	13.614	1285.4	1421.5	1.9059	780
14.949	1293.3	1431.7	1.9225	14.372	1293.3	1431.6	1.9182	13.838	1293.2	1431.5	1.9139	800
15.551	1312.9	1456.8	1.9421	14.951	1312.9	1456.7	1.9377	14.395	1312.8	1456.7	1.9335	850
16.152	1332.7	1482.2	1.9611	15.529	1332.7	1482.1	1.9567	14.952	1332.6	1482.0	1.9525	900
16.753	1352.7	1507.7	1.9796	16.107	1352.7	1507.7	1.9752	15.509	1352.6	1507.6	1.9710	950
17.352	1372.9	1533.5	1.9975	16.684	1372.9	1533.4	1.9932	16.064	1372.8	1533.4	1.9890	1000
18.551	1414.0	1585.6	2.0321	17.836	1414.0	1585.6	2.0277	17.174	1413.9	1585.5	2.0235	1100
19.747	1456.0	1638.7	2.0650	18.987	1455.9	1638.6	2.0607	18.283	1455.9	1638.6	2.0565	1200
20.943	1498.8	1692.6	2.0966	20.137	1498.8	1692.6	2.0922	19.390	1498.8	1692.5	2.0881	1300
22.138	1542.6	1747.4	2.1269	21.286	1542.6	1747.4	2.1225	20.497	1542.6	1747.4	2.1184	1400
23.332	1587.3	1803.2	2.1561	22.434	1587.3	1803.1	2.1517	21.603	1587.3	1803.1	2.1476	1500
24.53	1632.9	1859.8	2.1843	23.58	1632.9	1859.8	2.1799	22.71	1632.9	1859.8	2.1758	1600
26.91	1726.8	1975.8	2.2380	25.88	1726.8	1975.8	2.2337	24.92	1726.7	1975.7	2.2295	1800
29.30	1824.1	2095.2	2.2886	28.17	1824.1	2095.2	2.2843	27.13	1824.1	2095.2	2.2801	2000
31.68	1924.6	2217.8	2.3365	30.47	1924.6	2217.7	2.3322	29.34	1924.6	2217.7	2.3280	2200
34.07	2027.9	2343.1	2.3820	32.76	2027.9	2343.1	2.3776	31.55	2027.9	2343.1	2.3735	2400

Table 3. Vapor

p (t Sat.)	56 (288.26)				58 (290.53)				60 (292.73)			
t	v	u	h	s	v	u	h	s	v	u	h	s
Sat.	7.658	1097.3	1176.7	1.6499	7.410	1097.8	1177.4	1.6471	7.177	1098.3	1178.0	1.6444
200	6.520	1058.7	1126.2	1.5780	6.276	1057.9	1125.2	1.5729	6.047	1057.1	1124.2	1.5680
210	6.658	1063.3	1132.3	1.5872	6.410	1062.6	1131.4	1.5822	6.178	1061.9	1130.5	1.5774
220	6.792	1067.9	1138.3	1.5960	6.541	1067.3	1137.5	1.5912	6.307	1066.6	1136.6	1.5864
230	6.925	1072.4	1144.2	1.6046	6.670	1071.8	1143.4	1.5998	6.432	1071.2	1142.6	1.5952
240	7.055	1076.8	1150.0	1.6129	6.797	1076.3	1149.2	1.6082	6.556	1075.7	1148.5	1.6036
250	7.183	1081.2	1155.6	1.6210	6.921	1080.7	1155.0	1.6164	6.677	1080.1	1154.3	1.6118
260	7.309	1085.5	1161.2	1.6288	7.044	1085.0	1160.6	1.6243	6.797	1084.5	1160.0	1.6198
270	7.434	1089.7	1166.8	1.6365	7.165	1089.3	1166.2	1.6319	6.915	1088.8	1165.6	1.6275
280	7.557	1093.9	1172.2	1.6439	7.285	1093.5	1171.7	1.6394	7.031	1093.0	1171.1	1.6351
290	7.679	1098.0	1177.6	1.6511	7.404	1097.6	1177.1	1.6467	7.146	1097.2	1176.5	1.6424
300	7.800	1102.1	1182.9	1.6582	7.521	1101.7	1182.4	1.6538	7.260	1101.3	1181.9	1.6496
310	7.920	1106.1	1188.2	1.6651	7.637	1105.8	1187.8	1.6608	7.373	1105.4	1187.3	1.6565
320	8.039	1110.1	1193.5	1.6719	7.752	1109.8	1193.0	1.6675	7.485	1109.5	1192.6	1.6634
330	8.157	1114.1	1198.6	1.6785	7.866	1113.8	1198.2	1.6742	7.596	1113.5	1197.8	1.6700
340	8.274	1118.1	1203.8	1.6850	7.980	1117.8	1203.4	1.6807	7.706	1117.4	1203.0	1.6766
350	8.390	1122.0	1208.9	1.6913	8.093	1121.7	1208.5	1.6871	7.815	1121.4	1208.2	1.6830
360	8.506	1125.9	1214.0	1.6976	8.205	1125.6	1213.6	1.6934	7.924	1125.3	1213.3	1.6893
370	8.621	1129.7	1219.1	1.7037	8.316	1129.5	1218.7	1.6995	8.032	1129.2	1218.4	1.6955
380	8.735	1133.6	1224.1	1.7097	8.427	1133.3	1223.8	1.7056	8.139	1133.1	1223.5	1.7015
390	8.849	1137.4	1229.1	1.7157	8.538	1137.2	1228.8	1.7115	8.246	1136.9	1228.5	1.7075
400	8.963	1141.2	1234.1	1.7215	8.647	1141.0	1233.8	1.7174	8.353	1140.8	1233.5	1.7134
420	9.189	1148.8	1244.0	1.7329	8.866	1148.6	1243.8	1.7288	8.565	1148.4	1243.5	1.7249
440	9.413	1156.4	1253.9	1.7440	9.083	1156.2	1253.7	1.7400	8.775	1156.0	1253.4	1.7360
460	9.637	1163.9	1263.8	1.7549	9.299	1163.7	1263.6	1.7508	8.984	1163.6	1263.3	1.7469
480	9.859	1171.4	1273.6	1.7654	9.514	1171.3	1273.4	1.7614	9.192	1171.1	1273.2	1.7575
500	10.080	1178.9	1283.4	1.7758	9.728	1178.8	1283.2	1.7717	9.399	1178.6	1283.0	1.7678
520	10.300	1186.4	1293.2	1.7858	9.941	1186.3	1293.0	1.7818	9.606	1186.2	1292.8	1.7780
540	10.520	1193.9	1303.0	1.7957	10.154	1193.8	1302.8	1.7917	9.811	1193.7	1302.6	1.7879
560	10.739	1201.5	1312.7	1.8054	10.365	1201.3	1312.6	1.8014	10.016	1201.2	1312.4	1.7976
580	10.958	1209.0	1322.5	1.8149	10.577	1208.9	1322.4	1.8110	10.221	1208.8	1322.2	1.8071
600	11.176	1216.5	1332.3	1.8243	10.788	1216.4	1332.2	1.8203	10.425	1216.3	1332.1	1.8165
620	11.394	1224.1	1342.2	1.8334	10.998	1224.0	1342.0	1.8295	10.628	1223.9	1341.9	1.8257
640	11.612	1231.7	1352.0	1.8425	11.208	1231.6	1351.9	1.8385	10.832	1231.5	1351.7	1.8347
660	11.829	1239.3	1361.8	1.8513	11.418	1239.2	1361.7	1.8474	11.035	1239.1	1361.6	1.8436
680	12.045	1246.9	1371.7	1.8601	11.627	1246.8	1371.6	1.8561	11.237	1246.7	1371.5	1.8523
700	12.262	1254.5	1381.6	1.8687	11.837	1254.4	1381.5	1.8647	11.440	1254.4	1381.4	1.8609
720	12.478	1262.2	1391.5	1.8772	12.046	1262.1	1391.4	1.8732	11.642	1262.0	1391.3	1.8694
740	12.694	1269.9	1401.4	1.8855	12.254	1269.8	1401.3	1.8816	11.844	1269.7	1401.2	1.8778
760	12.910	1277.6	1411.4	1.8937	12.463	1277.5	1411.3	1.8898	12.045	1277.5	1411.2	1.8860
780	13.126	1285.4	1421.4	1.9019	12.671	1285.3	1421.3	1.8980	12.247	1285.2	1421.2	1.8942
800	13.341	1293.2	1431.4	1.9099	12.879	1293.1	1431.3	1.9060	12.448	1293.0	1431.2	1.9022
850	13.879	1312.8	1456.6	1.9295	13.399	1312.7	1456.5	1.9256	12.951	1312.7	1456.4	1.9218
900	14.417	1332.6	1482.0	1.9485	13.918	1332.5	1481.9	1.9446	13.452	1332.5	1481.8	1.9408
950	14.953	1352.6	1507.5	1.9670	14.436	1352.5	1507.5	1.9631	13.954	1352.5	1507.4	1.9593
1000	15.489	1372.8	1533.3	1.9849	14.954	1372.8	1533.3	1.9810	14.454	1372.7	1533.2	1.9773
1100	16.560	1413.9	1585.5	2.0195	15.988	1413.9	1585.4	2.0156	15.454	1413.8	1585.4	2.0119
1200	17.629	1455.9	1638.5	2.0525	17.020	1455.8	1638.5	2.0486	16.452	1455.8	1638.5	2.0448
1300	18.697	1498.7	1692.5	2.0840	18.051	1498.7	1692.5	2.0802	17.449	1498.7	1692.4	2.0764
1400	19.764	1542.5	1747.3	2.1144	19.082	1542.5	1747.3	2.1105	18.445	1542.5	1747.3	2.1067
1500	20.831	1587.2	1803.1	2.1436	20.112	1587.2	1803.1	2.1397	19.441	1587.2	1803.0	2.1359
1600	21.90	1632.8	1859.8	2.1718	21.14	1632.8	1859.7	2.1679	20.44	1632.8	1859.7	2.1641
1800	24.03	1726.7	1975.7	2.2255	23.20	1726.7	1975.7	2.2216	22.43	1726.7	1975.7	2.2179
2000	26.16	1824.1	2095.1	2.2761	25.26	1824.0	2095.1	2.2722	24.41	1824.0	2095.1	2.2685
2200	28.29	1924.6	2217.7	2.3240	27.31	1924.5	2217.7	2.3201	26.40	1924.5	2217.7	2.3164
2400	30.42	2027.9	2343.1	2.3695	29.37	2027.8	2343.1	2.3656	28.39	2027.8	2343.1	2.3618

	62 (294.88)				**64** (296.98)				**66** (299.02)			**p** (t Sat.)
v	u	h	s	v	u	h	s	v	u	h	s	t
6.960	1098.8	1178.7	1.6418	6.755	1099.3	1179.3	1.6392	6.562	1099.8	1179.9	1.6368	**Sat.**
5.834	*1056.3*	*1123.2*	*1.5631*	*5.633*	*1055.5*	*1122.2*	*1.5584*	*5.444*	*1054.7*	*1121.2*	*1.5537*	**200**
5.962	*1061.2*	*1129.6*	*1.5726*	*5.758*	*1060.4*	*1128.6*	*1.5680*	*5.567*	*1059.7*	*1127.7*	*1.5634*	**210**
6.087	*1065.9*	*1135.7*	*1.5818*	*5.881*	*1065.2*	*1134.9*	*1.5772*	*5.687*	*1064.5*	*1134.0*	*1.5728*	**220**
6.210	*1070.6*	*1141.8*	*1.5906*	*6.001*	*1069.9*	*1141.0*	*1.5862*	*5.805*	*1069.3*	*1140.2*	*1.5818*	**230**
6.330	*1075.1*	*1147.7*	*1.5992*	*6.118*	*1074.5*	*1147.0*	*1.5948*	*5.920*	*1073.9*	*1146.2*	*1.5905*	**240**
6.448	*1079.6*	*1153.6*	*1.6074*	*6.234*	*1079.0*	*1152.9*	*1.6032*	*6.033*	*1078.5*	*1152.2*	*1.5990*	**250**
6.565	*1084.0*	*1159.3*	*1.6155*	*6.348*	*1083.5*	*1158.7*	*1.6112*	*6.144*	*1083.0*	*1158.0*	*1.6071*	**260**
6.680	*1088.3*	*1165.0*	*1.6233*	*6.460*	*1087.8*	*1164.3*	*1.6191*	*6.253*	*1087.4*	*1163.7*	*1.6150*	**270**
6.793	*1092.6*	*1170.5*	*1.6308*	*6.570*	*1092.1*	*1169.9*	*1.6267*	*6.361*	*1091.7*	*1169.4*	*1.6227*	**280**
6.905	*1096.8*	*1176.0*	*1.6382*	*6.679*	*1096.4*	*1175.5*	*1.6342*	*6.467*	*1096.0*	*1174.9*	*1.6302*	**290**
7.016	1100.9	1181.4	1.6454	6.787	1100.6	1180.9	1.6414	6.572	1100.2	1180.4	1.6375	**300**
7.126	1105.1	1186.8	1.6524	6.894	1104.7	1186.3	1.6485	6.676	1104.3	1185.9	1.6446	**310**
7.234	1109.1	1192.1	1.6593	7.000	1108.8	1191.7	1.6553	6.779	1108.4	1191.2	1.6515	**320**
7.342	1113.2	1197.4	1.6660	7.105	1112.8	1197.0	1.6621	6.881	1112.5	1196.5	1.6583	**330**
7.449	1117.1	1202.6	1.6726	7.209	1116.8	1202.2	1.6687	6.983	1116.5	1201.8	1.6649	**340**
7.555	1121.1	1207.8	1.6790	7.312	1120.8	1207.4	1.6752	7.083	1120.5	1207.0	1.6714	**350**
7.661	1125.0	1212.9	1.6853	7.415	1124.8	1212.6	1.6815	7.183	1124.5	1212.2	1.6778	**360**
7.766	1129.0	1218.1	1.6915	7.516	1128.7	1217.7	1.6877	7.282	1128.4	1217.4	1.6840	**370**
7.870	1132.8	1223.1	1.6976	7.618	1132.6	1222.8	1.6938	7.381	1132.3	1222.5	1.6901	**380**
7.974	1136.7	1228.2	1.7036	7.719	1136.5	1227.9	1.6998	7.479	1136.2	1227.6	1.6962	**390**
8.077	1140.6	1233.2	1.7095	7.819	1140.3	1232.9	1.7057	7.576	1140.1	1232.6	1.7021	**400**
8.283	1148.2	1243.2	1.7210	8.019	1148.0	1243.0	1.7173	7.770	1147.8	1242.7	1.7137	**420**
8.487	1155.8	1253.2	1.7322	8.217	1155.6	1253.0	1.7285	7.963	1155.5	1252.7	1.7249	**440**
8.690	1163.4	1263.1	1.7431	8.413	1163.2	1262.9	1.7394	8.154	1163.1	1262.7	1.7358	**460**
8.891	1171.0	1273.0	1.7537	8.609	1170.8	1272.8	1.7501	8.344	1170.7	1272.6	1.7465	**480**
9.092	1178.5	1282.8	1.7641	8.804	1178.4	1282.6	1.7604	8.533	1178.2	1282.4	1.7569	**500**
9.292	1186.0	1292.6	1.7742	8.998	1185.9	1292.5	1.7706	8.721	1185.8	1292.3	1.7671	**520**
9.491	1193.6	1302.5	1.7841	9.191	1193.4	1302.3	1.7805	8.909	1193.3	1302.1	1.7770	**540**
9.690	1201.1	1312.3	1.7939	9.383	1201.0	1312.1	1.7902	9.096	1200.9	1312.0	1.7867	**560**
9.888	1208.7	1322.1	1.8034	9.576	1208.5	1321.9	1.7998	9.282	1208.4	1321.8	1.7963	**580**
10.085	1216.2	1331.9	1.8128	9.767	1216.1	1331.8	1.8092	9.468	1216.0	1331.6	1.8057	**600**
10.283	1223.8	1341.8	1.8220	9.958	1223.7	1341.6	1.8184	9.654	1223.6	1341.5	1.8149	**620**
10.480	1231.4	1351.6	1.8310	10.149	1231.3	1351.5	1.8274	9.839	1231.2	1351.4	1.8239	**640**
10.676	1239.0	1361.5	1.8399	10.340	1238.9	1361.4	1.8363	10.024	1238.8	1361.2	1.8328	**660**
10.872	1246.6	1371.4	1.8486	10.530	1246.5	1371.3	1.8451	10.209	1246.5	1371.1	1.8416	**680**
11.068	1254.3	1381.3	1.8573	10.720	1254.2	1381.2	1.8537	10.393	1254.1	1381.1	1.8502	**700**
11.264	1262.0	1391.2	1.8657	10.910	1261.9	1391.1	1.8622	10.577	1261.8	1391.0	1.8587	**720**
11.459	1269.7	1401.1	1.8741	11.099	1269.6	1401.1	1.8706	10.761	1269.5	1401.0	1.8671	**740**
11.655	1277.4	1411.1	1.8824	11.288	1277.3	1411.0	1.8788	10.944	1277.3	1410.9	1.8754	**760**
11.850	1285.2	1421.1	1.8905	11.478	1285.1	1421.0	1.8869	11.128	1285.0	1421.0	1.8835	**780**
12.045	1293.0	1431.2	1.8985	11.666	1292.9	1431.1	1.8950	11.311	1292.8	1431.0	1.8915	**800**
12.531	1312.6	1456.4	1.9181	12.138	1312.5	1456.3	1.9146	11.769	1312.5	1456.2	1.9112	**850**
13.017	1332.4	1481.8	1.9372	12.609	1332.4	1481.7	1.9336	12.225	1332.3	1481.6	1.9302	**900**
13.502	1352.4	1507.3	1.9557	13.079	1352.4	1507.3	1.9521	12.681	1352.3	1507.2	1.9487	**950**
13.987	1372.7	1533.1	1.9736	13.548	1372.6	1533.1	1.9701	13.137	1372.6	1533.0	1.9667	**1000**
14.954	1413.8	1585.4	2.0082	14.486	1413.8	1585.3	2.0047	14.046	1413.7	1585.3	2.0013	**1100**
15.920	1455.8	1638.4	2.0412	15.422	1455.7	1638.4	2.0377	14.954	1455.7	1638.4	2.0343	**1200**
16.885	1498.6	1692.4	2.0728	16.357	1498.6	1692.4	2.0693	15.861	1498.6	1692.3	2.0659	**1300**
17.850	1542.5	1747.3	2.1031	17.291	1542.4	1747.2	2.0996	16.767	1542.4	1747.2	2.0962	**1400**
18.813	1587.2	1803.0	2.1323	18.225	1587.2	1803.0	2.1288	17.672	1587.1	1803.0	2.1254	**1500**
19.777	1632.8	1859.7	2.1605	19.158	1632.8	1859.7	2.1570	18.577	1632.7	1859.6	2.1536	**1600**
21.702	1726.7	1975.7	2.2142	21.024	1726.7	1975.7	2.2107	20.387	1726.7	1975.6	2.2073	**1800**
23.627	1824.0	2095.1	2.2649	22.889	1824.0	2095.1	2.2614	22.195	1824.0	2095.1	2.2580	**2000**
25.552	1924.5	2217.7	2.3128	24.753	1924.5	2217.7	2.3093	24.003	1924.5	2217.6	2.3059	**2200**
27.476	2027.8	2343.0	2.3582	26.617	2027.8	2343.0	2.3547	25.811	2027.8	2343.0	2.3513	**2400**

Table 3. Vapor

p (t Sat.)	68 (301.01)				70 (302.96)				72 (304.86)			
t	v	u	h	s	v	u	h	s	v	u	h	s
Sat.	6.380	1100.2	1180.5	1.6344	6.209	1100.6	1181.0	1.6321	6.046	1101.0	1181.6	1.6298
250	*5.843*	*1077.9*	*1151.5*	*1.5949*	*5.664*	*1077.4*	*1150.7*	*1.5909*	*5.495*	*1076.8*	*1150.0*	*1.5869*
260	*5.951*	*1082.4*	*1157.3*	*1.6031*	*5.770*	*1081.9*	*1156.7*	*1.5992*	*5.599*	*1081.4*	*1156.0*	*1.5953*
270	*6.058*	*1086.9*	*1163.1*	*1.6111*	*5.875*	*1086.4*	*1162.5*	*1.6072*	*5.701*	*1085.9*	*1161.9*	*1.6034*
280	*6.164*	*1091.2*	*1168.8*	*1.6188*	*5.977*	*1090.8*	*1168.2*	*1.6150*	*5.802*	*1090.3*	*1167.6*	*1.6112*
290	*6.267*	*1095.5*	*1174.4*	*1.6263*	*6.079*	*1095.1*	*1173.9*	*1.6226*	*5.901*	*1094.7*	*1173.3*	*1.6189*
300	*6.370*	*1099.8*	*1179.9*	*1.6337*	*6.179*	*1099.4*	*1179.4*	*1.6299*	*5.999*	*1099.0*	*1178.9*	*1.6263*
310	6.471	1103.9	1185.4	1.6408	6.278	1103.6	1184.9	1.6371	6.096	1103.2	1184.4	1.6335
320	6.572	1108.1	1190.8	1.6478	6.376	1107.7	1190.3	1.6441	6.191	1107.4	1189.9	1.6405
330	6.671	1112.2	1196.1	1.6546	6.473	1111.8	1195.7	1.6509	6.286	1111.5	1195.3	1.6474
340	6.770	1116.2	1201.4	1.6612	6.569	1115.9	1201.0	1.6576	6.380	1115.6	1200.6	1.6541
350	6.868	1120.2	1206.7	1.6677	6.665	1119.9	1206.3	1.6642	6.473	1119.6	1205.9	1.6607
360	6.965	1124.2	1211.9	1.6741	6.760	1123.9	1211.5	1.6706	6.565	1123.7	1211.1	1.6671
370	7.062	1128.2	1217.0	1.6804	6.854	1127.9	1216.7	1.6769	6.657	1127.6	1216.3	1.6735
380	7.158	1132.1	1222.2	1.6866	6.947	1131.9	1221.8	1.6831	6.748	1131.6	1221.5	1.6796
390	7.253	1136.0	1227.3	1.6926	7.040	1135.8	1227.0	1.6891	6.839	1135.5	1226.7	1.6857
400	7.348	1139.9	1232.4	1.6985	7.133	1139.7	1232.1	1.6951	6.929	1139.4	1231.8	1.6917
410	7.442	1143.8	1237.4	1.7044	7.225	1143.5	1237.1	1.7009	7.019	1143.3	1236.8	1.6976
420	7.537	1147.6	1242.4	1.7101	7.316	1147.4	1242.2	1.7067	7.108	1147.2	1241.9	1.7034
430	7.630	1151.4	1247.5	1.7158	7.407	1151.3	1247.2	1.7124	7.197	1151.1	1247.0	1.7091
440	7.724	1155.3	1252.5	1.7214	7.498	1155.1	1252.2	1.7180	7.286	1154.9	1252.0	1.7147
450	7.817	1159.1	1257.5	1.7269	7.589	1158.9	1257.2	1.7235	7.374	1158.7	1257.0	1.7202
460	7.910	1162.9	1262.4	1.7324	7.679	1162.7	1262.2	1.7290	7.462	1162.6	1262.0	1.7257
470	8.002	1166.7	1267.4	1.7377	7.769	1166.5	1267.2	1.7344	7.549	1166.4	1267.0	1.7311
480	8.094	1170.5	1272.3	1.7430	7.859	1170.3	1272.1	1.7397	7.637	1170.2	1271.9	1.7364
490	8.186	1174.3	1277.3	1.7483	7.949	1174.1	1277.1	1.7449	7.724	1174.0	1276.9	1.7417
500	8.278	1178.1	1282.2	1.7534	8.038	1177.9	1282.0	1.7501	7.811	1177.8	1281.9	1.7468
520	8.461	1185.6	1292.1	1.7636	8.216	1185.5	1291.9	1.7603	7.984	1185.4	1291.7	1.7570
540	8.643	1193.2	1302.0	1.7736	8.393	1193.1	1301.8	1.7703	8.157	1192.9	1301.6	1.7670
560	8.825	1200.8	1311.8	1.7833	8.570	1200.6	1311.6	1.7800	8.329	1200.5	1311.5	1.7768
580	9.006	1208.3	1321.7	1.7929	8.746	1208.2	1321.5	1.7896	8.500	1208.1	1321.4	1.7864
600	9.187	1215.9	1331.5	1.8023	8.922	1215.8	1331.4	1.7990	8.671	1215.7	1331.2	1.7958
620	9.367	1223.5	1341.4	1.8115	9.097	1223.4	1341.2	1.8082	8.842	1223.3	1341.1	1.8050
640	9.547	1231.1	1351.2	1.8206	9.272	1231.0	1351.1	1.8173	9.012	1230.9	1351.0	1.8141
660	9.727	1238.7	1361.1	1.8295	9.447	1238.6	1361.0	1.8262	9.182	1238.6	1360.9	1.8230
680	9.906	1246.4	1371.0	1.8382	9.621	1246.3	1370.9	1.8350	9.351	1246.2	1370.8	1.8318
700	10.085	1254.0	1380.9	1.8469	9.795	1254.0	1380.8	1.8436	9.521	1253.9	1380.7	1.8404
720	10.264	1261.7	1390.9	1.8554	9.969	1261.7	1390.8	1.8521	9.690	1261.6	1390.7	1.8489
740	10.442	1269.5	1400.9	1.8638	10.142	1269.4	1400.8	1.8605	9.858	1269.3	1400.7	1.8573
760	10.621	1277.2	1410.9	1.8720	10.315	1277.1	1410.8	1.8688	10.027	1277.1	1410.7	1.8656
780	10.799	1285.0	1420.9	1.8802	10.488	1284.9	1420.8	1.8769	10.195	1284.9	1420.7	1.8737
800	10.977	1292.8	1430.9	1.8882	10.661	1292.7	1430.8	1.8849	10.364	1292.7	1430.7	1.8818
850	11.421	1312.4	1456.1	1.9078	11.093	1312.4	1456.1	1.9046	10.784	1312.3	1456.0	1.9014
900	11.864	1332.3	1481.6	1.9269	11.524	1332.2	1481.5	1.9236	11.203	1332.2	1481.4	1.9205
950	12.307	1352.3	1507.2	1.9454	11.954	1352.3	1507.1	1.9421	11.621	1352.2	1507.0	1.9390
1000	12.749	1372.5	1533.0	1.9634	12.384	1372.5	1532.9	1.9601	12.039	1372.5	1532.9	1.9570
1100	13.632	1413.7	1585.2	1.9980	13.242	1413.6	1585.2	1.9948	12.873	1413.6	1585.1	1.9916
1200	14.513	1455.7	1638.3	2.0310	14.098	1455.7	1638.3	2.0278	13.706	1455.6	1638.2	2.0246
1300	15.394	1498.6	1692.3	2.0626	14.953	1498.6	1692.3	2.0593	14.537	1498.5	1692.2	2.0562
1400	16.273	1542.4	1747.2	2.0929	15.808	1542.4	1747.1	2.0897	15.368	1542.4	1747.1	2.0866
1500	17.152	1587.1	1802.9	2.1221	16.662	1587.1	1802.9	2.1189	16.198	1587.1	1802.9	2.1158
1600	18.031	1632.7	1859.6	2.1503	17.515	1632.7	1859.6	2.1471	17.028	1632.7	1859.6	2.1440
1800	19.787	1726.6	1975.6	2.2040	19.221	1726.6	1975.6	2.2008	18.687	1726.6	1975.6	2.1977
2000	21.542	1824.0	2095.0	2.2547	20.927	1824.0	2095.0	2.2515	20.345	1823.9	2095.0	2.2484
2200	23.297	1924.5	2217.6	2.3026	22.631	1924.5	2217.6	2.2994	22.003	1924.4	2217.6	2.2963
2400	25.052	2027.8	2343.0	2.3480	24.336	2027.8	2343.0	2.3448	23.660	2027.7	2343.0	2.3417

	74 (306.72)				**76** (308.54)				**78** (310.32)			**p** (t Sat.)
v	**u**	**h**	**s**	**v**	**u**	**h**	**s**	**v**	**u**	**h**	**s**	**t**
5.892	1101.4	1182.1	1.6276	5.746	1101.8	1182.6	1.6255	5.607	1102.2	1183.1	1.6234	**Sat.**
5.335	*1076.2*	*1149.3*	*1.5831*	*5.183*	*1075.7*	*1148.6*	*1.5793*	*5.039*	*1075.1*	*1147.8*	*1.5756*	**250**
5.437	*1080.9*	*1155.3*	*1.5915*	*5.283*	*1080.3*	*1154.6*	*1.5878*	*5.137*	*1079.8*	*1154.0*	*1.5842*	**260**
5.537	*1085.4*	*1161.2*	*1.5997*	*5.381*	*1084.9*	*1160.6*	*1.5960*	*5.233*	*1084.4*	*1160.0*	*1.5925*	**270**
5.635	*1089.9*	*1167.0*	*1.6076*	*5.478*	*1089.4*	*1166.4*	*1.6040*	*5.328*	*1088.9*	*1165.8*	*1.6005*	**280**
5.732	*1094.3*	*1172.7*	*1.6153*	*5.573*	*1093.8*	*1172.2*	*1.6117*	*5.421*	*1093.4*	*1171.6*	*1.6083*	**290**
5.828	*1098.6*	*1178.4*	*1.6227*	*5.667*	*1098.2*	*1177.9*	*1.6192*	*5.513*	*1097.8*	*1177.3*	*1.6158*	**300**
5.923	1102.8	1183.9	1.6300	5.759	1102.4	1183.4	1.6265	*5.604*	*1102.1*	*1183.0*	*1.6232*	**310**
6.016	1107.0	1189.4	1.6371	5.851	1106.7	1189.0	1.6336	5.693	1106.3	1188.5	1.6303	**320**
6.109	1111.2	1194.8	1.6440	5.941	1110.8	1194.4	1.6406	5.782	1110.5	1194.0	1.6373	**330**
6.201	1115.3	1200.2	1.6507	6.031	1115.0	1199.8	1.6474	5.870	1114.7	1199.4	1.6441	**340**
6.292	1119.4	1205.5	1.6573	6.120	1119.1	1205.1	1.6540	5.957	1118.8	1204.7	1.6508	**350**
6.382	1123.4	1210.8	1.6638	6.208	1123.1	1210.4	1.6605	6.043	1122.8	1210.0	1.6573	**360**
6.471	1127.4	1216.0	1.6701	6.295	1127.1	1215.7	1.6668	6.128	1126.9	1215.3	1.6637	**370**
6.560	1131.4	1221.2	1.6763	6.382	1131.1	1220.9	1.6731	6.213	1130.8	1220.5	1.6699	**380**
6.649	1135.3	1226.3	1.6824	6.469	1135.1	1226.0	1.6792	6.298	1134.8	1225.7	1.6761	**390**
6.737	1139.2	1231.5	1.6884	6.554	1139.0	1231.2	1.6852	6.381	1138.8	1230.9	1.6821	**400**
6.824	1143.1	1236.6	1.6943	6.640	1142.9	1236.3	1.6911	6.465	1142.7	1236.0	1.6880	**410**
6.911	1147.0	1241.6	1.7001	6.725	1146.8	1241.4	1.6969	6.548	1146.6	1241.1	1.6938	**420**
6.998	1150.9	1246.7	1.7058	6.809	1150.7	1246.4	1.7027	6.630	1150.5	1246.2	1.6996	**430**
7.084	1154.7	1251.7	1.7115	6.894	1154.5	1251.5	1.7083	6.713	1154.3	1251.2	1.7052	**440**
7.170	1158.6	1256.7	1.7170	6.977	1158.4	1256.5	1.7139	6.794	1158.2	1256.3	1.7108	**450**
7.256	1162.4	1261.7	1.7225	7.061	1162.2	1261.5	1.7194	6.876	1162.0	1261.3	1.7163	**460**
7.341	1166.2	1266.7	1.7279	7.144	1166.0	1266.5	1.7248	6.957	1165.9	1266.3	1.7217	**470**
7.427	1170.0	1271.7	1.7332	7.227	1169.9	1271.5	1.7301	7.038	1169.7	1271.3	1.7271	**480**
7.511	1173.8	1276.7	1.7385	7.310	1173.7	1276.5	1.7354	7.119	1173.5	1276.3	1.7323	**490**
7.596	1177.6	1281.7	1.7437	7.393	1177.5	1281.5	1.7406	7.200	1177.3	1281.3	1.7376	**500**
7.765	1185.2	1291.6	1.7539	7.557	1185.1	1291.4	1.7508	7.361	1185.0	1291.2	1.7478	**520**
7.933	1192.8	1301.5	1.7639	7.721	1192.7	1301.3	1.7608	7.520	1192.6	1301.1	1.7578	**540**
8.101	1200.4	1311.3	1.7737	7.885	1200.3	1311.2	1.7706	7.680	1200.2	1311.0	1.7676	**560**
8.268	1208.0	1321.2	1.7833	8.047	1207.9	1321.1	1.7802	7.838	1207.8	1320.9	1.7772	**580**
8.434	1215.6	1331.1	1.7927	8.210	1215.5	1330.9	1.7896	7.997	1215.4	1330.8	1.7867	**600**
8.600	1223.2	1341.0	1.8019	8.372	1223.1	1340.8	1.7989	8.155	1223.0	1340.7	1.7959	**620**
8.766	1230.8	1350.9	1.8110	8.533	1230.7	1350.7	1.8080	8.312	1230.6	1350.6	1.8050	**640**
8.931	1238.5	1360.8	1.8199	8.694	1238.4	1360.7	1.8169	8.469	1238.3	1360.5	1.8140	**660**
9.097	1246.1	1370.7	1.8287	8.855	1246.0	1370.6	1.8257	8.626	1246.0	1370.5	1.8227	**680**
9.261	1253.8	1380.6	1.8373	9.016	1253.7	1380.5	1.8343	8.783	1253.7	1380.4	1.8314	**700**
9.426	1261.5	1390.6	1.8459	9.176	1261.4	1390.5	1.8429	8.939	1261.4	1390.4	1.8399	**720**
9.590	1269.2	1400.6	1.8542	9.336	1269.2	1400.5	1.8513	9.095	1269.1	1400.4	1.8483	**740**
9.754	1277.0	1410.6	1.8625	9.496	1276.9	1410.5	1.8595	9.251	1276.9	1410.4	1.8566	**760**
9.918	1284.8	1420.6	1.8707	9.656	1284.7	1420.5	1.8677	9.406	1284.7	1420.4	1.8648	**780**
10.082	1292.6	1430.7	1.8787	9.815	1292.5	1430.6	1.8757	9.562	1292.5	1430.5	1.8728	**800**
10.491	1312.3	1455.9	1.8984	10.213	1312.2	1455.8	1.8954	9.950	1312.1	1455.8	1.8925	**850**
10.899	1332.1	1481.4	1.9174	10.611	1332.1	1481.3	1.9145	10.337	1332.0	1481.2	1.9116	**900**
11.306	1352.2	1507.0	1.9360	11.007	1352.1	1506.9	1.9330	10.724	1352.1	1506.9	1.9301	**950**
11.712	1372.4	1532.8	1.9540	11.403	1372.4	1532.8	1.9510	11.110	1372.3	1532.7	1.9481	**1000**
12.524	1413.6	1585.1	1.9886	12.194	1413.5	1585.0	1.9856	11.880	1413.5	1585.0	1.9827	**1100**
13.335	1455.6	1638.2	2.0216	12.983	1455.6	1638.2	2.0186	12.649	1455.5	1638.1	2.0158	**1200**
14.144	1498.5	1692.2	2.0532	13.771	1498.5	1692.2	2.0502	13.417	1498.5	1692.1	2.0474	**1300**
14.952	1542.3	1747.1	2.0835	14.558	1542.3	1747.1	2.0806	14.185	1542.3	1747.0	2.0777	**1400**
15.760	1587.1	1802.9	2.1127	15.345	1587.0	1802.8	2.1098	14.951	1587.0	1802.8	2.1069	**1500**
16.568	1632.7	1859.6	2.1409	16.131	1632.7	1859.5	2.1380	15.717	1632.6	1859.5	2.1351	**1600**
18.182	1726.6	1975.6	2.1947	17.703	1726.6	1975.5	2.1917	17.249	1726.6	1975.5	2.1889	**1800**
19.795	1823.9	2095.0	2.2453	19.274	1823.9	2095.0	2.2424	18.780	1823.9	2095.0	2.2395	**2000**
21.408	1924.4	2217.6	2.2932	20.845	1924.4	2217.6	2.2903	20.310	1924.4	2217.6	2.2874	**2200**
23.021	2027.7	2343.0	2.3387	22.415	2027.7	2343.0	2.3357	21.840	2027.7	2342.9	2.3329	**2400**

Table 3. Vapor

p (t Sat.)	80 (312.07)				82 (313.78)				84 (315.46)			
t	v	u	h	s	v	u	h	s	v	u	h	s
Sat.	5.474	1102.6	1183.6	1.6214	5.348	1102.9	1184.1	1.6194	5.228	1103.3	1184.5	1.6175
250	*4.902*	*1074.5*	*1147.1*	*1.5720*	*4.772*	*1073.9*	*1146.3*	*1.5684*	*4.648*	*1073.3*	*1145.6*	*1.5649*
260	*4.998*	*1079.3*	*1153.3*	*1.5806*	*4.867*	*1078.7*	*1152.6*	*1.5771*	*4.741*	*1078.2*	*1151.9*	*1.5737*
270	*5.093*	*1083.9*	*1159.3*	*1.5890*	*4.959*	*1083.4*	*1158.7*	*1.5855*	*4.832*	*1082.9*	*1158.0*	*1.5822*
280	*5.186*	*1088.5*	*1165.2*	*1.5970*	*5.051*	*1088.0*	*1164.6*	*1.5937*	*4.922*	*1087.5*	*1164.0*	*1.5904*
290	*5.277*	*1092.9*	*1171.1*	*1.6049*	*5.140*	*1092.5*	*1170.5*	*1.6016*	*5.010*	*1092.1*	*1169.9*	*1.5983*
300	*5.367*	*1097.4*	*1176.8*	*1.6125*	*5.229*	*1096.9*	*1176.3*	*1.6092*	*5.097*	*1096.5*	*1175.7*	*1.6060*
310	*5.456*	*1101.7*	*1182.5*	*1.6199*	*5.316*	*1101.3*	*1182.0*	*1.6166*	*5.182*	*1100.9*	*1181.5*	*1.6135*
320	5.544	1106.0	1188.0	1.6271	5.402	1105.6	1187.6	1.6239	5.266	1105.2	1187.1	1.6207
330	5.631	1110.2	1193.5	1.6341	5.487	1109.8	1193.1	1.6309	5.350	1109.5	1192.6	1.6278
340	5.717	1114.3	1199.0	1.6409	5.571	1114.0	1198.6	1.6378	5.432	1113.7	1198.1	1.6347
350	5.802	1118.5	1204.3	1.6476	5.654	1118.2	1204.0	1.6445	5.514	1117.9	1203.6	1.6415
360	5.886	1122.5	1209.7	1.6541	5.737	1122.3	1209.3	1.6511	5.595	1122.0	1208.9	1.6480
370	5.970	1126.6	1215.0	1.6605	5.819	1126.3	1214.6	1.6575	5.675	1126.0	1214.3	1.6545
380	6.053	1130.6	1220.2	1.6668	5.900	1130.3	1219.9	1.6638	5.754	1130.1	1219.5	1.6608
390	6.135	1134.6	1225.4	1.6730	5.980	1134.3	1225.1	1.6700	5.833	1134.1	1224.8	1.6670
400	6.217	1138.5	1230.6	1.6790	6.061	1138.3	1230.3	1.6760	5.912	1138.1	1230.0	1.6731
410	6.299	1142.5	1235.7	1.6850	6.140	1142.3	1235.4	1.6820	5.990	1142.0	1235.1	1.6791
420	6.380	1146.4	1240.8	1.6908	6.220	1146.2	1240.6	1.6879	6.067	1146.0	1240.3	1.6850
430	6.460	1150.3	1245.9	1.6966	6.299	1150.1	1245.7	1.6936	6.144	1149.9	1245.4	1.6907
440	6.541	1154.2	1251.0	1.7022	6.377	1154.0	1250.7	1.6993	6.221	1153.8	1250.5	1.6964
450	6.621	1158.0	1256.0	1.7078	6.455	1157.8	1255.8	1.7049	6.298	1157.7	1255.6	1.7020
460	6.700	1161.9	1261.1	1.7133	6.533	1161.7	1260.8	1.7104	6.374	1161.5	1260.6	1.7076
470	6.780	1165.7	1266.1	1.7188	6.611	1165.6	1265.9	1.7158	6.450	1165.4	1265.6	1.7130
480	6.859	1169.6	1271.1	1.7241	6.688	1169.4	1270.9	1.7212	6.526	1169.2	1270.7	1.7184
490	6.938	1173.4	1276.1	1.7294	6.765	1173.2	1275.9	1.7265	6.601	1173.1	1275.7	1.7237
500	7.017	1177.2	1281.1	1.7346	6.842	1177.1	1280.9	1.7317	6.676	1176.9	1280.7	1.7289
520	7.173	1184.8	1291.0	1.7449	6.995	1184.7	1290.8	1.7420	6.826	1184.6	1290.7	1.7392
540	7.329	1192.4	1300.9	1.7549	7.148	1192.3	1300.8	1.7521	6.975	1192.2	1300.6	1.7493
560	7.485	1200.1	1310.9	1.7647	7.300	1199.9	1310.7	1.7619	7.123	1199.8	1310.5	1.7591
580	7.640	1207.7	1320.8	1.7743	7.451	1207.6	1320.6	1.7715	7.271	1207.4	1320.5	1.7688
600	7.794	1215.3	1330.7	1.7838	7.602	1215.2	1330.5	1.7810	7.419	1215.1	1330.4	1.7782
620	7.948	1222.9	1340.6	1.7930	7.752	1222.8	1340.4	1.7902	7.566	1222.7	1340.3	1.7875
640	8.102	1230.5	1350.5	1.8021	7.902	1230.5	1350.4	1.7993	7.712	1230.4	1350.2	1.7966
660	8.255	1238.2	1360.4	1.8111	8.052	1238.1	1360.3	1.8083	7.858	1238.0	1360.2	1.8056
680	8.408	1245.9	1370.4	1.8199	8.201	1245.8	1370.2	1.8171	8.004	1245.7	1370.1	1.8144
700	8.561	1253.6	1380.3	1.8285	8.351	1253.5	1380.2	1.8258	8.150	1253.4	1380.1	1.8230
720	8.714	1261.3	1390.3	1.8371	8.499	1261.2	1390.2	1.8343	8.295	1261.1	1390.1	1.8316
740	8.866	1269.0	1400.3	1.8455	8.648	1269.0	1400.2	1.8427	8.440	1268.9	1400.1	1.8400
760	9.018	1276.8	1410.3	1.8538	8.796	1276.7	1410.2	1.8510	8.585	1276.7	1410.1	1.8483
780	9.170	1284.6	1420.3	1.8619	8.945	1284.5	1420.3	1.8592	8.730	1284.5	1420.2	1.8564
800	9.321	1292.4	1430.4	1.8700	9.093	1292.4	1430.3	1.8672	8.875	1292.3	1430.2	1.8645
850	9.700	1312.1	1455.7	1.8897	9.462	1312.0	1455.6	1.8869	9.236	1312.0	1455.5	1.8842
900	10.078	1332.0	1481.2	1.9087	9.831	1331.9	1481.1	1.9060	9.596	1331.9	1481.0	1.9033
950	10.455	1352.0	1506.8	1.9273	10.199	1352.0	1506.7	1.9245	9.955	1351.9	1506.7	1.9218
1000	10.831	1372.3	1532.6	1.9453	10.566	1372.3	1532.6	1.9425	10.314	1372.2	1532.5	1.9398
1100	11.583	1413.5	1584.9	1.9799	11.299	1413.4	1584.9	1.9772	11.030	1413.4	1584.8	1.9745
1200	12.333	1455.5	1638.1	2.0130	12.031	1455.5	1638.0	2.0102	11.744	1455.5	1638.0	2.0075
1300	13.081	1498.4	1692.1	2.0446	12.762	1498.4	1692.1	2.0418	12.458	1498.4	1692.0	2.0391
1400	13.830	1542.3	1747.0	2.0749	13.492	1542.2	1747.0	2.0722	13.170	1542.2	1746.9	2.0695
1500	14.577	1587.0	1802.8	2.1041	14.221	1587.0	1802.8	2.1014	13.882	1587.0	1802.7	2.0987
1600	15.324	1632.6	1859.5	2.1323	14.950	1632.6	1859.5	2.1296	14.594	1632.6	1859.4	2.1269
1800	16.818	1726.5	1975.5	2.1861	16.407	1726.5	1975.5	2.1834	16.016	1726.5	1975.5	2.1807
2000	18.310	1823.9	2094.9	2.2367	17.863	1823.9	2094.9	2.2340	17.438	1823.9	2094.9	2.2313
2200	19.802	1924.4	2217.5	2.2846	19.319	1924.4	2217.5	2.2819	18.859	1924.4	2217.5	2.2792
2400	21.294	2027.7	2342.9	2.3301	20.775	2027.7	2342.9	2.3274	20.280	2027.7	2342.9	2.3247

v	u	h	s	v	u	h	s	v	u	h	s	t
5.113	1103.6	1185.0	1.6156	5.004	1104.0	1185.4	1.6137	4.898	1104.3	1185.9	1.6119	Sat.
4.529	*1072.7*	*1144.8*	*1.5614*	*4.416*	*1072.1*	*1144.0*	*1.5580*	*4.308*	*1071.5*	*1143.3*	*1.5546*	**250**
4.621	*1077.6*	*1151.2*	*1.5703*	*4.506*	*1077.1*	*1150.4*	*1.5670*	*4.397*	*1076.5*	*1149.7*	*1.5637*	**260**
4.711	*1082.4*	*1157.4*	*1.5788*	*4.595*	*1081.9*	*1156.7*	*1.5756*	*4.484*	*1081.4*	*1156.0*	*1.5724*	**270**
4.799	*1087.1*	*1163.4*	*1.5871*	*4.681*	*1086.6*	*1162.8*	*1.5839*	*4.569*	*1086.1*	*1162.2*	*1.5807*	**280**
4.885	*1091.6*	*1169.4*	*1.5951*	*4.766*	*1091.2*	*1168.8*	*1.5919*	*4.653*	*1090.7*	*1168.2*	*1.5888*	**290**
4.971	*1096.1*	*1175.2*	*1.6028*	*4.850*	*1095.7*	*1174.7*	*1.5997*	*4.735*	*1095.3*	*1174.1*	*1.5967*	**300**
5.054	*1100.5*	*1181.0*	*1.6103*	*4.933*	*1100.1*	*1180.5*	*1.6073*	*4.816*	*1099.7*	*1180.0*	*1.6043*	**310**
5.137	1104.9	1186.6	1.6177	5.014	1104.5	1186.2	1.6146	*4.896*	*1104.1*	*1185.7*	*1.6117*	**320**
5.219	1109.2	1192.2	1.6248	5.094	1108.8	1191.8	1.6218	4.975	1108.5	1191.3	1.6189	**330**
5.300	1113.4	1197.7	1.6317	5.173	1113.1	1197.3	1.6288	5.053	1112.7	1196.9	1.6259	**340**
5.380	1117.6	1203.2	1.6385	5.252	1117.3	1202.8	1.6356	5.130	1116.9	1202.4	1.6327	**350**
5.459	1121.7	1208.6	1.6451	5.330	1121.4	1208.2	1.6422	5.206	1121.1	1207.8	1.6394	**360**
5.538	1125.8	1213.9	1.6516	5.407	1125.5	1213.6	1.6487	5.281	1125.2	1213.2	1.6459	**370**
5.615	1129.8	1219.2	1.6579	5.483	1129.6	1218.9	1.6551	5.356	1129.3	1218.5	1.6523	**380**
5.693	1133.9	1224.5	1.6641	5.559	1133.6	1224.1	1.6613	5.431	1133.4	1223.8	1.6585	**390**
5.770	1137.8	1229.7	1.6702	5.634	1137.6	1229.4	1.6674	5.505	1137.4	1229.1	1.6647	**400**
5.846	1141.8	1234.9	1.6762	5.709	1141.6	1234.6	1.6734	5.578	1141.4	1234.3	1.6707	**410**
5.922	1145.8	1240.0	1.6821	5.783	1145.6	1239.7	1.6794	5.651	1145.3	1239.5	1.6766	**420**
5.998	1149.7	1245.1	1.6879	5.857	1149.5	1244.9	1.6852	5.723	1149.3	1244.6	1.6825	**430**
6.073	1153.6	1250.2	1.6936	5.931	1153.4	1250.0	1.6909	5.796	1153.2	1249.7	1.6882	**440**
6.148	1157.5	1255.3	1.6992	6.004	1157.3	1255.1	1.6965	5.867	1157.1	1254.8	1.6938	**450**
6.222	1161.4	1260.4	1.7048	6.077	1161.2	1260.1	1.7021	5.939	1161.0	1259.9	1.6994	**460**
6.296	1165.2	1265.4	1.7102	6.150	1165.1	1265.2	1.7075	6.010	1164.9	1265.0	1.7049	**470**
6.370	1169.1	1270.5	1.7156	6.222	1168.9	1270.2	1.7129	6.081	1168.8	1270.0	1.7103	**480**
6.444	1172.9	1275.5	1.7209	6.295	1172.8	1275.3	1.7182	6.152	1172.6	1275.1	1.7156	**490**
6.518	1176.8	1280.5	1.7262	6.367	1176.6	1280.3	1.7235	6.222	1176.5	1280.1	1.7209	**500**
6.664	1184.4	1290.5	1.7365	6.510	1184.3	1290.3	1.7338	6.363	1184.1	1290.1	1.7312	**520**
6.810	1192.1	1300.4	1.7466	6.653	1191.9	1300.3	1.7439	6.502	1191.8	1300.1	1.7413	**540**
6.955	1199.7	1310.4	1.7564	6.795	1199.6	1310.2	1.7538	6.641	1199.5	1310.1	1.7512	**560**
7.100	1207.3	1320.3	1.7661	6.936	1207.2	1320.2	1.7634	6.780	1207.1	1320.0	1.7608	**580**
7.244	1215.0	1330.3	1.7755	7.077	1214.9	1330.1	1.7729	6.918	1214.8	1330.0	1.7703	**600**
7.387	1222.6	1340.2	1.7848	7.217	1222.5	1340.0	1.7822	7.055	1222.4	1339.9	1.7796	**620**
7.531	1230.3	1350.1	1.7939	7.358	1230.2	1350.0	1.7913	7.192	1230.1	1349.9	1.7887	**640**
7.674	1237.9	1360.1	1.8029	7.497	1237.9	1359.9	1.8003	7.329	1237.8	1359.8	1.7977	**660**
7.816	1245.6	1370.0	1.8117	7.637	1245.5	1369.9	1.8091	7.465	1245.5	1369.8	1.8065	**680**
7.959	1253.3	1380.0	1.8204	7.776	1253.3	1379.9	1.8178	7.602	1253.2	1379.8	1.8152	**700**
8.101	1261.1	1390.0	1.8289	7.915	1261.0	1389.9	1.8263	7.738	1260.9	1389.8	1.8238	**720**
8.243	1268.8	1400.0	1.8373	8.054	1268.7	1399.9	1.8347	7.873	1268.7	1399.8	1.8322	**740**
8.384	1276.6	1410.0	1.8456	8.192	1276.5	1409.9	1.8430	8.009	1276.5	1409.8	1.8405	**760**
8.526	1284.4	1420.1	1.8538	8.331	1284.3	1420.0	1.8512	8.144	1284.3	1419.9	1.8487	**780**
8.667	1292.2	1430.2	1.8619	8.469	1292.2	1430.1	1.8593	8.279	1292.1	1430.0	1.8567	**800**
9.020	1311.9	1455.5	1.8816	8.813	1311.9	1455.4	1.8790	8.616	1311.8	1455.3	1.8765	**850**
9.371	1331.8	1480.9	1.9007	9.157	1331.8	1480.9	1.8981	8.953	1331.7	1480.8	1.8956	**900**
9.722	1351.9	1506.6	1.9192	9.500	1351.8	1506.6	1.9166	9.288	1351.8	1506.5	1.9141	**950**
10.073	1372.2	1532.5	1.9372	9.843	1372.1	1532.4	1.9347	9.623	1372.1	1532.4	1.9322	**1000**
10.772	1413.4	1584.8	1.9719	10.527	1413.3	1584.8	1.9693	10.292	1413.3	1584.7	1.9668	**1100**
11.470	1455.4	1638.0	2.0049	11.209	1455.4	1637.9	2.0024	10.959	1455.4	1637.9	1.9999	**1200**
12.167	1498.4	1692.0	2.0365	11.890	1498.3	1692.0	2.0340	11.626	1498.3	1691.9	2.0315	**1300**
12.864	1542.2	1746.9	2.0669	12.571	1542.2	1746.9	2.0643	12.291	1542.2	1746.9	2.0619	**1400**
13.559	1586.9	1802.7	2.0961	13.251	1586.9	1802.7	2.0936	12.956	1586.9	1802.7	2.0911	**1500**
14.254	1632.6	1859.4	2.1243	13.930	1632.6	1859.4	2.1218	13.620	1632.5	1859.4	2.1193	**1600**
15.644	1726.5	1975.5	2.1781	15.288	1726.5	1975.4	2.1755	14.948	1726.5	1975.4	2.1731	**1800**
17.032	1823.8	2094.9	2.2287	16.645	1823.8	2094.9	2.2262	16.275	1823.8	2094.9	2.2237	**2000**
18.421	1924.4	2217.5	2.2766	18.002	1924.3	2217.5	2.2741	17.602	1924.3	2217.5	2.2716	**2200**
19.809	2027.6	2342.9	2.3221	19.358	2027.6	2342.9	2.3196	18.928	2027.6	2342.9	2.3171	**2400**

Table 3. Vapor

t	v	u	h	s	v	u	h	s	v	u	h	s
Sat.	4.798	1104.6	1186.3	1.6101	4.701	1104.9	1186.7	1.6084	4.609	1105.2	1187.1	1.6067
300	4.625	1094.9	1173.6	1.5937	4.520	1094.4	1173.0	1.5907	4.419	1094.0	1172.5	1.5878
310	4.705	1099.3	1179.4	1.6013	4.598	1099.0	1178.9	1.5984	4.496	1098.6	1178.4	1.5956
320	4.783	1103.8	1185.2	1.6088	4.675	1103.4	1184.7	1.6059	4.572	1103.0	1184.2	1.6031
330	4.861	1108.1	1190.9	1.6160	4.751	1107.8	1190.4	1.6132	4.647	1107.4	1190.0	1.6104
340	4.937	1112.4	1196.5	1.6230	4.826	1112.1	1196.0	1.6202	4.720	1111.8	1195.6	1.6175
350	5.013	1116.6	1202.0	1.6299	4.901	1116.3	1201.6	1.6271	4.793	1116.0	1201.2	1.6244
360	5.088	1120.8	1207.4	1.6366	4.974	1120.5	1207.1	1.6339	4.866	1120.2	1206.7	1.6312
370	5.162	1125.0	1212.8	1.6431	5.047	1124.7	1212.5	1.6404	4.937	1124.4	1212.1	1.6378
380	5.235	1129.1	1218.2	1.6496	5.119	1128.8	1217.9	1.6469	5.008	1128.5	1217.5	1.6442
390	5.308	1133.1	1223.5	1.6558	5.191	1132.9	1223.2	1.6532	5.078	1132.6	1222.9	1.6506
400	5.381	1137.2	1228.8	1.6620	5.262	1136.9	1228.5	1.6593	5.148	1136.7	1228.1	1.6567
410	5.453	1141.2	1234.0	1.6680	5.333	1140.9	1233.7	1.6654	5.218	1140.7	1233.4	1.6628
420	5.524	1145.1	1239.2	1.6740	5.403	1144.9	1238.9	1.6714	5.286	1144.7	1238.6	1.6688
430	5.595	1149.1	1244.3	1.6798	5.472	1148.9	1244.1	1.6772	5.355	1148.7	1243.8	1.6747
440	5.666	1153.0	1249.5	1.6856	5.542	1152.8	1249.2	1.6830	5.423	1152.6	1249.0	1.6804
450	5.736	1156.9	1254.6	1.6912	5.611	1156.8	1254.4	1.6886	5.491	1156.6	1254.1	1.6861
460	5.806	1160.8	1259.7	1.6968	5.680	1160.7	1259.5	1.6942	5.558	1160.5	1259.2	1.6917
470	5.876	1164.7	1264.8	1.7023	5.748	1164.6	1264.5	1.6997	5.625	1164.4	1264.3	1.6972
480	5.946	1168.6	1269.8	1.7077	5.816	1168.4	1269.6	1.7051	5.692	1168.3	1269.4	1.7026
490	6.015	1172.5	1274.9	1.7130	5.884	1172.3	1274.7	1.7105	5.759	1172.2	1274.5	1.7080
500	6.084	1176.3	1279.9	1.7183	5.952	1176.2	1279.7	1.7158	5.825	1176.0	1279.5	1.7133
520	6.222	1184.0	1289.9	1.7286	6.086	1183.9	1289.7	1.7261	5.957	1183.7	1289.6	1.7237
540	6.358	1191.7	1299.9	1.7387	6.220	1191.6	1299.8	1.7362	6.088	1191.4	1299.6	1.7338
560	6.494	1199.3	1309.9	1.7486	6.354	1199.2	1309.7	1.7461	6.219	1199.1	1309.6	1.7437
580	6.630	1207.0	1319.9	1.7583	6.487	1206.9	1319.7	1.7558	6.349	1206.8	1319.6	1.7534
600	6.765	1214.7	1329.8	1.7678	6.619	1214.6	1329.7	1.7653	6.479	1214.4	1329.5	1.7629
620	6.900	1222.3	1339.8	1.7771	6.751	1222.2	1339.7	1.7746	6.608	1222.1	1339.5	1.7722
640	7.034	1230.0	1349.7	1.7862	6.882	1229.9	1349.6	1.7838	6.737	1229.8	1349.5	1.7814
660	7.168	1237.7	1359.7	1.7952	7.014	1237.6	1359.6	1.7928	6.866	1237.5	1359.5	1.7904
680	7.301	1245.4	1369.7	1.8040	7.144	1245.3	1369.6	1.8016	6.994	1245.2	1369.5	1.7992
700	7.435	1253.1	1379.7	1.8127	7.275	1253.0	1379.6	1.8103	7.122	1252.9	1379.5	1.8079
720	7.568	1260.8	1389.7	1.8213	7.405	1260.8	1389.6	1.8188	7.250	1260.7	1389.5	1.8165
740	7.701	1268.6	1399.7	1.8297	7.535	1268.5	1399.6	1.8273	7.377	1268.5	1399.5	1.8249
760	7.833	1276.4	1409.7	1.8380	7.665	1276.3	1409.7	1.8356	7.504	1276.3	1409.6	1.8332
780	7.966	1284.2	1419.8	1.8462	7.795	1284.1	1419.7	1.8438	7.631	1284.1	1419.6	1.8414
800	8.098	1292.0	1429.9	1.8543	7.924	1292.0	1429.8	1.8519	7.758	1291.9	1429.7	1.8495
820	8.230	1299.9	1440.0	1.8622	8.054	1299.8	1439.9	1.8598	7.885	1299.8	1439.9	1.8575
840	8.362	1307.8	1450.2	1.8701	8.183	1307.7	1450.1	1.8677	8.011	1307.7	1450.0	1.8653
860	8.494	1315.7	1460.3	1.8779	8.312	1315.7	1460.3	1.8755	8.138	1315.6	1460.2	1.8731
880	8.626	1323.7	1470.5	1.8855	8.441	1323.6	1470.4	1.8831	8.264	1323.6	1470.4	1.8808
900	8.757	1331.7	1480.7	1.8931	8.570	1331.6	1480.7	1.8907	8.390	1331.6	1480.6	1.8883
920	8.889	1339.7	1491.0	1.9006	8.699	1339.6	1490.9	1.8982	8.516	1339.6	1490.9	1.8958
940	9.020	1347.7	1501.3	1.9080	8.827	1347.7	1501.2	1.9056	8.642	1347.6	1501.2	1.9032
960	9.151	1355.8	1511.6	1.9153	8.956	1355.7	1511.5	1.9129	8.768	1355.7	1511.5	1.9106
980	9.282	1363.9	1521.9	1.9225	9.084	1363.9	1521.9	1.9201	8.894	1363.8	1521.8	1.9178
1000	9.413	1372.0	1532.3	1.9297	9.212	1372.0	1532.3	1.9273	9.020	1372.0	1532.2	1.9250
1020	9.544	1380.2	1542.7	1.9368	9.341	1380.2	1542.7	1.9344	9.145	1380.1	1542.6	1.9320
1040	9.675	1388.4	1553.2	1.9438	9.469	1388.4	1553.1	1.9414	9.271	1388.4	1553.0	1.9390
1060	9.806	1396.7	1563.6	1.9507	9.597	1396.6	1563.6	1.9483	9.396	1396.6	1563.5	1.9460
1080	9.937	1405.0	1574.1	1.9576	9.725	1404.9	1574.1	1.9552	9.522	1404.9	1574.0	1.9529
1100	10.068	1413.3	1584.7	1.9644	9.853	1413.2	1584.6	1.9620	9.647	1413.2	1584.6	1.9597
1200	10.721	1455.3	1637.9	1.9974	10.492	1455.3	1637.8	1.9951	10.273	1455.3	1637.8	1.9927
1300	11.372	1498.3	1691.9	2.0291	11.130	1498.3	1691.9	2.0267	10.898	1498.2	1691.8	2.0243
1400	12.023	1542.1	1746.8	2.0594	11.767	1542.1	1746.8	2.0570	11.522	1542.1	1746.8	2.0547
1500	12.674	1586.9	1802.6	2.0887	12.404	1586.9	1802.6	2.0863	12.145	1586.8	1802.6	2.0839
1600	13.324	1632.5	1859.4	2.1169	13.040	1632.5	1859.3	2.1145	12.768	1632.5	1859.3	2.1122
1800	14.623	1726.5	1975.4	2.1706	14.312	1726.4	1975.4	2.1683	14.013	1726.4	1975.4	2.1659
2000	15.921	1823.8	2094.9	2.2213	15.582	1823.8	2094.8	2.2189	15.258	1823.8	2094.8	2.2166
2200	17.219	1924.3	2217.5	2.2692	16.853	1924.3	2217.4	2.2668	16.502	1924.3	2217.4	2.2645
2400	18.517	2027.6	2342.8	2.3146	18.123	2027.6	2342.8	2.3123	17.745	2027.6	2342.8	2.3099

v	u	h	s	v	u	h	s	v	u	h	s	t
4.520	1105.5	1187.4	1.6050	4.434	1105.8	1187.8	1.6034	4.234	1106.5	1188.7	1.5995	**Sat.**
4.322	*1093.6*	*1171.9*	*1.5850*	*4.228*	*1093.1*	*1171.4*	*1.5822*	*4.011*	*1092.1*	*1170.0*	*1.5753*	**300**
4.398	*1098.2*	*1177.9*	*1.5928*	*4.303*	*1097.7*	*1177.4*	*1.5900*	*4.083*	*1096.7*	*1176.1*	*1.5833*	**310**
4.472	*1102.6*	*1183.8*	*1.6003*	*4.377*	*1102.3*	*1183.3*	*1.5976*	*4.154*	*1101.3*	*1182.0*	*1.5910*	**320**
4.546	1107.1	1189.5	1.6077	4.449	1106.7	1189.1	1.6050	*4.224*	*1105.8*	*1187.9*	*1.5984*	**330**
4.619	1111.4	1195.2	1.6148	4.521	1111.1	1194.8	1.6121	4.293	1110.3	1193.7	1.6057	**340**
4.691	1115.7	1200.8	1.6218	4.592	1115.4	1200.4	1.6191	4.361	1114.6	1199.4	1.6128	**350**
4.762	1120.0	1206.3	1.6285	4.662	1119.7	1205.9	1.6259	4.428	1118.9	1205.0	1.6196	**360**
4.832	1124.1	1211.8	1.6352	4.731	1123.9	1211.4	1.6326	4.495	1123.2	1210.5	1.6264	**370**
4.902	1128.3	1217.2	1.6416	4.799	1128.0	1216.8	1.6391	4.560	1127.4	1216.0	1.6329	**380**
4.971	1132.4	1222.5	1.6480	4.867	1132.1	1222.2	1.6455	4.625	1131.5	1221.4	1.6393	**390**
5.039	1136.5	1227.8	1.6542	4.934	1136.2	1227.5	1.6517	4.690	1135.6	1226.8	1.6456	**400**
5.107	1140.5	1233.1	1.6603	5.001	1140.3	1232.8	1.6578	4.754	1139.7	1232.1	1.6518	**410**
5.175	1144.5	1238.3	1.6663	5.068	1144.3	1238.1	1.6638	4.818	1143.8	1237.4	1.6578	**420**
5.242	1148.5	1243.5	1.6722	5.134	1148.3	1243.3	1.6697	4.881	1147.8	1242.6	1.6637	**430**
5.309	1152.4	1248.7	1.6779	5.199	1152.3	1248.5	1.6755	4.944	1151.8	1247.8	1.6696	**440**
5.375	1156.4	1253.9	1.6836	5.265	1156.2	1253.6	1.6812	5.006	1155.7	1253.0	1.6753	**450**
5.441	1160.3	1259.0	1.6892	5.330	1160.1	1258.8	1.6868	5.068	1159.7	1258.2	1.6810	**460**
5.507	1164.2	1264.1	1.6948	5.394	1164.1	1263.9	1.6923	5.130	1163.6	1263.3	1.6865	**470**
5.573	1168.1	1269.2	1.7002	5.459	1168.0	1269.0	1.6978	5.192	1167.6	1268.4	1.6920	**480**
5.638	1172.0	1274.3	1.7056	5.523	1171.8	1274.0	1.7032	5.253	1171.5	1273.5	1.6974	**490**
5.703	1175.9	1279.3	1.7109	5.587	1175.7	1279.1	1.7085	5.314	1175.4	1278.6	1.7027	**500**
5.833	1183.6	1289.4	1.7212	5.714	1183.5	1289.2	1.7189	5.436	1183.1	1288.7	1.7131	**520**
5.962	1191.3	1299.4	1.7314	5.840	1191.2	1299.2	1.7290	5.556	1190.9	1298.8	1.7233	**540**
6.090	1199.0	1309.4	1.7413	5.966	1198.9	1309.3	1.7390	5.677	1198.6	1308.9	1.7333	**560**
6.218	1206.7	1319.4	1.7510	6.091	1206.6	1319.3	1.7487	5.796	1206.3	1318.9	1.7430	**580**
6.345	1214.3	1329.4	1.7605	6.216	1214.2	1329.3	1.7582	5.915	1214.0	1328.9	1.7526	**600**
6.472	1222.0	1339.4	1.7699	6.340	1221.9	1339.3	1.7675	6.034	1221.7	1338.9	1.7619	**620**
6.598	1229.7	1349.4	1.7790	6.464	1229.6	1349.2	1.7767	6.152	1229.4	1348.9	1.7711	**640**
6.724	1237.4	1359.4	1.7880	6.588	1237.3	1359.2	1.7857	6.270	1237.1	1358.9	1.7801	**660**
6.850	1245.1	1369.3	1.7969	6.711	1245.0	1369.2	1.7946	6.388	1244.8	1369.0	1.7890	**680**
6.975	1252.9	1379.4	1.8056	6.834	1252.8	1379.2	1.8033	6.505	1252.6	1379.0	1.7977	**700**
7.100	1260.6	1389.4	1.8141	6.957	1260.5	1389.3	1.8118	6.622	1260.3	1389.0	1.8063	**720**
7.225	1268.4	1399.4	1.8226	7.079	1268.3	1399.3	1.8203	6.739	1268.1	1399.1	1.8148	**740**
7.350	1276.2	1409.5	1.8309	7.201	1276.1	1409.4	1.8286	6.855	1275.9	1409.1	1.8231	**760**
7.474	1284.0	1419.5	1.8391	7.324	1283.9	1419.5	1.8368	6.972	1283.8	1419.2	1.8313	**780**
7.599	1291.9	1429.7	1.8472	7.445	1291.8	1429.6	1.8449	7.088	1291.6	1429.4	1.8394	**800**
7.723	1299.7	1439.8	1.8551	7.567	1299.7	1439.7	1.8529	7.204	1299.5	1439.5	1.8474	**820**
7.847	1307.6	1449.9	1.8630	7.689	1307.6	1449.9	1.8607	7.320	1307.4	1449.7	1.8552	**840**
7.971	1315.6	1460.1	1.8708	7.810	1315.5	1460.0	1.8685	7.436	1315.4	1459.8	1.8630	**860**
8.094	1323.5	1470.3	1.8785	7.932	1323.5	1470.2	1.8762	7.552	1323.3	1470.1	1.8707	**880**
8.218	1331.5	1480.5	1.8860	8.053	1331.5	1480.5	1.8838	7.667	1331.3	1480.3	1.8783	**900**
8.342	1339.5	1490.8	1.8935	8.174	1339.5	1490.7	1.8913	7.783	1339.4	1490.6	1.8858	**920**
8.465	1347.6	1501.1	1.9009	8.295	1347.5	1501.0	1.8987	7.898	1347.4	1500.9	1.8932	**940**
8.588	1355.7	1511.4	1.9082	8.416	1355.6	1511.3	1.9060	8.013	1355.5	1511.2	1.9005	**960**
8.712	1363.8	1521.8	1.9155	8.537	1363.7	1521.7	1.9132	8.128	1363.6	1521.6	1.9078	**980**
8.835	1371.9	1532.1	1.9227	8.657	1371.9	1532.1	1.9204	8.243	1371.8	1531.9	1.9149	**1000**
8.958	1380.1	1542.6	1.9297	8.778	1380.1	1542.5	1.9275	8.358	1380.0	1542.4	1.9220	**1020**
9.081	1388.3	1553.0	1.9367	8.899	1388.3	1552.9	1.9345	8.473	1388.2	1552.8	1.9291	**1040**
9.204	1396.6	1563.5	1.9437	9.019	1396.5	1563.4	1.9414	8.588	1396.4	1563.3	1.9360	**1060**
9.327	1404.8	1574.0	1.9506	9.140	1404.8	1573.9	1.9483	8.703	1404.7	1573.8	1.9429	**1080**
9.450	1413.2	1584.5	1.9574	9.260	1413.1	1584.5	1.9551	8.817	1413.0	1584.4	1.9497	**1100**
10.063	1455.2	1637.7	1.9904	9.861	1455.2	1637.7	1.9882	9.390	1455.1	1637.6	1.9828	**1200**
10.675	1498.2	1691.8	2.0221	10.461	1498.2	1691.8	2.0198	9.962	1498.1	1691.7	2.0144	**1300**
11.286	1542.1	1746.7	2.0524	11.060	1542.0	1746.7	2.0502	10.533	1542.0	1746.6	2.0448	**1400**
11.897	1586.8	1802.6	2.0817	11.659	1586.8	1802.5	2.0794	11.103	1586.8	1802.5	2.0740	**1500**
12.507	1632.5	1859.3	2.1099	12.257	1632.4	1859.3	2.1076	11.673	1632.4	1859.2	2.1022	**1600**
13.727	1726.4	1975.3	2.1637	13.452	1726.4	1975.3	2.1614	12.812	1726.4	1975.3	2.1560	**1800**
14.946	1823.8	2094.8	2.2143	14.647	1823.7	2094.8	2.2121	13.950	1823.7	2094.8	2.2067	**2000**
16.165	1924.3	2217.4	2.2622	15.842	1924.3	2217.4	2.2600	15.087	1924.2	2217.4	2.2546	**2200**
17.383	2027.6	2342.8	2.3077	17.036	2027.5	2342.8	2.3054	16.224	2027.5	2342.8	2.3000	**2400**

Table 3. Vapor

p (t Sat.)	110 (334.82)				115 (338.12)				120 (341.30)			
t	v	u	h	s	v	u	h	s	v	u	h	s
Sat.	4.051	1107.1	1189.6	1.5957	3.884	1107.7	1190.4	1.5921	3.730	1108.3	1191.1	1.5886
300	*3.813*	*1090.9*	*1168.6*	*1.5687*	*3.632*	*1089.8*	*1167.1*	*1.5622*	*3.466*	*1088.7*	*1165.6*	*1.5560*
310	*3.883*	*1095.7*	*1174.7*	*1.5767*	*3.700*	*1094.7*	*1173.4*	*1.5704*	*3.532*	*1093.6*	*1172.0*	*1.5643*
320	*3.952*	*1100.4*	*1180.8*	*1.5846*	*3.766*	*1099.4*	*1179.5*	*1.5784*	*3.597*	*1098.4*	*1178.3*	*1.5724*
330	*4.019*	*1104.9*	*1186.7*	*1.5921*	*3.832*	*1104.0*	*1185.6*	*1.5860*	*3.660*	*1103.1*	*1184.4*	*1.5801*
340	4.086	1109.4	1192.6	1.5995	3.896	1108.6	1191.5	1.5935	*3.722*	*1107.7*	*1190.3*	*1.5877*
350	4.151	1113.8	1198.3	1.6066	3.959	1113.0	1197.3	1.6007	3.783	1112.2	1196.2	1.5950
360	4.216	1118.2	1204.0	1.6136	4.022	1117.4	1203.0	1.6077	3.844	1116.7	1202.0	1.6021
370	4.280	1122.5	1209.6	1.6204	4.084	1121.8	1208.7	1.6146	3.904	1121.0	1207.7	1.6090
380	4.343	1126.7	1215.1	1.6270	4.145	1126.0	1214.2	1.6213	3.963	1125.4	1213.3	1.6157
390	4.406	1130.9	1220.6	1.6335	4.205	1130.3	1219.7	1.6278	4.021	1129.6	1218.9	1.6223
400	4.468	1135.0	1226.0	1.6398	4.265	1134.4	1225.2	1.6342	4.079	1133.8	1224.4	1.6288
410	4.529	1139.1	1231.3	1.6460	4.324	1138.6	1230.6	1.6404	4.136	1138.0	1229.8	1.6351
420	4.590	1143.2	1236.7	1.6521	4.383	1142.7	1236.0	1.6465	4.192	1142.1	1235.2	1.6412
430	4.651	1147.3	1241.9	1.6580	4.441	1146.8	1241.3	1.6525	4.249	1146.2	1240.6	1.6473
440	4.711	1151.3	1247.2	1.6639	4.499	1150.8	1246.5	1.6584	4.305	1150.3	1245.9	1.6532
450	4.771	1155.3	1252.4	1.6697	4.557	1154.8	1251.8	1.6642	4.360	1154.3	1251.2	1.6590
460	4.831	1159.3	1257.6	1.6753	4.614	1158.8	1257.0	1.6699	4.415	1158.4	1256.4	1.6648
470	4.890	1163.2	1262.8	1.6809	4.671	1162.8	1262.2	1.6756	4.470	1162.4	1261.6	1.6704
480	4.949	1167.1	1267.9	1.6864	4.728	1166.7	1267.3	1.6811	4.525	1166.3	1266.8	1.6759
490	5.008	1171.1	1273.0	1.6918	4.784	1170.7	1272.5	1.6865	4.579	1170.3	1272.0	1.6814
500	5.066	1175.0	1278.1	1.6972	4.840	1174.6	1277.6	1.6919	4.633	1174.2	1277.1	1.6868
520	5.183	1182.8	1288.3	1.7077	4.952	1182.4	1287.8	1.7024	4.740	1182.1	1287.3	1.6973
540	5.299	1190.5	1298.4	1.7179	5.063	1190.2	1298.0	1.7127	4.847	1189.9	1297.5	1.7076
560	5.414	1198.3	1308.5	1.7279	5.173	1198.0	1308.1	1.7227	4.953	1197.7	1307.7	1.7177
580	5.528	1206.0	1318.5	1.7376	5.283	1205.7	1318.1	1.7325	5.059	1205.4	1317.8	1.7275
600	5.642	1213.7	1328.6	1.7472	5.392	1213.5	1328.2	1.7420	5.164	1213.2	1327.8	1.7371
620	5.755	1221.4	1338.6	1.7566	5.501	1221.2	1338.3	1.7514	5.268	1220.9	1337.9	1.7465
640	5.869	1229.2	1348.6	1.7658	5.610	1228.9	1348.3	1.7607	5.372	1228.7	1348.0	1.7557
660	5.981	1236.9	1358.6	1.7748	5.718	1236.7	1358.3	1.7697	5.476	1236.4	1358.0	1.7648
680	6.094	1244.6	1368.7	1.7837	5.825	1244.4	1368.4	1.7786	5.579	1244.2	1368.1	1.7737
700	6.206	1252.4	1378.7	1.7924	5.933	1252.2	1378.4	1.7873	5.682	1252.0	1378.2	1.7825
720	6.318	1260.2	1388.8	1.8010	6.040	1260.0	1388.5	1.7959	5.785	1259.8	1388.3	1.7911
740	6.429	1268.0	1398.8	1.8095	6.147	1267.8	1398.6	1.8044	5.888	1267.6	1398.3	1.7996
760	6.541	1275.8	1408.9	1.8178	6.254	1275.6	1408.7	1.8128	5.990	1275.4	1408.5	1.8079
780	6.652	1283.6	1419.0	1.8260	6.360	1283.4	1418.8	1.8210	6.093	1283.3	1418.6	1.8162
800	6.763	1291.5	1429.1	1.8341	6.467	1291.3	1428.9	1.8291	6.195	1291.2	1428.7	1.8243
820	6.874	1299.4	1439.3	1.8421	6.573	1299.2	1439.1	1.8371	6.297	1299.1	1438.9	1.8323
840	6.985	1307.3	1449.5	1.8500	6.679	1307.1	1449.3	1.8450	6.398	1307.0	1449.1	1.8402
860	7.096	1315.2	1459.7	1.8578	6.785	1315.1	1459.5	1.8528	6.500	1314.9	1459.3	1.8480
880	7.206	1323.2	1469.9	1.8655	6.891	1323.1	1469.7	1.8605	6.602	1322.9	1469.5	1.8557
900	7.317	1331.2	1480.1	1.8731	6.996	1331.1	1480.0	1.8681	6.703	1330.9	1479.8	1.8633
920	7.427	1339.2	1490.4	1.8806	7.102	1339.1	1490.2	1.8756	6.804	1339.0	1490.1	1.8708
940	7.537	1347.3	1500.7	1.8880	7.207	1347.2	1500.6	1.8830	6.905	1347.1	1500.4	1.8782
960	7.647	1355.4	1511.0	1.8953	7.313	1355.3	1510.9	1.8903	7.006	1355.2	1510.7	1.8856
980	7.757	1363.5	1521.4	1.9026	7.418	1363.4	1521.3	1.8976	7.107	1363.3	1521.1	1.8928
1000	7.867	1371.7	1531.8	1.9097	7.523	1371.6	1531.7	1.9048	7.208	1371.5	1531.5	1.9000
1020	7.977	1379.9	1542.2	1.9168	7.628	1379.8	1542.1	1.9119	7.309	1379.7	1542.0	1.9071
1040	8.086	1388.1	1552.7	1.9239	7.733	1388.0	1552.6	1.9189	7.410	1387.9	1552.4	1.9141
1060	8.196	1396.3	1563.2	1.9308	7.838	1396.2	1563.0	1.9258	7.510	1396.2	1562.9	1.9211
1080	8.306	1404.6	1573.7	1.9377	7.943	1404.5	1573.6	1.9327	7.611	1404.4	1573.5	1.9280
1100	8.415	1412.9	1584.2	1.9445	8.048	1412.9	1584.1	1.9395	7.711	1412.8	1584.0	1.9348
1200	8.962	1455.1	1637.5	1.9776	8.571	1455.0	1637.4	1.9726	8.213	1454.9	1637.3	1.9679
1300	9.508	1498.1	1691.6	2.0092	9.094	1498.0	1691.5	2.0043	8.714	1497.9	1691.4	1.9996
1400	10.053	1541.9	1746.6	2.0396	9.615	1541.9	1746.5	2.0347	9.214	1541.8	1746.4	2.0300
1500	10.598	1586.7	1802.4	2.0689	10.136	1586.7	1802.4	2.0639	9.713	1586.6	1802.3	2.0592
1600	11.142	1632.4	1859.2	2.0971	10.657	1632.3	1859.1	2.0922	10.212	1632.3	1859.0	2.0875
1800	12.229	1726.3	1975.2	2.1509	11.697	1726.3	1975.2	2.1460	11.209	1726.2	1975.1	2.1413
2000	13.315	1823.7	2094.7	2.2015	12.736	1823.6	2094.7	2.1966	12.205	1823.6	2094.6	2.1919
2200	14.401	1924.2	2217.3	2.2494	13.775	1924.1	2217.3	2.2445	13.201	1924.1	2217.3	2.2398
2400	15.487	2027.5	2342.7	2.2949	14.814	2027.4	2342.7	2.2900	14.197	2027.4	2342.7	2.2853

	125 (344.39)				130 (347.37)				135 (350.27)			p (t Sat.)
v	u	h	s	v	u	h	s	v	u	h	s	t
3.588	1108.8	1191.8	1.5853	3.457	1109.4	1192.5	1.5821	3.335	1109.8	1193.2	1.5790	**Sat.**
3.313	*1087.5*	*1164.2*	*1.5499*	*3.172*	*1086.3*	*1162.6*	*1.5439*	*3.040*	*1085.1*	*1161.1*	*1.5381*	**300**
3.377	*1092.5*	*1170.6*	*1.5584*	*3.234*	*1091.4*	*1169.2*	*1.5526*	*3.102*	*1090.3*	*1167.8*	*1.5469*	**310**
3.440	*1097.4*	*1177.0*	*1.5665*	*3.296*	*1096.4*	*1175.7*	*1.5609*	*3.162*	*1095.3*	*1174.3*	*1.5553*	**320**
3.502	*1102.2*	*1183.2*	*1.5744*	*3.355*	*1101.2*	*1181.9*	*1.5689*	*3.220*	*1100.2*	*1180.7*	*1.5634*	**330**
3.562	*1106.8*	*1189.2*	*1.5820*	*3.414*	*1105.9*	*1188.1*	*1.5766*	*3.277*	*1105.0*	*1186.9*	*1.5713*	**340**
3.622	1111.4	1195.2	1.5894	3.472	1110.6	1194.1	1.5841	*3.334*	*1109.7*	*1193.0*	*1.5788*	**350**
3.680	1115.9	1201.0	1.5966	3.529	1115.1	1200.0	1.5913	3.389	1114.3	1199.0	1.5862	**360**
3.738	1120.3	1206.8	1.6036	3.585	1119.6	1205.8	1.5984	3.443	1118.8	1204.9	1.5933	**370**
3.795	1124.7	1212.5	1.6104	3.640	1124.0	1211.6	1.6052	3.497	1123.3	1210.6	1.6002	**380**
3.851	1129.0	1218.1	1.6171	3.695	1128.3	1217.2	1.6119	3.550	1127.7	1216.4	1.6070	**390**
3.907	1133.2	1223.6	1.6235	3.749	1132.6	1222.8	1.6185	3.602	1132.0	1222.0	1.6136	**400**
3.962	1137.4	1229.1	1.6299	3.803	1136.8	1228.3	1.6249	3.654	1136.3	1227.6	1.6200	**410**
4.017	1141.6	1234.5	1.6361	3.856	1141.0	1233.8	1.6311	3.706	1140.5	1233.1	1.6263	**420**
4.072	1145.7	1239.9	1.6422	3.908	1145.2	1239.2	1.6372	3.757	1144.7	1238.5	1.6325	**430**
4.125	1149.8	1245.2	1.6481	3.960	1149.3	1244.6	1.6433	3.807	1148.8	1243.9	1.6385	**440**
4.179	1153.9	1250.5	1.6540	4.012	1153.4	1249.9	1.6491	3.857	1152.9	1249.3	1.6444	**450**
4.232	1157.9	1255.8	1.6598	4.063	1157.5	1255.2	1.6549	3.907	1157.0	1254.6	1.6503	**460**
4.285	1161.9	1261.0	1.6654	4.114	1161.5	1260.5	1.6606	3.956	1161.1	1259.9	1.6560	**470**
4.338	1165.9	1266.3	1.6710	4.165	1165.5	1265.7	1.6662	4.006	1165.1	1265.2	1.6616	**480**
4.390	1169.9	1271.4	1.6765	4.216	1169.5	1270.9	1.6717	4.054	1169.1	1270.4	1.6672	**490**
4.442	1173.9	1276.6	1.6819	4.266	1173.5	1276.1	1.6772	4.103	1173.1	1275.6	1.6726	**500**
4.546	1181.7	1286.9	1.6925	4.366	1181.4	1286.4	1.6878	4.200	1181.0	1285.9	1.6833	**520**
4.648	1189.6	1297.1	1.7028	4.465	1189.2	1296.7	1.6982	4.295	1188.9	1296.2	1.6937	**540**
4.751	1197.4	1307.3	1.7129	4.564	1197.1	1306.8	1.7083	4.390	1196.8	1306.4	1.7038	**560**
4.852	1205.2	1317.4	1.7227	4.661	1204.9	1317.0	1.7181	4.485	1204.6	1316.6	1.7137	**580**
4.953	1212.9	1327.5	1.7323	4.759	1212.7	1327.1	1.7278	4.579	1212.4	1326.8	1.7234	**600**
5.054	1220.7	1337.6	1.7418	4.856	1220.4	1337.2	1.7372	4.672	1220.2	1336.9	1.7328	**620**
5.154	1228.5	1347.7	1.7510	4.952	1228.2	1347.3	1.7465	4.765	1228.0	1347.0	1.7421	**640**
5.254	1236.2	1357.7	1.7601	5.048	1236.0	1357.4	1.7556	4.858	1235.8	1357.1	1.7512	**660**
5.353	1244.0	1367.8	1.7690	5.144	1243.8	1367.5	1.7645	4.951	1243.6	1367.3	1.7602	**680**
5.452	1251.8	1377.9	1.7778	5.240	1251.6	1377.6	1.7733	5.043	1251.4	1377.4	1.7690	**700**
5.551	1259.6	1388.0	1.7864	5.335	1259.4	1387.7	1.7819	5.135	1259.2	1387.5	1.7776	**720**
5.650	1267.4	1398.1	1.7949	5.430	1267.2	1397.9	1.7905	5.226	1267.1	1397.6	1.7861	**740**
5.748	1275.3	1408.2	1.8033	5.525	1275.1	1408.0	1.7988	5.318	1274.9	1407.8	1.7945	**760**
5.847	1283.1	1418.4	1.8115	5.619	1283.0	1418.1	1.8071	5.409	1282.8	1417.9	1.8028	**780**
5.945	1291.0	1428.5	1.8197	5.714	1290.8	1428.3	1.8152	5.500	1290.7	1428.1	1.8109	**800**
6.043	1298.9	1438.7	1.8277	5.808	1298.8	1438.5	1.8232	5.591	1298.6	1438.3	1.8190	**820**
6.140	1306.8	1448.9	1.8356	5.902	1306.7	1448.7	1.8311	5.682	1306.6	1448.5	1.8269	**840**
6.238	1314.8	1459.1	1.8434	5.996	1314.7	1458.9	1.8390	5.772	1314.5	1458.7	1.8347	**860**
6.336	1322.8	1469.3	1.8511	6.090	1322.7	1469.2	1.8467	5.863	1322.5	1469.0	1.8424	**880**
6.433	1330.8	1479.6	1.8587	6.184	1330.7	1479.5	1.8543	5.953	1330.6	1479.3	1.8500	**900**
6.530	1338.9	1489.9	1.8662	6.277	1338.7	1489.8	1.8618	6.043	1338.6	1489.6	1.8576	**920**
6.627	1346.9	1500.2	1.8737	6.371	1346.8	1500.1	1.8692	6.133	1346.7	1499.9	1.8650	**940**
6.724	1355.0	1510.6	1.8810	6.464	1354.9	1510.4	1.8766	6.223	1354.8	1510.3	1.8724	**960**
6.821	1363.2	1521.0	1.8883	6.558	1363.1	1520.8	1.8839	6.313	1363.0	1520.7	1.8796	**980**
6.918	1371.4	1531.4	1.8954	6.651	1371.3	1531.2	1.8910	6.403	1371.1	1531.1	1.8868	**1000**
7.015	1379.6	1541.8	1.9025	6.744	1379.5	1541.7	1.8982	6.493	1379.4	1541.6	1.8939	**1020**
7.112	1387.8	1552.3	1.9096	6.837	1387.7	1552.2	1.9052	6.583	1387.6	1552.0	1.9010	**1040**
7.209	1396.1	1562.8	1.9165	6.930	1396.0	1562.7	1.9121	6.672	1395.9	1562.6	1.9079	**1060**
7.305	1404.4	1573.3	1.9234	7.023	1404.3	1573.2	1.9190	6.762	1404.2	1573.1	1.9148	**1080**
7.402	1412.7	1583.9	1.9302	7.116	1412.6	1583.8	1.9259	6.851	1412.5	1583.7	1.9216	**1100**
7.884	1454.8	1637.2	1.9634	7.580	1454.8	1637.1	1.9590	7.298	1454.7	1637.0	1.9548	**1200**
8.365	1497.9	1691.4	1.9950	8.042	1497.8	1691.3	1.9907	7.743	1497.7	1691.2	1.9865	**1300**
8.845	1541.8	1746.4	2.0254	8.504	1541.7	1746.3	2.0211	8.188	1541.7	1746.2	2.0169	**1400**
9.324	1586.6	1802.2	2.0547	8.965	1586.5	1802.2	2.0504	8.632	1586.5	1802.1	2.0462	**1500**
9.803	1632.2	1859.0	2.0829	9.426	1632.2	1858.9	2.0786	9.076	1632.1	1858.9	2.0744	**1600**
10.760	1726.2	1975.1	2.1367	10.346	1726.2	1975.1	2.1324	9.963	1726.1	1975.0	2.1282	**1800**
11.717	1823.6	2094.6	2.1874	11.266	1823.5	2094.6	2.1831	10.849	1823.5	2094.5	2.1789	**2000**
12.673	1924.1	2217.2	2.2353	12.185	1924.0	2217.2	2.2310	11.734	1924.0	2217.1	2.2268	**2200**
13.629	2027.4	2342.6	2.2808	13.105	2027.3	2342.6	2.2764	12.619	2027.3	2342.5	2.2723	**2400**

Table 3. Vapor

p (t Sat.)	140 (353.08)				145 (355.82)				150 (358.48)			
t	v	u	h	s	v	u	h	s	v	u	h	s
Sat.	3.221	1110.3	1193.8	1.5761	3.115	1110.8	1194.4	1.5732	3.016	1111.2	1194.9	1.5704
310	*2.978*	*1089.2*	*1166.4*	*1.5414*	*2.864*	*1088.0*	*1164.9*	*1.5359*	*2.756*	*1086.9*	*1163.4*	*1.5306*
320	*3.037*	*1094.3*	*1173.0*	*1.5499*	*2.921*	*1093.2*	*1171.6*	*1.5446*	*2.812*	*1092.2*	*1170.2*	*1.5395*
330	*3.094*	*1099.3*	*1179.4*	*1.5581*	*2.977*	*1098.3*	*1178.2*	*1.5530*	*2.867*	*1097.3*	*1176.9*	*1.5479*
340	*3.150*	*1104.1*	*1185.7*	*1.5661*	*3.031*	*1103.2*	*1184.5*	*1.5610*	*2.920*	*1102.3*	*1183.3*	*1.5561*
350	*3.205*	*1108.9*	*1191.9*	*1.5737*	*3.085*	*1108.0*	*1190.8*	*1.5688*	*2.973*	*1107.2*	*1189.7*	*1.5639*
360	3.259	1113.5	1198.0	1.5812	3.137	1112.7	1196.9	1.5763	3.024	1111.9	1195.9	1.5715
370	3.312	1118.1	1203.9	1.5884	3.189	1117.3	1202.9	1.5836	3.074	1116.6	1201.9	1.5789
380	3.364	1122.6	1209.7	1.5954	3.240	1121.9	1208.8	1.5906	3.124	1121.2	1207.9	1.5860
390	3.415	1127.0	1215.5	1.6022	3.290	1126.3	1214.6	1.5975	3.173	1125.7	1213.7	1.5930
400	3.466	1131.4	1221.2	1.6088	3.340	1130.7	1220.3	1.6042	3.221	1130.1	1219.5	1.5997
410	3.517	1135.7	1226.8	1.6153	3.389	1135.1	1226.0	1.6107	3.269	1134.5	1225.2	1.6063
420	3.567	1139.9	1232.3	1.6217	3.437	1139.4	1231.6	1.6171	3.316	1138.8	1230.8	1.6127
430	3.616	1144.1	1237.8	1.6279	3.485	1143.6	1237.1	1.6234	3.363	1143.1	1236.4	1.6190
440	3.665	1148.3	1243.3	1.6339	3.533	1147.8	1242.6	1.6295	3.409	1147.3	1241.9	1.6252
450	3.713	1152.4	1248.6	1.6399	3.580	1152.0	1248.0	1.6355	3.455	1151.5	1247.4	1.6312
460	3.762	1156.5	1254.0	1.6458	3.626	1156.1	1253.4	1.6414	3.500	1155.6	1252.8	1.6371
470	3.810	1160.6	1259.3	1.6515	3.673	1160.2	1258.7	1.6472	3.545	1159.7	1258.1	1.6429
480	3.857	1164.7	1264.6	1.6572	3.719	1164.3	1264.0	1.6528	3.590	1163.8	1263.5	1.6486
490	3.904	1168.7	1269.9	1.6627	3.765	1168.3	1269.3	1.6584	3.635	1167.9	1268.8	1.6542
500	3.952	1172.7	1275.1	1.6682	3.811	1172.3	1274.6	1.6639	3.679	1171.9	1274.1	1.6598
520	4.045	1180.7	1285.5	1.6789	3.901	1180.3	1285.0	1.6747	3.767	1180.0	1284.5	1.6706
540	4.138	1188.6	1295.8	1.6893	3.991	1188.3	1295.3	1.6851	3.854	1187.9	1294.9	1.6811
560	4.230	1196.5	1306.0	1.6995	4.080	1196.2	1305.6	1.6953	3.940	1195.8	1305.2	1.6913
580	4.321	1204.3	1316.2	1.7094	4.168	1204.0	1315.9	1.7053	4.026	1203.7	1315.5	1.7012
600	4.412	1212.1	1326.4	1.7191	4.256	1211.9	1326.1	1.7150	4.111	1211.6	1325.7	1.7110
620	4.502	1219.9	1336.6	1.7286	4.344	1219.7	1336.2	1.7245	4.196	1219.4	1335.9	1.7205
640	4.592	1227.7	1346.7	1.7379	4.431	1227.5	1346.4	1.7338	4.280	1227.3	1346.1	1.7299
660	4.682	1235.6	1356.8	1.7470	4.517	1235.3	1356.5	1.7430	4.364	1235.1	1356.2	1.7390
680	4.771	1243.4	1367.0	1.7560	4.604	1243.2	1366.7	1.7519	4.448	1242.9	1366.4	1.7480
700	4.860	1251.2	1377.1	1.7648	4.690	1251.0	1376.8	1.7608	4.531	1250.8	1376.6	1.7568
720	4.949	1259.0	1387.2	1.7735	4.776	1258.8	1387.0	1.7694	4.614	1258.6	1386.7	1.7655
740	5.037	1266.9	1397.4	1.7820	4.861	1266.7	1397.1	1.7780	4.697	1266.5	1396.9	1.7741
760	5.125	1274.7	1407.5	1.7904	4.946	1274.6	1407.3	1.7864	4.779	1274.4	1407.1	1.7825
780	5.214	1282.6	1417.7	1.7986	5.032	1282.5	1417.5	1.7946	4.862	1282.3	1417.2	1.7908
800	5.301	1290.5	1427.9	1.8068	5.117	1290.4	1427.7	1.8028	4.944	1290.2	1427.5	1.7989
820	5.389	1298.5	1438.1	1.8148	5.201	1298.3	1437.9	1.8108	5.026	1298.2	1437.7	1.8070
840	5.477	1306.4	1448.3	1.8228	5.286	1306.3	1448.1	1.8188	5.108	1306.1	1447.9	1.8149
860	5.564	1314.4	1458.5	1.8306	5.371	1314.3	1458.4	1.8266	5.190	1314.1	1458.2	1.8228
880	5.651	1322.4	1468.8	1.8383	5.455	1322.3	1468.6	1.8343	5.271	1322.1	1468.5	1.8305
900	5.739	1330.4	1479.1	1.8459	5.539	1330.3	1478.9	1.8420	5.353	1330.2	1478.8	1.8381
920	5.826	1338.5	1489.4	1.8535	5.623	1338.4	1489.3	1.8495	5.434	1338.3	1489.1	1.8457
940	5.913	1346.6	1499.8	1.8609	5.707	1346.5	1499.6	1.8570	5.516	1346.4	1499.5	1.8531
960	6.000	1354.7	1510.1	1.8683	5.791	1354.6	1510.0	1.8643	5.597	1354.5	1509.8	1.8605
980	6.086	1362.9	1520.5	1.8755	5.875	1362.8	1520.4	1.8716	5.678	1362.6	1520.2	1.8678
1000	6.173	1371.0	1531.0	1.8827	5.959	1370.9	1530.8	1.8788	5.759	1370.8	1530.7	1.8750
1020	6.260	1379.3	1541.4	1.8898	6.043	1379.2	1541.3	1.8859	5.840	1379.1	1541.2	1.8821
1040	6.346	1387.5	1551.9	1.8969	6.126	1387.4	1551.8	1.8930	5.921	1387.3	1551.6	1.8891
1060	6.433	1395.8	1562.4	1.9039	6.210	1395.7	1562.3	1.8999	6.002	1395.6	1562.2	1.8961
1080	6.519	1404.1	1573.0	1.9107	6.293	1404.0	1572.9	1.9068	6.082	1403.9	1572.7	1.9030
1100	6.605	1412.4	1583.6	1.9176	6.377	1412.3	1583.4	1.9136	6.163	1412.2	1583.3	1.9099
1200	7.036	1454.6	1636.9	1.9507	6.793	1454.6	1636.8	1.9468	6.566	1454.5	1636.7	1.9430
1300	7.466	1497.7	1691.1	1.9824	7.208	1497.6	1691.0	1.9785	6.967	1497.6	1690.9	1.9748
1400	7.895	1541.6	1746.1	2.0129	7.622	1541.5	1746.1	2.0090	7.368	1541.5	1746.0	2.0052
1500	8.324	1586.4	1802.0	2.0421	8.036	1586.4	1802.0	2.0382	7.768	1586.3	1801.9	2.0345
1600	8.752	1632.1	1858.8	2.0704	8.449	1632.0	1858.8	2.0665	8.167	1632.0	1858.7	2.0627
1800	9.607	1726.1	1975.0	2.1242	9.275	1726.0	1974.9	2.1203	8.966	1726.0	1974.9	2.1166
2000	10.461	1823.5	2094.5	2.1749	10.100	1823.4	2094.4	2.1710	9.763	1823.4	2094.4	2.1672
2200	11.315	1924.0	2217.1	2.2228	10.925	1923.9	2217.1	2.2189	10.561	1923.9	2217.0	2.2152
2400	12.169	2027.3	2342.5	2.2682	11.749	2027.2	2342.5	2.2644	11.357	2027.2	2342.4	2.2606

	155 (361.07)				**160** (363.60)				**165** (366.07)			**p** (t Sat.)
v	u	h	s	v	u	h	s	v	u	h	s	t
2.923	1111.6	1195.5	1.5677	2.836	1112.0	1196.0	1.5651	2.754	1112.4	1196.4	1.5625	Sat.
												310
2.711	*1091.1*	*1168.8*	*1.5344*	*2.615*	*1090.0*	*1167.4*	*1.5294*					320
2.764	*1096.3*	*1175.6*	*1.5430*	*2.668*	*1095.3*	*1174.2*	*1.5381*	*2.577*	*1094.2*	*1172.9*	*1.5333*	330
2.817	*1101.3*	*1182.1*	*1.5512*	*2.719*	*1100.4*	*1180.9*	*1.5465*	*2.628*	*1099.4*	*1179.6*	*1.5418*	340
2.868	*1106.3*	*1188.5*	*1.5592*	*2.769*	*1105.4*	*1187.4*	*1.5545*	*2.677*	*1104.5*	*1186.2*	*1.5500*	350
2.918	*1111.1*	*1194.8*	*1.5669*	*2.818*	*1110.3*	*1193.7*	*1.5623*	*2.725*	*1109.4*	*1192.6*	*1.5579*	360
2.967	1115.8	1200.9	1.5743	2.867	1115.0	1199.9	1.5698	2.772	1114.3	1198.9	1.5655	370
3.016	1120.4	1206.9	1.5815	2.914	1119.7	1206.0	1.5771	2.818	1119.0	1205.0	1.5728	380
3.063	1125.0	1212.9	1.5885	2.961	1124.3	1212.0	1.5842	2.864	1123.6	1211.1	1.5800	390
3.110	1129.5	1218.7	1.5953	3.007	1128.8	1217.8	1.5911	2.909	1128.2	1217.0	1.5869	400
3.157	1133.9	1224.4	1.6020	3.052	1133.3	1223.6	1.5978	2.953	1132.7	1222.8	1.5936	410
3.203	1138.2	1230.1	1.6085	3.097	1137.6	1229.3	1.6043	2.997	1137.1	1228.6	1.6002	420
3.248	1142.5	1235.7	1.6148	3.141	1142.0	1235.0	1.6107	3.040	1141.4	1234.3	1.6066	430
3.293	1146.8	1241.2	1.6210	3.185	1146.3	1240.6	1.6169	3.083	1145.7	1239.9	1.6129	440
3.338	1151.0	1246.7	1.6270	3.228	1150.5	1246.1	1.6230	3.125	1150.0	1245.4	1.6190	450
3.382	1155.2	1252.2	1.6330	3.271	1154.7	1251.5	1.6290	3.167	1154.2	1250.9	1.6251	460
3.426	1159.3	1257.6	1.6388	3.314	1158.8	1257.0	1.6348	3.209	1158.4	1256.4	1.6310	470
3.469	1163.4	1262.9	1.6446	3.356	1163.0	1262.4	1.6406	3.250	1162.5	1261.8	1.6367	480
3.513	1167.5	1268.2	1.6502	3.398	1167.1	1267.7	1.6463	3.291	1166.7	1267.2	1.6424	490
3.556	1171.5	1273.5	1.6557	3.440	1171.2	1273.0	1.6518	3.332	1170.8	1272.5	1.6480	500
3.641	1179.6	1284.0	1.6666	3.523	1179.2	1283.6	1.6627	3.412	1178.9	1283.1	1.6589	520
3.726	1187.6	1294.5	1.6771	3.605	1187.3	1294.0	1.6733	3.492	1186.9	1293.6	1.6695	540
3.809	1195.5	1304.8	1.6874	3.687	1195.2	1304.4	1.6835	3.572	1194.9	1304.0	1.6798	560
3.893	1203.4	1315.1	1.6973	3.768	1203.2	1314.7	1.6936	3.650	1202.9	1314.3	1.6899	580
3.975	1211.3	1325.3	1.7071	3.848	1211.1	1325.0	1.7034	3.728	1210.8	1324.6	1.6997	600
4.057	1219.2	1335.6	1.7167	3.928	1218.9	1335.2	1.7129	3.806	1218.7	1334.9	1.7093	620
4.139	1227.0	1345.8	1.7260	4.007	1226.8	1345.4	1.7223	3.883	1226.6	1345.1	1.7187	640
4.220	1234.9	1355.9	1.7352	4.086	1234.7	1355.6	1.7315	3.960	1234.4	1355.3	1.7279	660
4.302	1242.7	1366.1	1.7442	4.165	1242.5	1365.8	1.7405	4.036	1242.3	1365.5	1.7369	680
4.382	1250.6	1376.3	1.7531	4.243	1250.4	1376.0	1.7494	4.112	1250.2	1375.7	1.7458	700
4.463	1258.4	1386.5	1.7617	4.321	1258.3	1386.2	1.7581	4.188	1258.1	1385.9	1.7545	720
4.543	1266.3	1396.6	1.7703	4.399	1266.1	1396.4	1.7667	4.264	1266.0	1396.1	1.7631	740
4.623	1274.2	1406.8	1.7787	4.477	1274.0	1406.6	1.7751	4.339	1273.9	1406.4	1.7716	760
4.703	1282.1	1417.0	1.7870	4.554	1282.0	1416.8	1.7834	4.414	1281.8	1416.6	1.7799	780
4.783	1290.1	1427.2	1.7952	4.631	1289.9	1427.0	1.7916	4.489	1289.7	1426.8	1.7881	800
4.862	1298.0	1437.5	1.8033	4.708	1297.9	1437.3	1.7996	4.564	1297.7	1437.1	1.7961	820
4.942	1306.0	1447.7	1.8112	4.785	1305.8	1447.5	1.8076	4.639	1305.7	1447.3	1.8041	840
5.021	1314.0	1458.0	1.8190	4.862	1313.8	1457.8	1.8154	4.713	1313.7	1457.6	1.8119	860
5.100	1322.0	1468.3	1.8268	4.939	1321.9	1468.1	1.8232	4.788	1321.7	1467.9	1.8197	880
5.179	1330.1	1478.6	1.8344	5.015	1329.9	1478.4	1.8308	4.862	1329.8	1478.3	1.8273	900
5.258	1338.1	1488.9	1.8420	5.092	1338.0	1488.8	1.8384	4.936	1337.9	1488.6	1.8349	920
5.336	1346.2	1499.3	1.8494	5.168	1346.1	1499.1	1.8459	5.010	1346.0	1499.0	1.8424	940
5.415	1354.4	1509.7	1.8568	5.244	1354.3	1509.5	1.8532	5.084	1354.1	1509.4	1.8498	960
5.494	1362.5	1520.1	1.8641	5.321	1362.4	1520.0	1.8605	5.158	1362.3	1519.8	1.8570	980
5.572	1370.7	1530.5	1.8713	5.397	1370.6	1530.4	1.8677	5.232	1370.5	1530.3	1.8643	1000
5.650	1379.0	1541.0	1.8784	5.473	1378.9	1540.9	1.8749	5.306	1378.8	1540.8	1.8714	1020
5.729	1387.2	1551.5	1.8855	5.549	1387.1	1551.4	1.8819	5.379	1387.0	1551.3	1.8784	1040
5.807	1395.5	1562.1	1.8924	5.624	1395.4	1561.9	1.8889	5.453	1395.3	1561.8	1.8854	1060
5.885	1403.8	1572.6	1.8993	5.700	1403.7	1572.5	1.8958	5.526	1403.6	1572.4	1.8923	1080
5.963	1412.2	1583.2	1.9062	5.776	1412.1	1583.1	1.9026	5.600	1412.0	1583.0	1.8992	1100
6.353	1454.4	1636.6	1.9394	6.154	1454.3	1636.5	1.9358	5.966	1454.3	1636.4	1.9324	1200
6.742	1497.5	1690.9	1.9711	6.530	1497.4	1690.8	1.9676	6.332	1497.4	1690.7	1.9641	1300
7.129	1541.4	1745.9	2.0015	6.906	1541.4	1745.9	1.9980	6.696	1541.3	1745.8	1.9946	1400
7.517	1586.3	1801.9	2.0308	7.281	1586.2	1801.8	2.0273	7.060	1586.2	1801.7	2.0239	1500
7.904	1632.0	1858.7	2.0591	7.656	1631.9	1858.6	2.0556	7.424	1631.9	1858.5	2.0522	1600
8.676	1726.0	1974.8	2.1129	8.405	1725.9	1974.8	2.1094	8.150	1725.9	1974.7	2.1060	1800
9.448	1823.4	2094.4	2.1636	9.153	1823.3	2094.3	2.1601	8.875	1823.3	2094.3	2.1567	2000
10.220	1923.9	2217.0	2.2115	9.900	1923.8	2217.0	2.2080	9.600	1923.8	2216.9	2.2046	2200
10.991	2027.2	2342.4	2.2570	10.648	2027.1	2342.4	2.2535	10.325	2027.1	2342.3	2.2501	2400

Table 3. Vapor

p (t Sat.)	170 (368.47)				175 (370.83)				180 (373.13)			
t	v	u	h	s	v	u	h	s	v	u	h	s
Sat.	2.676	1112.7	1196.9	1.5600	2.603	1113.1	1197.4	1.5576	2.533	1113.4	1197.8	1.5553
350	*2.590*	*1103.6*	*1185.0*	*1.5455*	*2.507*	*1102.7*	*1183.9*	*1.5412*	*2.429*	*1101.7*	*1182.6*	*1.5368*
360	*2.637*	*1108.6*	*1191.5*	*1.5535*	2.554	1107.7	1190.4	1.5492	2.475	1106.9	1189.3	1.5450
370	2.683	1113.5	1197.9	1.5612	*2.599*	*1112.7*	*1196.8*	*1.5570*	*2.520*	*1111.9*	*1195.8*	*1.5529*
380	2.728	1118.2	1204.1	1.5686	2.643	1117.5	1203.1	1.5645	2.563	1116.7	1202.1	1.5605
390	2.773	1122.9	1210.2	1.5758	2.687	1122.2	1209.2	1.5718	2.606	1121.5	1208.3	1.5678
400	2.817	1127.5	1216.1	1.5828	2.730	1126.9	1215.3	1.5788	2.648	1126.2	1214.4	1.5749
410	2.860	1132.0	1222.0	1.5896	2.773	1131.4	1221.2	1.5857	2.690	1130.8	1220.4	1.5818
420	2.903	1136.5	1227.8	1.5962	2.814	1135.9	1227.0	1.5924	2.731	1135.3	1226.3	1.5886
430	2.945	1140.9	1233.5	1.6027	2.856	1140.3	1232.8	1.5989	2.771	1139.8	1232.1	1.5951
440	2.987	1145.2	1239.2	1.6090	2.896	1144.7	1238.5	1.6052	2.811	1144.2	1237.8	1.6015
450	3.028	1149.5	1244.8	1.6152	2.937	1149.0	1244.1	1.6114	2.850	1148.5	1243.4	1.6078
460	3.069	1153.7	1250.3	1.6212	2.977	1153.3	1249.7	1.6175	2.889	1152.8	1249.0	1.6139
470	3.110	1157.9	1255.8	1.6272	3.016	1157.5	1255.2	1.6235	2.928	1157.0	1254.6	1.6199
480	3.150	1162.1	1261.2	1.6330	3.056	1161.7	1260.6	1.6293	2.966	1161.3	1260.1	1.6257
490	3.190	1166.3	1266.6	1.6387	3.095	1165.8	1266.1	1.6351	3.005	1165.4	1265.5	1.6315
500	3.230	1170.4	1272.0	1.6443	3.133	1170.0	1271.4	1.6407	3.042	1169.6	1270.9	1.6372
510	3.269	1174.5	1277.3	1.6498	3.172	1174.1	1276.8	1.6462	3.080	1173.7	1276.3	1.6427
520	3.308	1178.5	1282.6	1.6553	3.210	1178.2	1282.1	1.6517	3.117	1177.8	1281.6	1.6482
530	3.347	1182.6	1287.9	1.6606	3.248	1182.2	1287.4	1.6571	3.154	1181.9	1286.9	1.6536
540	3.386	1186.6	1293.1	1.6659	3.286	1186.3	1292.7	1.6624	3.191	1185.9	1292.2	1.6589
550	3.425	1190.6	1298.3	1.6711	3.323	1190.3	1297.9	1.6676	3.228	1190.0	1297.5	1.6642
560	3.463	1194.6	1303.6	1.6762	3.361	1194.3	1303.1	1.6727	3.264	1194.0	1302.7	1.6693
570	3.501	1198.6	1308.8	1.6813	3.398	1198.3	1308.3	1.6778	3.301	1198.0	1307.9	1.6744
580	3.540	1202.6	1313.9	1.6863	3.435	1202.3	1313.5	1.6828	3.337	1202.0	1313.1	1.6795
590	3.578	1206.5	1319.1	1.6913	3.472	1206.3	1318.7	1.6878	3.373	1206.0	1318.3	1.6844
600	3.616	1210.5	1324.3	1.6962	3.509	1210.2	1323.9	1.6927	3.409	1210.0	1323.5	1.6893
620	3.691	1218.4	1334.5	1.7058	3.583	1218.2	1334.2	1.7023	3.481	1217.9	1333.9	1.6990
640	3.766	1226.3	1344.8	1.7152	3.656	1226.1	1344.5	1.7118	3.552	1225.8	1344.1	1.7084
660	3.841	1234.2	1355.0	1.7244	3.728	1234.0	1354.7	1.7210	3.623	1233.8	1354.4	1.7177
680	3.915	1242.1	1365.3	1.7335	3.801	1241.9	1365.0	1.7301	3.693	1241.7	1364.7	1.7268
700	3.989	1250.0	1375.5	1.7423	3.873	1249.8	1375.2	1.7390	3.763	1249.6	1374.9	1.7357
720	4.063	1257.9	1385.7	1.7511	3.945	1257.7	1385.4	1.7477	3.833	1257.5	1385.2	1.7445
740	4.136	1265.8	1395.9	1.7597	4.016	1265.6	1395.7	1.7563	3.903	1265.4	1395.4	1.7531
760	4.209	1273.7	1406.1	1.7681	4.087	1273.5	1405.9	1.7648	3.972	1273.4	1405.7	1.7615
780	4.283	1281.6	1416.4	1.7764	4.158	1281.5	1416.1	1.7731	4.041	1281.3	1415.9	1.7699
800	4.355	1289.6	1426.6	1.7846	4.229	1289.4	1426.4	1.7813	4.110	1289.3	1426.2	1.7781
820	4.428	1297.6	1436.9	1.7927	4.300	1297.4	1436.7	1.7894	4.179	1297.3	1436.5	1.7862
840	4.501	1305.5	1447.1	1.8007	4.371	1305.4	1446.9	1.7974	4.248	1305.3	1446.7	1.7942
860	4.573	1313.6	1457.4	1.8085	4.441	1313.4	1457.2	1.8052	4.316	1313.3	1457.1	1.8020
880	4.645	1321.6	1467.7	1.8163	4.511	1321.5	1467.6	1.8130	4.385	1321.3	1467.4	1.8098
900	4.718	1329.7	1478.1	1.8240	4.582	1329.5	1477.9	1.8207	4.453	1329.4	1477.7	1.8175
920	4.790	1337.8	1488.4	1.8315	4.652	1337.6	1488.3	1.8282	4.521	1337.5	1488.1	1.8250
940	4.862	1345.9	1498.8	1.8390	4.722	1345.8	1498.7	1.8357	4.589	1345.6	1498.5	1.8325
960	4.934	1354.0	1509.2	1.8464	4.791	1353.9	1509.1	1.8431	4.657	1353.8	1508.9	1.8399
980	5.005	1362.2	1519.7	1.8537	4.861	1362.1	1519.5	1.8504	4.725	1362.0	1519.4	1.8472
1000	5.077	1370.4	1530.1	1.8609	4.931	1370.3	1530.0	1.8576	4.793	1370.2	1529.8	1.8545
1020	5.149	1378.6	1540.6	1.8680	5.000	1378.5	1540.5	1.8648	4.861	1378.4	1540.3	1.8616
1040	5.220	1386.9	1551.1	1.8751	5.070	1386.8	1551.0	1.8718	4.928	1386.7	1550.9	1.8687
1060	5.292	1395.2	1561.7	1.8821	5.139	1395.1	1561.6	1.8788	4.996	1395.0	1561.4	1.8757
1080	5.363	1403.5	1572.2	1.8890	5.209	1403.4	1572.1	1.8857	5.063	1403.4	1572.0	1.8826
1100	5.434	1411.9	1582.9	1.8958	5.278	1411.8	1582.7	1.8926	5.131	1411.7	1582.6	1.8894
1200	5.790	1454.2	1636.3	1.9291	5.624	1454.1	1636.2	1.9258	5.467	1454.0	1636.1	1.9227
1300	6.145	1497.3	1690.6	1.9608	5.969	1497.2	1690.5	1.9576	5.802	1497.2	1690.4	1.9544
1400	6.499	1541.3	1745.7	1.9913	6.313	1541.2	1745.6	1.9880	6.137	1541.2	1745.6	1.9849
1500	6.852	1586.1	1801.7	2.0206	6.656	1586.1	1801.6	2.0174	6.471	1586.0	1801.5	2.0142
1600	7.205	1631.8	1858.5	2.0489	6.999	1631.8	1858.4	2.0456	6.804	1631.7	1858.4	2.0425
1800	7.910	1725.9	1974.7	2.1027	7.684	1725.8	1974.6	2.0995	7.470	1725.8	1974.6	2.0964
2000	8.614	1823.2	2094.2	2.1534	8.368	1823.2	2094.2	2.1502	8.135	1823.2	2094.2	2.1470
2200	9.318	1923.8	2216.9	2.2013	9.052	1923.7	2216.9	2.1981	8.800	1923.7	2216.8	2.1950
2400	10.021	2027.1	2342.3	2.2468	9.735	2027.0	2342.3	2.2436	9.465	2027.0	2342.2	2.2404

v	u	h	s	v	u	h	s	v	u	h	s	t
2.468	1113.7	1198.2	1.5530	2.405	1114.0	1198.6	1.5507	2.346	1114.3	1198.9	1.5486	Sat.
2.356	*1100.8*	*1181.4*	*1.5326*	*2.286*	*1099.8*	*1180.2*	*1.5284*	*2.219*	*1098.9*	*1178.9*	*1.5243*	350
2.400	*1106.0*	*1188.2*	*1.5409*	*2.330*	*1105.1*	*1187.0*	*1.5368*	*2.263*	*1104.2*	*1185.8*	*1.5328*	360
2.444	*1111.0*	*1194.7*	*1.5488*	*2.373*	*1110.2*	*1193.6*	*1.5448*	*2.305*	*1109.4*	*1192.6*	*1.5409*	370
2.487	1116.0	1201.1	1.5565	2.415	1115.2	1200.1	1.5526	2.347	1114.4	1199.1	1.5488	380
2.529	1120.8	1207.4	1.5639	2.457	1120.1	1206.4	1.5601	2.387	1119.3	1205.5	1.5563	390
2.571	1125.5	1213.5	1.5711	2.497	1124.8	1212.6	1.5673	2.427	1124.1	1211.7	1.5636	400
2.611	1130.1	1219.5	1.5781	2.537	1129.5	1218.7	1.5744	2.466	1128.9	1217.9	1.5707	410
2.651	1134.7	1225.5	1.5848	2.576	1134.1	1224.7	1.5812	2.505	1133.5	1223.9	1.5776	420
2.691	1139.2	1231.3	1.5914	2.615	1138.6	1230.6	1.5878	2.543	1138.1	1229.8	1.5843	430
2.730	1143.6	1237.1	1.5979	2.653	1143.1	1236.4	1.5943	2.581	1142.5	1235.7	1.5908	440
2.769	1148.0	1242.8	1.6042	2.691	1147.5	1242.1	1.6007	2.618	1147.0	1241.4	1.5972	450
2.807	1152.3	1248.4	1.6103	2.729	1151.8	1247.8	1.6068	2.654	1151.3	1247.1	1.6034	460
2.845	1156.6	1254.0	1.6163	2.766	1156.1	1253.4	1.6129	2.691	1155.7	1252.7	1.6095	470
2.882	1160.8	1259.5	1.6222	2.802	1160.4	1258.9	1.6188	2.726	1159.9	1258.3	1.6155	480
2.919	1165.0	1265.0	1.6280	2.839	1164.6	1264.4	1.6247	2.762	1164.2	1263.8	1.6213	490
2.956	1169.2	1270.4	1.6337	2.875	1168.8	1269.9	1.6304	2.797	1168.4	1269.3	1.6271	500
2.993	1173.3	1275.8	1.6393	2.911	1172.9	1275.3	1.6360	2.832	1172.5	1274.8	1.6327	510
3.029	1177.4	1281.1	1.6448	2.946	1177.1	1280.6	1.6415	2.867	1176.7	1280.2	1.6383	520
3.066	1181.5	1286.5	1.6502	2.982	1181.2	1286.0	1.6469	2.902	1180.8	1285.5	1.6437	530
3.102	1185.6	1291.8	1.6556	3.017	1185.2	1291.3	1.6523	2.936	1184.9	1290.9	1.6491	540
3.137	1189.6	1297.0	1.6608	3.052	1189.3	1296.6	1.6575	2.970	1189.0	1296.2	1.6544	550
3.173	1193.7	1302.3	1.6660	3.087	1193.4	1301.9	1.6627	3.004	1193.0	1301.5	1.6596	560
3.209	1197.7	1307.5	1.6711	3.121	1197.4	1307.1	1.6679	3.038	1197.1	1306.7	1.6647	570
3.244	1201.7	1312.8	1.6762	3.156	1201.4	1312.4	1.6729	3.072	1201.1	1312.0	1.6698	580
3.279	1205.7	1318.0	1.6811	3.190	1205.4	1317.6	1.6779	3.106	1205.1	1317.2	1.6748	590
3.314	1209.7	1323.2	1.6861	3.224	1209.4	1322.8	1.6829	3.139	1209.2	1322.4	1.6797	600
3.384	1217.7	1333.5	1.6957	3.292	1217.4	1333.2	1.6926	3.206	1217.1	1332.8	1.6895	620
3.453	1225.6	1343.8	1.7052	3.360	1225.4	1343.5	1.7020	3.272	1225.1	1343.2	1.6990	640
3.522	1233.5	1354.1	1.7145	3.427	1233.3	1353.8	1.7113	3.337	1233.1	1353.5	1.7083	660
3.591	1241.5	1364.4	1.7236	3.494	1241.2	1364.1	1.7204	3.403	1241.0	1363.8	1.7174	680
3.659	1249.4	1374.6	1.7325	3.561	1249.2	1374.4	1.7294	3.468	1249.0	1374.1	1.7263	700
3.727	1257.3	1384.9	1.7413	3.627	1257.1	1384.6	1.7382	3.533	1256.9	1384.4	1.7351	720
3.795	1265.2	1395.2	1.7499	3.694	1265.0	1394.9	1.7468	3.597	1264.9	1394.7	1.7438	740
3.863	1273.2	1405.4	1.7584	3.759	1273.0	1405.2	1.7553	3.661	1272.8	1404.9	1.7523	760
3.930	1281.1	1415.7	1.7667	3.825	1281.0	1415.5	1.7636	3.726	1280.8	1415.2	1.7606	780
3.997	1289.1	1426.0	1.7749	3.891	1288.9	1425.7	1.7719	3.789	1288.8	1425.5	1.7689	800
4.065	1297.1	1436.2	1.7830	3.956	1296.9	1436.0	1.7800	3.853	1296.8	1435.8	1.7770	820
4.131	1305.1	1446.5	1.7910	4.021	1305.0	1446.4	1.7880	3.917	1304.8	1446.2	1.7850	840
4.198	1313.1	1456.9	1.7989	4.086	1313.0	1456.7	1.7959	3.980	1312.9	1456.5	1.7929	860
4.265	1321.2	1467.2	1.8067	4.151	1321.1	1467.0	1.8036	4.044	1320.9	1466.8	1.8007	880
4.331	1329.3	1477.6	1.8144	4.216	1329.2	1477.4	1.8113	4.107	1329.0	1477.2	1.8084	900
4.398	1337.4	1487.9	1.8219	4.281	1337.3	1487.8	1.8189	4.170	1337.1	1487.6	1.8160	920
4.464	1345.5	1498.3	1.8294	4.345	1345.4	1498.2	1.8264	4.233	1345.3	1498.0	1.8235	940
4.530	1353.7	1508.8	1.8368	4.410	1353.6	1508.6	1.8338	4.296	1353.5	1508.5	1.8309	960
4.596	1361.9	1519.2	1.8441	4.474	1361.8	1519.1	1.8411	4.359	1361.7	1518.9	1.8382	980
4.662	1370.1	1529.7	1.8514	4.539	1370.0	1529.6	1.8483	4.421	1369.9	1529.4	1.8454	1000
4.728	1378.3	1540.2	1.8585	4.603	1378.2	1540.1	1.8555	4.484	1378.1	1539.9	1.8526	1020
4.794	1386.6	1550.7	1.8656	4.667	1386.5	1550.6	1.8626	4.546	1386.4	1550.5	1.8596	1040
4.860	1394.9	1561.3	1.8726	4.731	1394.8	1561.2	1.8696	4.609	1394.7	1561.0	1.8666	1060
4.926	1403.3	1571.9	1.8795	4.795	1403.2	1571.8	1.8765	4.671	1403.1	1571.6	1.8736	1080
4.991	1411.6	1582.5	1.8863	4.859	1411.5	1582.4	1.8833	4.734	1411.5	1582.3	1.8804	1100
5.319	1454.0	1636.0	1.9196	5.178	1453.9	1635.9	1.9166	5.045	1453.8	1635.8	1.9137	1200
5.645	1497.1	1690.4	1.9514	5.496	1497.0	1690.3	1.9484	5.354	1497.0	1690.2	1.9455	1300
5.970	1541.1	1745.5	1.9819	5.813	1541.0	1745.4	1.9789	5.663	1541.0	1745.4	1.9760	1400
6.295	1586.0	1801.5	2.0112	6.129	1585.9	1801.4	2.0082	5.972	1585.9	1801.4	2.0053	1500
6.620	1631.7	1858.3	2.0395	6.445	1631.6	1858.3	2.0365	6.280	1631.6	1858.2	2.0336	1600
7.268	1725.7	1974.6	2.0933	7.077	1725.7	1974.5	2.0904	6.895	1725.7	1974.5	2.0875	1800
7.915	1823.1	2094.1	2.1440	7.707	1823.1	2094.1	2.1411	7.509	1823.1	2094.0	2.1382	2000
8.562	1923.7	2216.8	2.1919	8.337	1923.6	2216.7	2.1890	8.123	1923.6	2216.7	2.1861	2200
9.209	2026.9	2342.2	2.2374	8.967	2026.9	2342.2	2.2345	8.737	2026.9	2342.1	2.2316	2400

Table 3. Vapor

p (t Sat.)	200 (381.86)				205 (383.94)				210 (385.97)			
t	v	u	h	s	v	u	h	s	v	u	h	s
Sat.	2.289	1114.6	1199.3	1.5464	2.235	1114.8	1199.6	1.5444	2.184	1115.1	1200.0	1.5423
350	*2.156*	*1097.9*	*1177.7*	*1.5202*	*2.096*	*1096.9*	*1176.4*	*1.5162*				
360	*2.199*	*1103.3*	*1184.7*	*1.5288*	*2.138*	*1102.4*	*1183.5*	*1.5249*	*2.080*	*1101.4*	*1182.3*	*1.5211*
370	*2.241*	*1108.5*	*1191.5*	*1.5371*	*2.180*	*1107.7*	*1190.4*	*1.5333*	*2.121*	*1106.8*	*1189.2*	*1.5295*
380	*2.282*	*1113.6*	*1198.1*	*1.5450*	*2.220*	*1112.8*	*1197.0*	*1.5413*	*2.161*	*1112.0*	*1196.0*	*1.5376*
390	2.322	1118.6	1204.5	1.5526	2.259	1117.9	1203.6	1.5490	2.200	1117.1	1202.6	1.5454
400	2.361	1123.5	1210.8	1.5600	2.298	1122.8	1209.9	1.5565	2.238	1122.1	1209.0	1.5529
410	2.399	1128.2	1217.0	1.5672	2.336	1127.6	1216.2	1.5637	2.275	1126.9	1215.3	1.5602
420	2.437	1132.9	1223.1	1.5741	2.373	1132.3	1222.3	1.5707	2.311	1131.6	1221.5	1.5673
430	2.475	1137.5	1229.1	1.5809	2.409	1136.9	1228.3	1.5775	2.347	1136.3	1227.5	1.5741
440	2.511	1142.0	1234.9	1.5874	2.446	1141.4	1234.2	1.5841	2.383	1140.9	1233.5	1.5808
450	2.548	1146.4	1240.7	1.5938	2.481	1145.9	1240.1	1.5905	2.418	1145.4	1239.4	1.5873
460	2.584	1150.8	1246.5	1.6001	2.516	1150.4	1245.8	1.5968	2.452	1149.9	1245.2	1.5936
470	2.619	1155.2	1252.1	1.6062	2.551	1154.7	1251.5	1.6030	2.487	1154.2	1250.9	1.5998
480	2.654	1159.5	1257.7	1.6122	2.586	1159.0	1257.1	1.6090	2.521	1158.6	1256.5	1.6059
490	2.689	1163.7	1263.3	1.6181	2.620	1163.3	1262.7	1.6149	2.554	1162.9	1262.1	1.6118
500	2.724	1168.0	1268.8	1.6239	2.654	1167.6	1268.2	1.6207	2.587	1167.2	1267.7	1.6176
510	2.758	1172.2	1274.2	1.6295	2.688	1171.8	1273.7	1.6264	2.620	1171.4	1273.2	1.6233
520	2.792	1176.3	1279.7	1.6351	2.721	1175.9	1279.2	1.6320	2.653	1175.6	1278.7	1.6289
530	2.826	1180.5	1285.0	1.6405	2.754	1180.1	1284.6	1.6375	2.685	1179.7	1284.1	1.6344
540	2.860	1184.6	1290.4	1.6459	2.787	1184.2	1289.9	1.6429	2.718	1183.9	1289.5	1.6399
550	2.893	1188.7	1295.7	1.6512	2.820	1188.3	1295.3	1.6482	2.750	1188.0	1294.8	1.6452
560	2.926	1192.7	1301.0	1.6565	2.852	1192.4	1300.6	1.6534	2.782	1192.1	1300.2	1.6504
570	2.960	1196.8	1306.3	1.6616	2.885	1196.5	1305.9	1.6586	2.813	1196.2	1305.5	1.6556
580	2.993	1200.8	1311.6	1.6667	2.917	1200.5	1311.2	1.6637	2.845	1200.2	1310.8	1.6607
590	3.025	1204.9	1316.8	1.6717	2.949	1204.6	1316.4	1.6687	2.876	1204.3	1316.1	1.6658
600	3.058	1208.9	1322.1	1.6767	2.981	1208.6	1321.7	1.6737	2.908	1208.3	1321.3	1.6708
620	3.123	1216.9	1332.5	1.6864	3.045	1216.6	1332.1	1.6835	2.970	121ɔ.4	1331.8	1.6806
640	3.188	1224.9	1342.9	1.6959	3.108	1224.6	1342.5	1.6930	3.031	1224.4	1342.2	1.6901
660	3.252	1232.8	1353.2	1.7053	3.170	1232.6	1352.9	1.7023	3.093	1232.4	1352.6	1.6995
680	3.316	1240.8	1363.5	1.7144	3.233	1240.6	1363.2	1.7115	3.154	1240.4	1362.9	1.7086
700	3.379	1248.8	1373.8	1.7234	3.295	1248.6	1373.6	1.7205	3.215	1248.4	1373.3	1.7176
720	3.442	1256.7	1384.1	1.7322	3.357	1256.5	1383.9	1.7293	3.275	1256.3	1383.6	1.7265
740	3.505	1264.7	1394.4	1.7408	3.418	1264.5	1394.2	1.7380	3.335	1264.3	1393.9	1.7351
760	3.568	1272.7	1404.7	1.7493	3.480	1272.5	1404.5	1.7465	3.395	1272.3	1404.2	1.7437
780	3.631	1280.6	1415.0	1.7577	3.541	1280.5	1414.8	1.7549	3.455	1280.3	1414.6	1.7521
800	3.693	1288.6	1425.3	1.7660	3.602	1288.5	1425.1	1.7631	3.515	1288.3	1424.9	1.7603
820	3.755	1296.6	1435.6	1.7741	3.662	1296.5	1435.4	1.7712	3.574	1296.3	1435.2	1.7685
840	3.818	1304.7	1446.0	1.7821	3.723	1304.5	1445.8	1.7793	3.633	1304.4	1445.6	1.7765
860	3.879	1312.7	1456.3	1.7900	3.784	1312.6	1456.1	1.7872	3.692	1312.4	1455.9	1.7844
880	3.941	1320.8	1466.7	1.7978	3.844	1320.7	1466.5	1.7950	3.751	1320.5	1466.3	1.7922
900	4.003	1328.9	1477.1	1.8055	3.904	1328.8	1476.9	1.8027	3.810	1328.6	1476.7	1.7999
920	4.064	1337.0	1487.5	1.8131	3.964	1336.9	1487.3	1.8103	3.869	1336.8	1487.1	1.8075
940	4.126	1345.2	1497.9	1.8206	4.024	1345.1	1497.7	1.8178	3.927	1344.9	1497.6	1.8150
960	4.187	1353.3	1508.3	1.8280	4.084	1353.2	1508.2	1.8252	3.986	1353.1	1508.0	1.8224
980	4.249	1361.6	1518.8	1.8353	4.144	1361.4	1518.6	1.8325	4.044	1361.3	1518.5	1.8298
1000	4.310	1369.8	1529.3	1.8425	4.204	1369.7	1529.1	1.8398	4.103	1369.6	1529.0	1.8370
1020	4.371	1378.0	1539.8	1.8497	4.263	1377.9	1539.7	1.8469	4.161	1377.8	1539.5	1.8442
1040	4.432	1386.3	1550.3	1.8568	4.323	1386.2	1550.2	1.8540	4.219	1386.1	1550.1	1.8513
1060	4.493	1394.6	1560.9	1.8638	4.382	1394.5	1560.8	1.8610	4.277	1394.5	1560.7	1.8583
1080	4.554	1403.0	1571.5	1.8707	4.442	1402.9	1571.4	1.8679	4.335	1402.8	1571.3	1.8652
1100	4.615	1411.4	1582.2	1.8776	4.501	1411.3	1582.0	1.8748	4.393	1411.2	1581.9	1.8721
1200	4.918	1453.7	1635.7	1.9109	4.797	1453.7	1635.6	1.9081	4.682	1453.6	1635.5	1.9054
1300	5.220	1496.9	1690.1	1.9427	5.092	1496.9	1690.0	1.9399	4.970	1496.8	1689.9	1.9372
1400	5.521	1540.9	1745.3	1.9732	5.386	1540.9	1745.2	1.9704	5.258	1540.8	1745.1	1.9677
1500	5.822	1585.8	1801.3	2.0025	5.680	1585.8	1801.2	1.9998	5.544	1585.7	1801.2	1.9971
1600	6.123	1631.6	1858.2	2.0308	5.973	1631.5	1858.1	2.0281	5.830	1631.5	1858.0	2.0254
1800	6.722	1725.6	1974.4	2.0847	6.558	1725.6	1974.4	2.0819	6.402	1725.5	1974.3	2.0793
2000	7.321	1823.0	2094.0	2.1354	7.143	1823.0	2094.0	2.1326	6.973	1823.0	2093.9	2.1300
2200	7.920	1923.6	2216.7	2.1833	7.727	1923.5	2216.6	2.1806	7.543	1923.5	2216.6	2.1779
2400	8.518	2026.8	2342.1	2.2288	8.311	2026.8	2342.1	2.2260	8.113	2026.8	2342.0	2.2234

v	u	h	s	v	u	h	s	v	u	h	s	t
2.135	1115.3	1200.3	1.5403	2.088	1115.5	1200.6	1.5384	2.043	1115.8	1200.8	1.5365	Sat.
												350
2.0250	*1100.5*	*1181.1*	*1.5173*	*1.9721*	*1099.5*	*1179.8*	*1.5135*	*1.9215*	*1098.6*	*1178.6*	*1.5098*	360
2.0652	*1105.9*	*1188.1*	*1.5258*	*2.0119*	*1105.1*	*1187.0*	*1.5222*	*1.9609*	*1104.2*	*1185.8*	*1.5186*	370
2.1045	*1111.2*	*1194.9*	*1.5340*	*2.0506*	*1110.4*	*1193.9*	*1.5305*	*1.9991*	*1109.6*	*1192.8*	*1.5269*	380
2.1427	1116.3	1201.6	1.5419	2.0884	1115.6	1200.6	1.5384	*2.0364*	*1114.8*	*1199.6*	*1.5350*	390
2.180	1121.3	1208.1	1.5495	2.125	1120.6	1207.2	1.5461	2.073	1119.9	1206.2	1.5427	400
2.217	1126.2	1214.4	1.5568	2.161	1125.6	1213.6	1.5535	2.108	1124.9	1212.7	1.5502	410
2.253	1131.0	1220.7	1.5639	2.197	1130.4	1219.8	1.5607	2.143	1129.8	1219.0	1.5574	420
2.288	1135.7	1226.8	1.5708	2.232	1135.1	1226.0	1.5676	2.178	1134.5	1225.2	1.5644	430
2.323	1140.3	1232.8	1.5776	2.266	1139.8	1232.0	1.5744	2.211	1139.2	1231.3	1.5713	440
2.358	1144.9	1238.7	1.5841	2.300	1144.3	1238.0	1.5810	2.245	1143.8	1237.3	1.5779	450
2.391	1149.4	1244.5	1.5905	2.333	1148.9	1243.8	1.5874	2.277	1148.3	1243.2	1.5843	460
2.425	1153.8	1250.3	1.5967	2.366	1153.3	1249.6	1.5936	2.310	1152.8	1249.0	1.5906	470
2.458	1158.1	1255.9	1.6028	2.399	1157.7	1255.3	1.5997	2.342	1157.2	1254.7	1.5968	480
2.491	1162.5	1261.6	1.6087	2.431	1162.0	1261.0	1.6057	2.374	1161.6	1260.4	1.6028	490
2.524	1166.7	1267.2	1.6146	2.463	1166.3	1266.6	1.6116	2.405	1165.9	1266.1	1.6087	500
2.556	1171.0	1272.7	1.6203	2.495	1170.6	1272.2	1.6174	2.436	1170.2	1271.6	1.6145	510
2.588	1175.2	1278.2	1.6259	2.526	1174.8	1277.7	1.6230	2.467	1174.4	1277.2	1.6201	520
2.620	1179.4	1283.6	1.6315	2.557	1179.0	1283.1	1.6286	2.498	1178.7	1282.6	1.6257	530
2.652	1183.5	1289.0	1.6369	2.588	1183.2	1288.6	1.6340	2.528	1182.8	1288.1	1.6312	540
2.683	1187.7	1294.4	1.6423	2.619	1187.3	1294.0	1.6394	2.558	1187.0	1293.5	1.6366	550
2.714	1191.8	1299.8	1.6475	2.650	1191.5	1299.3	1.6447	2.588	1191.1	1298.9	1.6419	560
2.745	1195.9	1305.1	1.6527	2.680	1195.6	1304.7	1.6499	2.618	1195.2	1304.3	1.6471	570
2.776	1199.9	1310.4	1.6579	2.711	1199.6	1310.0	1.6550	2.648	1199.3	1309.6	1.6523	580
2.807	1204.0	1315.7	1.6629	2.741	1203.7	1315.3	1.6601	2.677	1203.4	1314.9	1.6574	590
2.838	1208.1	1320.9	1.6679	2.771	1207.8	1320.6	1.6651	2.707	1207.5	1320.2	1.6624	600
2.898	1216.1	1331.4	1.6777	2.830	1215.9	1331.1	1.6749	2.765	1215.6	1330.7	1.6722	620
2.959	1224.2	1341.9	1.6873	2.890	1223.9	1341.5	1.6845	2.823	1223.7	1341.2	1.6818	640
3.019	1232.2	1352.3	1.6967	2.948	1231.9	1352.0	1.6939	2.881	1231.7	1351.7	1.6913	660
3.079	1240.2	1362.6	1.7059	3.007	1239.9	1362.4	1.7031	2.938	1239.7	1362.1	1.7005	680
3.138	1248.2	1373.0	1.7149	3.065	1247.9	1372.7	1.7122	2.995	1247.7	1372.4	1.7095	700
3.197	1256.1	1383.3	1.7237	3.123	1255.9	1383.1	1.7210	3.052	1255.8	1382.8	1.7184	720
3.256	1264.1	1393.7	1.7324	3.180	1263.9	1393.4	1.7297	3.108	1263.8	1393.2	1.7271	740
3.315	1272.1	1404.0	1.7409	3.238	1272.0	1403.8	1.7383	3.164	1271.8	1403.5	1.7356	760
3.373	1280.1	1414.3	1.7493	3.295	1280.0	1414.1	1.7467	3.220	1279.8	1413.9	1.7441	780
3.431	1288.2	1424.7	1.7576	3.352	1288.0	1424.5	1.7549	3.276	1287.8	1424.2	1.7523	800
3.489	1296.2	1435.0	1.7658	3.409	1296.0	1434.8	1.7631	3.332	1295.9	1434.6	1.7605	820
3.547	1304.2	1445.4	1.7738	3.466	1304.1	1445.2	1.7711	3.387	1303.9	1445.0	1.7685	840
3.605	1312.3	1455.7	1.7817	3.522	1312.2	1455.6	1.7791	3.443	1312.0	1455.4	1.7765	860
3.663	1320.4	1466.1	1.7895	3.578	1320.3	1466.0	1.7869	3.498	1320.1	1465.8	1.7843	880
3.720	1328.5	1476.5	1.7972	3.635	1328.4	1476.4	1.7946	3.553	1328.3	1476.2	1.7920	900
3.778	1336.7	1487.0	1.8048	3.691	1336.5	1486.8	1.8022	3.608	1336.4	1486.6	1.7996	920
3.835	1344.8	1497.4	1.8123	3.747	1344.7	1497.2	1.8097	3.663	1344.6	1497.1	1.8072	940
3.892	1353.0	1507.9	1.8198	3.803	1352.9	1507.7	1.8172	3.717	1352.8	1507.6	1.8146	960
3.949	1361.2	1518.4	1.8271	3.859	1361.1	1518.2	1.8245	3.772	1361.0	1518.1	1.8219	980
4.006	1369.5	1528.9	1.8344	3.914	1369.4	1528.7	1.8318	3.827	1369.3	1528.6	1.8292	1000
4.063	1377.7	1539.4	1.8415	3.970	1377.6	1539.3	1.8389	3.881	1377.5	1539.1	1.8364	1020
4.120	1386.0	1550.0	1.8486	4.026	1385.9	1549.8	1.8460	3.936	1385.8	1549.7	1.8435	1040
4.177	1394.4	1560.5	1.8556	4.081	1394.3	1560.4	1.8530	3.990	1394.2	1560.3	1.8505	1060
4.234	1402.7	1571.2	1.8626	4.137	1402.6	1571.0	1.8600	4.044	1402.5	1570.9	1.8574	1080
4.290	1411.1	1581.8	1.8694	4.192	1411.0	1581.7	1.8668	4.098	1410.9	1581.6	1.8643	1100
4.573	1453.5	1635.4	1.9028	4.468	1453.4	1635.4	1.9002	4.369	1453.4	1635.3	1.8977	1200
4.854	1496.7	1689.9	1.9346	4.744	1496.7	1689.8	1.9320	4.638	1496.6	1689.7	1.9295	1300
5.135	1540.8	1745.1	1.9651	5.018	1540.7	1745.0	1.9626	4.906	1540.7	1744.9	1.9600	1400
5.415	1585.7	1801.1	1.9945	5.292	1585.6	1801.0	1.9919	5.174	1585.6	1801.0	1.9894	1500
5.695	1631.4	1858.0	2.0228	5.565	1631.4	1857.9	2.0202	5.441	1631.3	1857.9	2.0177	1600
6.253	1725.5	1974.3	2.0766	6.111	1725.5	1974.2	2.0741	5.975	1725.4	1974.2	2.0716	1800
6.810	1822.9	2093.9	2.1274	6.655	1822.9	2093.8	2.1248	6.507	1822.9	2093.8	2.1223	2000
7.367	1923.5	2216.6	2.1753	7.200	1923.4	2216.5	2.1728	7.040	1923.4	2216.5	2.1703	2200
7.924	2026.7	2342.0	2.2208	7.744	2026.7	2342.0	2.2182	7.572	2026.7	2341.9	2.2157	2400

Table 3. Vapor

t	230 (393.76)				235 (395.63)				240 (397.46)			
	v	u	h	s	v	u	h	s	v	u	h	s
Sat.	2.000	1116.0	1201.1	1.5346	1.9590	1116.2	1201.4	1.5327	1.9194	1116.4	1201.6	1.5309
370	*1.9120*	*1103.3*	*1184.6*	*1.5150*	*1.8651*	*1102.3*	*1183.4*	*1.5115*	*1.8201*	*1101.4*	*1182.2*	*1.5080*
380	*1.9498*	*1108.7*	*1191.7*	*1.5235*	*1.9025*	*1107.9*	*1190.6*	*1.5200*	*1.8572*	*1107.0*	*1189.5*	*1.5166*
390	*1.9866*	*1114.0*	*1198.6*	*1.5316*	*1.9389*	*1113.2*	*1197.6*	*1.5283*	*1.8932*	*1112.4*	*1196.5*	*1.5250*
400	2.023	1119.2	1205.3	1.5394	1.9744	1118.4	1204.3	1.5362	1.9283	1117.7	1203.3	1.5329
410	2.058	1124.2	1211.8	1.5470	2.0091	1123.5	1210.9	1.5438	1.9625	1122.8	1210.0	1.5406
420	2.092	1129.1	1218.2	1.5543	2.0431	1128.5	1217.3	1.5511	1.9960	1127.8	1216.5	1.5480
430	2.126	1133.9	1224.4	1.5613	2.0764	1133.3	1223.6	1.5582	2.0289	1132.7	1222.8	1.5552
440	2.159	1138.6	1230.5	1.5682	2.1091	1138.1	1229.8	1.5651	2.0611	1137.5	1229.0	1.5622
450	2.192	1143.3	1236.6	1.5748	2.141	1142.7	1235.9	1.5719	2.093	1142.2	1235.1	1.5689
460	2.224	1147.8	1242.5	1.5813	2.173	1147.3	1241.8	1.5784	2.124	1146.8	1241.2	1.5755
470	2.256	1152.3	1248.4	1.5877	2.205	1151.9	1247.7	1.5848	2.155	1151.4	1247.1	1.5819
480	2.288	1156.8	1254.1	1.5938	2.235	1156.3	1253.5	1.5910	2.186	1155.9	1252.9	1.5881
490	2.319	1161.2	1259.9	1.5999	2.266	1160.7	1259.3	1.5971	2.216	1160.3	1258.7	1.5943
500	2.350	1165.5	1265.5	1.6058	2.296	1165.1	1265.0	1.6030	2.246	1164.7	1264.4	1.6002
510	2.380	1169.8	1271.1	1.6116	2.327	1169.4	1270.6	1.6088	2.275	1169.0	1270.0	1.6061
520	2.410	1174.1	1276.7	1.6173	2.356	1173.7	1276.2	1.6145	2.304	1173.3	1275.6	1.6118
530	2.441	1178.3	1282.2	1.6229	2.386	1177.9	1281.7	1.6202	2.333	1177.6	1281.2	1.6175
540	2.470	1182.5	1287.6	1.6284	2.415	1182.1	1287.2	1.6257	2.362	1181.8	1286.7	1.6230
550	2.500	1186.7	1293.1	1.6338	2.444	1186.3	1292.6	1.6311	2.391	1186.0	1292.2	1.6284
560	2.530	1190.8	1298.5	1.6391	2.473	1190.5	1298.0	1.6365	2.419	1190.2	1297.6	1.6338
570	2.559	1194.9	1303.8	1.6444	2.502	1194.6	1303.4	1.6417	2.447	1194.3	1303.0	1.6391
580	2.588	1199.0	1309.2	1.6496	2.531	1198.7	1308.8	1.6469	2.476	1198.4	1308.4	1.6443
590	2.617	1203.1	1314.5	1.6547	2.559	1202.9	1314.1	1.6520	2.503	1202.6	1313.7	1.6494
600	2.646	1207.2	1319.8	1.6597	2.587	1206.9	1319.5	1.6570	2.531	1206.7	1319.1	1.6545
620	2.703	1215.3	1330.4	1.6696	2.644	1215.1	1330.0	1.6669	2.586	1214.8	1329.7	1.6644
640	2.760	1223.4	1340.9	1.6792	2.699	1223.2	1340.6	1.6766	2.641	1222.9	1340.2	1.6741
660	2.816	1231.5	1351.3	1.6886	2.755	1231.2	1351.0	1.6860	2.695	1231.0	1350.7	1.6835
680	2.873	1239.5	1361.8	1.6979	2.810	1239.3	1361.5	1.6953	2.749	1239.1	1361.2	1.6928
700	2.928	1247.5	1372.2	1.7069	2.864	1247.3	1371.9	1.7044	2.803	1247.1	1371.6	1.7018
720	2.984	1255.6	1382.6	1.7158	2.919	1255.4	1382.3	1.7132	2.857	1255.2	1382.0	1.7108
740	3.039	1263.6	1392.9	1.7245	2.973	1263.4	1392.7	1.7220	2.910	1263.2	1392.4	1.7195
760	3.094	1271.6	1403.3	1.7331	3.027	1271.4	1403.1	1.7305	2.963	1271.2	1402.8	1.7281
780	3.149	1279.6	1413.7	1.7415	3.081	1279.5	1413.4	1.7390	3.015	1279.3	1413.2	1.7365
800	3.204	1287.7	1424.0	1.7498	3.134	1287.5	1423.8	1.7473	3.068	1287.4	1423.6	1.7448
820	3.258	1295.7	1434.4	1.7580	3.188	1295.6	1434.2	1.7555	3.120	1295.4	1434.0	1.7530
840	3.313	1303.8	1444.8	1.7660	3.241	1303.7	1444.6	1.7635	3.172	1303.5	1444.4	1.7611
860	3.367	1311.9	1455.2	1.7740	3.294	1311.7	1455.0	1.7715	3.224	1311.6	1454.8	1.7690
880	3.421	1320.0	1465.6	1.7818	3.347	1319.9	1465.4	1.7793	3.276	1319.7	1465.2	1.7769
900	3.475	1328.1	1476.0	1.7895	3.400	1328.0	1475.8	1.7870	3.328	1327.9	1475.7	1.7846
920	3.528	1336.3	1486.5	1.7971	3.452	1336.2	1486.3	1.7947	3.380	1336.0	1486.1	1.7923
940	3.582	1344.5	1496.9	1.8047	3.505	1344.3	1496.8	1.8022	3.431	1344.2	1496.6	1.7998
960	3.636	1352.7	1507.4	1.8121	3.557	1352.5	1507.3	1.8096	3.483	1352.4	1507.1	1.8072
980	3.689	1360.9	1517.9	1.8194	3.610	1360.8	1517.8	1.8170	3.534	1360.7	1517.6	1.8146
1000	3.743	1369.1	1528.4	1.8267	3.662	1369.0	1528.3	1.8243	3.585	1368.9	1528.2	1.8219
1020	3.796	1377.4	1539.0	1.8339	3.714	1377.3	1538.9	1.8314	3.636	1377.2	1538.7	1.8291
1040	3.849	1385.7	1549.6	1.8410	3.767	1385.6	1549.4	1.8385	3.687	1385.5	1549.3	1.8362
1060	3.902	1394.1	1560.2	1.8480	3.819	1394.0	1560.0	1.8456	3.738	1393.9	1559.9	1.8432
1080	3.956	1402.4	1570.8	1.8550	3.871	1402.4	1570.7	1.8525	3.789	1402.3	1570.6	1.8501
1100	4.009	1410.8	1581.5	1.8618	3.923	1410.8	1581.3	1.8594	3.840	1410.7	1581.2	1.8570
1200	4.273	1453.3	1635.2	1.8952	4.182	1453.2	1635.1	1.8928	4.094	1453.1	1635.0	1.8904
1300	4.536	1496.5	1689.6	1.9271	4.439	1496.5	1689.5	1.9247	4.346	1496.4	1689.4	1.9223
1400	4.799	1540.6	1744.9	1.9576	4.696	1540.5	1744.8	1.9552	4.598	1540.5	1744.7	1.9528
1500	5.061	1585.5	1800.9	1.9870	4.953	1585.5	1800.9	1.9846	4.849	1585.4	1800.8	1.9822
1600	5.322	1631.3	1857.8	2.0153	5.209	1631.2	1857.8	2.0129	5.100	1631.2	1857.7	2.0105
1800	5.844	1725.4	1974.1	2.0692	5.720	1725.4	1974.1	2.0668	5.601	1725.3	1974.1	2.0644
2000	6.366	1822.8	2093.8	2.1199	6.230	1822.8	2093.7	2.1175	6.100	1822.7	2093.7	2.1152
2200	6.887	1923.4	2216.5	2.1678	6.740	1923.3	2216.4	2.1654	6.600	1923.3	2216.4	2.1631
2400	7.407	2026.6	2341.9	2.2133	7.250	2026.6	2341.9	2.2109	7.099	2026.6	2341.8	2.2086

	245 (399.26)				250 (401.04)				255 (402.79)			p (t Sat.)
v	u	h	s	v	u	h	s	v	u	h	s	t
1.8814	1116.6	1201.9	1.5292	1.8448	1116.7	1202.1	1.5274	1.8097	1116.9	1202.3	1.5257	Sat.
1.7769	*1100.5*	*1181.0*	*1.5045*									370
1.8136	*1106.1*	*1188.4*	*1.5133*	*1.7717*	*1105.3*	*1187.2*	*1.5100*	*1.7315*	*1104.4*	*1186.1*	*1.5067*	380
1.8492	*1111.6*	*1195.5*	*1.5217*	*1.8070*	*1110.8*	*1194.4*	*1.5185*	*1.7664*	*1110.0*	*1193.3*	*1.5153*	390
1.8839	1117.0	1202.4	1.5298	*1.8413*	*1116.2*	*1201.4*	*1.5266*	*1.8004*	*1115.4*	*1200.4*	*1.5235*	400
1.9178	1122.1	1209.1	1.5375	1.8748	1121.4	1208.1	1.5344	1.8334	1120.7	1207.2	1.5314	410
1.9508	1127.2	1215.6	1.5450	1.9075	1126.5	1214.7	1.5420	1.8658	1125.8	1213.9	1.5390	420
1.9833	1132.1	1222.0	1.5522	1.9395	1131.5	1221.2	1.5493	1.8974	1130.8	1220.4	1.5464	430
2.0151	1136.9	1228.3	1.5592	1.9709	1136.3	1227.5	1.5563	1.9284	1135.7	1226.7	1.5535	440
2.046	1141.6	1234.4	1.5660	2.002	1141.1	1233.7	1.5632	1.9588	1140.5	1233.0	1.5604	450
2.077	1146.3	1240.5	1.5726	2.032	1145.8	1239.8	1.5698	1.9888	1145.3	1239.1	1.5671	460
2.108	1150.9	1246.4	1.5791	2.062	1150.4	1245.8	1.5763	2.0183	1149.9	1245.1	1.5736	470
2.138	1155.4	1252.3	1.5854	2.092	1154.9	1251.7	1.5826	2.0475	1154.4	1251.1	1.5799	480
2.167	1159.8	1258.1	1.5915	2.121	1159.4	1257.5	1.5888	2.0763	1159.0	1256.9	1.5861	490
2.197	1164.2	1263.8	1.5975	2.150	1163.8	1263.3	1.5948	2.105	1163.4	1262.7	1.5922	500
2.226	1168.6	1269.5	1.6034	2.178	1168.2	1269.0	1.6007	2.133	1167.8	1268.4	1.5981	510
2.255	1172.9	1275.1	1.6092	2.207	1172.5	1274.6	1.6065	2.161	1172.1	1274.1	1.6039	520
2.283	1177.2	1280.7	1.6148	2.235	1176.8	1280.2	1.6122	2.188	1176.5	1279.7	1.6096	530
2.311	1181.4	1286.2	1.6204	2.263	1181.1	1285.8	1.6178	2.216	1180.7	1285.3	1.6152	540
2.340	1185.7	1291.7	1.6258	2.290	1185.3	1291.3	1.6233	2.243	1185.0	1290.8	1.6207	550
2.367	1189.8	1297.2	1.6312	2.318	1189.5	1296.7	1.6287	2.270	1189.2	1296.3	1.6261	560
2.395	1194.0	1302.6	1.6365	2.345	1193.7	1302.2	1.6340	2.297	1193.4	1301.8	1.6315	570
2.423	1198.1	1308.0	1.6417	2.372	1197.8	1307.6	1.6392	2.323	1197.5	1307.2	1.6367	580
2.450	1202.3	1313.4	1.6469	2.399	1202.0	1313.0	1.6443	2.350	1201.7	1312.6	1.6419	590
2.477	1206.4	1318.7	1.6519	2.426	1206.1	1318.3	1.6494	2.376	1205.8	1317.9	1.6470	600
2.532	1214.6	1329.3	1.6619	2.479	1214.3	1329.0	1.6594	2.429	1214.0	1328.6	1.6570	620
2.585	1222.7	1339.9	1.6716	2.532	1222.4	1339.6	1.6691	2.480	1222.2	1339.2	1.6667	640
2.639	1230.8	1350.4	1.6810	2.584	1230.6	1350.1	1.6786	2.532	1230.3	1349.8	1.6762	660
2.692	1238.9	1360.9	1.6903	2.636	1238.6	1360.6	1.6879	2.583	1238.4	1360.3	1.6855	680
2.744	1246.9	1371.3	1.6994	2.688	1246.7	1371.1	1.6970	2.634	1246.5	1370.8	1.6946	700
2.797	1255.0	1381.8	1.7083	2.739	1254.8	1381.5	1.7059	2.684	1254.6	1381.2	1.7036	720
2.849	1263.0	1392.2	1.7171	2.790	1262.8	1391.9	1.7147	2.734	1262.6	1391.7	1.7123	740
2.901	1271.1	1402.6	1.7257	2.841	1270.9	1402.3	1.7233	2.784	1270.7	1402.1	1.7210	760
2.952	1279.1	1413.0	1.7341	2.892	1279.0	1412.8	1.7318	2.834	1278.8	1412.5	1.7294	780
3.004	1287.2	1423.4	1.7424	2.943	1287.0	1423.2	1.7401	2.884	1286.9	1422.9	1.7378	800
3.055	1295.3	1433.8	1.7506	2.993	1295.1	1433.6	1.7483	2.933	1295.0	1433.4	1.7460	820
3.106	1303.4	1444.2	1.7587	3.043	1303.2	1444.0	1.7564	2.982	1303.1	1443.8	1.7541	840
3.157	1311.5	1454.6	1.7667	3.093	1311.3	1454.4	1.7643	3.032	1311.2	1454.2	1.7620	860
3.208	1319.6	1465.0	1.7745	3.143	1319.5	1464.9	1.7722	3.080	1319.3	1464.7	1.7699	880
3.259	1327.7	1475.5	1.7823	3.193	1327.6	1475.3	1.7799	3.129	1327.5	1475.2	1.7777	900
3.310	1335.9	1486.0	1.7899	3.243	1335.8	1485.8	1.7876	3.178	1335.7	1485.6	1.7853	920
3.360	1344.1	1496.4	1.7974	3.292	1344.0	1496.3	1.7951	3.227	1343.9	1496.1	1.7929	940
3.411	1352.3	1506.9	1.8049	3.342	1352.2	1506.8	1.8026	3.275	1352.1	1506.6	1.8003	960
3.461	1360.6	1517.5	1.8122	3.391	1360.5	1517.3	1.8099	3.324	1360.3	1517.2	1.8077	980
3.511	1368.8	1528.0	1.8195	3.440	1368.7	1527.9	1.8172	3.372	1368.6	1527.7	1.8150	1000
3.561	1377.1	1538.6	1.8267	3.489	1377.0	1538.4	1.8244	3.420	1376.9	1538.3	1.8222	1020
3.611	1385.4	1549.2	1.8338	3.538	1385.3	1549.0	1.8315	3.468	1385.2	1548.9	1.8293	1040
3.661	1393.8	1559.8	1.8409	3.588	1393.7	1559.7	1.8386	3.517	1393.6	1559.5	1.8363	1060
3.711	1402.2	1570.4	1.8478	3.636	1402.1	1570.3	1.8455	3.565	1402.0	1570.2	1.8433	1080
3.761	1410.6	1581.1	1.8547	3.685	1410.5	1581.0	1.8524	3.613	1410.4	1580.9	1.8502	1100
4.010	1453.1	1634.9	1.8881	3.929	1453.0	1634.8	1.8858	3.852	1452.9	1634.7	1.8836	1200
4.257	1496.3	1689.4	1.9200	4.172	1496.3	1689.3	1.9177	4.090	1496.2	1689.2	1.9155	1300
4.504	1540.4	1744.6	1.9505	4.414	1540.4	1744.6	1.9483	4.327	1540.3	1744.5	1.9461	1400
4.750	1585.4	1800.7	1.9799	4.655	1585.3	1800.7	1.9777	4.563	1585.3	1800.6	1.9755	1500
4.996	1631.2	1857.7	2.0082	4.896	1631.1	1857.6	2.0060	4.799	1631.1	1857.5	2.0038	1600
5.486	1725.3	1974.0	2.0621	5.376	1725.2	1974.0	2.0599	5.271	1725.2	1973.9	2.0577	1800
5.976	1822.7	2093.6	2.1129	5.856	1822.7	2093.6	2.1106	5.741	1822.6	2093.6	2.1084	2000
6.465	1923.2	2216.4	2.1608	6.336	1923.2	2216.3	2.1586	6.211	1923.2	2216.3	2.1564	2200
6.954	2026.5	2341.8	2.2063	6.815	2026.5	2341.8	2.2041	6.681	2026.5	2341.7	2.2019	2400

Table 3. Vapor

p (t Sat.)	260 (404.51)				265 (406.20)				270 (407.87)			
t	v	u	h	s	v	u	h	s	v	u	h	s
Sat.	1.7758	1117.1	1202.5	1.5240	1.7432	1117.2	1202.7	1.5224	1.7117	1117.4	1202.9	1.5208
400	*1.7609*	*1114.7*	*1199.4*	*1.5204*	*1.7230*	*1113.9*	*1198.4*	*1.5174*	*1.6863*	*1113.1*	*1197.3*	*1.5143*
410	1.7937	1120.0	1206.3	1.5284	1.7554	1119.3	1205.3	1.5254	1.7184	1118.5	1204.4	1.5225
420	1.8256	1125.2	1213.0	1.5361	1.7870	1124.5	1212.1	1.5332	1.7497	1123.8	1211.2	1.5303
430	1.8569	1130.2	1219.6	1.5435	1.8178	1129.6	1218.7	1.5406	1.7803	1128.9	1217.9	1.5378
440	1.8875	1135.1	1226.0	1.5506	1.8481	1134.6	1225.2	1.5478	1.8102	1134.0	1224.4	1.5451
450	1.9175	1140.0	1232.2	1.5576	1.8778	1139.4	1231.5	1.5548	1.8395	1138.9	1230.8	1.5521
460	1.9471	1144.7	1238.4	1.5643	1.9070	1144.2	1237.7	1.5616	1.8683	1143.7	1237.0	1.5590
470	1.9762	1149.4	1244.5	1.5709	1.9357	1148.9	1243.8	1.5682	1.8967	1148.4	1243.1	1.5656
480	2.0050	1154.0	1250.4	1.5773	1.9640	1153.5	1249.8	1.5746	1.9246	1153.0	1249.2	1.5721
490	2.0333	1158.5	1256.3	1.5835	1.9920	1158.1	1255.7	1.5809	1.9522	1157.6	1255.1	1.5784
500	2.061	1163.0	1262.1	1.5896	2.020	1162.5	1261.6	1.5870	1.9794	1162.1	1261.0	1.5845
510	2.089	1167.4	1267.9	1.5956	2.047	1167.0	1267.4	1.5930	2.0063	1166.6	1266.8	1.5905
520	2.117	1171.8	1273.6	1.6014	2.074	1171.4	1273.1	1.5989	2.0330	1171.0	1272.5	1.5964
530	2.144	1176.1	1279.2	1.6071	2.101	1175.7	1278.7	1.6046	2.0594	1175.3	1278.2	1.6022
540	2.171	1180.4	1284.8	1.6127	2.127	1180.0	1284.3	1.6103	2.0855	1179.7	1283.9	1.6078
550	2.197	1184.6	1290.4	1.6183	2.154	1184.3	1289.9	1.6158	2.111	1183.9	1289.4	1.6134
560	2.224	1188.9	1295.9	1.6237	2.180	1188.5	1295.4	1.6212	2.137	1188.2	1295.0	1.6189
570	2.250	1193.1	1301.3	1.6290	2.206	1192.7	1300.9	1.6266	2.163	1192.4	1300.5	1.6242
580	2.277	1197.2	1306.8	1.6343	2.232	1196.9	1306.4	1.6319	2.188	1196.6	1306.0	1.6295
590	2.303	1201.4	1312.2	1.6395	2.257	1201.1	1311.8	1.6371	2.213	1200.8	1311.4	1.6347
600	2.329	1205.5	1317.6	1.6446	2.283	1205.3	1317.2	1.6422	2.238	1205.0	1316.8	1.6399
620	2.380	1213.8	1328.3	1.6546	2.333	1213.5	1327.9	1.6522	2.288	1213.2	1327.6	1.6499
640	2.431	1221.9	1338.9	1.6643	2.383	1221.7	1338.6	1.6620	2.337	1221.5	1338.2	1.6597
660	2.481	1230.1	1349.5	1.6739	2.433	1229.9	1349.2	1.6715	2.386	1229.6	1348.9	1.6693
680	2.532	1238.2	1360.0	1.6832	2.482	1238.0	1359.7	1.6809	2.435	1237.8	1359.4	1.6786
700	2.582	1246.3	1370.5	1.6923	2.531	1246.1	1370.2	1.6900	2.483	1245.9	1369.9	1.6878
720	2.631	1254.4	1381.0	1.7013	2.580	1254.2	1380.7	1.6990	2.531	1254.0	1380.4	1.6968
740	2.680	1262.5	1391.4	1.7100	2.629	1262.3	1391.2	1.7078	2.579	1262.1	1390.9	1.7056
760	2.730	1270.5	1401.9	1.7187	2.677	1270.4	1401.6	1.7164	2.626	1270.2	1401.4	1.7142
780	2.778	1278.6	1412.3	1.7272	2.725	1278.5	1412.1	1.7249	2.673	1278.3	1411.8	1.7227
800	2.827	1286.7	1422.7	1.7355	2.773	1286.5	1422.5	1.7333	2.720	1286.4	1422.3	1.7311
820	2.876	1294.8	1433.2	1.7437	2.820	1294.7	1433.0	1.7415	2.767	1294.5	1432.8	1.7393
840	2.924	1302.9	1443.6	1.7518	2.868	1302.8	1443.4	1.7496	2.814	1302.6	1443.2	1.7474
860	2.972	1311.0	1454.0	1.7598	2.915	1310.9	1453.9	1.7576	2.860	1310.8	1453.7	1.7554
880	3.020	1319.2	1464.5	1.7677	2.962	1319.1	1464.3	1.7655	2.907	1318.9	1464.1	1.7633
900	3.068	1327.4	1475.0	1.7754	3.009	1327.2	1474.8	1.7732	2.953	1327.1	1474.6	1.7711
920	3.116	1335.5	1485.5	1.7831	3.056	1335.4	1485.3	1.7809	2.999	1335.3	1485.1	1.7787
940	3.164	1343.7	1496.0	1.7906	3.103	1343.6	1495.8	1.7884	3.045	1343.5	1495.6	1.7863
960	3.211	1352.0	1506.5	1.7981	3.150	1351.9	1506.3	1.7959	3.091	1351.7	1506.2	1.7938
980	3.259	1360.2	1517.0	1.8055	3.197	1360.1	1516.9	1.8033	3.137	1360.0	1516.7	1.8012
1000	3.306	1368.5	1527.6	1.8128	3.243	1368.4	1527.4	1.8106	3.183	1368.3	1527.3	1.8084
1020	3.354	1376.8	1538.2	1.8200	3.290	1376.7	1538.0	1.8178	3.228	1376.6	1537.9	1.8157
1040	3.401	1385.1	1548.8	1.8271	3.336	1385.0	1548.6	1.8249	3.274	1385.0	1548.5	1.8228
1060	3.448	1393.5	1559.4	1.8341	3.383	1393.4	1559.3	1.8320	3.319	1393.3	1559.2	1.8298
1080	3.495	1401.9	1570.1	1.8411	3.429	1401.8	1569.9	1.8389	3.365	1401.7	1569.8	1.8368
1100	3.542	1410.3	1580.8	1.8480	3.475	1410.2	1580.6	1.8458	3.410	1410.1	1580.5	1.8437
1150	3.660	1431.5	1607.6	1.8649	3.590	1431.4	1607.5	1.8628	3.523	1431.3	1607.4	1.8606
1200	3.777	1452.8	1634.6	1.8814	3.705	1452.8	1634.5	1.8793	3.636	1452.7	1634.4	1.8772
1250	3.894	1474.4	1661.7	1.8976	3.820	1474.3	1661.7	1.8954	3.749	1474.3	1661.6	1.8933
1300	4.011	1496.1	1689.1	1.9133	3.934	1496.1	1689.0	1.9112	3.861	1496.0	1688.9	1.9091
1350	4.127	1518.1	1716.7	1.9288	4.049	1518.0	1716.6	1.9266	3.973	1518.0	1716.5	1.9245
1400	4.243	1540.3	1744.4	1.9439	4.163	1540.2	1744.4	1.9418	4.085	1540.2	1744.3	1.9397
1450	4.359	1562.6	1772.4	1.9587	4.277	1562.6	1772.3	1.9566	4.197	1562.5	1772.2	1.9545
1500	4.475	1585.2	1800.5	1.9733	4.391	1585.2	1800.5	1.9712	4.309	1585.1	1800.4	1.9691
1600	4.707	1631.0	1857.5	2.0016	4.618	1631.0	1857.4	1.9995	4.532	1630.9	1857.4	1.9974
1800	5.169	1725.2	1973.9	2.0555	5.072	1725.1	1973.8	2.0534	4.977	1725.1	1973.8	2.0514
2000	5.631	1822.6	2093.5	2.1063	5.524	1822.6	2093.5	2.1042	5.422	1822.5	2093.4	2.1021
2200	6.092	1923.1	2216.2	2.1542	5.977	1923.1	2216.2	2.1521	5.866	1923.1	2216.2	2.1501
2400	6.553	2026.4	2341.7	2.1997	6.429	2026.4	2341.7	2.1976	6.310	2026.4	2341.6	2.1955

v	u	h	s	v	u	h	s	v	u	h	s	t
1.6813	1117.5	1203.1	1.5192	1.6520	1117.7	1203.3	1.5176	1.6237	1117.8	1203.4	1.5160	**Sat.**
1.6510	*1112.3*	*1196.3*	*1.5113*	*1.6169*	*1111.5*	*1195.3*	*1.5084*	*1.5840*	*1110.7*	*1194.2*	*1.5054*	**400**
1.6828	1117.8	1203.4	1.5196	*1.6484*	*1117.0*	*1202.5*	*1.5167*	*1.6152*	*1116.3*	*1201.5*	*1.5138*	**410**
1.7138	1123.1	1210.3	1.5274	1.6791	1122.4	1209.4	1.5246	1.6456	1121.7	1208.5	1.5218	**420**
1.7440	1128.3	1217.0	1.5350	1.7090	1127.6	1216.2	1.5323	1.6752	1127.0	1215.3	1.5296	**430**
1.7736	1133.3	1223.6	1.5424	1.7383	1132.7	1222.8	1.5397	1.7042	1132.1	1222.0	1.5370	**440**
1.8026	1138.3	1230.0	1.5495	1.7669	1137.7	1229.3	1.5468	1.7325	1137.1	1228.5	1.5442	**450**
1.8310	1143.1	1236.3	1.5563	1.7950	1142.6	1235.6	1.5537	1.7603	1142.0	1234.9	1.5512	**460**
1.8590	1147.9	1242.5	1.5630	1.8227	1147.4	1241.8	1.5605	1.7876	1146.8	1241.1	1.5579	**470**
1.8866	1152.5	1248.5	1.5695	1.8499	1152.1	1247.9	1.5670	1.8145	1151.6	1247.3	1.5645	**480**
1.9138	1157.1	1254.5	1.5758	1.8768	1156.7	1253.9	1.5734	1.8411	1156.2	1253.3	1.5709	**490**
1.9407	1161.7	1260.4	1.5820	1.9033	1161.2	1259.9	1.5796	1.8672	1160.8	1259.3	1.5771	**500**
1.9672	1166.1	1266.3	1.5881	1.9295	1165.7	1265.7	1.5856	1.8931	1165.3	1265.2	1.5832	**510**
1.9935	1170.6	1272.0	1.5940	1.9554	1170.2	1271.5	1.5916	1.9186	1169.8	1271.0	1.5892	**520**
2.0195	1175.0	1277.7	1.5998	1.9810	1174.6	1277.2	1.5974	1.9439	1174.2	1276.7	1.5950	**530**
2.0453	1179.3	1283.4	1.6054	2.0064	1178.9	1282.9	1.6031	1.9689	1178.6	1282.4	1.6008	**540**
2.071	1183.6	1289.0	1.6110	2.032	1183.3	1288.5	1.6087	1.9938	1182.9	1288.1	1.6064	**550**
2.096	1187.9	1294.5	1.6165	2.057	1187.5	1294.1	1.6142	2.0184	1187.2	1293.6	1.6119	**560**
2.121	1192.1	1300.1	1.6219	2.081	1191.8	1299.6	1.6196	2.0428	1191.5	1299.2	1.6173	**570**
2.146	1196.3	1305.5	1.6272	2.106	1196.0	1305.1	1.6249	2.0671	1195.7	1304.7	1.6227	**580**
2.171	1200.5	1311.0	1.6324	2.130	1200.2	1310.6	1.6301	2.0911	1199.9	1310.2	1.6279	**590**
2.196	1204.7	1316.4	1.6376	2.155	1204.4	1316.0	1.6353	2.115	1204.1	1315.7	1.6331	**600**
2.245	1213.0	1327.2	1.6476	2.203	1212.7	1326.8	1.6454	2.163	1212.4	1326.5	1.6432	**620**
2.293	1221.2	1337.9	1.6575	2.251	1221.0	1337.6	1.6552	2.210	1220.7	1337.2	1.6531	**640**
2.341	1229.4	1348.5	1.6670	2.298	1229.2	1348.2	1.6648	2.256	1228.9	1347.9	1.6627	**660**
2.389	1237.5	1359.1	1.6764	2.345	1237.3	1358.8	1.6742	2.302	1237.1	1358.5	1.6721	**680**
2.436	1245.7	1369.7	1.6856	2.392	1245.5	1369.4	1.6834	2.348	1245.3	1369.1	1.6813	**700**
2.484	1253.8	1380.2	1.6946	2.438	1253.6	1379.9	1.6924	2.394	1253.4	1379.7	1.6903	**720**
2.530	1261.9	1390.7	1.7034	2.484	1261.7	1390.4	1.7012	2.439	1261.5	1390.2	1.6991	**740**
2.577	1270.0	1401.2	1.7121	2.530	1269.8	1400.9	1.7099	2.484	1269.7	1400.7	1.7078	**760**
2.623	1278.1	1411.6	1.7206	2.575	1277.9	1411.4	1.7184	2.529	1277.8	1411.2	1.7164	**780**
2.670	1286.2	1422.1	1.7289	2.621	1286.1	1421.9	1.7268	2.574	1285.9	1421.6	1.7247	**800**
2.716	1294.3	1432.5	1.7372	2.666	1294.2	1432.3	1.7351	2.618	1294.0	1432.1	1.7330	**820**
2.762	1302.5	1443.0	1.7453	2.711	1302.3	1442.8	1.7432	2.663	1302.2	1442.6	1.7411	**840**
2.807	1310.6	1453.5	1.7533	2.756	1310.5	1453.3	1.7512	2.707	1310.3	1453.1	1.7491	**860**
2.853	1318.8	1464.0	1.7612	2.801	1318.7	1463.8	1.7591	2.751	1318.5	1463.6	1.7570	**880**
2.898	1327.0	1474.5	1.7689	2.846	1326.8	1474.3	1.7669	2.795	1326.7	1474.1	1.7648	**900**
2.944	1335.2	1485.0	1.7766	2.890	1335.0	1484.8	1.7745	2.839	1334.9	1484.6	1.7725	**920**
2.989	1343.4	1495.5	1.7842	2.935	1343.3	1495.3	1.7821	2.882	1343.2	1495.2	1.7801	**940**
3.034	1351.6	1506.0	1.7917	2.979	1351.5	1505.9	1.7896	2.926	1351.4	1505.7	1.7876	**960**
3.079	1359.9	1516.6	1.7991	3.023	1359.8	1516.4	1.7970	2.970	1359.7	1516.3	1.7950	**980**
3.124	1368.2	1527.2	1.8064	3.067	1368.1	1527.0	1.8043	3.013	1368.0	1526.9	1.8023	**1000**
3.169	1376.5	1537.8	1.8136	3.112	1376.4	1537.6	1.8115	3.056	1376.3	1537.5	1.8095	**1020**
3.214	1384.9	1548.4	1.8207	3.156	1384.8	1548.3	1.8186	3.100	1384.7	1548.1	1.8166	**1040**
3.258	1393.2	1559.0	1.8277	3.200	1393.1	1558.9	1.8257	3.143	1393.0	1558.8	1.8237	**1060**
3.303	1401.6	1569.7	1.8347	3.243	1401.5	1569.6	1.8327	3.186	1401.4	1569.5	1.8307	**1080**
3.348	1410.0	1580.4	1.8416	3.287	1410.0	1580.3	1.8396	3.229	1409.9	1580.2	1.8376	**1100**
3.459	1431.2	1607.2	1.8586	3.396	1431.2	1607.1	1.8565	3.336	1431.1	1607.0	1.8545	**1150**
3.570	1452.6	1634.3	1.8751	3.505	1452.5	1634.2	1.8731	3.443	1452.5	1634.1	1.8711	**1200**
3.680	1474.2	1661.5	1.8912	3.614	1474.1	1661.4	1.8892	3.550	1474.0	1661.3	1.8872	**1250**
3.791	1496.0	1688.9	1.9070	3.723	1495.9	1688.8	1.9050	3.657	1495.8	1688.7	1.9030	**1300**
3.901	1517.9	1716.4	1.9225	3.831	1517.9	1716.4	1.9205	3.763	1517.8	1716.3	1.9185	**1350**
4.011	1540.1	1744.2	1.9376	3.939	1540.0	1744.1	1.9356	3.869	1540.0	1744.1	1.9336	**1400**
4.121	1562.5	1772.2	1.9525	4.047	1562.4	1772.1	1.9505	3.976	1562.4	1772.0	1.9485	**1450**
4.230	1585.1	1800.4	1.9670	4.155	1585.0	1800.3	1.9650	4.081	1585.0	1800.2	1.9630	**1500**
4.450	1630.9	1857.3	1.9954	4.370	1630.8	1857.3	1.9934	4.293	1630.8	1857.2	1.9914	**1600**
4.887	1725.0	1973.7	2.0493	4.799	1725.0	1973.7	2.0473	4.715	1725.0	1973.6	2.0453	**1800**
5.323	1822.5	2093.4	2.1000	5.228	1822.5	2093.4	2.0980	5.136	1822.4	2093.3	2.0961	**2000**
5.760	1923.0	2216.1	2.1480	5.657	1923.0	2216.1	2.1460	5.557	1923.0	2216.1	2.1441	**2200**
6.196	2026.3	2341.6	2.1935	6.085	2026.3	2341.6	2.1915	5.978	2026.2	2341.5	2.1895	**2400**

Table 3. Vapor

p (t Sat.)	290 (414.33)				295 (415.89)				300 (417.43)			
t	v	u	h	s	v	u	h	s	v	u	h	s
Sat.	1.5963	1117.9	1203.6	1.5145	1.5698	1118.0	1203.7	1.5130	1.5442	1118.2	1203.9	1.5115
400	*1.5521*	*1109.9*	*1193.2*	*1.5025*	*1.5213*	*1109.0*	*1192.1*	*1.4996*	*1.4915*	*1108.2*	*1191.0*	*1.4967*
410	*1.5831*	*1115.5*	*1200.5*	*1.5110*	*1.5521*	*1114.8*	*1199.5*	*1.5081*	*1.5221*	*1114.0*	*1198.5*	*1.5054*
420	1.6133	1121.0	1207.6	1.5191	1.5820	1120.3	1206.7	1.5163	1.5517	1119.6	1205.7	1.5136
430	1.6426	1126.3	1214.5	1.5269	1.6110	1125.7	1213.6	1.5242	1.5805	1125.0	1212.7	1.5216
440	1.6713	1131.5	1221.2	1.5344	1.6394	1130.9	1220.4	1.5318	1.6086	1130.3	1219.6	1.5292
450	1.6993	1136.6	1227.7	1.5416	1.6672	1136.0	1227.0	1.5391	1.6361	1135.4	1226.2	1.5365
460	1.7268	1141.5	1234.2	1.5486	1.6944	1140.9	1233.4	1.5461	1.6630	1140.4	1232.7	1.5436
470	1.7538	1146.3	1240.4	1.5554	1.7211	1145.8	1239.8	1.5530	1.6894	1145.3	1239.1	1.5505
480	1.7804	1151.1	1246.6	1.5620	1.7473	1150.6	1246.0	1.5596	1.7154	1150.1	1245.3	1.5572
490	1.8066	1155.8	1252.7	1.5685	1.7732	1155.3	1252.1	1.5661	1.7410	1154.8	1251.5	1.5637
500	1.8324	1160.4	1258.7	1.5748	1.7987	1159.9	1258.1	1.5724	1.7662	1159.5	1257.5	1.5701
510	1.8579	1164.9	1264.6	1.5809	1.8239	1164.5	1264.0	1.5786	1.7910	1164.1	1263.5	1.5763
520	1.8831	1169.4	1270.4	1.5869	1.8488	1169.0	1269.9	1.5846	1.8156	1168.6	1269.4	1.5823
530	1.9080	1173.8	1276.2	1.5927	1.8734	1173.4	1275.7	1.5904	1.8399	1173.1	1275.2	1.5882
540	1.9327	1178.2	1281.9	1.5985	1.8978	1177.8	1281.4	1.5962	1.8640	1177.5	1281.0	1.5940
550	1.9572	1182.6	1287.6	1.6041	1.9219	1182.2	1287.1	1.6019	1.8878	1181.9	1286.7	1.5997
560	1.9815	1186.9	1293.2	1.6096	1.9458	1186.5	1292.8	1.6074	1.9114	1186.2	1292.3	1.6052
570	2.0056	1191.1	1298.8	1.6151	1.9696	1190.8	1298.3	1.6129	1.9348	1190.5	1297.9	1.6107
580	2.0295	1195.4	1304.3	1.6204	1.9932	1195.1	1303.9	1.6182	1.9580	1194.8	1303.5	1.6161
590	2.0532	1199.6	1309.8	1.6257	2.0166	1199.3	1309.4	1.6235	1.9811	1199.0	1309.0	1.6214
600	2.077	1203.8	1315.3	1.6309	2.040	1203.5	1314.9	1.6287	2.004	1203.2	1314.5	1.6266
620	2.124	1212.2	1326.1	1.6410	2.086	1211.9	1325.8	1.6389	2.049	1211.6	1325.4	1.6368
640	2.170	1220.5	1336.9	1.6509	2.132	1220.2	1336.6	1.6488	2.094	1220.0	1336.2	1.6467
660	2.216	1228.7	1347.6	1.6606	2.177	1228.5	1347.3	1.6585	2.139	1228.2	1347.0	1.6564
680	2.261	1236.9	1358.2	1.6700	2.222	1236.7	1357.9	1.6679	2.183	1236.4	1357.6	1.6658
700	2.306	1245.1	1368.8	1.6792	2.266	1244.8	1368.5	1.6771	2.227	1244.6	1368.3	1.6751
720	2.351	1253.2	1379.4	1.6882	2.310	1253.0	1379.1	1.6862	2.270	1252.8	1378.9	1.6841
740	2.396	1261.3	1389.9	1.6971	2.354	1261.2	1389.7	1.6950	2.314	1261.0	1389.4	1.6930
760	2.440	1269.5	1400.4	1.7058	2.398	1269.3	1400.2	1.7037	2.357	1269.1	1400.0	1.7017
780	2.484	1277.6	1410.9	1.7143	2.441	1277.4	1410.7	1.7123	2.400	1277.3	1410.5	1.7103
800	2.528	1285.7	1421.4	1.7227	2.485	1285.6	1421.2	1.7207	2.442	1285.4	1421.0	1.7187
820	2.572	1293.9	1431.9	1.7310	2.528	1293.7	1431.7	1.7290	2.485	1293.6	1431.5	1.7270
840	2.616	1302.0	1442.4	1.7391	2.571	1301.9	1442.2	1.7371	2.527	1301.7	1442.0	1.7351
860	2.659	1310.2	1452.9	1.7471	2.613	1310.1	1452.7	1.7451	2.569	1309.9	1452.5	1.7432
880	2.703	1318.4	1463.4	1.7550	2.656	1318.2	1463.2	1.7530	2.611	1318.1	1463.1	1.7511
900	2.746	1326.6	1473.9	1.7628	2.699	1326.4	1473.8	1.7608	2.653	1326.3	1473.6	1.7589
920	2.789	1334.8	1484.5	1.7705	2.741	1334.7	1484.3	1.7685	2.695	1334.5	1484.1	1.7666
940	2.832	1343.0	1495.0	1.7781	2.783	1342.9	1494.8	1.7761	2.736	1342.8	1494.7	1.7742
960	2.875	1351.3	1505.6	1.7856	2.825	1351.2	1505.4	1.7836	2.778	1351.1	1505.3	1.7817
980	2.918	1359.6	1516.1	1.7930	2.868	1359.5	1516.0	1.7910	2.819	1359.3	1515.8	1.7891
1000	2.960	1367.9	1526.7	1.8003	2.910	1367.8	1526.6	1.7983	2.860	1367.7	1526.5	1.7964
1020	3.003	1376.2	1537.4	1.8075	2.951	1376.1	1537.2	1.8055	2.902	1376.0	1537.1	1.8036
1040	3.046	1384.6	1548.0	1.8146	2.993	1384.5	1547.9	1.8127	2.943	1384.4	1547.7	1.8108
1060	3.088	1392.9	1558.7	1.8217	3.035	1392.8	1558.5	1.8198	2.984	1392.7	1558.4	1.8178
1080	3.130	1401.3	1569.3	1.8287	3.077	1401.3	1569.2	1.8267	3.025	1401.2	1569.1	1.8248
1100	3.173	1409.8	1580.0	1.8356	3.118	1409.7	1579.9	1.8337	3.066	1409.6	1579.8	1.8317
1150	3.278	1431.0	1606.9	1.8526	3.222	1430.9	1606.8	1.8506	3.168	1430.8	1606.7	1.8487
1200	3.384	1452.4	1634.0	1.8691	3.326	1452.3	1633.9	1.8672	3.270	1452.2	1633.8	1.8653
1250	3.489	1474.0	1661.2	1.8853	3.429	1473.9	1661.1	1.8833	3.372	1473.8	1661.0	1.8815
1300	3.593	1495.8	1688.6	1.9011	3.532	1495.7	1688.5	1.8992	3.473	1495.6	1688.4	1.8973
1350	3.698	1517.7	1716.2	1.9165	3.635	1517.7	1716.1	1.9146	3.574	1517.6	1716.0	1.9127
1400	3.802	1539.9	1744.0	1.9317	3.738	1539.9	1743.9	1.9298	3.675	1539.8	1743.8	1.9279
1450	3.907	1562.3	1772.0	1.9465	3.840	1562.3	1771.9	1.9446	3.776	1562.2	1771.8	1.9427
1500	4.011	1584.9	1800.2	1.9611	3.943	1584.9	1800.1	1.9592	3.877	1584.8	1800.0	1.9573
1600	4.219	1630.8	1857.2	1.9895	4.147	1630.7	1857.1	1.9876	4.078	1630.7	1857.0	1.9857
1800	4.634	1724.9	1973.6	2.0434	4.555	1724.9	1973.6	2.0415	4.479	1724.9	1973.5	2.0396
2000	5.048	1822.4	2093.3	2.0942	4.962	1822.4	2093.2	2.0923	4.879	1822.3	2093.2	2.0904
2200	5.462	1922.9	2216.0	2.1421	5.369	1922.9	2216.0	2.1402	5.280	1922.9	2216.0	2.1384
2400	5.875	2026.2	2341.5	2.1876	5.776	2026.2	2341.5	2.1857	5.679	2026.1	2341.4	2.1838

v	u	h	s	v	u	h	s	v	u	h	s	t
1.5193	1118.3	1204.0	1.5101	1.4953	1118.4	1204.1	1.5086	1.4720	1118.5	1204.3	1.5072	**Sat.**
1.4626	*1107.3*	*1189.9*	*1.4938*	*1.4346*	*1106.5*	*1188.8*	*1.4910*	*1.4075*	*1105.6*	*1187.7*	*1.4882*	**400**
1.4930	*1113.2*	*1197.5*	*1.5026*	*1.4648*	*1112.4*	*1196.4*	*1.4998*	*1.4375*	*1111.6*	*1195.4*	*1.4971*	**410**
1.5224	1118.8	1204.8	1.5109	*1.4940*	*1118.1*	*1203.8*	*1.5083*	*1.4665*	*1117.4*	*1202.9*	*1.5056*	**420**
1.5510	1124.3	1211.8	1.5189	1.5223	1123.6	1211.0	1.5163	1.4946	1122.9	1210.1	1.5138	**430**
1.5788	1129.6	1218.7	1.5266	1.5499	1129.0	1217.9	1.5241	1.5220	1128.3	1217.1	1.5216	**440**
1.6060	1134.8	1225.4	1.5340	1.5769	1134.2	1224.6	1.5316	1.5487	1133.6	1223.9	1.5291	**450**
1.6327	1139.8	1232.0	1.5412	1.6033	1139.3	1231.2	1.5388	1.5748	1138.7	1230.5	1.5364	**460**
1.6588	1144.8	1238.4	1.5481	1.6292	1144.2	1237.7	1.5457	1.6004	1143.7	1237.0	1.5434	**470**
1.6845	1149.6	1244.7	1.5549	1.6546	1149.1	1244.0	1.5525	1.6256	1148.6	1243.4	1.5502	**480**
1.7098	1154.4	1250.9	1.5614	1.6795	1153.9	1250.2	1.5591	1.6503	1153.4	1249.6	1.5568	**490**
1.7347	1159.0	1256.9	1.5678	1.7042	1158.6	1256.3	1.5655	1.6746	1158.1	1255.7	1.5632	**500**
1.7592	1163.6	1262.9	1.5740	1.7285	1163.2	1262.4	1.5717	1.6986	1162.8	1261.8	1.5695	**510**
1.7835	1168.2	1268.8	1.5800	1.7524	1167.8	1268.3	1.5778	1.7223	1167.4	1267.8	1.5756	**520**
1.8075	1172.7	1274.7	1.5860	1.7761	1172.3	1274.2	1.5838	1.7458	1171.9	1273.7	1.5816	**530**
1.8312	1177.1	1280.5	1.5918	1.7996	1176.7	1280.0	1.5896	1.7689	1176.4	1279.5	1.5875	**540**
1.8548	1181.5	1286.2	1.5975	1.8228	1181.1	1285.7	1.5953	1.7918	1180.8	1285.2	1.5932	**550**
1.8781	1185.9	1291.9	1.6031	1.8458	1185.5	1291.4	1.6009	1.8145	1185.2	1290.9	1.5988	**560**
1.9012	1190.2	1297.5	1.6086	1.8686	1189.9	1297.0	1.6064	1.8371	1189.5	1296.6	1.6044	**570**
1.9241	1194.5	1303.1	1.6140	1.8912	1194.2	1302.6	1.6119	1.8594	1193.8	1302.2	1.6098	**580**
1.9469	1198.7	1308.6	1.6193	1.9137	1198.4	1308.2	1.6172	1.8816	1198.1	1307.8	1.6151	**590**
1.9695	1203.0	1314.1	1.6245	1.9360	1202.7	1313.7	1.6224	1.9036	1202.4	1313.3	1.6204	**600**
2.0143	1211.4	1325.1	1.6347	1.9802	1211.1	1324.7	1.6327	1.9472	1210.8	1324.3	1.6307	**620**
2.0586	1219.7	1335.9	1.6447	2.0239	1219.4	1335.5	1.6426	1.9902	1219.2	1335.2	1.6406	**640**
2.1025	1228.0	1346.6	1.6544	2.0671	1227.7	1346.3	1.6523	2.0329	1227.5	1346.0	1.6504	**660**
2.1460	1236.2	1357.3	1.6638	2.1100	1236.0	1357.0	1.6618	2.0752	1235.8	1356.7	1.6599	**680**
2.189	1244.4	1368.0	1.6731	2.153	1244.2	1367.7	1.6711	2.117	1244.0	1367.4	1.6692	**700**
2.232	1252.6	1378.6	1.6821	2.195	1252.4	1378.3	1.6802	2.159	1252.2	1378.0	1.6782	**720**
2.275	1260.8	1389.2	1.6910	2.237	1260.6	1388.9	1.6891	2.200	1260.4	1388.7	1.6872	**740**
2.317	1268.9	1399.7	1.6998	2.279	1268.8	1399.5	1.6978	2.241	1268.6	1399.2	1.6959	**760**
2.359	1277.1	1410.2	1.7083	2.320	1276.9	1410.0	1.7064	2.282	1276.8	1409.8	1.7045	**780**
2.401	1285.3	1420.8	1.7167	2.361	1285.1	1420.6	1.7148	2.323	1284.9	1420.3	1.7129	**800**
2.443	1293.4	1431.3	1.7250	2.403	1293.3	1431.1	1.7231	2.363	1293.1	1430.9	1.7212	**820**
2.485	1301.6	1441.8	1.7332	2.444	1301.4	1441.6	1.7313	2.404	1301.3	1441.4	1.7294	**840**
2.526	1309.8	1452.3	1.7412	2.484	1309.6	1452.2	1.7393	2.444	1309.5	1452.0	1.7375	**860**
2.567	1318.0	1462.9	1.7491	2.525	1317.8	1462.7	1.7473	2.484	1317.7	1462.5	1.7454	**880**
2.609	1326.2	1473.4	1.7570	2.566	1326.1	1473.2	1.7551	2.524	1325.9	1473.1	1.7532	**900**
2.650	1334.4	1484.0	1.7647	2.606	1334.3	1483.8	1.7628	2.564	1334.2	1483.6	1.7609	**920**
2.691	1342.7	1494.5	1.7723	2.646	1342.6	1494.4	1.7704	2.604	1342.4	1494.2	1.7685	**940**
2.731	1350.9	1505.1	1.7798	2.687	1350.8	1505.0	1.7779	2.643	1350.7	1504.8	1.7760	**960**
2.772	1359.2	1515.7	1.7872	2.727	1359.1	1515.6	1.7853	2.683	1359.0	1515.4	1.7835	**980**
2.813	1367.6	1526.3	1.7945	2.767	1367.4	1526.2	1.7926	2.722	1367.3	1526.0	1.7908	**1000**
2.854	1375.9	1536.9	1.8017	2.807	1375.8	1536.8	1.7999	2.762	1375.7	1536.7	1.7980	**1020**
2.894	1384.3	1547.6	1.8089	2.847	1384.2	1547.5	1.8070	2.801	1384.1	1547.3	1.8052	**1040**
2.935	1392.7	1558.3	1.8160	2.887	1392.6	1558.1	1.8141	2.840	1392.5	1558.0	1.8123	**1060**
2.975	1401.1	1569.0	1.8230	2.926	1401.0	1568.8	1.8211	2.879	1400.9	1568.7	1.8193	**1080**
3.015	1409.5	1579.7	1.8299	2.966	1409.4	1579.6	1.8280	2.919	1409.3	1579.5	1.8262	**1100**
3.116	1430.7	1606.6	1.8469	3.065	1430.7	1606.5	1.8450	3.016	1430.6	1606.4	1.8432	**1150**
3.216	1452.2	1633.7	1.8634	3.164	1452.1	1633.6	1.8616	3.113	1452.0	1633.5	1.8598	**1200**
3.316	1473.8	1660.9	1.8796	3.262	1473.7	1660.8	1.8778	3.210	1473.6	1660.7	1.8760	**1250**
3.416	1495.6	1688.4	1.8954	3.360	1495.5	1688.3	1.8936	3.307	1495.4	1688.2	1.8918	**1300**
3.515	1517.6	1716.0	1.9109	3.458	1517.5	1715.9	1.9091	3.403	1517.4	1715.8	1.9073	**1350**
3.615	1539.8	1743.8	1.9260	3.556	1539.7	1743.7	1.9242	3.499	1539.7	1743.6	1.9224	**1400**
3.714	1562.2	1771.8	1.9409	3.654	1562.1	1771.7	1.9391	3.595	1562.1	1771.6	1.9373	**1450**
3.813	1584.8	1800.0	1.9555	3.751	1584.7	1799.9	1.9537	3.691	1584.7	1799.9	1.9519	**1500**
4.011	1630.6	1857.0	1.9838	3.946	1630.6	1856.9	1.9820	3.883	1630.5	1856.9	1.9802	**1600**
4.405	1724.9	1973.5	2.0378	4.334	1724.8	1973.4	2.0360	4.265	1724.7	1973.4	2.0342	**1800**
4.799	1822.3	2093.2	2.0886	4.722	1822.3	2093.1	2.0867	4.647	1822.2	2093.1	2.0850	**2000**
5.193	1922.8	2215.9	2.1365	5.109	1922.8	2215.9	2.1347	5.028	1922.8	2215.8	2.1329	**2200**
5.586	2026.1	2341.4	2.1820	5.496	2026.1	2341.4	2.1802	5.409	2026.0	2341.3	2.1784	**2400**

Table 3. Vapor

p (t Sat.)	320 (423.39)				325 (424.84)				330 (426.27)			
t	v	u	h	s	v	u	h	s	v	u	h	s
Sat.	1.4493	1118.6	1204.4	1.5058	1.4274	1118.6	1204.5	1.5044	1.4061	1118.7	1204.6	1.5031
410	*1.4110*	*1110.8*	*1194.4*	*1.4944*	*1.3852*	*1110.0*	*1193.3*	*1.4917*	*1.3602*	*1109.2*	*1192.2*	*1.4890*
420	*1.4398*	*1116.6*	*1201.9*	*1.5030*	*1.4139*	*1115.9*	*1200.9*	*1.5004*	*1.3887*	*1115.1*	*1199.9*	*1.4978*
430	1.4677	1122.2	1209.2	1.5112	1.4416	1121.5	1208.2	1.5087	1.4163	1120.8	1207.3	1.5061
440	1.4949	1127.7	1216.2	1.5191	1.4686	1127.0	1215.4	1.5166	1.4430	1126.4	1214.5	1.5142
450	1.5213	1133.0	1223.1	1.5267	1.4948	1132.4	1222.3	1.5243	1.4691	1131.8	1221.5	1.5219
460	1.5472	1138.1	1229.8	1.5340	1.5205	1137.6	1229.0	1.5316	1.4945	1137.0	1228.2	1.5293
470	1.5726	1143.2	1236.3	1.5411	1.5456	1142.6	1235.6	1.5387	1.5194	1142.1	1234.9	1.5364
480	1.5975	1148.1	1242.7	1.5479	1.5702	1147.6	1242.0	1.5456	1.5438	1147.1	1241.4	1.5434
490	1.6219	1152.9	1249.0	1.5545	1.5944	1152.4	1248.3	1.5523	1.5678	1152.0	1247.7	1.5501
500	1.6460	1157.7	1255.1	1.5610	1.6183	1157.2	1254.5	1.5588	1.5914	1156.8	1253.9	1.5566
510	1.6698	1162.4	1261.2	1.5673	1.6417	1161.9	1260.7	1.5651	1.6146	1161.5	1260.1	1.5630
520	1.6932	1167.0	1267.2	1.5735	1.6649	1166.5	1266.7	1.5713	1.6375	1166.1	1266.1	1.5692
530	1.7163	1171.5	1273.1	1.5795	1.6878	1171.1	1272.6	1.5774	1.6601	1170.7	1272.1	1.5753
540	1.7392	1176.0	1279.0	1.5854	1.7104	1175.6	1278.5	1.5833	1.6825	1175.2	1278.0	1.5812
550	1.7619	1180.4	1284.8	1.5911	1.7328	1180.1	1284.3	1.5890	1.7046	1179.7	1283.8	1.5870
560	1.7843	1184.8	1290.5	1.5968	1.7549	1184.5	1290.0	1.5947	1.7265	1184.2	1289.6	1.5927
570	1.8065	1189.2	1296.2	1.6023	1.7769	1188.9	1295.7	1.6003	1.7482	1188.5	1295.3	1.5983
580	1.8286	1193.5	1301.8	1.6077	1.7987	1193.2	1301.4	1.6057	1.7697	1192.9	1301.0	1.6037
590	1.8504	1197.8	1307.4	1.6131	1.8203	1197.5	1307.0	1.6111	1.7910	1197.2	1306.6	1.6091
600	1.8722	1202.1	1313.0	1.6184	1.8417	1201.8	1312.6	1.6164	1.8122	1201.5	1312.2	1.6144
620	1.9152	1210.6	1324.0	1.6287	1.8842	1210.3	1323.6	1.6267	1.8541	1210.0	1323.2	1.6248
640	1.9577	1218.9	1334.9	1.6387	1.9261	1218.7	1334.5	1.6367	1.8955	1218.4	1334.2	1.6348
660	1.9998	1227.3	1345.7	1.6484	1.9676	1227.0	1345.4	1.6465	1.9365	1226.8	1345.0	1.6446
680	2.0415	1235.6	1356.4	1.6579	2.0088	1235.3	1356.1	1.6560	1.9771	1235.1	1355.8	1.6541
700	2.083	1243.8	1367.1	1.6672	2.050	1243.6	1366.8	1.6653	2.017	1243.4	1366.6	1.6635
720	2.124	1252.0	1377.8	1.6763	2.090	1251.8	1377.5	1.6745	2.057	1251.6	1377.2	1.6726
740	2.165	1260.2	1388.4	1.6853	2.130	1260.0	1388.1	1.6834	2.097	1259.8	1387.9	1.6816
760	2.205	1268.4	1399.0	1.6940	2.170	1268.2	1398.7	1.6922	2.136	1268.0	1398.5	1.6903
780	2.246	1276.6	1409.6	1.7026	2.210	1276.4	1409.3	1.7008	2.176	1276.2	1409.1	1.6989
800	2.286	1284.8	1420.1	1.7111	2.250	1284.6	1419.9	1.7092	2.215	1284.4	1419.7	1.7074
820	2.326	1293.0	1430.7	1.7194	2.289	1292.8	1430.5	1.7175	2.253	1292.6	1430.3	1.7157
840	2.365	1301.1	1441.2	1.7276	2.328	1301.0	1441.0	1.7257	2.292	1300.9	1440.8	1.7239
860	2.405	1309.4	1451.8	1.7356	2.367	1309.2	1451.6	1.7338	2.331	1309.1	1451.4	1.7320
880	2.445	1317.6	1462.3	1.7436	2.406	1317.4	1462.1	1.7417	2.369	1317.3	1462.0	1.7400
900	2.484	1325.8	1472.9	1.7514	2.445	1325.7	1472.7	1.7496	2.407	1325.5	1472.5	1.7478
920	2.523	1334.0	1483.5	1.7591	2.484	1333.9	1483.3	1.7573	2.445	1333.8	1483.1	1.7555
940	2.562	1342.3	1494.0	1.7667	2.522	1342.2	1493.9	1.7649	2.483	1342.1	1493.7	1.7631
960	2.601	1350.6	1504.6	1.7742	2.561	1350.5	1504.5	1.7724	2.521	1350.4	1504.3	1.7707
980	2.640	1358.9	1515.3	1.7817	2.599	1358.8	1515.1	1.7799	2.559	1358.7	1515.0	1.7781
1000	2.679	1367.2	1525.9	1.7890	2.637	1367.1	1525.7	1.7872	2.597	1367.0	1525.6	1.7854
1020	2.718	1375.6	1536.5	1.7962	2.676	1375.5	1536.4	1.7945	2.635	1375.4	1536.3	1.7927
1040	2.757	1384.0	1547.2	1.8034	2.714	1383.9	1547.1	1.8016	2.672	1383.8	1546.9	1.7999
1060	2.795	1392.4	1557.9	1.8105	2.752	1392.3	1557.8	1.8087	2.710	1392.2	1557.6	1.8070
1080	2.834	1400.8	1568.6	1.8175	2.790	1400.7	1568.5	1.8157	2.747	1400.6	1568.4	1.8140
1100	2.872	1409.2	1579.3	1.8244	2.828	1409.2	1579.2	1.8226	2.784	1409.1	1579.1	1.8209
1150	2.968	1430.5	1606.3	1.8414	2.922	1430.4	1606.2	1.8396	2.878	1430.3	1606.1	1.8379
1200	3.064	1451.9	1633.4	1.8580	3.017	1451.9	1633.3	1.8562	2.970	1451.8	1633.2	1.8545
1250	3.159	1473.6	1660.7	1.8742	3.111	1473.5	1660.6	1.8724	3.063	1473.4	1660.5	1.8707
1300	3.255	1495.4	1688.1	1.8900	3.204	1495.3	1688.0	1.8883	3.155	1495.3	1687.9	1.8865
1350	3.350	1517.4	1715.7	1.9055	3.298	1517.3	1715.7	1.9037	3.247	1517.3	1715.6	1.9020
1400	3.444	1539.6	1743.6	1.9207	3.391	1539.5	1743.5	1.9189	3.339	1539.5	1743.4	1.9172
1450	3.539	1562.0	1771.6	1.9355	3.484	1562.0	1771.5	1.9338	3.431	1561.9	1771.4	1.9321
1500	3.633	1584.6	1799.8	1.9501	3.577	1584.6	1799.7	1.9484	3.523	1584.5	1799.7	1.9467
1600	3.822	1630.5	1856.8	1.9785	3.763	1630.4	1856.8	1.9768	3.706	1630.4	1856.7	1.9750
1800	4.199	1724.7	1973.3	2.0325	4.134	1724.7	1973.3	2.0307	4.071	1724.6	1973.2	2.0290
2000	4.574	1822.2	2093.0	2.0832	4.504	1822.1	2093.0	2.0815	4.435	1822.1	2093.0	2.0798
2200	4.949	1922.7	2215.8	2.1312	4.873	1922.7	2215.8	2.1295	4.799	1922.7	2215.7	2.1278
2400	5.324	2026.0	2341.3	2.1767	5.243	2026.0	2341.3	2.1750	5.163	2025.9	2341.2	2.1733

	335 (427.68)				340 (429.07)				345 (430.45)			p (t Sat.)
v	u	h	s	v	u	h	s	v	u	h	s	t
1.3854	1118.8	1204.7	1.5017	1.3653	1118.9	1204.8	1.5004	1.3457	1118.9	1204.9	1.4991	**Sat.**
1.3360	*1108.3*	*1191.2*	*1.4863*	*1.3124*	*1107.5*	*1190.1*	*1.4837*	*1.2894*	*1106.6*	*1189.0*	*1.4810*	**410**
1.3643	*1114.3*	*1198.9*	*1.4952*	*1.3406*	*1113.6*	*1197.9*	*1.4926*	*1.3175*	*1112.8*	*1196.9*	*1.4901*	**420**
1.3917	1120.1	1206.4	1.5036	1.3678	1119.4	1205.5	1.5012	*1.3445*	*1118.7*	*1204.5*	*1.4987*	**430**
1.4182	1125.7	1213.6	1.5117	1.3941	1125.0	1212.8	1.5093	1.3707	1124.4	1211.9	1.5069	**440**
1.4441	1131.1	1220.7	1.5195	1.4198	1130.5	1219.8	1.5171	1.3962	1129.9	1219.0	1.5148	**450**
1.4693	1136.4	1227.5	1.5270	1.4448	1135.8	1226.7	1.5247	1.4210	1135.2	1226.0	1.5224	**460**
1.4940	1141.5	1234.2	1.5342	1.4693	1141.0	1233.4	1.5319	1.4453	1140.4	1232.7	1.5297	**470**
1.5182	1146.6	1240.7	1.5412	1.4933	1146.0	1240.0	1.5389	1.4691	1145.5	1239.3	1.5368	**480**
1.5419	1151.5	1247.1	1.5479	1.5168	1151.0	1246.4	1.5458	1.4924	1150.5	1245.8	1.5436	**490**
1.5652	1156.3	1253.3	1.5545	1.5399	1155.8	1252.7	1.5524	1.5152	1155.4	1252.1	1.5502	**500**
1.5882	1161.0	1259.5	1.5609	1.5626	1160.6	1258.9	1.5588	1.5378	1160.2	1258.3	1.5567	**510**
1.6109	1165.7	1265.6	1.5671	1.5850	1165.3	1265.0	1.5650	1.5600	1164.9	1264.5	1.5630	**520**
1.6333	1170.3	1271.6	1.5732	1.6072	1169.9	1271.0	1.5712	1.5819	1169.5	1270.5	1.5691	**530**
1.6554	1174.9	1277.5	1.5791	1.6290	1174.5	1277.0	1.5771	1.6035	1174.1	1276.5	1.5751	**540**
1.6772	1179.4	1283.3	1.5850	1.6507	1179.0	1282.9	1.5830	1.6249	1178.6	1282.4	1.5810	**550**
1.6989	1183.8	1289.1	1.5907	1.6720	1183.5	1288.7	1.5887	1.6460	1183.1	1288.2	1.5867	**560**
1.7203	1188.2	1294.9	1.5963	1.6932	1187.9	1294.4	1.5943	1.6670	1187.5	1294.0	1.5924	**570**
1.7415	1192.6	1300.5	1.6018	1.7142	1192.3	1300.1	1.5998	1.6877	1191.9	1299.7	1.5979	**580**
1.7626	1196.9	1306.2	1.6072	1.7351	1196.6	1305.8	1.6052	1.7083	1196.3	1305.4	1.6033	**590**
1.7835	1201.2	1311.8	1.6125	1.7557	1200.9	1311.4	1.6106	1.7287	1200.6	1311.0	1.6087	**600**
1.8249	1209.7	1322.9	1.6228	1.7966	1209.5	1322.5	1.6209	1.7691	1209.2	1322.1	1.6191	**620**
1.8658	1218.2	1333.8	1.6329	1.8370	1217.9	1333.5	1.6310	1.8090	1217.7	1333.2	1.6292	**640**
1.9063	1226.6	1344.7	1.6427	1.8769	1226.3	1344.4	1.6409	1.8484	1226.1	1344.1	1.6390	**660**
1.9463	1234.9	1355.5	1.6523	1.9165	1234.7	1355.2	1.6505	1.8875	1234.4	1354.9	1.6486	**680**
1.9860	1243.2	1366.3	1.6616	1.9557	1243.0	1366.0	1.6598	1.9262	1242.7	1365.7	1.6580	**700**
2.0254	1251.4	1377.0	1.6708	1.9946	1251.2	1376.7	1.6690	1.9646	1251.0	1376.4	1.6672	**720**
2.0646	1259.6	1387.6	1.6797	2.0332	1259.5	1387.4	1.6779	2.0027	1259.3	1387.1	1.6762	**740**
2.1035	1267.9	1398.3	1.6885	2.0716	1267.7	1398.0	1.6867	2.0406	1267.5	1397.8	1.6850	**760**
2.1422	1276.1	1408.9	1.6971	2.1097	1275.9	1408.6	1.6954	2.0782	1275.7	1408.4	1.6936	**780**
2.181	1284.3	1419.5	1.7056	2.148	1284.1	1419.2	1.7039	2.116	1284.0	1419.0	1.7021	**800**
2.219	1292.5	1430.0	1.7140	2.185	1292.3	1429.8	1.7122	2.153	1292.2	1429.6	1.7105	**820**
2.257	1300.7	1440.6	1.7222	2.223	1300.6	1440.4	1.7204	2.190	1300.4	1440.2	1.7187	**840**
2.295	1308.9	1451.2	1.7302	2.261	1308.8	1451.0	1.7285	2.227	1308.6	1450.8	1.7268	**860**
2.333	1317.2	1461.8	1.7382	2.298	1317.0	1461.6	1.7365	2.264	1316.9	1461.4	1.7347	**880**
2.371	1325.4	1472.4	1.7460	2.335	1325.3	1472.2	1.7443	2.301	1325.1	1472.0	1.7426	**900**
2.408	1333.7	1483.0	1.7538	2.372	1333.5	1482.8	1.7520	2.337	1333.4	1482.6	1.7503	**920**
2.446	1342.0	1493.6	1.7614	2.409	1341.8	1493.4	1.7597	2.374	1341.7	1493.2	1.7580	**940**
2.483	1350.2	1504.2	1.7689	2.446	1350.1	1504.0	1.7672	2.410	1350.0	1503.9	1.7655	**960**
2.520	1358.6	1514.8	1.7764	2.483	1358.5	1514.7	1.7747	2.446	1358.3	1514.5	1.7730	**980**
2.558	1366.9	1525.5	1.7837	2.519	1366.8	1525.3	1.7820	2.482	1366.7	1525.2	1.7803	**1000**
2.595	1375.3	1536.1	1.7910	2.556	1375.2	1536.0	1.7893	2.518	1375.1	1535.8	1.7876	**1020**
2.632	1383.7	1546.8	1.7981	2.592	1383.6	1546.7	1.7964	2.554	1383.5	1546.5	1.7948	**1040**
2.669	1392.1	1557.5	1.8052	2.629	1392.0	1557.4	1.8035	2.590	1391.9	1557.3	1.8019	**1060**
2.706	1400.5	1568.2	1.8122	2.665	1400.4	1568.1	1.8106	2.626	1400.3	1568.0	1.8089	**1080**
2.742	1409.0	1579.0	1.8192	2.702	1408.9	1578.9	1.8175	2.662	1408.8	1578.8	1.8158	**1100**
2.834	1430.3	1606.0	1.8362	2.792	1430.2	1605.9	1.8345	2.751	1430.1	1605.7	1.8329	**1150**
2.926	1451.7	1633.1	1.8528	2.882	1451.6	1633.0	1.8511	2.840	1451.6	1632.9	1.8495	**1200**
3.017	1473.4	1660.4	1.8690	2.972	1473.3	1660.3	1.8673	2.929	1473.2	1660.2	1.8657	**1250**
3.108	1495.2	1687.9	1.8848	3.062	1495.1	1687.8	1.8832	3.017	1495.1	1687.7	1.8815	**1300**
3.199	1517.2	1715.5	1.9003	3.151	1517.1	1715.4	1.8987	3.105	1517.1	1715.3	1.8970	**1350**
3.289	1539.4	1743.3	1.9155	3.241	1539.4	1743.3	1.9138	3.194	1539.3	1743.2	1.9122	**1400**
3.380	1561.9	1771.4	1.9304	3.330	1561.8	1771.3	1.9287	3.281	1561.8	1771.2	1.9271	**1450**
3.470	1584.5	1799.6	1.9450	3.419	1584.4	1799.5	1.9433	3.369	1584.4	1799.5	1.9417	**1500**
3.651	1630.4	1856.7	1.9734	3.597	1630.3	1856.6	1.9717	3.544	1630.3	1856.5	1.9701	**1600**
4.010	1724.6	1973.2	2.0274	3.951	1724.5	1973.1	2.0257	3.894	1724.5	1973.1	2.0241	**1800**
4.369	1822.1	2092.9	2.0781	4.305	1822.0	2092.9	2.0765	4.242	1822.0	2092.8	2.0749	**2000**
4.728	1922.6	2215.7	2.1261	4.658	1922.6	2215.7	2.1245	4.591	1922.6	2215.6	2.1228	**2200**
5.086	2025.9	2341.2	2.1716	5.011	2025.9	2341.2	2.1699	4.939	2025.8	2341.1	2.1683	**2400**

Table 3. Vapor

p (t Sat.)	350 (431.82)				355 (433.17)				360 (434.50)			
t	v	u	h	s	v	u	h	s	v	u	h	s
Sat.	1.3267	1119.0	1204.9	1.4978	1.3083	1119.1	1205.0	1.4965	1.2903	1119.1	1205.1	1.4952
420	*1.2950*	*1112.0*	*1195.9*	*1.4875*	*1.2732*	*1111.2*	*1194.8*	*1.4850*	*1.2519*	*1110.4*	*1193.8*	*1.4825*
430	*1.3219*	*1117.9*	*1203.6*	*1.4962*	*1.2999*	*1117.2*	*1202.6*	*1.4938*	*1.2786*	*1116.5*	*1201.6*	*1.4914*
440	1.3480	1123.7	1211.0	1.5045	1.3258	1123.0	1210.1	1.5022	1.3043	1122.3	1209.2	1.4998
450	1.3733	1129.2	1218.2	1.5125	1.3510	1128.6	1217.4	1.5102	1.3293	1128.0	1216.5	1.5079
460	1.3979	1134.6	1225.2	1.5201	1.3754	1134.0	1224.4	1.5179	1.3536	1133.4	1223.6	1.5157
470	1.4220	1139.9	1232.0	1.5275	1.3993	1139.3	1231.2	1.5253	1.3773	1138.8	1230.5	1.5231
480	1.4455	1145.0	1238.6	1.5346	1.4227	1144.5	1237.9	1.5325	1.4004	1143.9	1237.2	1.5303
490	1.4686	1150.0	1245.1	1.5415	1.4456	1149.5	1244.5	1.5394	1.4231	1149.0	1243.8	1.5373
500	1.4913	1154.9	1251.5	1.5482	1.4680	1154.4	1250.9	1.5461	1.4454	1154.0	1250.3	1.5440
510	1.5136	1159.7	1257.8	1.5546	1.4901	1159.3	1257.2	1.5526	1.4673	1158.8	1256.6	1.5506
520	1.5356	1164.5	1263.9	1.5610	1.5119	1164.0	1263.4	1.5590	1.4888	1163.6	1262.8	1.5570
530	1.5572	1169.1	1270.0	1.5671	1.5333	1168.7	1269.5	1.5651	1.5101	1168.3	1268.9	1.5632
540	1.5786	1173.7	1276.0	1.5731	1.5545	1173.3	1275.5	1.5712	1.5310	1173.0	1275.0	1.5692
550	1.5998	1178.3	1281.9	1.5790	1.5754	1177.9	1281.4	1.5771	1.5517	1177.5	1280.9	1.5752
560	1.6207	1182.8	1287.7	1.5848	1.5961	1182.4	1287.3	1.5829	1.5722	1182.1	1286.8	1.5810
570	1.6414	1187.2	1293.5	1.5904	1.6166	1186.9	1293.1	1.5885	1.5925	1186.5	1292.6	1.5867
580	1.6619	1191.6	1299.3	1.5960	1.6369	1191.3	1298.8	1.5941	1.6126	1191.0	1298.4	1.5922
590	1.6823	1196.0	1304.9	1.6014	1.6570	1195.7	1304.5	1.5996	1.6324	1195.4	1304.1	1.5977
600	1.7025	1200.3	1310.6	1.6068	1.6770	1200.0	1310.2	1.6049	1.6522	1199.7	1309.8	1.6031
620	1.7424	1208.9	1321.8	1.6172	1.7164	1208.6	1321.4	1.6154	1.6912	1208.4	1321.0	1.6136
640	1.7818	1217.4	1332.8	1.6274	1.7554	1217.2	1332.5	1.6256	1.7297	1216.9	1332.1	1.6238
660	1.8207	1225.8	1343.8	1.6372	1.7938	1225.6	1343.4	1.6355	1.7677	1225.4	1343.1	1.6337
680	1.8593	1234.2	1354.6	1.6469	1.8319	1234.0	1354.3	1.6451	1.8053	1233.8	1354.0	1.6433
700	1.8975	1242.5	1365.4	1.6562	1.8697	1242.3	1365.1	1.6545	1.8426	1242.1	1364.9	1.6528
720	1.9354	1250.8	1376.2	1.6654	1.9071	1250.6	1375.9	1.6637	1.8796	1250.4	1375.6	1.6620
740	1.9731	1259.1	1386.9	1.6744	1.9443	1258.9	1386.6	1.6727	1.9163	1258.7	1386.4	1.6710
760	2.0104	1267.3	1397.5	1.6832	1.9812	1267.1	1397.3	1.6815	1.9527	1267.0	1397.0	1.6798
780	2.0476	1275.6	1408.2	1.6919	2.0179	1275.4	1407.9	1.6902	1.9889	1275.2	1407.7	1.6885
800	2.085	1283.8	1418.8	1.7004	2.054	1283.6	1418.6	1.6987	2.025	1283.5	1418.4	1.6970
820	2.121	1292.0	1429.4	1.7088	2.091	1291.9	1429.2	1.7071	2.061	1291.7	1429.0	1.7054
840	2.158	1300.3	1440.0	1.7170	2.127	1300.1	1439.8	1.7153	2.096	1300.0	1439.6	1.7136
860	2.194	1308.5	1450.6	1.7251	2.163	1308.4	1450.4	1.7234	2.132	1308.2	1450.2	1.7218
880	2.231	1316.7	1461.2	1.7331	2.199	1316.6	1461.0	1.7314	2.167	1316.5	1460.9	1.7297
900	2.267	1325.0	1471.8	1.7409	2.234	1324.9	1471.7	1.7393	2.203	1324.8	1471.5	1.7376
920	2.303	1333.3	1482.5	1.7487	2.270	1333.2	1482.3	1.7470	2.238	1333.0	1482.1	1.7454
940	2.339	1341.6	1493.1	1.7563	2.305	1341.5	1492.9	1.7547	2.273	1341.3	1492.8	1.7530
960	2.375	1349.9	1503.7	1.7639	2.341	1349.8	1503.6	1.7622	2.308	1349.7	1503.4	1.7606
980	2.411	1358.2	1514.4	1.7713	2.376	1358.1	1514.2	1.7697	2.343	1358.0	1514.1	1.7680
1000	2.446	1366.6	1525.0	1.7787	2.411	1366.5	1524.9	1.7770	2.377	1366.4	1524.7	1.7754
1020	2.482	1375.0	1535.7	1.7859	2.446	1374.9	1535.6	1.7843	2.412	1374.8	1535.4	1.7827
1040	2.517	1383.4	1546.4	1.7931	2.481	1383.3	1546.3	1.7915	2.447	1383.2	1546.1	1.7899
1060	2.553	1391.8	1557.1	1.8002	2.516	1391.7	1557.0	1.7986	2.481	1391.6	1556.9	1.7970
1080	2.588	1400.2	1567.9	1.8072	2.551	1400.1	1567.7	1.8056	2.515	1400.1	1567.6	1.8040
1100	2.624	1408.7	1578.6	1.8142	2.586	1408.6	1578.5	1.8126	2.550	1408.5	1578.4	1.8110
1150	2.712	1430.0	1605.6	1.8312	2.673	1429.9	1605.5	1.8296	2.635	1429.9	1605.4	1.8280
1200	2.799	1451.5	1632.8	1.8478	2.759	1451.4	1632.7	1.8462	2.721	1451.3	1632.6	1.8446
1250	2.887	1473.1	1660.1	1.8641	2.846	1473.1	1660.0	1.8625	2.806	1473.0	1659.9	1.8609
1300	2.974	1495.0	1687.6	1.8799	2.932	1494.9	1687.5	1.8783	2.891	1494.9	1687.4	1.8767
1350	3.061	1517.0	1715.3	1.8954	3.017	1517.0	1715.2	1.8938	2.975	1516.9	1715.1	1.8922
1400	3.148	1539.3	1743.1	1.9106	3.103	1539.2	1743.1	1.9090	3.060	1539.2	1743.0	1.9074
1450	3.234	1561.7	1771.2	1.9255	3.189	1561.6	1771.1	1.9239	3.144	1561.6	1771.0	1.9223
1500	3.321	1584.3	1799.4	1.9401	3.274	1584.3	1799.4	1.9385	3.228	1584.2	1799.3	1.9369
1600	3.494	1630.2	1856.5	1.9685	3.444	1630.2	1856.4	1.9669	3.396	1630.1	1856.4	1.9653
1800	3.838	1724.5	1973.1	2.0225	3.784	1724.4	1973.0	2.0209	3.731	1724.4	1973.0	2.0193
2000	4.182	1822.0	2092.8	2.0733	4.123	1821.9	2092.8	2.0717	4.065	1821.9	2092.7	2.0701
2200	4.525	1922.5	2215.6	2.1212	4.461	1922.5	2215.6	2.1197	4.399	1922.5	2215.5	2.1181
2400	4.868	2025.8	2341.1	2.1667	4.800	2025.8	2341.1	2.1652	4.733	2025.7	2341.0	2.1636

	365 (435.82)				**370** (437.13)				**375** (438.42)		**p** (t Sat.)
v	u	h	s	v	u	h	s	v	u	h	s	t
1.2727	1119.2	1205.1	1.4940	1.2557	1119.2	1205.2	1.4927	1.2391	1119.3	1205.3	1.4915	Sat.
1.2312	*1109.6*	*1192.7*	*1.4800*	*1.2110*	*1108.7*	*1191.7*	*1.4775*	*1.1913*	*1107.9*	*1190.6*	*1.4750*	420
1.2577	*1115.7*	*1200.7*	*1.4889*	*1.2374*	*1115.0*	*1199.7*	*1.4865*	*1.2176*	*1114.2*	*1198.7*	*1.4842*	430
1.2833	1121.6	1208.3	1.4975	1.2629	1120.9	1207.4	1.4952	1.2430	1120.2	1206.5	1.4928	440
1.3081	1127.3	1215.7	1.5056	1.2876	1126.7	1214.8	1.5034	1.2675	1126.0	1214.0	1.5011	450
1.3323	1132.8	1222.8	1.5134	1.3115	1132.2	1222.0	1.5113	1.2913	1131.6	1221.2	1.5091	460
1.3558	1138.2	1229.8	1.5210	1.3349	1137.6	1229.0	1.5188	1.3145	1137.1	1228.3	1.5167	470
1.3788	1143.4	1236.5	1.5282	1.3577	1142.9	1235.8	1.5261	1.3372	1142.3	1235.1	1.5240	480
1.4013	1148.5	1243.2	1.5352	1.3800	1148.0	1242.5	1.5332	1.3593	1147.5	1241.8	1.5311	490
1.4234	1153.5	1249.6	1.5420	1.4019	1153.0	1249.0	1.5400	1.3810	1152.5	1248.4	1.5380	500
1.4450	1158.4	1256.0	1.5486	1.4234	1157.9	1255.4	1.5466	1.4024	1157.5	1254.8	1.5446	510
1.4664	1163.2	1262.2	1.5550	1.4446	1162.8	1261.7	1.5530	1.4233	1162.3	1261.1	1.5511	520
1.4874	1167.9	1268.4	1.5612	1.4654	1167.5	1267.8	1.5593	1.4440	1167.1	1267.3	1.5574	530
1.5082	1172.6	1274.4	1.5673	1.4860	1172.2	1273.9	1.5654	1.4643	1171.8	1273.4	1.5636	540
1.5287	1177.2	1280.4	1.5733	1.5063	1176.8	1279.9	1.5714	1.4844	1176.4	1279.4	1.5696	550
1.5490	1181.7	1286.3	1.5791	1.5263	1181.4	1285.9	1.5773	1.5043	1181.0	1285.4	1.5754	560
1.5690	1186.2	1292.2	1.5848	1.5462	1185.9	1291.7	1.5830	1.5239	1185.5	1291.3	1.5812	570
1.5889	1190.7	1298.0	1.5904	1.5658	1190.3	1297.5	1.5886	1.5434	1190.0	1297.1	1.5868	580
1.6085	1195.1	1303.7	1.5959	1.5853	1194.8	1303.3	1.5941	1.5626	1194.4	1302.9	1.5923	590
1.6281	1199.4	1309.4	1.6013	1.6046	1199.1	1309.0	1.5995	1.5817	1198.8	1308.6	1.5977	600
1.6666	1208.1	1320.7	1.6118	1.6427	1207.8	1320.3	1.6101	1.6195	1207.5	1319.9	1.6083	620
1.7047	1216.6	1331.8	1.6220	1.6803	1216.4	1331.4	1.6203	1.6567	1216.1	1331.1	1.6186	640
1.7422	1225.1	1342.8	1.6320	1.7175	1224.9	1342.5	1.6302	1.6934	1224.6	1342.1	1.6285	660
1.7794	1233.5	1353.7	1.6416	1.7542	1233.3	1353.4	1.6399	1.7297	1233.1	1353.1	1.6382	680
1.8163	1241.9	1364.6	1.6511	1.7906	1241.7	1364.3	1.6494	1.7657	1241.5	1364.0	1.6477	700
1.8528	1250.2	1375.4	1.6603	1.8267	1250.0	1375.1	1.6586	1.8014	1249.8	1374.8	1.6570	720
1.8890	1258.5	1386.1	1.6693	1.8625	1258.3	1385.8	1.6677	1.8368	1258.1	1385.6	1.6660	740
1.9250	1266.8	1396.8	1.6782	1.8981	1266.6	1396.6	1.6765	1.8719	1266.4	1396.3	1.6749	760
1.9608	1275.0	1407.5	1.6869	1.9334	1274.9	1407.3	1.6852	1.9068	1274.7	1407.0	1.6836	780
1.9964	1283.3	1418.1	1.6954	1.9686	1283.1	1417.9	1.6938	1.9415	1283.0	1417.7	1.6921	800
2.0318	1291.6	1428.8	1.7038	2.0035	1291.4	1428.6	1.7021	1.9760	1291.2	1428.4	1.7005	820
2.0670	1299.8	1439.4	1.7120	2.0383	1299.7	1439.2	1.7104	2.0104	1299.5	1439.0	1.7088	840
2.1021	1308.1	1450.1	1.7201	2.0730	1307.9	1449.9	1.7185	2.0446	1307.8	1449.7	1.7169	860
2.1370	1316.3	1460.7	1.7281	2.1075	1316.2	1460.5	1.7265	2.0787	1316.1	1460.3	1.7249	880
2.172	1324.6	1471.3	1.7360	2.142	1324.5	1471.1	1.7344	2.113	1324.4	1471.0	1.7328	900
2.206	1332.9	1481.9	1.7438	2.176	1332.8	1481.8	1.7422	2.146	1332.7	1481.6	1.7406	920
2.241	1341.2	1492.6	1.7514	2.210	1341.1	1492.4	1.7498	2.180	1341.0	1492.3	1.7483	940
2.276	1349.6	1503.3	1.7590	2.244	1349.4	1503.1	1.7574	2.214	1349.3	1502.9	1.7558	960
2.310	1357.9	1513.9	1.7665	2.278	1357.8	1513.8	1.7649	2.247	1357.7	1513.6	1.7633	980
2.344	1366.3	1524.6	1.7738	2.312	1366.2	1524.5	1.7722	2.281	1366.1	1524.3	1.7707	1000
2.378	1374.7	1535.3	1.7811	2.346	1374.6	1535.2	1.7795	2.314	1374.4	1535.0	1.7780	1020
2.413	1383.1	1546.0	1.7883	2.379	1383.0	1545.9	1.7867	2.347	1382.9	1545.7	1.7852	1040
2.447	1391.5	1556.7	1.7954	2.413	1391.4	1556.6	1.7938	2.380	1391.3	1556.5	1.7923	1060
2.481	1400.0	1567.5	1.8024	2.447	1399.9	1567.4	1.8009	2.414	1399.8	1567.3	1.7993	1080
2.514	1408.4	1578.3	1.8094	2.480	1408.4	1578.2	1.8078	2.447	1408.3	1578.0	1.8063	1100
2.599	1429.8	1605.3	1.8264	2.563	1429.7	1605.2	1.8249	2.529	1429.6	1605.1	1.8234	1150
2.683	1451.3	1632.5	1.8431	2.647	1451.2	1632.4	1.8415	2.611	1451.1	1632.3	1.8400	1200
2.767	1472.9	1659.8	1.8593	2.729	1472.9	1659.7	1.8578	2.693	1472.8	1659.7	1.8562	1250
2.851	1494.8	1687.4	1.8752	2.812	1494.7	1687.3	1.8736	2.774	1494.7	1687.2	1.8721	1300
2.934	1516.9	1715.0	1.8907	2.894	1516.8	1715.0	1.8892	2.856	1516.7	1714.9	1.8876	1350
3.018	1539.1	1742.9	1.9059	2.977	1539.0	1742.8	1.9043	2.937	1539.0	1742.8	1.9028	1400
3.101	1561.5	1771.0	1.9208	3.059	1561.5	1770.9	1.9192	3.018	1561.4	1770.8	1.9177	1450
3.184	1584.2	1799.2	1.9354	3.141	1584.1	1799.2	1.9338	3.098	1584.1	1799.1	1.9323	1500
3.349	1630.1	1856.3	1.9638	3.304	1630.0	1856.3	1.9623	3.260	1630.0	1856.2	1.9608	1600
3.680	1724.4	1972.9	2.0178	3.630	1724.3	1972.9	2.0163	3.582	1724.3	1972.8	2.0148	1800
4.010	1821.9	2092.7	2.0686	3.955	1821.8	2092.7	2.0671	3.903	1821.8	2092.6	2.0656	2000
4.339	1922.4	2215.5	2.1166	4.280	1922.4	2215.5	2.1151	4.223	1922.3	2215.4	2.1136	2200
4.668	2025.7	2341.0	2.1621	4.605	2025.7	2341.0	2.1606	4.544	2025.6	2340.9	2.1591	2400

Table 3. Vapor

p (t Sat.)	380 (439.70)				385 (440.97)				390 (442.23)			
t	v	u	h	s	v	u	h	s	v	u	h	s
Sat.	1.2229	1119.3	1205.3	1.4903	1.2071	1119.4	1205.4	1.4891	1.1916	1119.4	1205.4	1.4879
420	*1.1721*	*1107.1*	*1189.5*	*1.4725*								
430	*1.1983*	*1113.4*	*1197.7*	*1.4818*	*1.1795*	*1112.6*	*1196.7*	*1.4794*	*1.1612*	*1111.8*	*1195.6*	*1.4770*
440	1.2236	1119.5	1205.5	1.4905	*1.2047*	*1118.8*	*1204.6*	*1.4883*	*1.1862*	*1118.0*	*1203.7*	*1.4860*
450	1.2480	1125.3	1213.1	1.4989	1.2289	1124.7	1212.2	1.4967	1.2104	1124.0	1211.4	1.4945
460	1.2717	1131.0	1220.4	1.5069	1.2525	1130.4	1219.6	1.5048	1.2338	1129.7	1218.8	1.5026
470	1.2947	1136.5	1227.5	1.5146	1.2754	1135.9	1226.8	1.5125	1.2565	1135.3	1226.0	1.5104
480	1.3172	1141.8	1234.4	1.5220	1.2977	1141.3	1233.7	1.5199	1.2787	1140.7	1233.0	1.5179
490	1.3392	1147.0	1241.2	1.5291	1.3195	1146.5	1240.5	1.5271	1.3004	1146.0	1239.8	1.5251
500	1.3607	1152.1	1247.7	1.5360	1.3409	1151.6	1247.1	1.5340	1.3216	1151.1	1246.5	1.5321
510	1.3818	1157.0	1254.2	1.5427	1.3618	1156.6	1253.6	1.5408	1.3424	1156.1	1253.0	1.5388
520	1.4026	1161.9	1260.5	1.5492	1.3824	1161.5	1260.0	1.5473	1.3628	1161.0	1259.4	1.5454
530	1.4231	1166.7	1266.8	1.5555	1.4027	1166.3	1266.2	1.5537	1.3829	1165.9	1265.7	1.5518
540	1.4432	1171.4	1272.9	1.5617	1.4227	1171.0	1272.4	1.5598	1.4027	1170.6	1271.9	1.5580
550	1.4632	1176.1	1279.0	1.5677	1.4424	1175.7	1278.5	1.5659	1.4222	1175.3	1278.0	1.5641
560	1.4828	1180.7	1284.9	1.5736	1.4619	1180.3	1284.5	1.5718	1.4415	1179.9	1284.0	1.5700
570	1.5023	1185.2	1290.8	1.5794	1.4812	1184.9	1290.4	1.5776	1.4606	1184.5	1289.9	1.5758
580	1.5215	1189.7	1296.7	1.5850	1.5002	1189.4	1296.2	1.5832	1.4795	1189.0	1295.8	1.5815
590	1.5406	1194.1	1302.5	1.5906	1.5191	1193.8	1302.0	1.5888	1.4982	1193.5	1301.6	1.5871
600	1.5595	1198.5	1308.2	1.5960	1.5378	1198.2	1307.8	1.5943	1.5167	1197.9	1307.4	1.5925
620	1.5968	1207.3	1319.5	1.6066	1.5747	1207.0	1319.2	1.6049	1.5532	1206.7	1318.8	1.6032
640	1.6336	1215.9	1330.7	1.6169	1.6111	1215.6	1330.4	1.6152	1.5892	1215.4	1330.0	1.6135
660	1.6699	1224.4	1341.8	1.6269	1.6471	1224.2	1341.5	1.6252	1.6248	1223.9	1341.2	1.6236
680	1.7058	1232.8	1352.8	1.6366	1.6826	1232.6	1352.5	1.6349	1.6599	1232.4	1352.2	1.6333
700	1.7414	1241.2	1363.7	1.6461	1.7178	1241.0	1363.4	1.6444	1.6947	1240.8	1363.1	1.6428
720	1.7767	1249.6	1374.5	1.6553	1.7526	1249.4	1374.3	1.6537	1.7292	1249.2	1374.0	1.6521
740	1.8116	1257.9	1385.3	1.6644	1.7872	1257.7	1385.1	1.6628	1.7634	1257.5	1384.8	1.6612
760	1.8464	1266.2	1396.1	1.6733	1.8215	1266.1	1395.8	1.6717	1.7973	1265.9	1395.6	1.6701
780	1.8809	1274.5	1406.8	1.6820	1.8556	1274.4	1406.6	1.6804	1.8310	1274.2	1406.3	1.6788
800	1.9152	1282.8	1417.5	1.6906	1.8895	1282.6	1417.3	1.6890	1.8645	1282.5	1417.0	1.6874
820	1.9493	1291.1	1428.2	1.6990	1.9232	1290.9	1427.9	1.6974	1.8978	1290.8	1427.7	1.6958
840	1.9832	1299.4	1438.8	1.7072	1.9567	1299.2	1438.6	1.7057	1.9309	1299.1	1438.4	1.7041
860	2.0170	1307.6	1449.5	1.7154	1.9901	1307.5	1449.3	1.7138	1.9639	1307.4	1449.1	1.7123
880	2.0507	1315.9	1460.1	1.7234	2.0234	1315.8	1459.9	1.7218	1.9968	1315.7	1459.8	1.7203
900	2.084	1324.2	1470.8	1.7313	2.056	1324.1	1470.6	1.7297	2.030	1324.0	1470.4	1.7282
920	2.118	1332.5	1481.4	1.7391	2.089	1332.4	1481.3	1.7375	2.062	1332.3	1481.1	1.7360
940	2.151	1340.9	1492.1	1.7467	2.122	1340.7	1491.9	1.7452	2.095	1340.6	1491.8	1.7437
960	2.184	1349.2	1502.8	1.7543	2.155	1349.1	1502.6	1.7528	2.127	1349.0	1502.5	1.7513
980	2.217	1357.6	1513.5	1.7618	2.188	1357.5	1513.3	1.7603	2.159	1357.3	1513.2	1.7588
1000	2.250	1365.9	1524.2	1.7692	2.220	1365.8	1524.0	1.7676	2.191	1365.7	1523.9	1.7661
1020	2.283	1374.3	1534.9	1.7764	2.253	1374.2	1534.7	1.7749	2.224	1374.1	1534.6	1.7734
1040	2.316	1382.8	1545.6	1.7837	2.285	1382.7	1545.5	1.7821	2.256	1382.6	1545.4	1.7807
1060	2.349	1391.2	1556.4	1.7908	2.318	1391.1	1556.2	1.7893	2.288	1391.0	1556.1	1.7878
1080	2.381	1399.7	1567.1	1.7978	2.350	1399.6	1567.0	1.7963	2.319	1399.5	1566.9	1.7948
1100	2.414	1408.2	1577.9	1.8048	2.382	1408.1	1577.8	1.8033	2.351	1408.0	1577.7	1.8018
1150	2.495	1429.5	1605.0	1.8219	2.462	1429.4	1604.9	1.8204	2.431	1429.4	1604.8	1.8189
1200	2.576	1451.0	1632.2	1.8385	2.542	1451.0	1632.1	1.8370	2.510	1450.9	1632.0	1.8355
1250	2.657	1472.7	1659.6	1.8547	2.622	1472.7	1659.5	1.8533	2.588	1472.6	1659.4	1.8518
1300	2.737	1494.6	1687.1	1.8706	2.702	1494.5	1687.0	1.8691	2.667	1494.5	1686.9	1.8677
1350	2.818	1516.7	1714.8	1.8861	2.781	1516.6	1714.7	1.8847	2.745	1516.6	1714.7	1.8832
1400	2.898	1538.9	1742.7	1.9013	2.860	1538.9	1742.6	1.8999	2.823	1538.8	1742.6	1.8984
1450	2.978	1561.4	1770.8	1.9162	2.939	1561.3	1770.7	1.9148	2.901	1561.3	1770.6	1.9133
1500	3.058	1584.0	1799.0	1.9309	3.018	1584.0	1799.0	1.9294	2.979	1583.9	1798.9	1.9279
1600	3.217	1630.0	1856.2	1.9593	3.175	1629.9	1856.1	1.9578	3.134	1629.9	1856.0	1.9564
1800	3.534	1724.2	1972.8	2.0133	3.488	1724.2	1972.7	2.0119	3.444	1724.2	1972.7	2.0104
2000	3.851	1821.8	2092.6	2.0641	3.801	1821.7	2092.5	2.0626	3.752	1821.7	2092.5	2.0612
2200	4.168	1922.3	2215.4	2.1121	4.114	1922.3	2215.3	2.1106	4.061	1922.2	2215.3	2.1092
2400	4.484	2025.6	2340.9	2.1576	4.426	2025.5	2340.9	2.1561	4.369	2025.5	2340.8	2.1547

v	u	h	s	v	u	h	s	v	u	h	s	t
1.1766	1119.4	1205.4	1.4867	1.1620	1119.5	1205.5	1.4856	1.1476	1119.5	1205.5	1.4844	**Sat.**
												420
1.1432	*1111.1*	*1194.6*	*1.4747*	*1.1257*	*1110.3*	*1193.6*	*1.4723*	*1.1086*	*1109.4*	*1192.5*	*1.4700*	**430**
1.1682	*1117.3*	*1202.7*	*1.4837*	*1.1506*	*1116.6*	*1201.7*	*1.4814*	*1.1334*	*1115.8*	*1200.8*	*1.4792*	**440**
1.1922	1123.3	1210.5	1.4923	1.1745	1122.6	1209.6	1.4901	1.1573	1121.9	1208.7	1.4879	**450**
1.2155	1129.1	1218.0	1.5005	1.1977	1128.5	1217.1	1.4984	1.1803	1127.8	1216.3	1.4963	**460**
1.2381	1134.7	1225.2	1.5083	1.2202	1134.1	1224.4	1.5063	1.2027	1133.5	1223.7	1.5042	**470**
1.2602	1140.2	1232.3	1.5159	1.2421	1139.6	1231.5	1.5139	1.2244	1139.0	1230.8	1.5119	**480**
1.2817	1145.4	1239.1	1.5231	1.2634	1144.9	1238.4	1.5212	1.2457	1144.4	1237.8	1.5192	**490**
1.3027	1150.6	1245.8	1.5302	1.2843	1150.1	1245.2	1.5282	1.2664	1149.6	1244.5	1.5263	**500**
1.3234	1155.7	1252.4	1.5370	1.3048	1155.2	1251.8	1.5351	1.2867	1154.7	1251.2	1.5332	**510**
1.3436	1160.6	1258.8	1.5435	1.3249	1160.2	1258.2	1.5417	1.3067	1159.7	1257.6	1.5399	**520**
1.3635	1165.5	1265.1	1.5500	1.3447	1165.0	1264.6	1.5481	1.3263	1164.6	1264.0	1.5463	**530**
1.3832	1170.2	1271.3	1.5562	1.3642	1169.8	1270.8	1.5544	1.3456	1169.4	1270.3	1.5526	**540**
1.4025	1174.9	1277.5	1.5623	1.3833	1174.6	1277.0	1.5605	1.3646	1174.2	1276.5	1.5588	**550**
1.4217	1179.6	1283.5	1.5682	1.4023	1179.2	1283.0	1.5665	1.3834	1178.9	1282.5	1.5648	**560**
1.4406	1184.2	1289.5	1.5741	1.4210	1183.8	1289.0	1.5723	1.4020	1183.5	1288.5	1.5706	**570**
1.4593	1188.7	1295.4	1.5798	1.4395	1188.4	1294.9	1.5781	1.4203	1188.0	1294.5	1.5764	**580**
1.4778	1193.2	1301.2	1.5854	1.4579	1192.9	1300.8	1.5837	1.4384	1192.6	1300.4	1.5820	**590**
1.4961	1197.6	1307.0	1.5909	1.4760	1197.3	1306.6	1.5892	1.4564	1197.0	1306.2	1.5875	**600**
1.5323	1206.4	1318.4	1.6015	1.5118	1206.1	1318.0	1.5999	1.4919	1205.9	1317.7	1.5983	**620**
1.5679	1215.1	1329.7	1.6119	1.5471	1214.8	1329.3	1.6103	1.5268	1214.6	1329.0	1.6086	**640**
1.6031	1223.7	1340.8	1.6219	1.5819	1223.4	1340.5	1.6203	1.5613	1223.2	1340.2	1.6187	**660**
1.6378	1232.2	1351.9	1.6317	1.6163	1231.9	1351.6	1.6301	1.5953	1231.7	1351.3	1.6285	**680**
1.6722	1240.6	1362.8	1.6412	1.6503	1240.4	1362.5	1.6397	1.6290	1240.2	1362.3	1.6381	**700**
1.7063	1249.0	1373.7	1.6505	1.6840	1248.8	1373.4	1.6490	1.6623	1248.6	1373.2	1.6474	**720**
1.7401	1257.4	1384.6	1.6596	1.7175	1257.2	1384.3	1.6581	1.6954	1257.0	1384.0	1.6566	**740**
1.7737	1265.7	1395.3	1.6686	1.7506	1265.5	1395.1	1.6670	1.7282	1265.3	1394.8	1.6655	**760**
1.8070	1274.0	1406.1	1.6773	1.7836	1273.8	1405.9	1.6758	1.7608	1273.7	1405.6	1.6743	**780**
1.8401	1282.3	1416.8	1.6859	1.8163	1282.1	1416.6	1.6844	1.7932	1282.0	1416.4	1.6829	**800**
1.8730	1290.6	1427.5	1.6943	1.8489	1290.5	1427.3	1.6928	1.8253	1290.3	1427.1	1.6913	**820**
1.9058	1298.9	1438.2	1.7026	1.8813	1298.8	1438.0	1.7011	1.8574	1298.6	1437.8	1.6996	**840**
1.9384	1307.2	1448.9	1.7108	1.9135	1307.1	1448.7	1.7093	1.8892	1306.9	1448.5	1.7078	**860**
1.9709	1315.5	1459.6	1.7188	1.9456	1315.4	1459.4	1.7173	1.9210	1315.2	1459.2	1.7158	**880**
2.003	1323.8	1470.3	1.7267	1.9776	1323.7	1470.1	1.7252	1.9525	1323.6	1469.9	1.7238	**900**
2.035	1332.2	1480.9	1.7345	2.0094	1332.0	1480.8	1.7330	1.9840	1331.9	1480.6	1.7316	**920**
2.068	1340.5	1491.6	1.7422	2.0411	1340.4	1491.5	1.7407	2.0154	1340.3	1491.3	1.7393	**940**
2.100	1348.9	1502.3	1.7498	2.0727	1348.7	1502.2	1.7483	2.0466	1348.6	1502.0	1.7469	**960**
2.131	1357.2	1513.0	1.7573	2.1043	1357.1	1512.9	1.7558	2.0778	1357.0	1512.7	1.7544	**980**
2.163	1365.6	1523.7	1.7647	2.136	1365.5	1523.6	1.7632	2.109	1365.4	1523.5	1.7618	**1000**
2.195	1374.0	1534.5	1.7720	2.167	1373.9	1534.3	1.7705	2.140	1373.8	1534.2	1.7691	**1020**
2.227	1382.5	1545.2	1.7792	2.198	1382.4	1545.1	1.7777	2.171	1382.3	1545.0	1.7763	**1040**
2.258	1390.9	1556.0	1.7863	2.229	1390.8	1555.9	1.7849	2.202	1390.7	1555.7	1.7834	**1060**
2.290	1399.4	1566.8	1.7934	2.261	1399.3	1566.6	1.7919	2.232	1399.2	1566.5	1.7905	**1080**
2.321	1407.9	1577.6	1.8003	2.292	1407.8	1577.4	1.7989	2.263	1407.7	1577.3	1.7975	**1100**
2.399	1429.3	1604.7	1.8174	2.369	1429.2	1604.6	1.8160	2.339	1429.1	1604.4	1.8146	**1150**
2.477	1450.8	1631.9	1.8341	2.446	1450.7	1631.8	1.8327	2.416	1450.7	1631.7	1.8312	**1200**
2.555	1472.5	1659.3	1.8504	2.523	1472.5	1659.2	1.8489	2.492	1472.4	1659.1	1.8475	**1250**
2.633	1494.4	1686.8	1.8662	2.599	1494.3	1686.8	1.8648	2.567	1494.3	1686.7	1.8634	**1300**
2.710	1516.5	1714.6	1.8818	2.676	1516.4	1714.5	1.8804	2.643	1516.4	1714.4	1.8790	**1350**
2.787	1538.8	1742.5	1.8970	2.752	1538.7	1742.4	1.8956	2.718	1538.6	1742.3	1.8942	**1400**
2.864	1561.2	1770.6	1.9119	2.828	1561.2	1770.5	1.9105	2.793	1561.1	1770.4	1.9091	**1450**
2.941	1583.9	1798.8	1.9265	2.904	1583.8	1798.8	1.9251	2.868	1583.8	1798.7	1.9237	**1500**
3.094	1629.8	1856.0	1.9549	3.055	1629.8	1855.9	1.9535	3.017	1629.7	1855.9	1.9522	**1600**
3.400	1724.1	1972.6	2.0090	3.357	1724.1	1972.6	2.0076	3.316	1724.1	1972.6	2.0062	**1800**
3.705	1821.7	2092.5	2.0598	3.658	1821.6	2092.4	2.0584	3.613	1821.6	2092.4	2.0570	**2000**
4.009	1922.2	2215.3	2.1078	3.959	1922.2	2215.2	2.1064	3.910	1922.1	2215.2	2.1050	**2200**
4.314	2025.5	2340.8	2.1533	4.260	2025.4	2340.8	2.1519	4.207	2025.4	2340.7	2.1505	**2400**

Table 3. Vapor

p (t Sat.)	410 (447.12)				415 (448.32)				420 (449.50)			
t	v	u	h	s	v	u	h	s	v	u	h	s
Sat.	1.1337	1119.5	1205.5	1.4833	1.1200	1119.5	1205.5	1.4822	1.1067	1119.5	1205.5	1.4810
430	*1.0919*	*1108.6*	*1191.5*	*1.4676*	*1.0756*	*1107.8*	*1190.4*	*1.4653*				
440	*1.1166*	*1115.1*	*1199.8*	*1.4770*	*1.1002*	*1114.3*	*1198.8*	*1.4747*	*1.0842*	*1113.5*	*1197.8*	*1.4725*
450	1.1404	1121.3	1207.8	1.4858	1.1239	1120.6	1206.9	1.4836	1.1078	1119.8	1205.9	1.4815
460	1.1633	1127.2	1215.5	1.4942	1.1468	1126.5	1214.6	1.4921	1.1305	1125.9	1213.8	1.4900
470	1.1856	1132.9	1222.9	1.5022	1.1689	1132.3	1222.1	1.5002	1.1526	1131.7	1221.3	1.4982
480	1.2072	1138.5	1230.1	1.5099	1.1904	1137.9	1229.3	1.5079	1.1740	1137.3	1228.6	1.5060
490	1.2283	1143.9	1237.1	1.5173	1.2113	1143.3	1236.4	1.5154	1.1948	1142.8	1235.7	1.5135
500	1.2489	1149.1	1243.9	1.5244	1.2318	1148.6	1243.2	1.5226	1.2151	1148.1	1242.6	1.5207
510	1.2691	1154.3	1250.5	1.5313	1.2518	1153.8	1249.9	1.5295	1.2350	1153.3	1249.3	1.5277
520	1.2889	1159.3	1257.1	1.5380	1.2715	1158.8	1256.5	1.5362	1.2545	1158.4	1255.9	1.5344
530	1.3083	1164.2	1263.5	1.5445	1.2908	1163.8	1262.9	1.5428	1.2737	1163.4	1262.3	1.5410
540	1.3275	1169.0	1269.8	1.5509	1.3098	1168.6	1269.2	1.5491	1.2925	1168.2	1268.7	1.5474
550	1.3463	1173.8	1276.0	1.5570	1.3285	1173.4	1275.4	1.5553	1.3111	1173.0	1274.9	1.5536
560	1.3650	1178.5	1282.1	1.5630	1.3470	1178.1	1281.6	1.5613	1.3294	1177.8	1281.1	1.5597
570	1.3833	1183.1	1288.1	1.5689	1.3652	1182.8	1287.6	1.5672	1.3474	1182.4	1287.2	1.5656
580	1.4015	1187.7	1294.0	1.5747	1.3832	1187.4	1293.6	1.5730	1.3653	1187.0	1293.2	1.5714
590	1.4195	1192.2	1299.9	1.5803	1.4010	1191.9	1299.5	1.5787	1.3830	1191.6	1299.1	1.5770
600	1.4373	1196.7	1305.8	1.5859	1.4187	1196.4	1305.4	1.5842	1.4004	1196.1	1305.0	1.5826
620	1.4724	1205.6	1317.3	1.5966	1.4535	1205.3	1316.9	1.5950	1.4349	1205.0	1316.5	1.5934
640	1.5070	1214.3	1328.6	1.6070	1.4877	1214.0	1328.3	1.6055	1.4688	1213.8	1327.9	1.6039
660	1.5411	1222.9	1339.9	1.6172	1.5215	1222.7	1339.5	1.6156	1.5023	1222.4	1339.2	1.6140
680	1.5748	1231.5	1351.0	1.6270	1.5548	1231.2	1350.6	1.6254	1.5353	1231.0	1350.3	1.6239
700	1.6081	1240.0	1362.0	1.6366	1.5878	1239.7	1361.7	1.6350	1.5679	1239.5	1361.4	1.6335
720	1.6411	1248.4	1372.9	1.6459	1.6204	1248.2	1372.6	1.6444	1.6002	1248.0	1372.4	1.6429
740	1.6738	1256.8	1383.8	1.6550	1.6528	1256.6	1383.5	1.6535	1.6323	1256.4	1383.3	1.6521
760	1.7063	1265.1	1394.6	1.6640	1.6849	1265.0	1394.4	1.6625	1.6640	1264.8	1394.1	1.6610
780	1.7385	1273.5	1405.4	1.6728	1.7168	1273.3	1405.2	1.6713	1.6956	1273.1	1404.9	1.6698
800	1.7705	1281.8	1416.1	1.6814	1.7485	1281.7	1415.9	1.6799	1.7269	1281.5	1415.7	1.6785
820	1.8024	1290.1	1426.9	1.6898	1.7799	1290.0	1426.7	1.6884	1.7581	1289.8	1426.5	1.6869
840	1.8340	1298.5	1437.6	1.6981	1.8113	1298.3	1437.4	1.6967	1.7890	1298.2	1437.2	1.6953
860	1.8655	1306.8	1448.3	1.7063	1.8424	1306.6	1448.1	1.7049	1.8199	1306.5	1447.9	1.7035
880	1.8969	1315.1	1459.0	1.7144	1.8734	1315.0	1458.8	1.7129	1.8505	1314.8	1458.7	1.7115
900	1.9281	1323.4	1469.7	1.7223	1.9043	1323.3	1469.5	1.7209	1.8811	1323.2	1469.4	1.7195
920	1.9593	1331.8	1480.4	1.7301	1.9351	1331.7	1480.3	1.7287	1.9115	1331.5	1480.1	1.7273
940	1.9903	1340.1	1491.1	1.7378	1.9657	1340.0	1491.0	1.7364	1.9418	1339.9	1490.8	1.7350
960	2.0211	1348.5	1501.9	1.7454	1.9963	1348.4	1501.7	1.7440	1.9720	1348.3	1501.5	1.7426
980	2.0519	1356.9	1512.6	1.7529	2.0267	1356.8	1512.4	1.7515	2.0021	1356.7	1512.3	1.7501
1000	2.083	1365.3	1523.3	1.7603	2.057	1365.2	1523.2	1.7589	2.032	1365.1	1523.0	1.7575
1020	2.113	1373.7	1534.1	1.7677	2.087	1373.6	1533.9	1.7662	2.062	1373.5	1533.8	1.7649
1040	2.144	1382.2	1544.8	1.7749	2.118	1382.1	1544.7	1.7735	2.092	1382.0	1544.6	1.7721
1060	2.174	1390.6	1555.6	1.7820	2.148	1390.5	1555.5	1.7806	2.122	1390.4	1555.3	1.7792
1080	2.205	1399.1	1566.4	1.7891	2.178	1399.0	1566.3	1.7877	2.151	1398.9	1566.1	1.7863
1100	2.235	1407.6	1577.2	1.7961	2.208	1407.5	1577.1	1.7947	2.181	1407.5	1577.0	1.7933
1150	2.311	1429.0	1604.3	1.8132	2.282	1428.9	1604.2	1.8118	2.255	1428.9	1604.1	1.8104
1200	2.386	1450.6	1631.6	1.8299	2.357	1450.5	1631.5	1.8285	2.328	1450.4	1631.4	1.8271
1250	2.461	1472.3	1659.0	1.8461	2.431	1472.2	1658.9	1.8447	2.402	1472.2	1658.8	1.8434
1300	2.536	1494.2	1686.6	1.8620	2.505	1494.2	1686.5	1.8607	2.475	1494.1	1686.4	1.8593
1350	2.610	1516.3	1714.3	1.8776	2.578	1516.2	1714.3	1.8762	2.548	1516.2	1714.2	1.8748
1400	2.684	1538.6	1742.3	1.8928	2.652	1538.5	1742.2	1.8914	2.620	1538.5	1742.1	1.8901
1450	2.759	1561.1	1770.4	1.9077	2.725	1561.0	1770.3	1.9063	2.693	1561.0	1770.2	1.9050
1500	2.833	1583.7	1798.7	1.9223	2.798	1583.7	1798.6	1.9210	2.765	1583.6	1798.5	1.9196
1600	2.981	1629.7	1855.8	1.9508	2.944	1629.6	1855.8	1.9494	2.909	1629.6	1855.7	1.9481
1800	3.275	1724.0	1972.5	2.0048	3.236	1724.0	1972.5	2.0035	3.197	1723.9	1972.4	2.0021
2000	3.569	1821.5	2092.3	2.0556	3.526	1821.5	2092.3	2.0543	3.484	1821.5	2092.3	2.0530
2200	3.863	1922.1	2215.2	2.1036	3.816	1922.1	2215.1	2.1023	3.771	1922.0	2215.1	2.1010
2400	4.156	2025.4	2340.7	2.1491	4.106	2025.3	2340.7	2.1478	4.057	2025.3	2340.6	2.1465

v	u	h	s	v	u	h	s	v	u	h	s	t
1.0936	1119.6	1205.6	1.4799	1.0809	1119.6	1205.6	1.4788	1.0684	1119.6	1205.6	1.4778	Sat.
												430
1.0685	*1112.8*	*1196.8*	*1.4703*	*1.0532*	*1112.0*	*1195.8*	*1.4680*	*1.0382*	*1111.2*	*1194.8*	*1.4658*	440
1.0921	*1119.1*	*1205.0*	*1.4793*	*1.0766*	*1118.4*	*1204.1*	*1.4772*	*1.0616*	*1117.7*	*1203.1*	*1.4751*	450
1.1147	1125.2	1212.9	1.4879	1.0992	1124.6	1212.0	1.4859	1.0841	1123.9	1211.2	1.4838	460
1.1366	1131.1	1220.5	1.4962	1.1210	1130.5	1219.7	1.4942	1.1058	1129.9	1218.9	1.4922	470
1.1579	1136.8	1227.8	1.5040	1.1422	1136.2	1227.1	1.5021	1.1268	1135.6	1226.3	1.5002	480
1.1786	1142.3	1235.0	1.5116	1.1628	1141.7	1234.3	1.5097	1.1473	1141.2	1233.5	1.5078	490
1.1988	1147.6	1241.9	1.5188	1.1829	1147.1	1241.2	1.5170	1.1673	1146.6	1240.6	1.5152	500
1.2186	1152.8	1248.7	1.5259	1.2025	1152.4	1248.0	1.5240	1.1868	1151.9	1247.4	1.5223	510
1.2379	1157.9	1255.3	1.5326	1.2217	1157.5	1254.7	1.5309	1.2059	1157.0	1254.1	1.5291	520
1.2570	1162.9	1261.8	1.5392	1.2406	1162.5	1261.2	1.5375	1.2246	1162.1	1260.7	1.5358	530
1.2756	1167.8	1268.2	1.5457	1.2592	1167.4	1267.6	1.5439	1.2431	1167.0	1267.1	1.5422	540
1.2941	1172.7	1274.4	1.5519	1.2774	1172.3	1273.9	1.5502	1.2612	1171.9	1273.4	1.5485	550
1.3122	1177.4	1280.6	1.5580	1.2954	1177.0	1280.1	1.5563	1.2790	1176.7	1279.6	1.5547	560
1.3301	1182.1	1286.7	1.5639	1.3132	1181.7	1286.2	1.5623	1.2967	1181.4	1285.8	1.5607	570
1.3478	1186.7	1292.7	1.5697	1.3308	1186.4	1292.3	1.5681	1.3141	1186.0	1291.8	1.5665	580
1.3653	1191.3	1298.7	1.5754	1.3481	1191.0	1298.2	1.5738	1.3313	1190.6	1297.8	1.5722	590
1.3827	1195.8	1304.5	1.5810	1.3653	1195.5	1304.1	1.5794	1.3483	1195.2	1303.7	1.5778	600
1.4168	1204.7	1316.2	1.5919	1.3991	1204.4	1315.8	1.5903	1.3818	1204.2	1315.4	1.5888	620
1.4504	1213.5	1327.6	1.6024	1.4324	1213.3	1327.2	1.6008	1.4148	1213.0	1326.9	1.5993	640
1.4835	1222.2	1338.9	1.6125	1.4652	1221.9	1338.5	1.6110	1.4473	1221.7	1338.2	1.6095	660
1.5162	1230.8	1350.0	1.6224	1.4976	1230.6	1349.7	1.6209	1.4794	1230.3	1349.4	1.6194	680
1.5485	1239.3	1361.1	1.6320	1.5296	1239.1	1360.8	1.6305	1.5111	1238.9	1360.5	1.6291	700
1.5805	1247.8	1372.1	1.6414	1.5612	1247.6	1371.8	1.6399	1.5424	1247.4	1371.5	1.6385	720
1.6122	1256.2	1383.0	1.6506	1.5926	1256.0	1382.7	1.6491	1.5735	1255.8	1382.5	1.6477	740
1.6436	1264.6	1393.9	1.6596	1.6237	1264.4	1393.6	1.6581	1.6043	1264.2	1393.4	1.6567	760
1.6749	1273.0	1404.7	1.6684	1.6546	1272.8	1404.5	1.6670	1.6349	1272.6	1404.2	1.6655	780
1.7059	1281.3	1415.5	1.6770	1.6853	1281.2	1415.3	1.6756	1.6652	1281.0	1415.0	1.6742	800
1.7367	1289.7	1426.3	1.6855	1.7158	1289.5	1426.0	1.6841	1.6954	1289.4	1425.8	1.6827	820
1.7673	1298.0	1437.0	1.6938	1.7461	1297.9	1436.8	1.6924	1.7254	1297.7	1436.6	1.6910	840
1.7978	1306.3	1447.7	1.7020	1.7763	1306.2	1447.5	1.7006	1.7553	1306.1	1447.4	1.6993	860
1.8282	1314.7	1458.5	1.7101	1.8063	1314.6	1458.3	1.7087	1.7850	1314.4	1458.1	1.7073	880
1.8584	1323.0	1469.2	1.7181	1.8362	1322.9	1469.0	1.7167	1.8145	1322.8	1468.8	1.7153	900
1.8885	1331.4	1479.9	1.7259	1.8660	1331.3	1479.8	1.7245	1.8440	1331.1	1479.6	1.7231	920
1.9184	1339.8	1490.7	1.7336	1.8956	1339.7	1490.5	1.7322	1.8733	1339.5	1490.3	1.7309	940
1.9483	1348.2	1501.4	1.7412	1.9252	1348.0	1501.2	1.7399	1.9026	1347.9	1501.1	1.7385	960
1.9781	1356.6	1512.1	1.7487	1.9546	1356.4	1512.0	1.7474	1.9317	1356.3	1511.8	1.7460	980
2.008	1365.0	1522.9	1.7562	1.9840	1364.9	1522.7	1.7548	1.9607	1364.8	1522.6	1.7534	1000
2.037	1373.4	1533.6	1.7635	2.0133	1373.3	1533.5	1.7621	1.9897	1373.2	1533.4	1.7608	1020
2.067	1381.9	1544.4	1.7707	2.0425	1381.8	1544.3	1.7694	2.0186	1381.7	1544.2	1.7680	1040
2.096	1390.3	1555.2	1.7779	2.0716	1390.2	1555.1	1.7765	2.0474	1390.2	1555.0	1.7752	1060
2.126	1398.8	1566.0	1.7849	2.1006	1398.8	1565.9	1.7836	2.0761	1398.7	1565.8	1.7822	1080
2.155	1407.4	1576.9	1.7919	2.130	1407.3	1576.7	1.7906	2.105	1407.2	1576.6	1.7892	1100
2.228	1428.8	1604.0	1.8091	2.202	1428.7	1603.9	1.8077	2.176	1428.6	1603.8	1.8064	1150
2.301	1450.4	1631.3	1.8258	2.274	1450.3	1631.2	1.8244	2.247	1450.2	1631.1	1.8231	1200
2.373	1472.1	1658.7	1.8420	2.345	1472.0	1658.7	1.8407	2.318	1472.0	1658.6	1.8394	1250
2.445	1494.0	1686.3	1.8580	2.417	1494.0	1686.3	1.8566	2.389	1493.9	1686.2	1.8553	1300
2.517	1516.1	1714.1	1.8735	2.488	1516.1	1714.0	1.8722	2.459	1516.0	1714.0	1.8709	1350
2.589	1538.4	1742.0	1.8887	2.559	1538.4	1742.0	1.8874	2.529	1538.3	1741.9	1.8861	1400
2.661	1560.9	1770.2	1.9037	2.630	1560.9	1770.1	1.9023	2.599	1560.8	1770.0	1.9010	1450
2.732	1583.6	1798.5	1.9183	2.700	1583.5	1798.4	1.9170	2.669	1583.5	1798.3	1.9157	1500
2.875	1629.6	1855.7	1.9468	2.841	1629.5	1855.6	1.9454	2.809	1629.5	1855.5	1.9441	1600
3.159	1723.9	1972.4	2.0008	3.123	1723.9	1972.3	1.9995	3.087	1723.8	1972.3	1.9982	1800
3.443	1821.4	2092.2	2.0516	3.403	1821.4	2092.2	2.0503	3.364	1821.4	2092.1	2.0490	2000
3.726	1922.0	2215.1	2.0996	3.683	1922.0	2215.0	2.0983	3.641	1921.9	2215.0	2.0971	2200
4.009	2025.3	2340.6	2.1451	3.963	2025.2	2340.5	2.1438	3.917	2025.2	2340.5	2.1426	2400

Table 3. Vapor

p (t Sat.)	440 (454.14)				445 (455.27)				450 (456.39)			
t	v	u	h	s	v	u	h	s	v	u	h	s
Sat.	1.0562	1119.6	1205.6	1.4767	1.0443	1119.6	1205.6	1.4756	1.0326	1119.6	1205.6	1.4746
450	*1.0468*	*1117.0*	*1202.2*	*1.4730*	*1.0324*	*1116.2*	*1201.2*	*1.4709*	*1.0183*	*1115.5*	*1200.3*	*1.4687*
460	1.0692	1123.2	1210.3	1.4818	1.0547	1122.5	1209.4	1.4798	1.0405	1121.8	1208.5	1.4777
470	1.0909	1129.2	1218.0	1.4902	1.0763	1128.6	1217.2	1.4882	1.0620	1128.0	1216.4	1.4863
480	1.1118	1135.0	1225.6	1.4982	1.0972	1134.4	1224.8	1.4963	1.0828	1133.8	1224.0	1.4944
490	1.1322	1140.6	1232.8	1.5059	1.1174	1140.1	1232.1	1.5041	1.1029	1139.5	1231.4	1.5022
500	1.1520	1146.1	1239.9	1.5133	1.1371	1145.6	1239.2	1.5115	1.1226	1145.1	1238.5	1.5097
510	1.1714	1151.4	1246.8	1.5205	1.1564	1150.9	1246.1	1.5187	1.1417	1150.4	1245.5	1.5170
520	1.1904	1156.6	1253.5	1.5274	1.1753	1156.1	1252.9	1.5256	1.1605	1155.7	1252.3	1.5239
530	1.2090	1161.6	1260.1	1.5341	1.1938	1161.2	1259.5	1.5324	1.1788	1160.8	1258.9	1.5307
540	1.2273	1166.6	1266.5	1.5406	1.2119	1166.2	1266.0	1.5389	1.1969	1165.8	1265.5	1.5372
550	1.2453	1171.5	1272.9	1.5469	1.2298	1171.1	1272.4	1.5452	1.2146	1170.7	1271.9	1.5436
560	1.2630	1176.3	1279.1	1.5530	1.2474	1175.9	1278.6	1.5514	1.2320	1175.6	1278.2	1.5498
570	1.2805	1181.0	1285.3	1.5590	1.2647	1180.7	1284.8	1.5574	1.2492	1180.3	1284.4	1.5559
580	1.2978	1185.7	1291.4	1.5649	1.2818	1185.4	1290.9	1.5633	1.2662	1185.0	1290.5	1.5618
590	1.3148	1190.3	1297.4	1.5707	1.2987	1190.0	1296.9	1.5691	1.2830	1189.7	1296.5	1.5675
600	1.3317	1194.9	1303.3	1.5763	1.3155	1194.6	1302.9	1.5747	1.2996	1194.3	1302.5	1.5732
610	1.3484	1199.4	1309.2	1.5818	1.3320	1199.1	1308.8	1.5803	1.3160	1198.8	1308.4	1.5788
620	1.3650	1203.9	1315.0	1.5872	1.3485	1203.6	1314.6	1.5857	1.3323	1203.3	1314.2	1.5842
630	1.3814	1208.3	1320.8	1.5926	1.3647	1208.0	1320.4	1.5910	1.3484	1207.8	1320.0	1.5896
640	1.3976	1212.7	1326.5	1.5978	1.3808	1212.5	1326.2	1.5963	1.3644	1212.2	1325.8	1.5948
650	1.4138	1217.1	1332.2	1.6029	1.3968	1216.8	1331.9	1.6015	1.3803	1216.6	1331.5	1.6000
660	1.4298	1221.5	1337.9	1.6080	1.4127	1221.2	1337.5	1.6066	1.3960	1221.0	1337.2	1.6051
670	1.4458	1225.8	1343.5	1.6130	1.4285	1225.5	1343.2	1.6116	1.4116	1225.3	1342.9	1.6101
680	1.4616	1230.1	1349.1	1.6180	1.4442	1229.9	1348.8	1.6165	1.4272	1229.6	1348.5	1.6151
690	1.4773	1234.4	1354.7	1.6228	1.4598	1234.2	1354.4	1.6214	1.4426	1233.9	1354.1	1.6199
700	1.4930	1238.7	1360.2	1.6276	1.4753	1238.4	1359.9	1.6262	1.4580	1238.2	1359.6	1.6248
720	1.5240	1247.2	1371.2	1.6371	1.5060	1247.0	1371.0	1.6356	1.4884	1246.7	1370.7	1.6342
740	1.5548	1255.6	1382.2	1.6463	1.5365	1255.4	1382.0	1.6449	1.5186	1255.2	1381.7	1.6435
760	1.5853	1264.0	1393.1	1.6553	1.5667	1263.9	1392.9	1.6539	1.5485	1263.7	1392.6	1.6525
780	1.6155	1272.4	1404.0	1.6641	1.5967	1272.3	1403.7	1.6627	1.5782	1272.1	1403.5	1.6614
800	1.6456	1280.8	1414.8	1.6728	1.6264	1280.7	1414.6	1.6714	1.6077	1280.5	1414.4	1.6701
820	1.6755	1289.2	1425.6	1.6813	1.6560	1289.0	1425.4	1.6799	1.6369	1288.9	1425.2	1.6786
840	1.7052	1297.6	1436.4	1.6897	1.6854	1297.4	1436.2	1.6883	1.6660	1297.3	1436.0	1.6870
860	1.7347	1305.9	1447.2	1.6979	1.7146	1305.8	1447.0	1.6965	1.6950	1305.6	1446.8	1.6952
880	1.7641	1314.3	1457.9	1.7060	1.7437	1314.1	1457.7	1.7046	1.7238	1314.0	1457.5	1.7033
900	1.7934	1322.6	1468.7	1.7139	1.7727	1322.5	1468.5	1.7126	1.7524	1322.4	1468.3	1.7113
920	1.8225	1331.0	1479.4	1.7218	1.8015	1330.9	1479.2	1.7204	1.7810	1330.8	1479.1	1.7191
940	1.8515	1339.4	1490.2	1.7295	1.8302	1339.3	1490.0	1.7282	1.8094	1339.2	1489.8	1.7269
960	1.8805	1347.8	1500.9	1.7372	1.8589	1347.7	1500.8	1.7358	1.8377	1347.6	1500.6	1.7345
980	1.9093	1356.2	1511.7	1.7447	1.8874	1356.1	1511.5	1.7434	1.8660	1356.0	1511.4	1.7420
1000	1.9380	1364.7	1522.4	1.7521	1.9158	1364.5	1522.3	1.7508	1.8941	1364.4	1522.2	1.7495
1020	1.9667	1373.1	1533.2	1.7594	1.9441	1373.0	1533.1	1.7581	1.9221	1372.9	1533.0	1.7568
1040	1.9952	1381.6	1544.0	1.7667	1.9724	1381.5	1543.9	1.7654	1.9501	1381.4	1543.8	1.7641
1060	2.0237	1390.1	1554.8	1.7738	2.0006	1390.0	1554.7	1.7725	1.9780	1389.9	1554.6	1.7712
1080	2.0521	1398.6	1565.7	1.7809	2.0287	1398.5	1565.5	1.7796	2.0058	1398.4	1565.4	1.7783
1100	2.081	1407.1	1576.5	1.7879	2.057	1407.0	1576.4	1.7866	2.034	1406.9	1576.3	1.7853
1150	2.151	1428.5	1603.7	1.8051	2.127	1428.5	1603.6	1.8038	2.103	1428.4	1603.5	1.8025
1200	2.221	1450.1	1631.0	1.8218	2.196	1450.1	1630.9	1.8205	2.172	1450.0	1630.8	1.8192
1250	2.291	1471.9	1658.5	1.8381	2.265	1471.8	1658.4	1.8368	2.240	1471.8	1658.3	1.8355
1300	2.361	1493.8	1686.1	1.8540	2.335	1493.8	1686.0	1.8527	2.308	1493.7	1685.9	1.8515
1350	2.431	1515.9	1713.9	1.8696	2.403	1515.9	1713.8	1.8683	2.376	1515.8	1713.7	1.8670
1400	2.500	1538.3	1741.8	1.8848	2.472	1538.2	1741.8	1.8835	2.444	1538.1	1741.7	1.8823
1450	2.570	1560.8	1770.0	1.8998	2.540	1560.7	1769.9	1.8985	2.512	1560.6	1769.8	1.8972
1500	2.639	1583.4	1798.3	1.9144	2.609	1583.4	1798.2	1.9131	2.580	1583.3	1798.2	1.9119
1600	2.777	1629.4	1855.5	1.9429	2.745	1629.4	1855.4	1.9416	2.715	1629.3	1855.4	1.9403
1800	3.051	1723.8	1972.2	1.9969	3.017	1723.7	1972.2	1.9957	2.983	1723.7	1972.1	1.9944
2000	3.325	1821.3	2092.1	2.0478	3.288	1821.3	2092.1	2.0465	3.251	1821.3	2092.0	2.0453
2200	3.599	1921.9	2215.0	2.0958	3.559	1921.9	2214.9	2.0945	3.519	1921.8	2214.9	2.0933
2400	3.873	2025.2	2340.5	2.1413	3.829	2025.1	2340.4	2.1400	3.787	2025.1	2340.4	2.1388

v	u	h	s	v	u	h	s	v	u	h	s	t
1.0212	1119.6	1205.6	1.4735	1.0100	1119.6	1205.5	1.4725	.9990	1119.6	1205.5	1.4714	**Sat.**
1.0044	*1114.7*	*1199.3*	*1.4666*	*.9909*	*1114.0*	*1198.3*	*1.4646*	*.9776*	*1113.2*	*1197.3*	*1.4625*	**450**
1.0266	1121.2	1207.6	1.4757	1.0130	1120.5	1206.7	1.4737	.9997	1119.7	1205.8	1.4717	**460**
1.0480	1127.3	1215.6	1.4843	1.0343	1126.7	1214.7	1.4824	1.0209	1126.0	1213.9	1.4805	**470**
1.0687	1133.2	1223.2	1.4925	1.0549	1132.6	1222.4	1.4907	1.0415	1132.0	1221.7	1.4888	**480**
1.0888	1139.0	1230.6	1.5004	1.0749	1138.4	1229.9	1.4986	1.0613	1137.9	1229.2	1.4968	**490**
1.1083	1144.5	1237.8	1.5079	1.0943	1144.0	1237.2	1.5062	1.0807	1143.5	1236.5	1.5044	**500**
1.1273	1149.9	1244.9	1.5152	1.1133	1149.4	1244.2	1.5135	1.0995	1148.9	1243.6	1.5117	**510**
1.1460	1155.2	1251.7	1.5222	1.1318	1154.7	1251.1	1.5205	1.1179	1154.3	1250.5	1.5188	**520**
1.1642	1160.3	1258.4	1.5290	1.1499	1159.9	1257.8	1.5273	1.1359	1159.5	1257.2	1.5257	**530**
1.1821	1165.4	1264.9	1.5356	1.1677	1165.0	1264.4	1.5339	1.1536	1164.5	1263.8	1.5323	**540**
1.1997	1170.3	1271.3	1.5420	1.1852	1169.9	1270.8	1.5404	1.1709	1169.5	1270.3	1.5388	**550**
1.2170	1175.2	1277.7	1.5482	1.2024	1174.8	1277.2	1.5466	1.1880	1174.4	1276.7	1.5450	**560**
1.2341	1180.0	1283.9	1.5543	1.2193	1179.6	1283.4	1.5527	1.2048	1179.3	1282.9	1.5512	**570**
1.2510	1184.7	1290.0	1.5602	1.2361	1184.3	1289.6	1.5587	1.2214	1184.0	1289.1	1.5571	**580**
1.2676	1189.3	1296.1	1.5660	1.2526	1189.0	1295.6	1.5645	1.2378	1188.7	1295.2	1.5630	**590**
1.2841	1193.9	1302.1	1.5717	1.2689	1193.6	1301.6	1.5702	1.2540	1193.3	1301.2	1.5687	**600**
1.3004	1198.5	1308.0	1.5773	1.2851	1198.2	1307.6	1.5758	1.2701	1197.9	1307.2	1.5743	**610**
1.3165	1203.0	1313.9	1.5827	1.3011	1202.7	1313.5	1.5812	1.2859	1202.4	1313.1	1.5798	**620**
1.3325	1207.5	1319.7	1.5881	1.3169	1207.2	1319.3	1.5866	1.3016	1206.9	1318.9	1.5852	**630**
1.3483	1211.9	1325.4	1.5934	1.3326	1211.7	1325.1	1.5919	1.3172	1211.4	1324.7	1.5905	**640**
1.3641	1216.3	1331.2	1.5985	1.3482	1216.1	1330.8	1.5971	1.3327	1215.8	1330.5	1.5957	**650**
1.3797	1220.7	1336.9	1.6036	1.3637	1220.5	1336.5	1.6022	1.3480	1220.2	1336.2	1.6008	**660**
1.3952	1225.1	1342.5	1.6087	1.3790	1224.8	1342.2	1.6073	1.3632	1224.6	1341.9	1.6058	**670**
1.4106	1229.4	1348.2	1.6136	1.3943	1229.2	1347.8	1.6122	1.3784	1228.9	1347.5	1.6108	**680**
1.4259	1233.7	1353.8	1.6185	1.4094	1233.5	1353.5	1.6171	1.3934	1233.3	1353.2	1.6157	**690**
1.4411	1238.0	1359.3	1.6234	1.4245	1237.8	1359.0	1.6220	1.4083	1237.6	1358.7	1.6206	**700**
1.4712	1246.5	1370.4	1.6328	1.4544	1246.3	1370.1	1.6315	1.4379	1246.1	1369.9	1.6301	**720**
1.5011	1255.0	1381.4	1.6421	1.4840	1254.8	1381.2	1.6407	1.4673	1254.6	1380.9	1.6394	**740**
1.5307	1263.5	1392.4	1.6511	1.5134	1263.3	1392.1	1.6498	1.4963	1263.1	1391.9	1.6484	**760**
1.5601	1271.9	1403.3	1.6600	1.5425	1271.7	1403.0	1.6587	1.5252	1271.6	1402.8	1.6573	**780**
1.5893	1280.3	1414.1	1.6687	1.5714	1280.2	1413.9	1.6674	1.5538	1280.0	1413.7	1.6660	**800**
1.6183	1288.7	1425.0	1.6772	1.6001	1288.6	1424.8	1.6759	1.5822	1288.4	1424.5	1.6746	**820**
1.6471	1297.1	1435.8	1.6856	1.6286	1296.9	1435.6	1.6843	1.6105	1296.8	1435.4	1.6830	**840**
1.6758	1305.5	1446.6	1.6939	1.6570	1305.3	1446.4	1.6925	1.6386	1305.2	1446.2	1.6912	**860**
1.7043	1313.9	1457.4	1.7020	1.6852	1313.7	1457.2	1.7007	1.6665	1313.6	1457.0	1.6994	**880**
1.7326	1322.2	1468.1	1.7099	1.7133	1322.1	1468.0	1.7086	1.6943	1322.0	1467.8	1.7074	**900**
1.7609	1330.6	1478.9	1.7178	1.7413	1330.5	1478.7	1.7165	1.7220	1330.4	1478.6	1.7152	**920**
1.7890	1339.0	1489.7	1.7256	1.7691	1338.9	1489.5	1.7243	1.7496	1338.8	1489.3	1.7230	**940**
1.8171	1347.5	1500.4	1.7332	1.7969	1347.3	1500.3	1.7319	1.7771	1347.2	1500.1	1.7306	**960**
1.8450	1355.9	1511.2	1.7407	1.8245	1355.8	1511.1	1.7395	1.8044	1355.7	1510.9	1.7382	**980**
1.8728	1364.3	1522.0	1.7482	1.8521	1364.2	1521.9	1.7469	1.8317	1364.1	1521.7	1.7456	**1000**
1.9006	1372.8	1532.8	1.7555	1.8795	1372.7	1532.7	1.7543	1.8589	1372.6	1532.5	1.7530	**1020**
1.9283	1381.3	1543.6	1.7628	1.9069	1381.2	1543.5	1.7615	1.8860	1381.1	1543.4	1.7603	**1040**
1.9559	1389.8	1554.4	1.7700	1.9342	1389.7	1554.3	1.7687	1.9131	1389.6	1554.2	1.7674	**1060**
1.9834	1398.3	1565.3	1.7770	1.9615	1398.2	1565.2	1.7758	1.9401	1398.1	1565.0	1.7745	**1080**
2.011	1406.8	1576.1	1.7841	1.9887	1406.7	1576.0	1.7828	1.9670	1406.6	1575.9	1.7815	**1100**
2.079	1428.3	1603.4	1.8012	2.0564	1428.2	1603.3	1.8000	2.0340	1428.1	1603.1	1.7987	**1150**
2.147	1449.9	1630.7	1.8180	2.1238	1449.8	1630.6	1.8167	2.1006	1449.8	1630.5	1.8155	**1200**
2.215	1471.7	1658.2	1.8343	2.1908	1471.6	1658.1	1.8330	2.1670	1471.5	1658.0	1.8318	**1250**
2.283	1493.6	1685.8	1.8502	2.258	1493.6	1685.8	1.8490	2.233	1493.5	1685.7	1.8477	**1300**
2.350	1515.8	1713.6	1.8658	2.324	1515.7	1713.6	1.8646	2.299	1515.6	1713.5	1.8633	**1350**
2.417	1538.1	1741.6	1.8810	2.391	1538.0	1741.5	1.8798	2.365	1538.0	1741.5	1.8786	**1400**
2.484	1560.6	1769.8	1.8960	2.457	1560.5	1769.7	1.8947	2.431	1560.5	1769.6	1.8935	**1450**
2.551	1583.3	1798.1	1.9106	2.523	1583.2	1798.0	1.9094	2.496	1583.2	1798.0	1.9082	**1500**
2.685	1629.3	1855.3	1.9391	2.655	1629.2	1855.3	1.9379	2.627	1629.2	1855.2	1.9367	**1600**
2.951	1723.7	1972.1	1.9932	2.918	1723.6	1972.1	1.9920	2.887	1723.6	1972.0	1.9908	**1800**
3.216	1821.2	2092.0	2.0440	3.181	1821.2	2091.9	2.0428	3.146	1821.2	2091.9	2.0416	**2000**
3.480	1921.8	2214.8	2.0920	3.443	1921.8	2214.8	2.0908	3.406	1921.7	2214.8	2.0896	**2200**
3.745	2025.1	2340.4	2.1375	3.704	2025.0	2340.3	2.1363	3.665	2025.0	2340.3	2.1351	**2400**

Table 3. Vapor

p (t Sat.)	470 (460.79)				475 (461.87)				480 (462.94)			
t	v	u	h	s	v	u	h	s	v	u	h	s
Sat.	.9883	1119.6	1205.5	1.4704	.9778	1119.5	1205.5	1.4694	.9675	1119.5	1205.5	1.4684
450	*.9645*	*1112.4*	*1196.3*	*1.4604*	*.9518*	*1111.6*	*1195.3*	*1.4583*	.9392	*1110.9*	*1194.3*	*1.4562*
460	*.9866*	*1119.0*	*1204.8*	*1.4697*	*.9738*	*1118.3*	*1203.9*	*1.4677*	*.9612*	*1117.6*	*1203.0*	*1.4657*
470	1.0078	1125.4	1213.0	1.4785	.9949	1124.7	1212.2	1.4766	.9823	1124.0	1211.3	1.4747
480	1.0282	1131.4	1220.9	1.4869	1.0153	1130.8	1220.1	1.4851	1.0026	1130.2	1219.3	1.4832
490	1.0480	1137.3	1228.4	1.4950	1.0350	1136.7	1227.7	1.4932	1.0222	1136.1	1226.9	1.4914
500	1.0673	1142.9	1235.8	1.5026	1.0541	1142.4	1235.1	1.5009	1.0413	1141.9	1234.4	1.4991
510	1.0860	1148.4	1242.9	1.5100	1.0728	1147.9	1242.2	1.5083	1.0598	1147.4	1241.6	1.5066
520	1.1043	1153.8	1249.8	1.5171	1.0910	1153.3	1249.2	1.5155	1.0779	1152.9	1248.6	1.5138
530	1.1222	1159.0	1256.6	1.5240	1.1088	1158.6	1256.0	1.5224	1.0956	1158.1	1255.4	1.5208
540	1.1398	1164.1	1263.3	1.5307	1.1262	1163.7	1262.7	1.5291	1.1130	1163.3	1262.1	1.5275
550	1.1570	1169.1	1269.8	1.5372	1.1434	1168.7	1269.2	1.5356	1.1300	1168.3	1268.7	1.5340
560	1.1740	1174.1	1276.2	1.5435	1.1602	1173.7	1275.7	1.5419	1.1467	1173.3	1275.2	1.5404
570	1.1907	1178.9	1282.4	1.5496	1.1768	1178.5	1282.0	1.5481	1.1632	1178.2	1281.5	1.5466
580	1.2071	1183.7	1288.6	1.5556	1.1931	1183.3	1288.2	1.5541	1.1794	1183.0	1287.7	1.5526
590	1.2234	1188.4	1294.8	1.5615	1.2093	1188.0	1294.3	1.5600	1.1955	1187.7	1293.9	1.5585
600	1.2395	1193.0	1300.8	1.5672	1.2252	1192.7	1300.4	1.5657	1.2113	1192.4	1300.0	1.5643
610	1.2554	1197.6	1306.8	1.5728	1.2410	1197.3	1306.4	1.5714	1.2269	1197.0	1306.0	1.5699
620	1.2711	1202.1	1312.7	1.5783	1.2566	1201.9	1312.3	1.5769	1.2424	1201.6	1311.9	1.5754
630	1.2867	1206.6	1318.6	1.5837	1.2721	1206.4	1318.2	1.5823	1.2578	1206.1	1317.8	1.5809
640	1.3022	1211.1	1324.4	1.5890	1.2874	1210.8	1324.0	1.5876	1.2730	1210.6	1323.6	1.5862
650	1.3175	1215.6	1330.1	1.5942	1.3026	1215.3	1329.8	1.5928	1.2880	1215.0	1329.4	1.5914
660	1.3327	1220.0	1335.9	1.5994	1.3177	1219.7	1335.5	1.5980	1.3030	1219.5	1335.2	1.5966
670	1.3478	1224.3	1341.6	1.6044	1.3326	1224.1	1341.2	1.6031	1.3178	1223.9	1340.9	1.6017
680	1.3628	1228.7	1347.2	1.6094	1.3475	1228.5	1346.9	1.6081	1.3326	1228.2	1346.6	1.6067
690	1.3777	1233.0	1352.8	1.6144	1.3623	1232.8	1352.5	1.6130	1.3472	1232.6	1352.2	1.6116
700	1.3925	1237.3	1358.4	1.6192	1.3769	1237.1	1358.2	1.6178	1.3617	1236.9	1357.9	1.6165
720	1.4218	1245.9	1369.6	1.6287	1.4060	1245.7	1369.3	1.6274	1.3906	1245.5	1369.0	1.6260
740	1.4509	1254.5	1380.6	1.6380	1.4348	1254.3	1380.4	1.6367	1.4191	1254.1	1380.1	1.6354
760	1.4797	1262.9	1391.6	1.6471	1.4634	1262.8	1391.4	1.6458	1.4474	1262.6	1391.1	1.6445
780	1.5083	1271.4	1402.6	1.6560	1.4917	1271.2	1402.3	1.6547	1.4755	1271.0	1402.1	1.6534
800	1.5366	1279.8	1413.5	1.6647	1.5198	1279.7	1413.2	1.6634	1.5033	1279.5	1413.0	1.6621
820	1.5648	1288.2	1424.3	1.6733	1.5477	1288.1	1424.1	1.6720	1.5309	1287.9	1423.9	1.6707
840	1.5927	1296.6	1435.2	1.6817	1.5754	1296.5	1435.0	1.6804	1.5584	1296.3	1434.8	1.6791
860	1.6206	1305.0	1446.0	1.6899	1.6029	1304.9	1445.8	1.6887	1.5857	1304.8	1445.6	1.6874
880	1.6482	1313.4	1456.8	1.6981	1.6304	1313.3	1456.6	1.6968	1.6128	1313.2	1456.4	1.6955
900	1.6758	1321.8	1467.6	1.7061	1.6576	1321.7	1467.4	1.7048	1.6399	1321.6	1467.2	1.7036
920	1.7032	1330.3	1478.4	1.7140	1.6848	1330.1	1478.2	1.7127	1.6668	1330.0	1478.0	1.7114
940	1.7305	1338.7	1489.2	1.7217	1.7118	1338.6	1489.0	1.7205	1.6935	1338.4	1488.9	1.7192
960	1.7577	1347.1	1500.0	1.7294	1.7388	1347.0	1499.8	1.7281	1.7202	1346.9	1499.7	1.7269
980	1.7848	1355.5	1510.8	1.7369	1.7656	1355.4	1510.6	1.7357	1.7468	1355.3	1510.5	1.7345
1000	1.8118	1364.0	1521.6	1.7444	1.7923	1363.9	1521.4	1.7431	1.7733	1363.8	1521.3	1.7419
1020	1.8388	1372.5	1532.4	1.7517	1.8190	1372.4	1532.3	1.7505	1.7997	1372.3	1532.1	1.7493
1040	1.8656	1381.0	1543.2	1.7590	1.8456	1380.9	1543.1	1.7578	1.8260	1380.8	1543.0	1.7566
1060	1.8924	1389.5	1554.1	1.7662	1.8721	1389.4	1553.9	1.7650	1.8522	1389.3	1553.8	1.7637
1080	1.9191	1398.0	1564.9	1.7733	1.8985	1397.9	1564.8	1.7721	1.8784	1397.8	1564.7	1.7708
1100	1.9457	1406.6	1575.8	1.7803	1.9249	1406.5	1575.7	1.7791	1.9045	1406.4	1575.5	1.7779
1150	2.0120	1428.0	1603.0	1.7975	1.9905	1428.0	1602.9	1.7963	1.9695	1427.9	1602.8	1.7951
1200	2.0780	1449.7	1630.4	1.8142	2.0559	1449.6	1630.3	1.8130	2.0342	1449.5	1630.2	1.8118
1250	2.1437	1471.5	1657.9	1.8306	2.1209	1471.4	1657.8	1.8294	2.0986	1471.3	1657.7	1.8282
1300	2.209	1493.4	1685.6	1.8465	2.186	1493.4	1685.5	1.8453	2.163	1493.3	1685.4	1.8441
1350	2.274	1515.6	1713.4	1.8621	2.250	1515.5	1713.3	1.8609	2.227	1515.5	1713.2	1.8597
1400	2.340	1537.9	1741.4	1.8774	2.315	1537.9	1741.3	1.8762	2.290	1537.8	1741.3	1.8750
1450	2.404	1560.4	1769.6	1.8923	2.379	1560.4	1769.5	1.8911	2.354	1560.3	1769.4	1.8899
1500	2.469	1583.1	1797.9	1.9070	2.443	1583.1	1797.8	1.9058	2.418	1583.0	1797.8	1.9046
1600	2.599	1629.2	1855.2	1.9355	2.571	1629.1	1855.1	1.9343	2.544	1629.1	1855.0	1.9331
1800	2.856	1723.6	1972.0	1.9896	2.826	1723.5	1971.9	1.9884	2.796	1723.5	1971.9	1.9872
2000	3.113	1821.1	2091.9	2.0404	3.080	1821.1	2091.8	2.0392	3.048	1821.1	2091.8	2.0381
2200	3.369	1921.7	2214.7	2.0884	3.334	1921.7	2214.7	2.0873	3.299	1921.6	2214.7	2.0861
2400	3.626	2025.0	2340.3	2.1339	3.587	2024.9	2340.2	2.1328	3.550	2024.9	2340.2	2.1316

v	u	h	s	v	u	h	s	v	u	h	s	t
.9574	1119.5	1205.4	1.4674	.9475	1119.5	1205.4	1.4664	.9378	1119.5	1205.4	1.4654	**Sat.**
.9270	*1110.1*	*1193.3*	*1.4541*	*.9149*	*1109.3*	*1192.2*	*1.4520*					**450**
.9489	*1116.9*	*1202.0*	*1.4637*	*.9368*	*1116.1*	*1201.1*	*1.4617*	*.9250*	*1115.4*	*1200.1*	*1.4598*	**460**
.9699	1123.4	1210.4	1.4728	.9578	1122.7	1209.5	1.4709	.9459	1122.0	1208.7	1.4690	**470**
.9902	1129.6	1218.5	1.4814	.9780	1129.0	1217.6	1.4795	.9660	1128.3	1216.8	1.4777	**480**
1.0097	1135.6	1226.2	1.4896	.9975	1135.0	1225.4	1.4878	.9854	1134.4	1224.7	1.4860	**490**
1.0287	1141.3	1233.7	1.4974	1.0163	1140.8	1233.0	1.4957	1.0042	1140.2	1232.2	1.4940	**500**
1.0472	1146.9	1240.9	1.5049	1.0347	1146.4	1240.3	1.5032	1.0225	1145.9	1239.6	1.5016	**510**
1.0651	1152.4	1248.0	1.5122	1.0526	1151.9	1247.3	1.5105	1.0403	1151.4	1246.7	1.5089	**520**
1.0827	1157.7	1254.9	1.5192	1.0701	1157.2	1254.3	1.5176	1.0577	1156.8	1253.7	1.5160	**530**
1.1000	1162.9	1261.6	1.5259	1.0873	1162.4	1261.0	1.5243	1.0748	1162.0	1260.5	1.5228	**540**
1.1169	1167.9	1268.2	1.5325	1.1041	1167.5	1267.6	1.5309	1.0915	1167.1	1267.1	1.5294	**550**
1.1335	1172.9	1274.6	1.5389	1.1206	1172.5	1274.1	1.5373	1.1079	1172.1	1273.6	1.5358	**560**
1.1499	1177.8	1281.0	1.5451	1.1368	1177.4	1280.5	1.5436	1.1240	1177.1	1280.0	1.5421	**570**
1.1660	1182.6	1287.3	1.5511	1.1528	1182.3	1286.8	1.5496	1.1399	1181.9	1286.3	1.5482	**580**
1.1819	1187.4	1293.4	1.5570	1.1686	1187.0	1293.0	1.5556	1.1556	1186.7	1292.6	1.5541	**590**
1.1976	1192.1	1299.5	1.5628	1.1842	1191.7	1299.1	1.5614	1.1711	1191.4	1298.7	1.5599	**600**
1.2132	1196.7	1305.6	1.5685	1.1997	1196.4	1305.2	1.5670	1.1864	1196.1	1304.8	1.5656	**610**
1.2285	1201.3	1311.5	1.5740	1.2149	1201.0	1311.1	1.5726	1.2016	1200.7	1310.7	1.5712	**620**
1.2438	1205.8	1317.4	1.5795	1.2300	1205.5	1317.1	1.5781	1.2165	1205.2	1316.7	1.5767	**630**
1.2588	1210.3	1323.3	1.5848	1.2450	1210.0	1322.9	1.5834	1.2314	1209.8	1322.6	1.5821	**640**
1.2738	1214.8	1329.1	1.5901	1.2598	1214.5	1328.7	1.5887	1.2461	1214.3	1328.4	1.5873	**650**
1.2886	1219.2	1334.9	1.5952	1.2745	1219.0	1334.5	1.5939	1.2607	1218.7	1334.2	1.5925	**660**
1.3033	1223.6	1340.6	1.6003	1.2891	1223.4	1340.3	1.5990	1.2752	1223.1	1339.9	1.5976	**670**
1.3179	1228.0	1346.3	1.6053	1.3036	1227.8	1346.0	1.6040	1.2895	1227.5	1345.6	1.6027	**680**
1.3324	1232.3	1351.9	1.6103	1.3180	1232.1	1351.6	1.6090	1.3038	1231.9	1351.3	1.6076	**690**
1.3469	1236.7	1357.6	1.6152	1.3323	1236.5	1357.3	1.6138	1.3180	1236.2	1357.0	1.6125	**700**
1.3755	1245.3	1368.7	1.6247	1.3606	1245.1	1368.5	1.6234	1.3461	1244.9	1368.2	1.6221	**720**
1.4038	1253.9	1379.8	1.6341	1.3887	1253.7	1379.6	1.6328	1.3739	1253.5	1379.3	1.6315	**740**
1.4318	1262.4	1390.9	1.6432	1.4165	1262.2	1390.6	1.6419	1.4015	1262.0	1390.4	1.6406	**760**
1.4596	1270.9	1401.9	1.6521	1.4440	1270.7	1401.6	1.6508	1.4288	1270.5	1401.4	1.6496	**780**
1.4872	1279.3	1412.8	1.6609	1.4713	1279.2	1412.6	1.6596	1.4558	1279.3	1412.3	1.6583	**800**
1.5145	1287.8	1423.7	1.6694	1.4985	1287.6	1423.5	1.6682	1.4827	1287.4	1423.3	1.6669	**820**
1.5417	1296.2	1434.6	1.6779	1.5254	1296.0	1434.4	1.6766	1.5095	1295.9	1434.2	1.6754	**840**
1.5688	1304.6	1445.4	1.6862	1.5522	1304.5	1445.2	1.6849	1.5360	1304.3	1445.0	1.6837	**860**
1.5957	1313.0	1456.2	1.6943	1.5789	1312.9	1456.1	1.6931	1.5624	1312.7	1455.9	1.6918	**880**
1.6225	1321.4	1467.1	1.7023	1.6054	1321.3	1466.9	1.7011	1.5887	1321.2	1466.7	1.6999	**900**
1.6491	1329.9	1477.9	1.7102	1.6318	1329.7	1477.7	1.7090	1.6148	1329.6	1477.5	1.7078	**920**
1.6756	1338.3	1488.7	1.7180	1.6581	1338.2	1488.5	1.7168	1.6409	1338.1	1488.4	1.7156	**940**
1.7020	1346.7	1499.5	1.7257	1.6842	1346.6	1499.3	1.7245	1.6668	1346.5	1499.2	1.7233	**960**
1.7284	1355.2	1510.3	1.7332	1.7103	1355.1	1510.2	1.7320	1.6926	1355.0	1510.0	1.7308	**980**
1.7546	1363.7	1521.1	1.7407	1.7363	1363.6	1521.0	1.7395	1.7184	1363.5	1520.9	1.7383	**1000**
1.7807	1372.2	1532.0	1.7481	1.7622	1372.1	1531.8	1.7469	1.7440	1372.0	1531.7	1.7457	**1020**
1.8068	1380.7	1542.8	1.7553	1.7880	1380.6	1542.7	1.7541	1.7696	1380.5	1542.6	1.7530	**1040**
1.8328	1389.2	1553.7	1.7625	1.8137	1389.1	1553.5	1.7613	1.7951	1389.0	1553.4	1.7602	**1060**
1.8587	1397.7	1564.5	1.7696	1.8394	1397.6	1564.4	1.7685	1.8205	1397.5	1564.3	1.7673	**1080**
1.8845	1406.3	1575.4	1.7767	1.8650	1406.2	1575.3	1.7755	1.8458	1406.1	1575.2	1.7743	**1100**
1.9489	1427.8	1602.7	1.7939	1.9287	1427.7	1602.6	1.7927	1.9090	1427.6	1602.5	1.7915	**1150**
2.0130	1449.4	1630.1	1.8106	1.9922	1449.4	1630.0	1.8095	1.9718	1449.3	1629.9	1.8083	**1200**
2.0767	1471.3	1657.6	1.8270	2.0553	1471.2	1657.6	1.8258	2.0343	1471.1	1657.5	1.8247	**1250**
2.140	1493.2	1685.3	1.8430	2.118	1493.2	1685.2	1.8418	2.097	1493.1	1685.2	1.8406	**1300**
2.204	1515.4	1713.2	1.8586	2.181	1515.3	1713.1	1.8574	2.159	1515.3	1713.0	1.8562	**1350**
2.267	1537.7	1741.2	1.8738	2.243	1537.7	1741.1	1.8727	2.221	1537.6	1741.0	1.8715	**1400**
2.330	1560.3	1769.4	1.8888	2.306	1560.2	1769.3	1.8876	2.282	1560.2	1769.2	1.8865	**1450**
2.392	1583.0	1797.7	1.9034	2.368	1582.9	1797.7	1.9023	2.344	1582.9	1797.6	1.9011	**1500**
2.518	1629.0	1855.0	1.9319	2.492	1629.0	1854.9	1.9308	2.467	1628.9	1854.9	1.9296	**1600**
2.768	1723.4	1971.8	1.9861	2.739	1723.4	1971.8	1.9849	2.711	1723.4	1971.7	1.9838	**1800**
3.017	1821.0	2091.7	2.0369	2.986	1821.0	2091.7	2.0358	2.956	1820.9	2091.7	2.0346	**2000**
3.265	1921.6	2214.6	2.0849	3.232	1921.6	2214.6	2.0838	3.199	1921.5	2214.6	2.0827	**2200**
3.513	2024.8	2340.2	2.1304	3.478	2024.8	2340.1	2.1293	3.442	2024.8	2340.1	2.1282	**2400**

Table 3. Vapor

p (t Sat.)	500 (467.13)				510 (469.17)				520 (471.19)			
t	v	u	h	s	v	u	h	s	v	u	h	s
Sat.	.9283	1119.4	1205.3	1.4645	.9098	1119.4	1205.2	1.4626	.8920	1119.3	1205.2	1.4607
460	*.9133*	*1114.7*	*1199.2*	*1.4578*	*.8907*	*1113.1*	*1197.2*	*1.4538*	.8688	*1111.6*	*1195.2*	*1.4499*
470	.9342	1121.3	1207.8	1.4671	.9115	1119.9	1206.0	1.4633	*.8896*	*1118.5*	*1204.1*	*1.4595*
480	.9543	1127.7	1216.0	1.4759	.9315	1126.4	1214.3	1.4723	.9095	1125.1	1212.6	1.4686
490	.9736	1133.8	1223.9	1.4843	.9507	1132.6	1222.3	1.4808	.9286	1131.4	1220.8	1.4773
500	.9924	1139.7	1231.5	1.4923	.9693	1138.6	1230.1	1.4889	.9471	1137.5	1228.6	1.4855
510	1.0106	1145.4	1238.9	1.4999	.9873	1144.4	1237.5	1.4966	.9650	1143.3	1236.2	1.4933
520	1.0283	1150.9	1246.1	1.5073	1.0049	1150.0	1244.8	1.5040	.9824	1149.0	1243.5	1.5008
530	1.0456	1156.3	1253.1	1.5144	1.0220	1155.4	1251.9	1.5112	.9993	1154.5	1250.6	1.5081
540	1.0625	1161.6	1259.9	1.5212	1.0388	1160.7	1258.7	1.5181	1.0159	1159.8	1257.6	1.5151
550	1.0792	1166.7	1266.6	1.5279	1.0552	1165.9	1265.5	1.5248	1.0322	1165.1	1264.4	1.5219
560	1.0955	1171.8	1273.1	1.5343	1.0713	1171.0	1272.1	1.5314	1.0481	1170.2	1271.1	1.5284
570	1.1115	1176.7	1279.5	1.5406	1.0872	1176.0	1278.6	1.5377	1.0637	1175.2	1277.6	1.5348
580	1.1273	1181.6	1285.9	1.5467	1.1027	1180.9	1284.9	1.5438	1.0791	1180.2	1284.0	1.5410
590	1.1429	1186.4	1292.1	1.5527	1.1181	1185.7	1291.2	1.5498	1.0943	1185.0	1290.3	1.5470
600	1.1583	1191.1	1298.3	1.5585	1.1333	1190.5	1297.4	1.5557	1.1093	1189.8	1296.5	1.5529
610	1.1735	1195.8	1304.3	1.5642	1.1483	1195.2	1303.5	1.5615	1.1240	1194.5	1302.7	1.5587
620	1.1885	1200.4	1310.4	1.5698	1.1631	1199.8	1309.6	1.5671	1.1386	1199.2	1308.8	1.5644
630	1.2033	1205.0	1316.3	1.5753	1.1777	1204.4	1315.5	1.5726	1.1531	1203.8	1314.8	1.5699
640	1.2181	1209.5	1322.2	1.5807	1.1922	1208.9	1321.5	1.5780	1.1674	1208.4	1320.7	1.5753
650	1.2327	1214.0	1328.0	1.5860	1.2066	1213.5	1327.3	1.5833	1.1815	1212.9	1326.6	1.5807
660	1.2472	1218.4	1333.8	1.5912	1.2209	1217.9	1333.2	1.5885	1.1956	1217.4	1332.5	1.5859
670	1.2615	1222.9	1339.6	1.5963	1.2350	1222.4	1338.9	1.5937	1.2095	1221.9	1338.3	1.5911
680	1.2758	1227.3	1345.3	1.6014	1.2490	1226.8	1344.7	1.5987	1.2233	1226.3	1344.0	1.5962
690	1.2899	1231.7	1351.0	1.6063	1.2630	1231.2	1350.4	1.6037	1.2370	1230.7	1349.8	1.6012
700	1.3040	1236.0	1356.7	1.6112	1.2768	1235.6	1356.1	1.6086	1.2507	1235.1	1355.5	1.6061
720	1.3319	1244.7	1367.9	1.6208	1.3042	1244.3	1367.3	1.6183	1.2776	1243.8	1366.8	1.6158
740	1.3594	1253.3	1379.1	1.6302	1.3313	1252.9	1378.5	1.6277	1.3043	1252.5	1378.0	1.6252
760	1.3867	1261.8	1390.1	1.6394	1.3582	1261.5	1389.6	1.6369	1.3307	1261.1	1389.1	1.6344
780	1.4138	1270.3	1401.1	1.6483	1.3848	1270.0	1400.7	1.6458	1.3569	1269.6	1400.2	1.6434
800	1.4407	1278.8	1412.1	1.6571	1.4112	1278.5	1411.7	1.6546	1.3828	1278.1	1411.2	1.6522
820	1.4673	1287.3	1423.0	1.6657	1.4374	1287.0	1422.6	1.6633	1.4086	1286.6	1422.2	1.6609
840	1.4938	1295.7	1433.9	1.6742	1.4634	1295.4	1433.5	1.6717	1.4342	1295.1	1433.1	1.6694
860	1.5201	1304.2	1444.8	1.6825	1.4892	1303.9	1444.4	1.6801	1.4596	1303.6	1444.0	1.6777
880	1.5463	1312.6	1455.7	1.6906	1.5150	1312.3	1455.3	1.6882	1.4848	1312.1	1454.9	1.6859
900	1.5723	1321.0	1466.5	1.6987	1.5405	1320.8	1466.2	1.6963	1.5100	1320.5	1465.8	1.6939
920	1.5982	1329.5	1477.4	1.7066	1.5660	1329.2	1477.0	1.7042	1.5350	1329.0	1476.7	1.7019
940	1.6240	1337.9	1488.2	1.7144	1.5913	1337.7	1487.9	1.7120	1.5598	1337.4	1487.5	1.7097
960	1.6497	1346.4	1499.0	1.7221	1.6165	1346.2	1498.7	1.7197	1.5846	1345.9	1498.4	1.7174
980	1.6753	1354.9	1509.9	1.7296	1.6416	1354.6	1509.6	1.7273	1.6093	1354.4	1509.3	1.7250
1000	1.7008	1363.3	1520.7	1.7371	1.6667	1363.1	1520.4	1.7348	1.6339	1362.9	1520.1	1.7325
1020	1.7262	1371.8	1531.6	1.7445	1.6916	1371.6	1531.3	1.7422	1.6584	1371.4	1531.0	1.7399
1040	1.7515	1380.4	1542.4	1.7518	1.7165	1380.2	1542.2	1.7495	1.6828	1380.0	1541.9	1.7472
1060	1.7768	1388.9	1553.3	1.7590	1.7413	1388.7	1553.0	1.7567	1.7071	1388.5	1552.8	1.7544
1080	1.8020	1397.4	1564.2	1.7661	1.7660	1397.3	1563.9	1.7638	1.7314	1397.1	1563.7	1.7615
1100	1.8271	1406.0	1575.1	1.7731	1.7906	1405.8	1574.8	1.7708	1.7556	1405.7	1574.6	1.7686
1150	1.8896	1427.5	1602.4	1.7904	1.8520	1427.4	1602.2	1.7881	1.8158	1427.2	1601.9	1.7858
1200	1.9518	1449.2	1629.8	1.8072	1.9130	1449.1	1629.6	1.8049	1.8757	1448.9	1629.4	1.8026
1250	2.0137	1471.1	1657.4	1.8235	1.9738	1470.9	1657.2	1.8212	1.9354	1470.8	1657.0	1.8190
1300	2.075	1493.1	1685.1	1.8395	2.034	1492.9	1684.9	1.8372	1.9948	1492.8	1684.7	1.8350
1350	2.137	1515.2	1712.9	1.8551	2.095	1515.1	1712.8	1.8528	2.0539	1515.0	1712.6	1.8506
1400	2.198	1537.6	1741.0	1.8704	2.155	1537.5	1740.8	1.8681	2.1129	1537.4	1740.7	1.8659
1450	2.259	1560.1	1769.2	1.8853	2.215	1560.0	1769.0	1.8831	2.1718	1559.9	1768.9	1.8809
1500	2.320	1582.8	1797.5	1.9000	2.274	1582.7	1797.4	1.8978	2.2305	1582.6	1797.3	1.8956
1600	2.442	1628.9	1854.8	1.9285	2.394	1628.8	1854.7	1.9263	2.348	1628.7	1854.6	1.9241
1800	2.684	1723.3	1971.7	1.9827	2.632	1723.2	1971.6	1.9804	2.581	1723.2	1971.5	1.9783
2000	2.926	1820.9	2091.6	2.0335	2.868	1820.8	2091.6	2.0313	2.813	1820.8	2091.5	2.0291
2200	3.167	1921.5	2214.5	2.0815	3.105	1921.4	2214.5	2.0793	3.045	1921.4	2214.4	2.0772
2400	3.408	2024.7	2340.1	2.1270	3.341	2024.7	2340.0	2.1248	3.277	2024.6	2339.9	2.1227

v	u	h	s	v	u	h	s	v	u	h	s	t
.8748	1119.3	1205.1	1.4588	.8583	1119.2	1204.9	1.4570	.8423	1119.1	1204.8	1.4551	**Sat.**
												460
.8477	*1110.0*	*1193.2*	*1.4460*									**460**
.8684	*1117.1*	*1202.3*	*1.4558*	*.8480*	*1115.6*	*1200.4*	*1.4521*	*.8283*	*1114.1*	*1198.4*	*1.4483*	**470**
.8883	1123.8	1210.9	1.4650	.8678	1122.5	1209.2	1.4615	.8480	1121.1	1207.4	1.4579	**480**
.9073	1130.2	1219.2	1.4738	.8867	1129.0	1217.6	1.4704	.8669	1127.7	1215.9	1.4669	**490**
.9256	1136.3	1227.1	1.4821	.9050	1135.2	1225.6	1.4788	.8850	1134.0	1224.1	1.4755	**500**
.9434	1142.3	1234.8	1.4901	.9226	1141.2	1233.4	1.4868	.9026	1140.1	1232.0	1.4836	**510**
.9607	1148.0	1242.2	1.4977	.9397	1147.0	1240.9	1.4945	.9196	1146.0	1239.6	1.4914	**520**
.9775	1153.6	1249.4	1.5050	.9564	1152.6	1248.2	1.5019	.9361	1151.7	1246.9	1.4989	**530**
.9939	1159.0	1256.4	1.5121	.9727	1158.1	1255.3	1.5091	.9522	1157.2	1254.1	1.5061	**540**
1.0100	1164.2	1263.3	1.5189	.9886	1163.4	1262.2	1.5160	.9679	1162.6	1261.1	1.5131	**550**
1.0257	1169.4	1270.0	1.5255	1.0042	1168.6	1269.0	1.5226	.9834	1167.8	1267.9	1.5198	**560**
1.0412	1174.5	1276.6	1.5319	1.0194	1173.7	1275.6	1.5291	.9985	1173.0	1274.6	1.5263	**570**
1.0564	1179.4	1283.1	1.5382	1.0345	1178.7	1282.1	1.5354	1.0133	1178.0	1281.1	1.5327	**580**
1.0714	1184.3	1289.4	1.5443	1.0493	1183.7	1288.5	1.5415	1.0280	1183.0	1287.6	1.5388	**590**
1.0861	1189.2	1295.7	1.5502	1.0639	1188.5	1294.8	1.5475	1.0424	1187.9	1293.9	1.5448	**600**
1.1007	1193.9	1301.9	1.5560	1.0782	1193.3	1301.0	1.5534	1.0566	1192.7	1300.2	1.5507	**610**
1.1151	1198.6	1308.0	1.5617	1.0925	1198.0	1307.2	1.5591	1.0706	1197.4	1306.4	1.5565	**620**
1.1294	1203.3	1314.0	1.5673	1.1065	1202.7	1313.2	1.5647	1.0845	1202.1	1312.5	1.5621	**630**
1.1435	1207.8	1320.0	1.5727	1.1204	1207.3	1319.3	1.5702	1.0982	1206.7	1318.5	1.5676	**640**
1.1574	1212.4	1325.9	1.5781	1.1342	1211.9	1325.2	1.5755	1.1118	1211.3	1324.5	1.5730	**650**
1.1712	1216.9	1331.8	1.5834	1.1478	1216.4	1331.1	1.5808	1.1252	1215.9	1330.4	1.5783	**660**
1.1850	1221.4	1337.6	1.5885	1.1613	1220.9	1337.0	1.5860	1.1386	1220.4	1336.3	1.5836	**670**
1.1986	1225.9	1343.4	1.5936	1.1747	1225.4	1342.8	1.5912	1.1518	1224.9	1342.1	1.5887	**680**
1.2121	1230.3	1349.2	1.5987	1.1881	1229.8	1348.5	1.5962	1.1649	1229.4	1347.9	1.5938	**690**
1.2255	1234.7	1354.9	1.6036	1.2013	1234.2	1354.3	1.6012	1.1779	1233.8	1353.7	1.5987	**700**
1.2521	1243.4	1366.2	1.6133	1.2274	1243.0	1365.7	1.6109	1.2037	1242.6	1365.1	1.6085	**720**
1.2783	1252.1	1377.5	1.6228	1.2533	1251.7	1376.9	1.6204	1.2291	1251.3	1376.4	1.6180	**740**
1.3043	1260.7	1388.6	1.6320	1.2789	1260.3	1388.1	1.6296	1.2543	1260.0	1387.6	1.6273	**760**
1.3300	1269.3	1399.7	1.6410	1.3042	1268.9	1399.2	1.6387	1.2793	1268.6	1398.8	1.6364	**780**
1.3556	1277.8	1410.8	1.6499	1.3293	1277.5	1410.3	1.6475	1.3040	1277.1	1409.8	1.6452	**800**
1.3809	1286.3	1421.8	1.6585	1.3542	1286.0	1421.3	1.6562	1.3285	1285.7	1420.9	1.6539	**820**
1.4060	1294.8	1432.7	1.6670	1.3789	1294.5	1432.3	1.6647	1.3528	1294.2	1431.9	1.6625	**840**
1.4310	1303.3	1443.6	1.6754	1.4035	1303.0	1443.3	1.6731	1.3770	1302.7	1442.9	1.6708	**860**
1.4558	1311.8	1454.6	1.6836	1.4279	1311.5	1454.2	1.6813	1.4010	1311.2	1453.8	1.6791	**880**
1.4805	1320.2	1465.5	1.6916	1.4522	1320.0	1465.1	1.6894	1.4249	1319.7	1464.7	1.6872	**900**
1.5051	1328.7	1476.3	1.6996	1.4764	1328.5	1476.0	1.6973	1.4487	1328.2	1475.6	1.6951	**920**
1.5296	1337.2	1487.2	1.7074	1.5004	1337.0	1486.9	1.7052	1.4723	1336.7	1486.6	1.7030	**940**
1.5539	1345.7	1498.1	1.7151	1.5243	1345.5	1497.8	1.7129	1.4958	1345.2	1497.5	1.7107	**960**
1.5782	1354.2	1509.0	1.7227	1.5482	1354.0	1508.7	1.7205	1.5193	1353.7	1508.4	1.7183	**980**
1.6023	1362.7	1519.8	1.7302	1.5719	1362.5	1519.6	1.7280	1.5426	1362.3	1519.3	1.7259	**1000**
1.6264	1371.2	1530.7	1.7376	1.5956	1371.0	1530.4	1.7354	1.5659	1370.8	1530.2	1.7333	**1020**
1.6504	1379.8	1541.6	1.7450	1.6191	1379.6	1541.3	1.7428	1.5890	1379.3	1541.1	1.7406	**1040**
1.6743	1388.3	1552.5	1.7522	1.6426	1388.1	1552.3	1.7500	1.6121	1387.9	1552.0	1.7478	**1060**
1.6981	1396.9	1563.4	1.7593	1.6660	1396.7	1563.2	1.7571	1.6352	1396.5	1562.9	1.7550	**1080**
1.7219	1405.5	1574.3	1.7664	1.6894	1405.3	1574.1	1.7642	1.6581	1405.1	1573.9	1.7620	**1100**
1.7810	1427.0	1601.7	1.7836	1.7475	1426.9	1601.5	1.7815	1.7152	1426.7	1601.3	1.7793	**1150**
1.8399	1448.8	1629.2	1.8005	1.8053	1448.6	1629.0	1.7983	1.7720	1448.5	1628.8	1.7962	**1200**
1.8984	1470.6	1656.8	1.8168	1.8628	1470.5	1656.6	1.8147	1.8285	1470.3	1656.5	1.8126	**1250**
1.9567	1492.7	1684.6	1.8328	1.9201	1492.5	1684.4	1.8307	1.8848	1492.4	1684.2	1.8286	**1300**
2.0148	1514.9	1712.5	1.8485	1.9772	1514.7	1712.3	1.8463	1.9409	1514.6	1712.2	1.8443	**1350**
2.0727	1537.2	1740.5	1.8638	2.0340	1537.1	1740.4	1.8616	1.9967	1537.0	1740.2	1.8596	**1400**
2.1305	1559.8	1768.8	1.8787	2.0908	1559.7	1768.6	1.8766	2.0525	1559.6	1768.5	1.8746	**1450**
2.1881	1582.5	1797.2	1.8934	2.1473	1582.4	1797.0	1.8913	2.1080	1582.3	1796.9	1.8892	**1500**
2.303	1628.6	1854.5	1.9220	2.260	1628.5	1854.4	1.9199	2.219	1628.4	1854.3	1.9178	**1600**
2.532	1723.1	1971.4	1.9761	2.485	1723.0	1971.3	1.9740	2.440	1722.9	1971.2	1.9720	**1800**
2.760	1820.7	2091.4	2.0270	2.709	1820.6	2091.3	2.0249	2.660	1820.6	2091.2	2.0229	**2000**
2.988	1921.3	2214.3	2.0750	2.932	1921.2	2214.2	2.0729	2.879	1921.1	2214.2	2.0709	**2200**
3.215	2024.5	2339.9	2.1205	3.156	2024.5	2339.8	2.1185	3.098	2024.4	2339.7	2.1164	**2400**

Table 3. Vapor

p (t Sat.)	560 (478.97)				570 (480.85)				580 (482.70)			
t	v	u	h	s	v	u	h	s	v	u	h	s
Sat.	.8269	1119.0	1204.7	1.4534	.8120	1118.9	1204.5	1.4516	.7976	1118.8	1204.4	1.4499
470	*.8092*	*1112.6*	*1196.5*	*1.4446*	*.7907*	*1111.1*	*1194.5*	*1.4408*	*.7728*	*1109.5*	*1192.5*	*1.4371*
480	.8289	1119.7	1205.6	1.4543	*.8104*	*1118.3*	*1203.8*	*1.4508*	*.7924*	*1116.9*	*1201.9*	*1.4472*
490	.8477	1126.4	1214.3	1.4635	.8291	1125.1	1212.6	1.4601	.8112	1123.8	1210.9	1.4567
500	.8657	1132.8	1222.6	1.4722	.8471	1131.7	1221.0	1.4689	.8291	1130.4	1219.4	1.4657
510	.8832	1139.0	1230.5	1.4805	.8645	1137.9	1229.1	1.4773	.8463	1136.8	1227.6	1.4741
520	.9001	1144.9	1238.2	1.4883	.8812	1143.9	1236.9	1.4853	.8630	1142.9	1235.5	1.4822
530	.9165	1150.7	1245.7	1.4959	.8975	1149.7	1244.4	1.4929	.8792	1148.8	1243.1	1.4900
540	.9324	1156.3	1252.9	1.5032	.9133	1155.4	1251.7	1.5003	.8949	1154.4	1250.5	1.4974
550	.9480	1161.7	1260.0	1.5102	.9288	1160.9	1258.8	1.5074	.9102	1160.0	1257.7	1.5045
560	.9633	1167.0	1266.8	1.5170	.9439	1166.2	1265.8	1.5142	.9252	1165.4	1264.7	1.5114
570	.9783	1172.2	1273.6	1.5236	.9587	1171.4	1272.6	1.5208	.9399	1170.7	1271.5	1.5181
580	.9930	1177.3	1280.2	1.5299	.9733	1176.6	1279.2	1.5273	.9543	1175.8	1278.2	1.5246
590	1.0074	1182.3	1286.7	1.5362	.9876	1181.6	1285.8	1.5335	.9684	1180.9	1284.8	1.5309
600	1.0216	1187.2	1293.1	1.5422	1.0016	1186.5	1292.2	1.5396	.9823	1185.9	1291.3	1.5370
610	1.0357	1192.0	1299.4	1.5481	1.0155	1191.4	1298.5	1.5456	.9960	1190.8	1297.7	1.5430
620	1.0495	1196.8	1305.6	1.5539	1.0292	1196.2	1304.8	1.5514	1.0096	1195.6	1303.9	1.5489
630	1.0632	1201.5	1311.7	1.5596	1.0427	1200.9	1310.9	1.5571	1.0229	1200.4	1310.1	1.5546
640	1.0768	1206.2	1317.8	1.5651	1.0561	1205.6	1317.0	1.5626	1.0361	1205.1	1316.3	1.5602
650	1.0902	1210.8	1323.8	1.5705	1.0693	1210.3	1323.1	1.5681	1.0492	1209.7	1322.3	1.5657
660	1.1034	1215.4	1329.7	1.5759	1.0824	1214.9	1329.0	1.5735	1.0621	1214.3	1328.3	1.5711
670	1.1166	1219.9	1335.6	1.5811	1.0954	1219.4	1335.0	1.5787	1.0749	1218.9	1334.3	1.5763
680	1.1296	1224.4	1341.5	1.5863	1.1082	1223.9	1340.8	1.5839	1.0876	1223.4	1340.2	1.5815
690	1.1425	1228.9	1347.3	1.5914	1.1210	1228.4	1346.7	1.5890	1.1002	1228.0	1346.0	1.5867
700	1.1554	1233.3	1353.1	1.5964	1.1336	1232.9	1352.5	1.5940	1.1126	1232.4	1351.8	1.5917
720	1.1808	1242.2	1364.5	1.6062	1.1587	1241.7	1363.9	1.6038	1.1373	1241.3	1363.4	1.6016
740	1.2059	1250.9	1375.9	1.6157	1.1834	1250.5	1375.3	1.6134	1.1617	1250.1	1374.8	1.6111
760	1.2307	1259.6	1387.1	1.6250	1.2079	1259.2	1386.6	1.6227	1.1858	1258.8	1386.1	1.6205
780	1.2552	1268.2	1398.3	1.6341	1.2320	1267.8	1397.8	1.6318	1.2097	1267.5	1397.3	1.6296
800	1.2796	1276.8	1409.4	1.6430	1.2560	1276.5	1408.9	1.6407	1.2333	1276.1	1408.5	1.6386
820	1.3037	1285.4	1420.5	1.6517	1.2798	1285.0	1420.0	1.6495	1.2567	1284.7	1419.6	1.6473
840	1.3277	1293.9	1431.5	1.6602	1.3034	1293.6	1431.1	1.6580	1.2799	1293.3	1430.6	1.6559
860	1.3515	1302.4	1442.5	1.6686	1.3268	1302.1	1442.1	1.6664	1.3030	1301.8	1441.7	1.6643
880	1.3751	1310.9	1453.4	1.6769	1.3501	1310.7	1453.1	1.6747	1.3259	1310.4	1452.7	1.6726
900	1.3986	1319.4	1464.4	1.6850	1.3732	1319.2	1464.0	1.6828	1.3487	1318.9	1463.7	1.6807
920	1.4220	1328.0	1475.3	1.6930	1.3962	1327.7	1475.0	1.6908	1.3713	1327.4	1474.6	1.6887
940	1.4452	1336.5	1486.2	1.7008	1.4191	1336.2	1485.9	1.6987	1.3938	1336.0	1485.6	1.6966
960	1.4684	1345.0	1497.1	1.7086	1.4419	1344.7	1496.8	1.7064	1.4163	1344.5	1496.5	1.7044
980	1.4914	1353.5	1508.1	1.7162	1.4645	1353.3	1507.8	1.7141	1.4386	1353.0	1507.4	1.7120
1000	1.5144	1362.0	1519.0	1.7237	1.4871	1361.8	1518.7	1.7216	1.4608	1361.6	1518.4	1.7195
1020	1.5372	1370.6	1529.9	1.7311	1.5096	1370.4	1529.6	1.7291	1.4829	1370.2	1529.3	1.7270
1040	1.5600	1379.1	1540.8	1.7385	1.5320	1378.9	1540.5	1.7364	1.5050	1378.7	1540.3	1.7343
1060	1.5827	1387.7	1551.7	1.7457	1.5544	1387.5	1551.5	1.7436	1.5270	1387.3	1551.2	1.7416
1080	1.6054	1396.3	1562.7	1.7529	1.5766	1396.1	1562.4	1.7508	1.5489	1395.9	1562.2	1.7488
1100	1.6279	1404.9	1573.6	1.7599	1.5988	1404.7	1573.4	1.7579	1.5707	1404.6	1573.1	1.7558
1150	1.6841	1426.5	1601.1	1.7773	1.6541	1426.4	1600.8	1.7752	1.6250	1426.2	1600.6	1.7732
1200	1.7399	1448.3	1628.6	1.7941	1.7090	1448.2	1628.4	1.7921	1.6791	1448.0	1628.2	1.7901
1250	1.7955	1470.2	1656.3	1.8105	1.7636	1470.1	1656.1	1.8085	1.7328	1469.9	1655.9	1.8065
1300	1.8508	1492.3	1684.1	1.8265	1.8179	1492.1	1683.9	1.8245	1.7862	1492.0	1683.7	1.8225
1350	1.9059	1514.5	1712.0	1.8422	1.8721	1514.4	1711.8	1.8402	1.8395	1514.3	1711.7	1.8382
1400	1.9608	1536.9	1740.1	1.8575	1.9261	1536.8	1739.9	1.8555	1.8926	1536.7	1739.8	1.8535
1450	2.0155	1559.5	1768.3	1.8725	1.9799	1559.4	1768.2	1.8705	1.9455	1559.3	1768.1	1.8685
1500	2.0702	1582.3	1796.8	1.8872	2.0336	1582.2	1796.6	1.8852	1.9983	1582.1	1796.5	1.8832
1600	2.179	1628.3	1854.2	1.9158	2.141	1628.3	1854.0	1.9138	2.103	1628.2	1853.9	1.9118
1800	2.396	1722.9	1971.1	1.9700	2.354	1722.8	1971.1	1.9680	2.313	1722.7	1971.0	1.9660
2000	2.612	1820.5	2091.2	2.0208	2.566	1820.4	2091.1	2.0189	2.522	1820.3	2091.0	2.0169
2200	2.828	1921.1	2214.1	2.0689	2.778	1921.0	2214.0	2.0669	2.730	1920.9	2214.0	2.0650
2400	3.043	2024.3	2339.7	2.1144	2.990	2024.3	2339.6	2.1124	2.938	2024.2	2339.5	2.1105

	590 (484.53)				**600** (486.33)				**610** (488.12)			**p** (t Sat.)
v	u	h	s	v	u	h	s	v	u	h	s	t
.7837	1118.7	1204.2	1.4481	.7702	1118.6	1204.1	1.4464	.7571	1118.4	1203.9	1.4448	Sat.
												470
.7751	*1115.4*	*1200.0*	*1.4437*	*.7582*	*1113.9*	*1198.1*	*1.4401*	*.7419*	*1112.4*	*1196.2*	*1.4366*	480
.7938	1122.5	1209.2	1.4533	.7769	1121.1	1207.4	1.4500	.7606	1119.8	1205.6	1.4466	490
.8116	1129.2	1217.8	1.4624	.7947	1128.0	1216.2	1.4592	.7783	1126.7	1214.6	1.4560	500
.8288	1135.6	1226.1	1.4710	.8118	1134.5	1224.6	1.4679	.7954	1133.3	1223.1	1.4648	510
.8454	1141.8	1234.1	1.4792	.8283	1140.7	1232.7	1.4762	.8118	1139.7	1231.3	1.4732	520
.8614	1147.8	1241.8	1.4870	.8443	1146.8	1240.5	1.4841	.8277	1145.8	1239.2	1.4812	530
.8771	1153.5	1249.3	1.4945	.8598	1152.6	1248.0	1.4917	.8431	1151.6	1246.8	1.4889	540
.8923	1159.1	1256.5	1.5018	.8749	1158.2	1255.4	1.4990	.8580	1157.3	1254.2	1.4962	550
.9071	1164.6	1263.6	1.5087	.8896	1163.7	1262.5	1.5060	.8726	1162.9	1261.4	1.5033	560
.9216	1169.9	1270.5	1.5155	.9040	1169.1	1269.5	1.5128	.8869	1168.3	1268.4	1.5102	570
.9359	1175.1	1277.3	1.5220	.9181	1174.3	1276.3	1.5194	.9009	1173.6	1275.3	1.5168	580
.9499	1180.2	1283.9	1.5283	.9320	1179.5	1282.9	1.5258	.9146	1178.8	1282.0	1.5232	590
.9637	1185.2	1290.4	1.5345	.9456	1184.5	1289.5	1.5320	.9281	1183.8	1288.6	1.5295	600
.9772	1190.1	1296.8	1.5405	.9590	1189.5	1295.9	1.5381	.9414	1188.8	1295.1	1.5356	610
.9906	1195.0	1303.1	1.5464	.9722	1194.4	1302.3	1.5440	.9545	1193.7	1301.5	1.5415	620
1.0038	1199.8	1309.4	1.5521	.9853	1199.2	1308.6	1.5497	.9674	1198.6	1307.8	1.5473	630
1.0168	1204.5	1315.5	1.5578	.9982	1203.9	1314.8	1.5554	.9801	1203.4	1314.0	1.5530	640
1.0297	1209.2	1321.6	1.5633	1.0109	1208.6	1320.9	1.5609	.9927	1208.1	1320.1	1.5586	650
1.0425	1213.8	1327.6	1.5687	1.0235	1213.3	1326.9	1.5664	1.0051	1212.8	1326.2	1.5641	660
1.0551	1218.4	1333.6	1.5740	1.0360	1217.9	1332.9	1.5717	1.0175	1217.4	1332.2	1.5694	670
1.0676	1223.0	1339.5	1.5792	1.0483	1222.5	1338.9	1.5769	1.0297	1222.0	1338.2	1.5747	680
1.0800	1227.5	1345.4	1.5844	1.0606	1227.0	1344.8	1.5821	1.0418	1226.5	1344.1	1.5798	690
1.0924	1232.0	1351.2	1.5894	1.0727	1231.5	1350.6	1.5872	1.0538	1231.1	1350.0	1.5849	700
1.1167	1240.9	1362.8	1.5993	1.0968	1240.4	1362.2	1.5971	1.0775	1240.0	1361.6	1.5949	720
1.1408	1249.7	1374.2	1.6089	1.1205	1249.3	1373.7	1.6067	1.1009	1248.9	1373.2	1.6046	740
1.1645	1258.4	1385.6	1.6183	1.1439	1258.1	1385.1	1.6161	1.1240	1257.7	1384.6	1.6140	760
1.1880	1267.1	1396.8	1.6274	1.1671	1266.8	1396.3	1.6253	1.1469	1266.4	1395.9	1.6232	780
1.2113	1275.8	1408.0	1.6364	1.1900	1275.4	1407.6	1.6343	1.1695	1275.1	1407.1	1.6322	800
1.2344	1284.4	1419.1	1.6452	1.2128	1284.1	1418.7	1.6430	1.1919	1283.7	1418.3	1.6410	820
1.2572	1293.0	1430.2	1.6538	1.2353	1292.7	1429.8	1.6517	1.2141	1292.3	1429.4	1.6496	840
1.2800	1301.5	1441.3	1.6622	1.2577	1301.2	1440.9	1.6601	1.2362	1300.9	1440.5	1.6581	860
1.3025	1310.1	1452.3	1.6705	1.2800	1309.8	1451.9	1.6684	1.2581	1309.5	1451.5	1.6664	880
1.3250	1318.6	1463.3	1.6786	1.3021	1318.4	1462.9	1.6766	1.2799	1318.1	1462.6	1.6745	900
1.3473	1327.2	1474.3	1.6866	1.3240	1326.9	1473.9	1.6846	1.3015	1326.7	1473.6	1.6826	920
1.3695	1335.7	1485.2	1.6945	1.3459	1335.5	1484.9	1.6925	1.3231	1335.2	1484.6	1.6905	940
1.3915	1344.3	1496.2	1.7023	1.3676	1344.0	1495.9	1.7003	1.3445	1343.8	1495.6	1.6983	960
1.4135	1352.8	1507.1	1.7100	1.3893	1352.6	1506.8	1.7079	1.3658	1352.4	1506.5	1.7060	980
1.4354	1361.4	1518.1	1.7175	1.4108	1361.2	1517.8	1.7155	1.3870	1360.9	1517.5	1.7135	1000
1.4572	1370.0	1529.0	1.7250	1.4322	1369.7	1528.8	1.7230	1.4082	1369.5	1528.5	1.7210	1020
1.4789	1378.5	1540.0	1.7323	1.4536	1378.3	1539.7	1.7303	1.4292	1378.1	1539.5	1.7284	1040
1.5005	1387.1	1551.0	1.7396	1.4749	1386.9	1550.7	1.7376	1.4502	1386.7	1550.4	1.7356	1060
1.5221	1395.8	1561.9	1.7468	1.4961	1395.6	1561.7	1.7448	1.4711	1395.4	1561.4	1.7428	1080
1.5436	1404.4	1572.9	1.7538	1.5173	1404.2	1572.7	1.7519	1.4919	1404.0	1572.4	1.7499	1100
1.5970	1426.0	1600.4	1.7712	1.5699	1425.9	1600.2	1.7692	1.5437	1425.7	1600.0	1.7673	1150
1.6502	1447.8	1628.0	1.7881	1.6222	1447.7	1627.8	1.7861	1.5952	1447.5	1627.6	1.7842	1200
1.7030	1469.8	1655.7	1.8045	1.6742	1469.6	1655.5	1.8026	1.6464	1469.5	1655.3	1.8007	1250
1.7556	1491.9	1683.6	1.8206	1.7260	1491.7	1683.4	1.8186	1.6974	1491.6	1683.2	1.8167	1300
1.8080	1514.1	1711.5	1.8363	1.7775	1514.0	1711.4	1.8343	1.7481	1513.9	1711.2	1.8324	1350
1.8602	1536.6	1739.7	1.8516	1.8289	1536.5	1739.5	1.8497	1.7986	1536.3	1739.4	1.8478	1400
1.9122	1559.2	1767.9	1.8666	1.8801	1559.1	1767.8	1.8647	1.8490	1559.0	1767.7	1.8628	1450
1.9642	1582.0	1796.4	1.8813	1.9312	1581.9	1796.3	1.8794	1.8993	1581.8	1796.1	1.8775	1500
2.068	1628.1	1853.8	1.9099	2.033	1628.0	1853.7	1.9080	1.9995	1627.9	1853.6	1.9061	1600
2.274	1722.6	1970.9	1.9641	2.236	1722.6	1970.8	1.9622	2.1989	1722.5	1970.7	1.9604	1800
2.479	1820.3	2090.9	2.0150	2.438	1820.2	2090.8	2.0131	2.3975	1820.1	2090.8	2.0113	2000
2.684	1920.9	2213.9	2.0631	2.639	1920.8	2213.8	2.0612	2.5957	1920.7	2213.7	2.0593	2200
2.888	2024.1	2339.5	2.1086	2.840	2024.0	2339.4	2.1067	2.7937	2024.0	2339.3	2.1049	2400

Table 3. Vapor

t	v	u	h	s	v	u	h	s	v	u	h	s
Sat.	.7445	1118.3	1203.7	1.4431	.7322	1118.2	1203.5	1.4415	.7203	1118.0	1203.3	1.4398
480	*.7260*	*1110.9*	*1194.2*	*1.4330*	*.7106*	*1109.4*	*1192.2*	*1.4295*				
490	.7447	1118.4	1203.8	1.4432	*.7293*	*1117.0*	*1202.0*	*1.4399*	*.7143*	*1115.5*	*1200.1*	*1.4365*
500	.7624	1125.4	1212.9	1.4527	.7470	1124.1	1211.2	1.4495	.7320	1122.8	1209.5	1.4463
510	.7794	1132.2	1221.6	1.4617	.7640	1131.0	1220.0	1.4586	.7489	1129.7	1218.4	1.4556
520	.7958	1138.6	1229.9	1.4702	.7802	1137.5	1228.4	1.4673	.7652	1136.3	1227.0	1.4643
530	.8116	1144.7	1237.8	1.4783	.7959	1143.7	1236.5	1.4755	.7808	1142.7	1235.1	1.4726
540	.8269	1150.7	1245.6	1.4861	.8112	1149.7	1244.3	1.4833	.7959	1148.8	1243.0	1.4805
550	.8417	1156.5	1253.0	1.4935	.8259	1155.5	1251.8	1.4908	.8106	1154.6	1250.6	1.4881
560	.8562	1162.0	1260.3	1.5007	.8403	1161.2	1259.2	1.4980	.8249	1160.3	1258.0	1.4954
570	.8704	1167.5	1267.4	1.5076	.8543	1166.7	1266.3	1.5050	.8388	1165.9	1265.2	1.5024
580	.8842	1172.8	1274.3	1.5143	.8681	1172.1	1273.3	1.5117	.8524	1171.3	1272.3	1.5092
590	.8978	1178.0	1281.0	1.5207	.8816	1177.3	1280.1	1.5183	.8658	1176.6	1279.1	1.5158
600	.9112	1183.1	1287.7	1.5270	.8948	1182.5	1286.8	1.5246	.8789	1181.8	1285.9	1.5222
610	.9243	1188.2	1294.2	1.5332	.9078	1187.5	1293.3	1.5308	.8918	1186.9	1292.5	1.5284
620	.9373	1193.1	1300.6	1.5392	.9206	1192.5	1299.8	1.5368	.9045	1191.9	1299.0	1.5344
630	.9500	1198.0	1307.0	1.5450	.9332	1197.4	1306.2	1.5427	.9169	1196.8	1305.4	1.5404
640	.9626	1202.8	1313.2	1.5507	.9457	1202.2	1312.5	1.5484	.9293	1201.6	1311.7	1.5461
650	.9751	1207.5	1319.4	1.5563	.9580	1207.0	1318.7	1.5540	.9414	1206.4	1317.9	1.5518
660	.9874	1212.2	1325.5	1.5618	.9702	1211.7	1324.8	1.5595	.9535	1211.2	1324.1	1.5573
670	.9995	1216.9	1331.6	1.5672	.9822	1216.4	1330.9	1.5649	.9654	1215.9	1330.2	1.5627
680	1.0116	1221.5	1337.6	1.5724	.9941	1221.0	1336.9	1.5702	.9772	1220.5	1336.2	1.5680
690	1.0236	1226.1	1343.5	1.5776	1.0059	1225.6	1342.9	1.5754	.9888	1225.1	1342.2	1.5733
700	1.0354	1230.6	1349.4	1.5827	1.0176	1230.1	1348.8	1.5806	1.0004	1229.7	1348.2	1.5784
720	1.0588	1239.6	1361.1	1.5927	1.0408	1239.2	1360.5	1.5906	1.0232	1238.7	1359.9	1.5885
740	1.0819	1248.5	1372.6	1.6024	1.0636	1248.1	1372.1	1.6003	1.0458	1247.7	1371.5	1.5982
760	1.1047	1257.3	1384.0	1.6119	1.0861	1256.9	1383.5	1.6098	1.0680	1256.5	1383.0	1.6077
780	1.1273	1266.0	1395.4	1.6211	1.1083	1265.7	1394.9	1.6190	1.0899	1265.3	1394.4	1.6170
800	1.1496	1274.7	1406.6	1.6301	1.1303	1274.4	1406.2	1.6281	1.1117	1274.1	1405.7	1.6260
820	1.1717	1283.4	1417.8	1.6389	1.1522	1283.1	1417.4	1.6369	1.1332	1282.8	1417.0	1.6349
840	1.1936	1292.0	1429.0	1.6476	1.1738	1291.7	1428.6	1.6456	1.1545	1291.4	1428.1	1.6436
860	1.2154	1300.6	1440.1	1.6560	1.1952	1300.3	1439.7	1.6540	1.1757	1300.1	1439.3	1.6521
880	1.2370	1309.2	1451.2	1.6644	1.2165	1309.0	1450.8	1.6624	1.1967	1308.7	1450.4	1.6604
900	1.2585	1317.8	1462.2	1.6726	1.2377	1317.6	1461.8	1.6706	1.2176	1317.3	1461.5	1.6686
920	1.2798	1326.4	1473.2	1.6806	1.2587	1326.1	1472.9	1.6786	1.2383	1325.9	1472.5	1.6767
940	1.3010	1335.0	1484.2	1.6885	1.2796	1334.7	1483.9	1.6866	1.2589	1334.5	1483.6	1.6847
960	1.3221	1343.6	1495.2	1.6963	1.3004	1343.3	1494.9	1.6944	1.2794	1343.1	1494.6	1.6925
980	1.3431	1352.1	1506.2	1.7040	1.3211	1351.9	1505.9	1.7021	1.2998	1351.7	1505.6	1.7002
1000	1.3640	1360.7	1517.2	1.7116	1.3417	1360.5	1516.9	1.7097	1.3202	1360.3	1516.6	1.7078
1020	1.3848	1369.3	1528.2	1.7191	1.3623	1369.1	1527.9	1.7172	1.3404	1368.9	1527.6	1.7153
1040	1.4056	1377.9	1539.2	1.7264	1.3827	1377.7	1538.9	1.7245	1.3605	1377.5	1538.6	1.7227
1060	1.4262	1386.5	1550.2	1.7337	1.4030	1386.3	1549.9	1.7318	1.3806	1386.2	1549.7	1.7300
1080	1.4468	1395.2	1561.2	1.7409	1.4233	1395.0	1560.9	1.7390	1.4006	1394.8	1560.7	1.7372
1100	1.4673	1403.8	1572.2	1.7480	1.4435	1403.7	1571.9	1.7461	1.4205	1403.5	1571.7	1.7443
1150	1.5184	1425.5	1599.8	1.7654	1.4938	1425.4	1599.5	1.7635	1.4700	1425.2	1599.3	1.7617
1200	1.5691	1447.4	1627.4	1.7823	1.5438	1447.2	1627.2	1.7805	1.5192	1447.1	1627.0	1.7786
1250	1.6195	1469.4	1655.2	1.7988	1.5934	1469.2	1655.0	1.7970	1.5681	1469.1	1654.8	1.7951
1300	1.6696	1491.5	1683.0	1.8149	1.6428	1491.4	1682.9	1.8130	1.6168	1491.2	1682.7	1.8112
1350	1.7196	1513.8	1711.1	1.8306	1.6920	1513.6	1710.9	1.8288	1.6653	1513.5	1710.7	1.8269
1400	1.7694	1536.2	1739.2	1.8459	1.7410	1536.1	1739.1	1.8441	1.7135	1536.0	1738.9	1.8423
1450	1.8190	1558.8	1767.5	1.8610	1.7898	1558.7	1767.4	1.8591	1.7616	1558.6	1767.3	1.8573
1500	1.8684	1581.7	1796.0	1.8757	1.8385	1581.6	1795.9	1.8739	1.8096	1581.5	1795.8	1.8721
1600	1.9670	1627.8	1853.5	1.9043	1.9356	1627.7	1853.4	1.9025	1.9052	1627.6	1853.3	1.9007
1800	2.1633	1722.4	1970.6	1.9585	2.1288	1722.3	1970.5	1.9567	2.0955	1722.3	1970.4	1.9550
2000	2.3588	1820.1	2090.7	2.0095	2.3213	1820.0	2090.6	2.0077	2.2849	1819.9	2090.5	2.0059
2200	2.5538	1920.7	2213.7	2.0575	2.5133	1920.6	2213.6	2.0557	2.4740	1920.5	2213.5	2.0540
2400	2.7487	2023.9	2339.3	2.1030	2.7051	2023.8	2339.2	2.1013	2.6628	2023.8	2339.1	2.0995

v	u	h	s	v	u	h	s	v	u	h	s	t
.7088	1117.9	1203.1	1.4382	.6975	1117.7	1202.9	1.4367	.6867	1117.5	1202.7	1.4351	Sat.
												480
.6997	*1114.1*	*1198.2*	*1.4331*	*.6855*	*1112.6*	*1196.3*	*1.4297*	*.6717*	*1111.1*	*1194.4*	*1.4264*	**490**
.7175	1121.5	1207.8	1.4431	.7033	1120.1	1206.0	1.4399	.6895	1118.8	1204.3	1.4367	**500**
.7343	1128.5	1216.9	1.4525	.7202	1127.3	1215.2	1.4495	.7064	1126.0	1213.6	1.4464	**510**
.7505	1135.2	1225.5	1.4614	.7363	1134.1	1224.0	1.4584	.7225	1132.9	1222.5	1.4555	**520**
.7661	1141.6	1233.8	1.4698	.7518	1140.6	1232.4	1.4670	.7379	1139.5	1231.0	1.4641	**530**
.7811	1147.8	1241.7	1.4778	.7668	1146.8	1240.4	1.4751	.7528	1145.8	1239.1	1.4723	**540**
.7957	1153.7	1249.4	1.4855	.7813	1152.8	1248.2	1.4828	.7672	1151.9	1247.0	1.4802	**550**
.8099	1159.5	1256.9	1.4928	.7954	1158.6	1255.7	1.4902	.7813	1157.7	1254.6	1.4877	**560**
.8237	1165.1	1264.2	1.4999	.8091	1164.3	1263.1	1.4974	.7949	1163.4	1262.0	1.4949	**570**
.8373	1170.5	1271.2	1.5067	.8225	1169.7	1270.2	1.5043	.8082	1169.0	1269.2	1.5018	**580**
.8505	1175.9	1278.2	1.5134	.8356	1175.1	1277.2	1.5110	.8212	1174.4	1276.2	1.5086	**590**
.8635	1181.1	1284.9	1.5198	.8485	1180.4	1284.0	1.5174	.8340	1179.7	1283.1	1.5151	**600**
.8762	1186.2	1291.6	1.5260	.8612	1185.5	1290.7	1.5237	.8465	1184.9	1289.8	1.5214	**610**
.8888	1191.2	1298.1	1.5321	.8736	1190.6	1297.3	1.5298	.8589	1189.9	1296.4	1.5276	**620**
.9012	1196.2	1304.6	1.5381	.8859	1195.6	1303.8	1.5358	.8710	1195.0	1302.9	1.5336	**630**
.9134	1201.1	1310.9	1.5439	.8979	1200.5	1310.1	1.5416	.8829	1199.9	1309.4	1.5394	**640**
.9254	1205.9	1317.2	1.5495	.9098	1205.3	1316.4	1.5473	.8947	1204.7	1315.7	1.5452	**650**
.9373	1210.6	1323.4	1.5551	.9216	1210.1	1322.7	1.5529	.9064	1209.6	1321.9	1.5508	**660**
.9491	1215.3	1329.5	1.5605	.9333	1214.8	1328.8	1.5584	.9179	1214.3	1328.1	1.5563	**670**
.9607	1220.0	1335.6	1.5659	.9448	1219.5	1334.9	1.5638	.9293	1219.0	1334.2	1.5616	**680**
.9723	1224.6	1341.6	1.5711	.9562	1224.1	1340.9	1.5690	.9406	1223.7	1340.3	1.5669	**690**
.9837	1229.2	1347.5	1.5763	.9675	1228.8	1346.9	1.5742	.9518	1228.3	1346.3	1.5721	**700**
1.0063	1238.3	1359.3	1.5864	.9898	1237.9	1358.7	1.5843	.9738	1237.4	1358.2	1.5823	**720**
1.0285	1247.3	1371.0	1.5962	1.0118	1246.9	1370.4	1.5941	.9955	1246.4	1369.9	1.5921	**740**
1.0505	1256.1	1382.5	1.6057	1.0335	1255.8	1382.0	1.6037	1.0170	1255.4	1381.5	1.6017	**760**
1.0721	1265.0	1393.9	1.6150	1.0549	1264.6	1393.4	1.6130	1.0381	1264.2	1392.9	1.6110	**780**
1.0936	1273.7	1405.3	1.6241	1.0760	1273.4	1404.8	1.6221	1.0590	1273.0	1404.3	1.6202	**800**
1.1148	1282.4	1416.5	1.6329	1.0970	1282.1	1416.1	1.6310	1.0797	1281.8	1415.6	1.6291	**820**
1.1359	1291.1	1427.7	1.6416	1.1178	1290.8	1427.3	1.6397	1.1003	1290.5	1426.9	1.6378	**840**
1.1568	1299.8	1438.9	1.6501	1.1384	1299.5	1438.5	1.6482	1.1206	1299.2	1438.1	1.6463	**860**
1.1775	1308.4	1450.0	1.6585	1.1589	1308.1	1449.6	1.6566	1.1408	1307.8	1449.3	1.6547	**880**
1.1981	1317.0	1461.1	1.6667	1.1792	1316.7	1460.8	1.6649	1.1608	1316.5	1460.4	1.6630	**900**
1.2185	1325.6	1472.2	1.6748	1.1994	1325.4	1471.8	1.6729	1.1807	1325.1	1471.5	1.6711	**920**
1.2389	1334.2	1483.2	1.6828	1.2194	1334.0	1482.9	1.6809	1.2005	1333.7	1482.6	1.6791	**940**
1.2591	1342.8	1494.3	1.6906	1.2394	1342.6	1494.0	1.6888	1.2202	1342.4	1493.6	1.6869	**960**
1.2792	1351.4	1505.3	1.6983	1.2592	1351.2	1505.0	1.6965	1.2398	1351.0	1504.7	1.6947	**980**
1.2992	1360.1	1516.3	1.7059	1.2790	1359.8	1516.0	1.7041	1.2593	1359.6	1515.7	1.7023	**1000**
1.3192	1368.7	1527.4	1.7134	1.2986	1368.5	1527.1	1.7116	1.2787	1368.3	1526.8	1.7098	**1020**
1.3390	1377.3	1538.4	1.7208	1.3182	1377.1	1538.1	1.7190	1.2980	1376.9	1537.8	1.7172	**1040**
1.3588	1386.0	1549.4	1.7281	1.3377	1385.8	1549.1	1.7263	1.3172	1385.6	1548.9	1.7245	**1060**
1.3785	1394.6	1560.4	1.7353	1.3571	1394.4	1560.2	1.7335	1.3364	1394.2	1559.9	1.7317	**1080**
1.3982	1403.3	1571.5	1.7425	1.3765	1403.1	1571.2	1.7407	1.3555	1402.9	1571.0	1.7389	**1100**
1.4470	1425.0	1599.1	1.7599	1.4246	1424.9	1598.9	1.7581	1.4029	1424.7	1598.6	1.7563	**1150**
1.4955	1446.9	1626.8	1.7768	1.4724	1446.8	1626.6	1.7751	1.4501	1446.6	1626.4	1.7733	**1200**
1.5437	1468.9	1654.6	1.7934	1.5199	1468.8	1654.4	1.7916	1.4969	1468.7	1654.2	1.7898	**1250**
1.5916	1491.1	1682.5	1.8094	1.5672	1491.0	1682.4	1.8077	1.5435	1490.8	1682.2	1.8060	**1300**
1.6393	1513.4	1710.6	1.8252	1.6142	1513.3	1710.4	1.8234	1.5898	1513.2	1710.3	1.8217	**1350**
1.6869	1535.9	1738.8	1.8405	1.6611	1535.8	1738.6	1.8388	1.6360	1535.7	1738.5	1.8371	**1400**
1.7343	1558.5	1767.1	1.8556	1.7078	1558.4	1767.0	1.8538	1.6820	1558.3	1766.9	1.8521	**1450**
1.7815	1581.4	1795.6	1.8703	1.7543	1581.3	1795.5	1.8686	1.7279	1581.2	1795.4	1.8669	**1500**
1.8757	1627.5	1853.2	1.8989	1.8471	1627.5	1853.0	1.8972	1.8194	1627.4	1852.9	1.8955	**1600**
2.0631	1722.2	1970.3	1.9532	2.0317	1722.1	1970.2	1.9515	2.0013	1722.0	1970.2	1.9498	**1800**
2.2497	1819.9	2090.5	2.0042	2.2156	1819.8	2090.4	2.0024	2.1824	1819.7	2090.3	2.0008	**2000**
2.4359	1920.5	2213.5	2.0522	2.3990	1920.4	2213.4	2.0505	2.3631	1920.3	2213.3	2.0488	**2200**
2.6219	2023.7	2339.1	2.0978	2.5822	2023.6	2339.0	2.0961	2.5436	2023.6	2338.9	2.0944	**2400**

Table 3. Vapor

p (t Sat.)	680 (500.00)				690 (501.62)				700 (503.23)			
t	v	u	h	s	v	u	h	s	v	u	h	s
Sat.	.6761	1117.4	1202.5	1.4335	.6658	1117.2	1202.2	1.4320	.6558	1117.0	1202.0	1.4305
490	.6583	1109.5	1192.4	1.4230	.6452	1108.0	1190.4	1.4196				
500	.6761	1117.4	1202.5	1.4335	.6630	1116.0	1200.6	1.4303	.6503	1114.5	1198.8	1.4271
510	.6929	1124.8	1211.9	1.4434	.6799	1123.5	1210.3	1.4403	.6671	1122.1	1208.6	1.4373
520	.7090	1131.7	1221.0	1.4526	.6959	1130.5	1219.4	1.4497	.6832	1129.3	1217.8	1.4468
530	.7244	1138.4	1229.5	1.4613	.7113	1137.3	1228.1	1.4586	.6985	1136.2	1226.7	1.4558
540	.7393	1144.8	1237.8	1.4696	.7261	1143.7	1236.5	1.4669	.7132	1142.7	1235.1	1.4643
550	.7536	1150.9	1245.7	1.4775	.7404	1150.0	1244.5	1.4749	.7275	1149.0	1243.2	1.4723
560	.7675	1156.8	1253.4	1.4851	.7542	1155.9	1252.2	1.4826	.7412	1155.0	1251.1	1.4801
570	.7811	1162.6	1260.9	1.4924	.7677	1161.7	1259.8	1.4899	.7546	1160.9	1258.7	1.4875
580	.7943	1168.2	1268.1	1.4994	.7808	1167.4	1267.1	1.4970	.7677	1166.6	1266.0	1.4946
590	.8072	1173.6	1275.2	1.5062	.7936	1172.9	1274.2	1.5038	.7804	1172.1	1273.2	1.5015
600	.8199	1179.0	1282.1	1.5127	.8062	1178.2	1281.2	1.5104	.7929	1177.5	1280.2	1.5081
610	.8323	1184.2	1288.9	1.5191	.8185	1183.5	1288.0	1.5168	.8051	1182.8	1287.1	1.5146
620	.8445	1189.3	1295.6	1.5253	.8306	1188.7	1294.7	1.5231	.8171	1188.0	1293.9	1.5209
630	.8566	1194.3	1302.1	1.5313	.8425	1193.7	1301.3	1.5292	.8289	1193.1	1300.5	1.5270
640	.8684	1199.3	1308.6	1.5372	.8543	1198.7	1307.8	1.5351	.8405	1198.1	1307.0	1.5329
650	.8801	1204.2	1314.9	1.5430	.8658	1203.6	1314.2	1.5409	.8520	1203.1	1313.4	1.5387
660	.8916	1209.0	1321.2	1.5486	.8773	1208.5	1320.5	1.5465	.8633	1207.9	1319.8	1.5444
670	.9030	1213.8	1327.4	1.5541	.8885	1213.3	1326.7	1.5521	.8745	1212.7	1326.0	1.5500
680	.9143	1218.5	1333.6	1.5596	.8997	1218.0	1332.9	1.5575	.8855	1217.5	1332.2	1.5554
690	.9254	1223.2	1339.6	1.5649	.9107	1222.7	1339.0	1.5628	.8965	1222.2	1338.3	1.5608
700	.9365	1227.8	1345.7	1.5701	.9217	1227.3	1345.0	1.5681	.9073	1226.9	1344.4	1.5661
720	.9583	1237.0	1357.6	1.5803	.9432	1236.5	1357.0	1.5783	.9286	1236.1	1356.4	1.5763
740	.9798	1246.0	1369.3	1.5901	.9645	1245.6	1368.8	1.5882	.9496	1245.2	1368.2	1.5863
760	1.0009	1255.0	1380.9	1.5998	.9854	1254.6	1380.4	1.5978	.9703	1254.2	1379.9	1.5959
780	1.0218	1263.9	1392.4	1.6091	1.0061	1263.5	1392.0	1.6072	.9907	1263.1	1391.5	1.6053
800	1.0425	1272.7	1403.9	1.6182	1.0265	1272.3	1403.4	1.6164	1.0109	1272.0	1402.9	1.6145
820	1.0630	1281.4	1415.2	1.6272	1.0467	1281.1	1414.8	1.6253	1.0308	1280.8	1414.3	1.6235
840	1.0832	1290.2	1426.5	1.6359	1.0667	1289.8	1426.0	1.6341	1.0506	1289.5	1425.6	1.6322
860	1.1033	1298.9	1437.7	1.6445	1.0865	1298.6	1437.3	1.6426	1.0702	1298.3	1436.9	1.6408
880	1.1232	1307.5	1448.9	1.6529	1.1062	1307.2	1448.5	1.6511	1.0896	1307.0	1448.1	1.6493
900	1.1430	1316.2	1460.0	1.6612	1.1257	1315.9	1459.7	1.6593	1.1089	1315.6	1459.3	1.6576
920	1.1627	1324.8	1471.1	1.6693	1.1451	1324.6	1470.8	1.6675	1.1281	1324.3	1470.4	1.6657
940	1.1822	1333.5	1482.2	1.6773	1.1644	1333.2	1481.9	1.6755	1.1471	1333.0	1481.6	1.6737
960	1.2016	1342.1	1493.3	1.6851	1.1836	1341.9	1493.0	1.6833	1.1661	1341.6	1492.7	1.6816
980	1.2210	1350.7	1504.4	1.6929	1.2027	1350.5	1504.1	1.6911	1.1849	1350.3	1503.8	1.6893
1000	1.2402	1359.4	1515.4	1.7005	1.2216	1359.2	1515.2	1.6987	1.2036	1358.9	1514.9	1.6970
1020	1.2593	1368.0	1526.5	1.7080	1.2405	1367.8	1526.2	1.7063	1.2223	1367.6	1525.9	1.7045
1040	1.2784	1376.7	1537.6	1.7154	1.2593	1376.5	1537.3	1.7137	1.2408	1376.3	1537.0	1.7120
1060	1.2973	1385.4	1548.6	1.7228	1.2780	1385.2	1548.4	1.7210	1.2593	1385.0	1548.1	1.7193
1080	1.3162	1394.0	1559.7	1.7300	1.2967	1393.9	1559.4	1.7283	1.2777	1393.7	1559.2	1.7265
1100	1.3351	1402.7	1570.7	1.7371	1.3153	1402.6	1570.5	1.7354	1.2960	1402.4	1570.2	1.7337
1150	1.3819	1424.5	1598.4	1.7546	1.3614	1424.4	1598.2	1.7529	1.3416	1424.2	1598.0	1.7512
1200	1.4283	1446.5	1626.2	1.7716	1.4073	1446.3	1626.0	1.7699	1.3868	1446.2	1625.8	1.7682
1250	1.4745	1468.5	1654.1	1.7881	1.4528	1468.4	1653.9	1.7864	1.4317	1468.2	1653.7	1.7848
1300	1.5205	1490.7	1682.0	1.8043	1.4981	1490.6	1681.9	1.8026	1.4764	1490.4	1681.7	1.8009
1350	1.5662	1513.0	1710.1	1.8200	1.5432	1512.9	1710.0	1.8183	1.5209	1512.8	1709.8	1.8167
1400	1.6117	1535.5	1738.4	1.8354	1.5881	1535.4	1738.2	1.8337	1.5652	1535.3	1738.1	1.8321
1450	1.6571	1558.2	1766.7	1.8504	1.6328	1558.1	1766.6	1.8488	1.6093	1558.0	1766.5	1.8471
1500	1.7023	1581.1	1795.3	1.8652	1.6774	1581.0	1795.1	1.8635	1.6533	1580.9	1795.0	1.8619
1600	1.7924	1627.3	1852.8	1.8938	1.7663	1627.2	1852.7	1.8922	1.7409	1627.1	1852.6	1.8906
1800	1.9717	1722.0	1970.1	1.9482	1.9431	1721.9	1970.0	1.9465	1.9152	1721.8	1969.9	1.9449
2000	2.1503	1819.6	2090.2	1.9991	2.1190	1819.6	2090.1	1.9975	2.0887	1819.5	2090.1	1.9958
2200	2.3284	1920.3	2213.2	2.0472	2.2946	1920.2	2213.2	2.0455	2.2618	1920.1	2213.1	2.0439
2400	2.5063	2023.5	2338.9	2.0927	2.4699	2023.4	2338.8	2.0911	2.4347	2023.4	2338.7	2.0895

v	u	h	s	v	u	h	s	v	u	h	s	t
.6461	1116.9	1201.7	1.4290	.6366	1116.7	1201.5	1.4275	.6274	1116.5	1201.2	1.4260	**Sat.**
												490
.6378	*1113.1*	*1196.9*	*1.4239*	*.6257*	*1111.6*	*1194.9*	*1.4207*	*.6138*	*1110.1*	*1193.0*	*1.4175*	**500**
.6547	1120.8	1206.8	1.4343	.6426	1119.5	1205.1	1.4312	.6308	1118.1	1203.3	1.4282	**510**
.6708	1128.1	1216.3	1.4439	.6586	1126.9	1214.6	1.4410	.6468	1125.6	1213.0	1.4381	**520**
.6861	1135.1	1225.2	1.4530	.6739	1133.9	1223.7	1.4502	.6621	1132.8	1222.2	1.4475	**530**
.7008	1141.7	1233.7	1.4616	.6886	1140.6	1232.4	1.4589	.6767	1139.6	1231.0	1.4563	**540**
.7149	1148.0	1241.9	1.4698	.7027	1147.0	1240.7	1.4672	.6908	1146.0	1239.4	1.4646	**550**
.7286	1154.1	1249.9	1.4776	.7164	1153.2	1248.7	1.4751	.7044	1152.3	1247.4	1.4726	**560**
.7419	1160.0	1257.5	1.4850	.7296	1159.2	1256.4	1.4826	.7176	1158.3	1255.3	1.4802	**570**
.7549	1165.8	1265.0	1.4922	.7425	1165.0	1263.9	1.4899	.7304	1164.2	1262.8	1.4875	**580**
.7676	1171.4	1272.2	1.4992	.7550	1170.6	1271.2	1.4969	.7429	1169.8	1270.2	1.4946	**590**
.7799	1176.8	1279.3	1.5059	.7673	1176.1	1278.3	1.5036	.7551	1175.4	1277.4	1.5014	**600**
.7921	1182.1	1286.2	1.5124	.7794	1181.4	1285.3	1.5101	.7670	1180.8	1284.4	1.5080	**610**
.8040	1187.4	1293.0	1.5187	.7912	1186.7	1292.1	1.5165	.7787	1186.0	1291.2	1.5143	**620**
.8157	1192.5	1299.6	1.5248	.8028	1191.9	1298.8	1.5227	.7903	1191.2	1298.0	1.5206	**630**
.8272	1197.5	1306.2	1.5308	.8142	1196.9	1305.4	1.5287	.8016	1196.3	1304.6	1.5266	**640**
.8386	1202.5	1312.7	1.5366	.8255	1201.9	1311.9	1.5346	.8127	1201.3	1311.1	1.5325	**650**
.8497	1207.4	1319.0	1.5424	.8366	1206.8	1318.3	1.5403	.8237	1206.3	1317.5	1.5383	**660**
.8608	1212.2	1325.3	1.5479	.8475	1211.7	1324.6	1.5459	.8346	1211.1	1323.9	1.5439	**670**
.8717	1217.0	1331.5	1.5534	.8583	1216.5	1330.8	1.5514	.8453	1216.0	1330.2	1.5494	**680**
.8826	1221.7	1337.7	1.5588	.8691	1221.2	1337.0	1.5568	.8559	1220.7	1336.4	1.5549	**690**
.8933	1226.4	1343.8	1.5641	.8797	1225.9	1343.1	1.5621	.8664	1225.5	1342.5	1.5602	**700**
.9144	1235.7	1355.8	1.5744	.9006	1235.2	1355.2	1.5724	.8871	1234.8	1354.6	1.5705	**720**
.9351	1244.8	1367.7	1.5843	.9211	1244.4	1367.1	1.5824	.9074	1244.0	1366.5	1.5806	**740**
.9556	1253.8	1379.4	1.5940	.9413	1253.4	1378.8	1.5921	.9274	1253.0	1378.3	1.5903	**760**
.9758	1262.8	1391.0	1.6034	.9613	1262.4	1390.5	1.6016	.9472	1262.0	1390.0	1.5998	**780**
.9957	1271.6	1402.5	1.6126	.9810	1271.3	1402.0	1.6108	.9667	1270.9	1401.5	1.6090	**800**
1.0155	1280.4	1413.9	1.6216	1.0005	1280.1	1413.4	1.6198	.9860	1279.8	1413.0	1.6180	**820**
1.0350	1289.2	1425.2	1.6304	1.0198	1288.9	1424.8	1.6286	1.0050	1288.6	1424.4	1.6269	**840**
1.0544	1298.0	1436.5	1.6390	1.0389	1297.7	1436.1	1.6373	1.0240	1297.4	1435.7	1.6355	**860**
1.0736	1306.7	1447.7	1.6475	1.0579	1306.4	1447.3	1.6457	1.0427	1306.1	1447.0	1.6440	**880**
1.0926	1315.4	1458.9	1.6558	1.0767	1315.1	1458.6	1.6540	1.0613	1314.8	1458.2	1.6523	**900**
1.1115	1324.0	1470.1	1.6639	1.0954	1323.8	1469.7	1.6622	1.0798	1323.5	1469.4	1.6605	**920**
1.1303	1332.7	1481.2	1.6720	1.1140	1332.5	1480.9	1.6702	1.0981	1332.2	1480.6	1.6685	**940**
1.1490	1341.4	1492.4	1.6798	1.1325	1341.1	1492.0	1.6781	1.1164	1340.9	1491.7	1.6764	**960**
1.1676	1350.1	1503.5	1.6876	1.1508	1349.8	1503.2	1.6859	1.1345	1349.6	1502.8	1.6842	**980**
1.1861	1358.7	1514.6	1.6953	1.1691	1358.5	1514.3	1.6936	1.1525	1358.3	1514.0	1.6919	**1000**
1.2045	1367.4	1525.7	1.7028	1.1873	1367.2	1525.4	1.7011	1.1705	1367.0	1525.1	1.6995	**1020**
1.2228	1376.1	1536.7	1.7103	1.2053	1375.9	1536.5	1.7086	1.1883	1375.7	1536.2	1.7069	**1040**
1.2411	1384.8	1547.8	1.7176	1.2233	1384.6	1547.6	1.7159	1.2061	1384.4	1547.3	1.7143	**1060**
1.2592	1393.5	1558.9	1.7249	1.2413	1393.3	1558.7	1.7232	1.2238	1393.1	1558.4	1.7215	**1080**
1.2773	1402.2	1570.0	1.7320	1.2591	1402.0	1569.8	1.7304	1.2415	1401.8	1569.5	1.7287	**1100**
1.3223	1424.0	1597.8	1.7495	1.3035	1423.9	1597.5	1.7479	1.2853	1423.7	1597.3	1.7463	**1150**
1.3669	1446.0	1625.6	1.7666	1.3476	1445.8	1625.4	1.7649	1.3288	1445.7	1625.2	1.7633	**1200**
1.4112	1468.1	1653.5	1.7831	1.3913	1467.9	1653.3	1.7815	1.3719	1467.8	1653.1	1.7799	**1250**
1.4553	1490.3	1681.5	1.7993	1.4348	1490.2	1681.3	1.7977	1.4149	1490.0	1681.2	1.7961	**1300**
1.4992	1512.7	1709.6	1.8150	1.4781	1512.6	1709.5	1.8134	1.4576	1512.4	1709.3	1.8118	**1350**
1.5429	1535.2	1737.9	1.8304	1.5212	1535.1	1737.8	1.8288	1.5001	1535.0	1737.6	1.8273	**1400**
1.5864	1557.9	1766.3	1.8455	1.5642	1557.8	1766.2	1.8439	1.5425	1557.7	1766.1	1.8423	**1450**
1.6298	1580.8	1794.9	1.8603	1.6070	1580.7	1794.8	1.8587	1.5847	1580.6	1794.6	1.8571	**1500**
1.7162	1627.0	1852.5	1.8890	1.6922	1626.9	1852.4	1.8874	1.6689	1626.8	1852.3	1.8858	**1600**
1.8881	1721.7	1969.8	1.9433	1.8618	1721.6	1969.7	1.9417	1.8362	1721.6	1969.6	1.9402	**1800**
2.0592	1819.4	2090.0	1.9943	2.0306	1819.4	2089.9	1.9927	2.0027	1819.3	2089.8	1.9911	**2000**
2.2299	1920.1	2213.0	2.0423	2.1989	1920.0	2213.0	2.0408	2.1688	1919.9	2212.9	2.0392	**2200**
2.4004	2023.3	2338.7	2.0879	2.3671	2023.2	2338.6	2.0863	2.3347	2023.1	2338.5	2.0848	**2400**

Table 3. Vapor

p (t Sat.)	740 (509.47)				750 (510.99)				775 (514.72)			
t	v	u	h	s	v	u	h	s	v	u	h	s
Sat.	.6184	1116.3	1201.0	1.4246	.6097	1116.1	1200.7	1.4231	.5887	1115.6	1200.0	1.4195
500	*.6023*	*1108.5*	*1191.0*	*1.4142*	*.5910*	*1107.0*	*1189.0*	*1.4110*				
510	.6193	1116.7	1201.5	1.4251	*.6080*	*1115.3*	*1199.7*	*1.4221*	*.5810*	*1111.7*	*1195.0*	*1.4144*
520	.6353	1124.4	1211.4	1.4352	.6241	1123.1	1209.7	1.4323	.5971	1119.8	1205.4	1.4251
530	.6506	1131.6	1220.7	1.4447	.6394	1130.4	1219.2	1.4419	.6124	1127.4	1215.2	1.4351
540	.6652	1138.5	1229.6	1.4536	.6539	1137.4	1228.1	1.4510	.6269	1134.6	1224.5	1.4444
550	.6792	1145.0	1238.1	1.4621	.6679	1144.0	1236.7	1.4595	.6408	1141.5	1233.4	1.4532
560	.6927	1151.4	1246.2	1.4701	.6814	1150.4	1245.0	1.4677	.6542	1148.0	1241.8	1.4616
570	.7059	1157.4	1254.1	1.4778	.6944	1156.6	1252.9	1.4754	.6671	1154.3	1250.0	1.4695
580	.7186	1163.3	1261.7	1.4852	.7071	1162.5	1260.6	1.4829	.6797	1160.4	1257.9	1.4771
590	.7310	1169.1	1269.2	1.4923	.7195	1168.3	1268.1	1.4900	.6918	1166.3	1265.5	1.4844
600	.7431	1174.6	1276.4	1.4991	.7315	1173.9	1275.4	1.4969	.7037	1172.0	1272.9	1.4915
610	.7550	1180.1	1283.4	1.5058	.7433	1179.4	1282.5	1.5036	.7153	1177.6	1280.2	1.4983
620	.7666	1185.4	1290.4	1.5122	.7548	1184.7	1289.5	1.5101	.7266	1183.0	1287.2	1.5048
630	.7781	1190.6	1297.1	1.5185	.7662	1190.0	1296.3	1.5164	.7378	1188.4	1294.2	1.5112
640	.7893	1195.7	1303.8	1.5245	.7773	1195.1	1303.0	1.5225	.7487	1193.6	1300.9	1.5174
650	.8003	1200.8	1310.3	1.5305	.7883	1200.2	1309.6	1.5284	.7594	1198.7	1307.6	1.5235
660	.8112	1205.7	1316.8	1.5363	.7991	1205.2	1316.1	1.5343	.7700	1203.8	1314.2	1.5294
670	.8220	1210.6	1323.2	1.5419	.8097	1210.1	1322.5	1.5400	.7805	1208.7	1320.7	1.5351
680	.8326	1215.5	1329.5	1.5475	.8203	1214.9	1328.8	1.5455	.7908	1213.6	1327.1	1.5407
690	.8431	1220.2	1335.7	1.5529	.8307	1219.7	1335.0	1.5510	.8009	1218.5	1333.4	1.5463
700	.8535	1225.0	1341.9	1.5583	.8410	1224.5	1341.2	1.5563	.8110	1223.3	1339.6	1.5517
720	.8740	1234.3	1354.0	1.5686	.8613	1233.9	1353.4	1.5668	.8308	1232.8	1351.9	1.5622
740	.8941	1243.5	1366.0	1.5787	.8812	1243.1	1365.4	1.5769	.8502	1242.1	1364.0	1.5724
760	.9139	1252.6	1377.8	1.5885	.9008	1252.2	1377.3	1.5867	.8694	1251.3	1375.9	1.5822
780	.9335	1261.6	1389.5	1.5980	.9201	1261.3	1389.0	1.5962	.8882	1260.3	1387.7	1.5918
800	.9527	1270.6	1401.0	1.6072	.9392	1270.2	1400.6	1.6055	.9068	1269.3	1399.4	1.6011
820	.9718	1279.4	1412.5	1.6163	.9580	1279.1	1412.1	1.6145	.9251	1278.3	1411.0	1.6102
840	.9907	1288.3	1423.9	1.6251	.9767	1288.0	1423.5	1.6234	.9433	1287.2	1422.4	1.6191
860	1.0094	1297.1	1435.3	1.6338	.9952	1296.8	1434.9	1.6321	.9613	1296.0	1433.9	1.6279
880	1.0279	1305.8	1446.6	1.6423	1.0135	1305.5	1446.2	1.6406	.9791	1304.8	1445.2	1.6364
900	1.0463	1314.5	1457.8	1.6506	1.0317	1314.3	1457.5	1.6489	.9968	1313.6	1456.5	1.6448
920	1.0645	1323.3	1469.0	1.6588	1.0497	1323.0	1468.7	1.6571	1.0143	1322.3	1467.8	1.6530
940	1.0827	1332.0	1480.2	1.6668	1.0676	1331.7	1479.9	1.6652	1.0317	1331.1	1479.0	1.6611
960	1.1007	1340.7	1491.4	1.6748	1.0854	1340.4	1491.1	1.6731	1.0490	1339.8	1490.3	1.6691
980	1.1186	1349.4	1502.5	1.6826	1.1031	1349.1	1502.2	1.6809	1.0662	1348.5	1501.5	1.6769
1000	1.1364	1358.1	1513.7	1.6902	1.1207	1357.8	1513.4	1.6886	1.0833	1357.3	1512.6	1.6846
1020	1.1541	1366.8	1524.8	1.6978	1.1382	1366.5	1524.5	1.6962	1.1003	1366.0	1523.8	1.6922
1040	1.1718	1375.5	1535.9	1.7053	1.1557	1375.3	1535.6	1.7037	1.1172	1374.7	1535.0	1.6997
1060	1.1893	1384.2	1547.0	1.7126	1.1730	1384.0	1546.8	1.7110	1.1341	1383.5	1546.1	1.7071
1080	1.2068	1392.9	1558.2	1.7199	1.1903	1392.7	1557.9	1.7183	1.1508	1392.2	1557.3	1.7144
1100	1.2242	1401.6	1569.3	1.7271	1.2075	1401.4	1569.0	1.7255	1.1675	1401.0	1568.4	1.7216
1120	1.2416	1410.4	1580.4	1.7342	1.2246	1410.2	1580.2	1.7326	1.1841	1409.8	1579.6	1.7287
1140	1.2589	1419.1	1591.5	1.7412	1.2417	1419.0	1591.3	1.7396	1.2007	1418.5	1590.7	1.7357
1160	1.2761	1427.9	1602.7	1.7481	1.2587	1427.8	1602.5	1.7465	1.2172	1427.3	1601.9	1.7426
1180	1.2933	1436.7	1613.8	1.7549	1.2757	1436.6	1613.6	1.7534	1.2337	1436.2	1613.1	1.7495
1200	1.3105	1445.5	1625.0	1.7617	1.2926	1445.4	1624.8	1.7601	1.2501	1445.0	1624.3	1.7563
1220	1.3275	1454.4	1636.2	1.7684	1.3095	1454.2	1636.0	1.7668	1.2665	1453.8	1635.5	1.7630
1240	1.3446	1463.2	1647.3	1.7750	1.3263	1463.1	1647.2	1.7735	1.2828	1462.7	1646.7	1.7696
1260	1.3616	1472.1	1658.5	1.7816	1.3431	1472.0	1658.4	1.7800	1.2991	1471.6	1657.9	1.7762
1280	1.3785	1481.0	1669.8	1.7881	1.3599	1480.9	1669.6	1.7865	1.3153	1480.5	1669.1	1.7827
1300	1.3955	1489.9	1681.0	1.7945	1.3766	1489.8	1680.8	1.7929	1.3315	1489.4	1680.4	1.7891
1350	1.4376	1512.3	1709.2	1.8103	1.4182	1512.2	1709.0	1.8087	1.3719	1511.9	1708.6	1.8049
1400	1.4796	1534.9	1737.5	1.8257	1.4597	1534.7	1737.3	1.8242	1.4120	1534.5	1737.0	1.8204
1450	1.5215	1557.6	1765.9	1.8408	1.5010	1557.5	1765.8	1.8393	1.4520	1557.2	1765.4	1.8355
1500	1.5631	1580.5	1794.5	1.8556	1.5421	1580.4	1794.4	1.8540	1.4919	1580.1	1794.1	1.8503
1600	1.6462	1626.7	1852.2	1.8843	1.6241	1626.6	1852.0	1.8827	1.5713	1626.4	1851.8	1.8790
1800	1.8113	1721.5	1969.5	1.9386	1.7870	1721.4	1969.4	1.9371	1.7291	1721.2	1969.2	1.9334
2000	1.9756	1819.2	2089.8	1.9896	1.9492	1819.2	2089.7	1.9881	1.8862	1819.0	2089.5	1.9844
2200	2.1395	1919.8	2212.8	2.0377	2.1109	1919.8	2212.7	2.0362	2.0428	1919.6	2212.6	2.0325
2400	2.3031	2023.1	2338.5	2.0832	2.2724	2023.0	2338.4	2.0817	2.1992	2022.8	2338.2	2.0781

	800 (518.36)				825 (521.92)				850 (525.39)			p (t Sat.)
v	u	h	s	v	u	h	s	v	u	h	s	t
.5691	1115.0	1199.3	1.4160	.5506	1114.5	1198.5	1.4126	.5331	1113.9	1197.7	1.4093	Sat.
												500
												510
.5554	*1107.9*	*1190.1*	*1.4066*									
.5717	1116.4	1201.0	1.4178	*.5475*	*1112.8*	*1196.4*	*1.4105*	*.5246*	*1109.1*	*1191.6*	*1.4031*	520
.5870	1124.3	1211.2	1.4282	.5629	1121.1	1207.0	1.4213	.5401	1117.8	1202.7	1.4143	530
.6015	1131.8	1220.8	1.4378	.5775	1128.8	1217.0	1.4313	.5547	1125.8	1213.0	1.4247	540
.6154	1138.8	1229.9	1.4469	.5913	1136.1	1226.4	1.4407	.5686	1133.3	1222.8	1.4344	550
.6287	1145.6	1238.6	1.4555	.6045	1143.1	1235.4	1.4495	.5817	1140.5	1232.0	1.4435	560
.6415	1152.0	1247.0	1.4637	.6173	1149.7	1243.9	1.4579	.5944	1147.3	1240.8	1.4521	570
.6539	1158.3	1255.1	1.4714	.6295	1156.1	1252.2	1.4658	.6066	1153.9	1249.3	1.4603	580
.6659	1164.3	1262.9	1.4789	.6414	1162.2	1260.2	1.4734	.6184	1160.1	1257.4	1.4681	590
.6776	1170.1	1270.4	1.4861	.6530	1168.2	1267.9	1.4808	.6298	1166.2	1265.3	1.4755	600
.6890	1175.8	1277.8	1.4930	.6643	1174.0	1275.4	1.4878	.6409	1172.1	1272.9	1.4827	610
.7002	1181.3	1285.0	1.4997	.6753	1179.6	1282.7	1.4946	.6518	1177.8	1280.3	1.4896	620
.7111	1186.7	1292.0	1.5062	.6860	1185.1	1289.8	1.5012	.6624	1183.4	1287.6	1.4963	630
.7218	1192.0	1298.9	1.5125	.6966	1190.5	1296.8	1.5076	.6728	1188.9	1294.7	1.5028	640
.7324	1197.2	1305.6	1.5186	.7070	1195.7	1303.7	1.5138	.6830	1194.2	1301.6	1.5091	650
.7428	1202.3	1312.3	1.5245	.7171	1200.9	1310.4	1.5198	.6930	1199.5	1308.5	1.5152	660
.7530	1207.4	1318.8	1.5304	.7272	1206.0	1317.0	1.5257	.7028	1204.6	1315.2	1.5211	670
.7631	1212.3	1325.3	1.5361	.7371	1211.0	1323.5	1.5315	.7126	1209.7	1321.8	1.5270	680
.7730	1217.2	1331.7	1.5416	.7468	1216.0	1330.0	1.5371	.7221	1214.7	1328.3	1.5327	690
.7829	1222.1	1338.0	1.5471	.7565	1220.9	1336.4	1.5426	.7316	1219.6	1334.7	1.5382	700
.8023	1231.6	1350.4	1.5577	.7754	1230.5	1348.9	1.5533	.7501	1229.4	1347.3	1.5490	720
.8212	1241.0	1362.6	1.5680	.7940	1240.0	1361.2	1.5637	.7683	1238.9	1359.7	1.5594	740
.8399	1250.3	1374.6	1.5779	.8122	1249.3	1373.3	1.5737	.7861	1248.3	1371.9	1.5695	760
.8583	1259.4	1386.5	1.5875	.8301	1258.5	1385.2	1.5834	.8037	1257.5	1383.9	1.5793	780
.8764	1268.5	1398.2	1.5969	.8478	1267.6	1397.0	1.5928	.8210	1266.7	1395.8	1.5888	800
.8943	1277.4	1409.8	1.6061	.8653	1276.6	1408.7	1.6020	.8380	1275.7	1407.6	1.5980	820
.9120	1286.4	1421.4	1.6150	.8825	1285.6	1420.3	1.6110	.8549	1284.8	1419.2	1.6071	840
.9295	1295.2	1432.8	1.6238	.8996	1294.5	1431.8	1.6198	.8715	1293.7	1430.8	1.6159	860
.9468	1304.1	1444.2	1.6324	.9165	1303.3	1443.3	1.6284	.8880	1302.6	1442.3	1.6246	880
.9640	1312.9	1455.6	1.6408	.9333	1312.2	1454.7	1.6369	.9043	1311.5	1453.7	1.6331	900
.9811	1321.7	1466.9	1.6490	.9499	1321.0	1466.0	1.6452	.9206	1320.3	1465.1	1.6414	920
.9980	1330.4	1478.2	1.6572	.9664	1329.8	1477.3	1.6533	.9366	1329.2	1476.5	1.6495	940
1.0149	1339.2	1489.4	1.6651	.9828	1338.6	1488.6	1.6613	.9526	1338.0	1487.8	1.6576	960
1.0316	1348.0	1500.7	1.6730	.9991	1347.4	1499.9	1.6692	.9684	1346.8	1499.1	1.6655	980
1.0482	1356.7	1511.9	1.6807	1.0152	1356.2	1511.1	1.6769	.9842	1355.6	1510.4	1.6733	1000
1.0647	1365.5	1523.1	1.6883	1.0313	1364.9	1522.4	1.6846	.9999	1364.4	1521.7	1.6809	1020
1.0812	1374.2	1534.3	1.6959	1.0473	1373.7	1533.6	1.6921	1.0154	1373.2	1532.9	1.6885	1040
1.0975	1383.0	1545.5	1.7033	1.0632	1382.5	1544.8	1.6995	1.0309	1382.0	1544.1	1.6959	1060
1.1138	1391.7	1556.6	1.7106	1.0791	1391.3	1556.0	1.7069	1.0463	1390.8	1555.4	1.7033	1080
1.1300	1400.5	1567.8	1.7178	1.0948	1400.1	1567.2	1.7141	1.0617	1399.6	1566.6	1.7105	1100
1.1462	1409.3	1579.0	1.7249	1.1105	1408.9	1578.4	1.7212	1.0770	1408.4	1577.8	1.7177	1120
1.1623	1418.1	1590.2	1.7319	1.1262	1417.7	1589.6	1.7283	1.0922	1417.2	1589.0	1.7247	1140
1.1783	1426.9	1601.4	1.7389	1.1418	1426.5	1600.8	1.7352	1.1074	1426.1	1600.3	1.7317	1160
1.1943	1435.8	1612.6	1.7458	1.1573	1435.4	1612.0	1.7421	1.1225	1434.9	1611.5	1.7386	1180
1.2102	1444.6	1623.8	1.7526	1.1728	1444.2	1623.3	1.7489	1.1375	1443.8	1622.7	1.7454	1200
1.2261	1453.5	1635.0	1.7593	1.1882	1453.1	1634.5	1.7557	1.1525	1452.7	1634.0	1.7521	1220
1.2420	1462.3	1646.2	1.7659	1.2036	1462.0	1645.7	1.7623	1.1675	1461.6	1645.3	1.7588	1240
1.2578	1471.3	1657.4	1.7725	1.2190	1470.9	1657.0	1.7689	1.1824	1470.5	1656.5	1.7654	1260
1.2735	1480.2	1668.7	1.7790	1.2343	1479.8	1668.3	1.7754	1.1973	1479.5	1667.8	1.7719	1280
1.2892	1489.1	1680.0	1.7854	1.2495	1488.8	1679.5	1.7819	1.2122	1488.5	1679.1	1.7784	1300
1.3284	1511.6	1708.2	1.8013	1.2876	1511.3	1707.8	1.7977	1.2491	1511.0	1707.4	1.7943	1350
1.3674	1534.2	1736.6	1.8167	1.3254	1533.9	1736.2	1.8132	1.2859	1533.6	1735.9	1.8097	1400
1.4062	1556.9	1765.1	1.8319	1.3631	1556.7	1764.8	1.8283	1.3225	1556.4	1764.4	1.8249	1450
1.4448	1579.9	1793.7	1.8467	1.4006	1579.6	1793.4	1.8431	1.3590	1579.4	1793.1	1.8397	1500
1.5218	1626.2	1851.5	1.8754	1.4754	1626.0	1851.2	1.8719	1.4316	1625.8	1850.9	1.8685	1600
1.6749	1721.0	1969.0	1.9298	1.6239	1720.8	1968.8	1.9264	1.5759	1720.7	1968.5	1.9230	1800
1.8271	1818.8	2089.3	1.9808	1.7716	1818.6	2089.1	1.9774	1.7194	1818.5	2088.9	1.9740	2000
1.9789	1919.4	2212.4	2.0290	1.9189	1919.3	2212.2	2.0255	1.8624	1919.1	2212.0	2.0221	2200
2.1305	2022.7	2338.1	2.0745	2.0660	2022.5	2337.9	2.0711	2.0052	2022.3	2337.7	2.0677	2400

Table 3. Vapor

p (t Sat.)	875 (528.79)				900 (532.12)				950 (538.56)			
t	v	u	h	s	v	u	h	s	v	u	h	s
Sat.	.5166	1113.3	1196.9	1.4060	.5009	1112.6	1196.0	1.4027	.4720	1111.3	1194.3	1.3964
520	*.5027*	*1105.3*	*1186.7*	*1.3956*								
530	.5184	1114.3	1198.2	1.4073	*.4977*	*1110.7*	*1193.6*	*1.4003*	*.4589*	*1103.1*	*1183.8*	*1.3859*
540	.5331	1122.7	1209.0	1.4181	.5125	1119.4	1204.8	1.4115	.4742	1112.6	1196.0	1.3981
550	.5470	1130.5	1219.1	1.4282	.5265	1127.5	1215.2	1.4219	.4883	1121.4	1207.2	1.4093
560	.5602	1137.9	1228.6	1.4375	.5397	1135.2	1225.0	1.4316	.5016	1129.5	1217.7	1.4197
570	.5728	1144.9	1237.6	1.4464	.5522	1142.4	1234.4	1.4407	.5142	1137.2	1227.6	1.4293
580	.5849	1151.6	1246.3	1.4547	.5643	1149.3	1243.2	1.4493	.5262	1144.5	1237.0	1.4384
590	.5966	1158.0	1254.6	1.4627	.5759	1155.8	1251.8	1.4574	.5377	1151.4	1245.9	1.4469
600	.6079	1164.2	1262.6	1.4703	.5871	1162.2	1260.0	1.4652	.5487	1158.0	1254.5	1.4551
610	.6189	1170.2	1270.4	1.4776	.5980	1168.3	1267.9	1.4727	.5594	1164.4	1262.7	1.4628
620	.6296	1176.0	1278.0	1.4847	.6086	1174.2	1275.6	1.4798	.5698	1170.5	1270.7	1.4702
630	.6401	1181.7	1285.4	1.4915	.6189	1180.0	1283.1	1.4867	.5799	1176.5	1278.5	1.4774
640	.6503	1187.3	1292.6	1.4981	.6290	1185.6	1290.4	1.4934	.5898	1182.3	1286.0	1.4843
650	.6603	1192.7	1299.6	1.5044	.6389	1191.1	1297.5	1.4999	.5994	1188.0	1293.4	1.4909
660	.6702	1198.0	1306.5	1.5106	.6486	1196.5	1304.5	1.5061	.6089	1193.5	1300.5	1.4974
670	.6799	1203.2	1313.3	1.5167	.6582	1201.8	1311.4	1.5123	.6181	1198.9	1307.6	1.5036
680	.6894	1208.3	1320.0	1.5226	.6675	1207.0	1318.2	1.5182	.6272	1204.2	1314.5	1.5097
690	.6988	1213.4	1326.6	1.5283	.6768	1212.1	1324.8	1.5240	.6361	1209.5	1321.3	1.5157
700	.7081	1218.4	1333.1	1.5339	.6859	1217.1	1331.4	1.5297	.6449	1214.6	1328.0	1.5215
720	.7263	1228.2	1345.8	1.5448	.7037	1227.0	1344.2	1.5407	.6622	1224.7	1341.1	1.5327
740	.7441	1237.8	1358.3	1.5553	.7212	1236.7	1356.8	1.5513	.6790	1234.5	1353.9	1.5434
760	.7616	1247.3	1370.6	1.5655	.7383	1246.2	1369.2	1.5615	.6955	1244.2	1366.5	1.5538
780	.7787	1256.6	1382.7	1.5753	.7551	1255.6	1381.4	1.5714	.7116	1253.7	1378.8	1.5639
800	.7956	1265.8	1394.6	1.5849	.7717	1264.9	1393.4	1.5810	.7275	1263.1	1391.0	1.5736
820	.8123	1274.9	1406.4	1.5942	.7880	1274.0	1405.3	1.5904	.7432	1272.3	1403.0	1.5830
840	.8287	1283.9	1418.1	1.6033	.8041	1283.1	1417.0	1.5995	.7586	1281.5	1414.9	1.5923
860	.8450	1292.9	1429.8	1.6121	.8200	1292.2	1428.7	1.6084	.7738	1290.6	1426.7	1.6013
880	.8611	1301.9	1441.3	1.6208	.8357	1301.1	1440.3	1.6172	.7889	1299.7	1438.4	1.6101
900	.8771	1310.8	1452.8	1.6293	.8513	1310.1	1451.9	1.6257	.8038	1308.7	1450.0	1.6187
920	.8929	1319.7	1464.2	1.6377	.8667	1319.0	1463.4	1.6341	.8185	1317.7	1461.6	1.6271
940	.9085	1328.5	1475.6	1.6459	.8820	1327.9	1474.8	1.6423	.8332	1326.6	1473.1	1.6354
960	.9241	1337.4	1487.0	1.6539	.8972	1336.8	1486.2	1.6504	.8477	1335.5	1484.5	1.6436
980	.9396	1346.2	1498.3	1.6619	.9123	1345.6	1497.6	1.6584	.8621	1344.4	1496.0	1.6516
1000	.9549	1355.0	1509.6	1.6697	.9273	1354.5	1508.9	1.6662	.8764	1353.3	1507.4	1.6594
1020	.9702	1363.8	1520.9	1.6774	.9422	1363.3	1520.2	1.6739	.8906	1362.2	1518.8	1.6672
1040	.9854	1372.7	1532.2	1.6849	.9570	1372.1	1531.5	1.6815	.9047	1371.1	1530.1	1.6748
1060	1.0005	1381.5	1543.5	1.6924	.9717	1381.0	1542.8	1.6889	.9187	1380.0	1541.5	1.6823
1080	1.0155	1390.3	1554.7	1.6997	.9863	1389.8	1554.1	1.6963	.9327	1388.8	1552.8	1.6897
1100	1.0304	1399.1	1566.0	1.7070	1.0009	1398.7	1565.4	1.7036	.9466	1397.7	1564.1	1.6970
1120	1.0453	1408.0	1577.2	1.7142	1.0154	1407.5	1576.6	1.7108	.9604	1406.6	1575.4	1.7042
1140	1.0601	1416.8	1588.5	1.7212	1.0299	1416.4	1587.9	1.7179	.9741	1415.5	1586.8	1.7114
1160	1.0749	1425.7	1599.7	1.7282	1.0443	1425.3	1599.2	1.7249	.9878	1424.4	1598.1	1.7184
1180	1.0896	1434.5	1611.0	1.7351	1.0586	1434.1	1610.4	1.7318	1.0015	1433.3	1609.4	1.7253
1200	1.1043	1443.4	1622.2	1.7420	1.0729	1443.0	1621.7	1.7386	1.0151	1442.3	1620.7	1.7322
1220	1.1189	1452.3	1633.5	1.7487	1.0871	1452.0	1633.0	1.7454	1.0286	1451.2	1632.0	1.7390
1240	1.1335	1461.3	1644.8	1.7554	1.1013	1460.9	1644.3	1.7521	1.0421	1460.2	1643.4	1.7457
1260	1.1480	1470.2	1656.1	1.7620	1.1155	1469.8	1655.6	1.7587	1.0556	1469.1	1654.7	1.7523
1280	1.1625	1479.1	1667.4	1.7685	1.1296	1478.8	1666.9	1.7652	1.0690	1478.1	1666.0	1.7589
1300	1.1769	1488.1	1678.7	1.7750	1.1437	1487.8	1678.3	1.7717	1.0824	1487.1	1677.4	1.7654
1350	1.2129	1510.7	1707.0	1.7909	1.1787	1510.3	1706.6	1.7876	1.1156	1509.7	1705.9	1.7813
1400	1.2487	1533.3	1735.5	1.8064	1.2135	1533.0	1735.1	1.8031	1.1488	1532.5	1734.4	1.7969
1450	1.2843	1556.1	1764.1	1.8216	1.2482	1555.9	1763.7	1.8183	1.1817	1555.3	1763.1	1.8121
1500	1.3198	1579.1	1792.8	1.8364	1.2827	1578.8	1792.5	1.8332	1.2145	1578.3	1791.9	1.8270
1600	1.3904	1625.5	1850.7	1.8652	1.3515	1625.3	1850.4	1.8620	1.2797	1624.9	1849.8	1.8558
1800	1.5307	1720.5	1968.3	1.9197	1.4879	1720.3	1968.1	1.9165	1.4092	1719.9	1967.6	1.9104
2000	1.6701	1818.3	2088.7	1.9708	1.6236	1818.1	2088.5	1.9676	1.5379	1817.8	2088.1	1.9615
2200	1.8092	1918.9	2211.9	2.0189	1.7589	1918.8	2211.7	2.0157	1.6662	1918.4	2211.3	2.0096
2400	1.9480	2022.1	2337.6	2.0644	1.8939	2022.0	2337.4	2.0613	1.7943	2021.6	2337.1	2.0552

1000 (544.75)				**1050** (550.71)				**1100** (556.45)				**p** (t Sat.)
v	u	h	s	v	u	h	s	v	u	h	s	t
.4459	1109.9	1192.4	1.3903	.4222	1108.4	1190.4	1.3844	.4005	1106.8	1188.3	1.3786	Sat.
												520
												530
.4389	*1105.2*	*1186.4*	*1.3844*									540
.4534	1114.8	1198.7	1.3966	*.4212*	*1107.7*	*1189.5*	*1.3835*	*.3912*	*1100.0*	*1179.6*	*1.3699*	550
.4669	1123.6	1210.0	1.4077	.4350	1117.2	1201.7	1.3955	.4054	1110.4	1192.9	1.3831	560
.4795	1131.7	1220.5	1.4179	.4478	1126.0	1213.0	1.4065	.4185	1119.9	1205.1	1.3949	570
.4915	1139.4	1230.4	1.4275	.4599	1134.2	1223.5	1.4167	.4308	1128.6	1216.3	1.4058	580
.5030	1146.7	1239.8	1.4365	.4713	1141.9	1233.4	1.4262	.4423	1136.8	1226.8	1.4158	590
.5140	1153.7	1248.8	1.4450	.4823	1149.2	1242.9	1.4351	.4532	1144.5	1236.7	1.4252	600
.5245	1160.3	1257.4	1.4531	.4928	1156.1	1251.9	1.4436	.4637	1151.7	1246.1	1.4341	610
.5348	1166.7	1265.7	1.4609	.5029	1162.8	1260.5	1.4516	.4737	1158.7	1255.1	1.4425	620
.5447	1172.9	1273.7	1.4682	.5126	1169.2	1268.8	1.4593	.4834	1165.4	1263.8	1.4504	630
.5543	1178.9	1281.5	1.4754	.5221	1175.4	1276.9	1.4666	.4927	1171.8	1272.1	1.4580	640
.5637	1184.7	1289.1	1.4822	.5314	1181.4	1284.7	1.4737	.5018	1178.0	1280.2	1.4653	650
.5730	1190.4	1296.5	1.4889	.5404	1187.3	1292.3	1.4805	.5107	1184.1	1288.0	1.4724	660
.5820	1196.0	1303.7	1.4953	.5492	1193.0	1299.7	1.4871	.5193	1189.9	1295.6	1.4791	670
.5908	1201.4	1310.8	1.5015	.5578	1198.6	1307.0	1.4935	.5277	1195.7	1303.1	1.4857	680
.5995	1206.8	1317.7	1.5076	.5663	1204.0	1314.1	1.4997	.5360	1201.3	1310.4	1.4921	690
.6080	1212.0	1324.6	1.5135	.5746	1209.4	1321.1	1.5058	.5441	1206.7	1317.5	1.4982	700
.6247	1222.3	1337.9	1.5249	.5908	1219.9	1334.7	1.5174	.5599	1217.4	1331.4	1.5101	720
.6410	1232.3	1350.9	1.5359	.6066	1230.1	1347.9	1.5286	.5752	1227.8	1344.9	1.5215	740
.6569	1242.1	1363.7	1.5464	.6220	1240.0	1360.9	1.5393	.5902	1237.9	1358.0	1.5323	760
.6725	1251.7	1376.2	1.5566	.6370	1249.8	1373.5	1.5496	.6048	1247.8	1370.9	1.5428	780
.6878	1261.2	1388.5	1.5664	.6518	1259.4	1386.0	1.5595	.6190	1257.5	1383.5	1.5529	800
.7028	1270.6	1400.6	1.5760	.6663	1268.8	1398.3	1.5692	.6331	1267.1	1395.9	1.5627	820
.7176	1279.9	1412.7	1.5853	.6806	1278.2	1410.4	1.5786	.6469	1276.5	1408.2	1.5722	840
.7323	1289.1	1424.6	1.5944	.6947	1287.5	1422.5	1.5878	.6605	1285.9	1420.3	1.5815	860
.7467	1298.2	1436.4	1.6033	.7086	1296.7	1434.4	1.5968	.6739	1295.2	1432.3	1.5905	880
.7610	1307.3	1448.1	1.6120	.7223	1305.8	1446.2	1.6055	.6871	1304.4	1444.3	1.5993	900
.7752	1316.3	1459.7	1.6205	.7359	1314.9	1457.9	1.6141	.7002	1313.6	1456.1	1.6080	920
.7892	1325.3	1471.3	1.6288	.7494	1324.0	1469.6	1.6225	.7132	1322.7	1467.9	1.6164	940
.8031	1334.3	1482.9	1.6370	.7627	1333.0	1481.2	1.6308	.7260	1331.8	1479.6	1.6247	960
.8169	1343.2	1494.4	1.6451	.7759	1342.0	1492.8	1.6388	.7387	1340.8	1491.2	1.6329	980
.8305	1352.2	1505.9	1.6530	.7891	1351.0	1504.3	1.6468	.7513	1349.9	1502.8	1.6409	1000
.8441	1361.1	1517.3	1.6608	.8021	1360.0	1515.8	1.6546	.7639	1358.9	1514.4	1.6488	1020
.8576	1370.0	1528.7	1.6684	.8150	1369.0	1527.3	1.6623	.7763	1367.9	1525.9	1.6565	1040
.8710	1378.9	1540.1	1.6760	.8279	1377.9	1538.8	1.6699	.7886	1376.9	1537.4	1.6641	1060
.8844	1387.9	1551.5	1.6834	.8406	1386.9	1550.2	1.6774	.8009	1385.9	1548.9	1.6716	1080
.8976	1396.8	1562.9	1.6908	.8533	1395.8	1561.6	1.6848	.8131	1394.9	1560.4	1.6790	1100
.9108	1405.7	1574.3	1.6980	.8660	1404.8	1573.1	1.6921	.8252	1403.9	1571.9	1.6863	1120
.9240	1414.6	1585.6	1.7052	.8786	1413.8	1584.5	1.6992	.8373	1412.9	1583.3	1.6935	1140
.9370	1423.6	1597.0	1.7122	.8911	1422.7	1595.9	1.7063	.8493	1421.9	1594.7	1.7007	1160
.9500	1432.5	1608.3	1.7192	.9035	1431.7	1607.3	1.7133	.8612	1430.9	1606.2	1.7077	1180
.9630	1441.5	1619.7	1.7261	.9159	1440.7	1618.6	1.7202	.8731	1439.9	1617.6	1.7146	1200
.9759	1450.4	1631.0	1.7329	.9283	1449.7	1630.0	1.7270	.8850	1448.9	1629.0	1.7214	1220
.9888	1459.4	1642.4	1.7396	.9406	1458.7	1641.4	1.7338	.8968	1457.9	1640.5	1.7282	1240
1.0016	1468.4	1653.8	1.7462	.9529	1467.7	1652.8	1.7404	.9085	1467.0	1651.9	1.7349	1260
1.0144	1477.4	1665.1	1.7528	.9651	1476.7	1664.3	1.7470	.9202	1476.0	1663.4	1.7415	1280
1.0272	1486.5	1676.5	1.7593	.9773	1485.8	1675.7	1.7536	.9319	1485.1	1674.8	1.7481	1300
1.0589	1509.1	1705.1	1.7753	1.0076	1508.5	1704.3	1.7696	.9609	1507.9	1703.5	1.7641	1350
1.0905	1531.9	1733.7	1.7909	1.0377	1531.3	1732.9	1.7852	.9898	1530.7	1732.2	1.7798	1400
1.1218	1554.8	1762.4	1.8061	1.0677	1554.3	1761.7	1.8005	1.0184	1553.7	1761.0	1.7951	1450
1.1531	1577.8	1791.2	1.8210	1.0975	1577.3	1790.6	1.8154	1.0470	1576.8	1789.9	1.8100	1500
1.2152	1624.4	1849.3	1.8499	1.1568	1624.0	1848.7	1.8443	1.1037	1623.5	1848.2	1.8390	1600
1.3384	1719.5	1967.2	1.9046	1.2743	1719.1	1966.7	1.8990	1.2161	1718.8	1966.3	1.8937	1800
1.4608	1817.4	2087.7	1.9557	1.3911	1817.1	2087.4	1.9502	1.3276	1816.7	2087.0	1.9449	2000
1.5828	1918.1	2211.0	2.0038	1.5074	1917.7	2210.6	1.9983	1.4388	1917.4	2210.3	1.9931	2200
1.7046	2021.3	2336.7	2.0494	1.6235	2020.9	2336.4	2.0439	1.5498	2020.6	2336.1	2.0387	2400

Table 3. Vapor

p (t Sat.)	1150 (562.00)				1200 (567.37)				1250 (572.56)			
t	v	u	h	s	v	u	h	s	v	u	h	s
Sat.	.3806	1105.2	1186.2	1.3729	.3623	1103.5	1183.9	1.3673	.3454	1101.7	1181.6	1.3619
560	*.3778*	*1103.0*	*1183.4*	*1.3702*	*.3518*	*1095.0*	*1173.1*	*1.3568*				
570	.3914	1113.3	1196.6	1.3831	.3659	1106.3	1187.6	1.3709	*.3419*	*1098.7*	*1177.8*	*1.3582*
580	.4038	1122.7	1208.7	1.3947	.3787	1116.5	1200.6	1.3835	.3552	1109.8	1192.0	1.3719
590	.4155	1131.4	1219.8	1.4054	.3906	1125.8	1212.5	1.3949	.3673	1119.8	1204.8	1.3842
600	.4265	1139.5	1230.3	1.4153	.4017	1134.4	1223.6	1.4054	.3786	1129.0	1216.6	1.3954
610	.4369	1147.2	1240.2	1.4246	.4122	1142.5	1234.0	1.4152	.3892	1137.5	1227.6	1.4057
620	.4469	1154.5	1249.6	1.4334	.4222	1150.1	1243.9	1.4243	.3992	1145.6	1237.9	1.4153
630	.4565	1161.4	1258.6	1.4417	.4317	1157.4	1253.2	1.4330	.4088	1153.1	1247.7	1.4243
640	.4658	1168.1	1267.2	1.4496	.4409	1164.3	1262.2	1.4412	.4179	1160.3	1257.0	1.4329
650	.4747	1174.5	1275.6	1.4571	.4498	1170.9	1270.8	1.4490	.4267	1167.2	1266.0	1.4410
660	.4834	1180.8	1283.6	1.4644	.4584	1177.4	1279.2	1.4565	.4352	1173.9	1274.6	1.4487
670	.4919	1186.8	1291.5	1.4713	.4667	1183.6	1287.2	1.4636	.4435	1180.3	1282.9	1.4561
680	.5002	1192.7	1299.1	1.4781	.4749	1189.6	1295.1	1.4706	.4515	1186.5	1290.9	1.4632
690	.5083	1198.4	1306.6	1.4846	.4828	1195.5	1302.7	1.4772	.4593	1192.5	1298.8	1.4700
700	.5162	1204.0	1313.9	1.4909	.4906	1201.3	1310.2	1.4837	.4670	1198.4	1306.4	1.4767
710	.5240	1209.5	1321.0	1.4970	.4982	1206.9	1317.5	1.4900	.4744	1204.2	1313.9	1.4831
720	.5316	1214.9	1328.1	1.5030	.5057	1212.4	1324.7	1.4961	.4818	1209.8	1321.2	1.4893
730	.5392	1220.2	1335.0	1.5089	.5130	1217.8	1331.7	1.5020	.4890	1215.3	1328.4	1.4954
740	.5466	1225.5	1341.8	1.5146	.5203	1223.1	1338.7	1.5078	.4960	1220.7	1335.5	1.5013
750	.5539	1230.6	1348.5	1.5201	.5274	1228.4	1345.5	1.5135	.5030	1226.1	1342.4	1.5070
760	.5611	1235.7	1355.1	1.5256	.5344	1233.6	1352.2	1.5191	.5098	1231.3	1349.3	1.5127
770	.5682	1240.8	1361.7	1.5310	.5414	1238.7	1358.9	1.5245	.5166	1236.5	1356.0	1.5182
780	.5753	1245.8	1368.2	1.5362	.5482	1243.7	1365.5	1.5298	.5233	1241.7	1362.7	1.5236
790	.5822	1250.7	1374.6	1.5414	.5550	1248.7	1372.0	1.5351	.5299	1246.7	1369.3	1.5289
800	.5891	1255.6	1381.0	1.5464	.5617	1253.7	1378.4	1.5402	.5364	1251.8	1375.8	1.5341
820	.6027	1265.3	1393.5	1.5564	.5749	1263.5	1391.1	1.5502	.5492	1261.7	1388.7	1.5443
840	.6161	1274.8	1405.9	1.5660	.5878	1273.1	1403.7	1.5599	.5618	1271.4	1401.4	1.5541
860	.6292	1284.3	1418.2	1.5753	.6006	1282.7	1416.0	1.5694	.5742	1281.1	1413.9	1.5636
880	.6422	1293.7	1430.3	1.5844	.6131	1292.1	1428.3	1.5786	.5864	1290.6	1426.2	1.5729
900	.6550	1302.9	1442.3	1.5933	.6255	1301.5	1440.4	1.5876	.5984	1300.0	1438.4	1.5820
920	.6676	1312.2	1454.3	1.6020	.6377	1310.8	1452.4	1.5963	.6102	1309.4	1450.5	1.5908
940	.6801	1321.4	1466.1	1.6106	.6498	1320.0	1464.3	1.6049	.6219	1318.7	1462.6	1.5994
960	.6925	1330.5	1477.9	1.6189	.6618	1329.2	1476.2	1.6133	.6335	1328.0	1474.5	1.6079
980	.7047	1339.6	1489.6	1.6271	.6736	1338.4	1488.0	1.6216	.6449	1337.2	1486.4	1.6162
1000	.7169	1348.7	1501.3	1.6352	.6853	1347.5	1499.7	1.6297	.6563	1346.4	1498.2	1.6244
1020	.7290	1357.8	1512.9	1.6431	.6970	1356.7	1511.4	1.6376	.6675	1355.5	1509.9	1.6324
1040	.7409	1366.8	1524.5	1.6509	.7085	1365.8	1523.1	1.6455	.6787	1364.7	1521.7	1.6402
1060	.7528	1375.9	1536.1	1.6585	.7200	1374.8	1534.7	1.6532	.6897	1373.8	1533.4	1.6480
1080	.7646	1384.9	1547.6	1.6661	.7313	1383.9	1546.3	1.6607	.7007	1382.9	1545.0	1.6556
1100	.7763	1393.9	1559.1	1.6735	.7426	1393.0	1557.9	1.6682	.7116	1392.0	1556.6	1.6631
1120	.7880	1403.0	1570.7	1.6809	.7539	1402.0	1569.4	1.6756	.7225	1401.1	1568.2	1.6705
1140	.7996	1412.0	1582.1	1.6881	.7650	1411.1	1581.0	1.6828	.7333	1410.2	1579.8	1.6778
1160	.8111	1421.0	1593.6	1.6952	.7761	1420.2	1592.5	1.6900	.7440	1419.3	1591.4	1.6850
1180	.8226	1430.1	1605.1	1.7023	.7872	1429.2	1604.0	1.6971	.7546	1428.4	1602.9	1.6921
1200	.8340	1439.1	1616.6	1.7092	.7982	1438.3	1615.5	1.7040	.7652	1437.5	1614.5	1.6991
1220	.8454	1448.1	1628.0	1.7161	.8091	1447.4	1627.0	1.7109	.7758	1446.6	1626.0	1.7060
1240	.8567	1457.2	1639.5	1.7229	.8200	1456.5	1638.6	1.7177	.7863	1455.7	1637.6	1.7128
1260	.8680	1466.3	1651.0	1.7296	.8309	1465.5	1650.1	1.7245	.7968	1464.8	1649.1	1.7196
1280	.8793	1475.3	1662.5	1.7362	.8417	1474.7	1661.6	1.7311	.8072	1474.0	1660.7	1.7262
1300	.8905	1484.4	1673.9	1.7428	.8525	1483.8	1673.1	1.7377	.8176	1483.1	1672.2	1.7328
1350	.9183	1507.2	1702.7	1.7589	.8793	1506.6	1701.9	1.7538	.8433	1506.0	1701.1	1.7490
1400	.9460	1530.2	1731.5	1.7746	.9059	1529.6	1730.7	1.7696	.8689	1529.0	1730.0	1.7648
1450	.9735	1553.2	1760.3	1.7899	.9323	1552.6	1759.7	1.7849	.8944	1552.1	1759.0	1.7801
1500	1.0008	1576.3	1789.3	1.8049	.9586	1575.8	1788.7	1.7999	.9197	1575.3	1788.0	1.7952
1600	1.0552	1623.1	1847.6	1.8339	1.0107	1622.6	1847.1	1.8290	.9699	1622.2	1846.5	1.8243
1800	1.1629	1718.4	1965.8	1.8887	1.1141	1718.0	1965.4	1.8838	1.0693	1717.6	1965.0	1.8791
2000	1.2697	1816.4	2086.6	1.9399	1.2167	1816.0	2086.2	1.9350	1.1678	1815.7	2085.8	1.9304
2200	1.3762	1917.1	2209.9	1.9881	1.3188	1916.7	2209.6	1.9832	1.2660	1916.4	2209.2	1.9786
2400	1.4824	2020.3	2335.7	2.0337	1.4207	2019.9	2335.4	2.0289	1.3639	2019.6	2335.1	2.0242

1300 (577.60)				1350 (582.50)				1400 (587.25)				p (t Sat.)
v	u	h	s	v	u	h	s	v	u	h	s	t
.3297	1099.8	1179.2	1.3565	.3152	1097.9	1176.7	1.3513	.3016	1096.0	1174.1	1.3461	Sat.
												560
												570
.3329	1102.6	1182.7	1.3600	*.3118*	*1094.8*	*1172.7*	*1.3475*					580
.3455	1113.5	1196.6	1.3732	.3248	1106.7	1187.8	1.3620	.3052	1099.4	1178.4	1.3502	590
.3570	1123.3	1209.2	1.3852	.3367	1117.3	1201.4	1.3748	.3175	1110.9	1193.1	1.3642	600
.3678	1132.4	1220.8	1.3961	.3477	1126.9	1213.8	1.3865	.3287	1121.2	1206.4	1.3766	610
.3779	1140.8	1231.7	1.4063	.3579	1135.9	1225.3	1.3971	.3391	1130.7	1218.5	1.3879	620
.3874	1148.8	1242.0	1.4157	.3675	1144.2	1236.0	1.4070	.3488	1139.5	1229.8	1.3983	630
.3966	1156.3	1251.7	1.4246	.3766	1152.1	1246.2	1.4163	.3580	1147.7	1240.5	1.4080	640
.4053	1163.4	1261.0	1.4330	.3854	1159.5	1255.8	1.4250	.3668	1155.5	1250.5	1.4171	650
.4138	1170.3	1269.9	1.4410	.3938	1166.6	1265.0	1.4333	.3751	1162.9	1260.1	1.4257	660
.4219	1176.9	1278.4	1.4486	.4019	1173.5	1273.9	1.4412	.3832	1169.9	1269.2	1.4338	670
.4299	1183.3	1286.7	1.4559	.4097	1180.1	1282.4	1.4487	.3910	1176.7	1278.0	1.4416	680
.4376	1189.5	1294.8	1.4629	.4174	1186.4	1290.7	1.4559	.3985	1183.3	1286.5	1.4490	690
.4451	1195.6	1302.6	1.4697	.4248	1192.6	1298.7	1.4629	.4059	1189.6	1294.8	1.4562	700
.4524	1201.4	1310.3	1.4763	.4320	1198.6	1306.6	1.4696	.4130	1195.8	1302.8	1.4631	710
.4596	1207.2	1317.7	1.4827	.4391	1204.5	1314.2	1.4761	.4200	1201.8	1310.6	1.4697	720
.4667	1212.8	1325.1	1.4888	.4460	1210.3	1321.7	1.4824	.4268	1207.7	1318.2	1.4761	730
.4736	1218.3	1332.3	1.4949	.4528	1215.9	1329.0	1.4886	.4335	1213.4	1325.7	1.4824	740
.4804	1223.8	1339.3	1.5007	.4595	1221.4	1336.2	1.4945	.4400	1219.0	1333.0	1.4885	750
.4871	1229.1	1346.3	1.5065	.4661	1226.8	1343.3	1.5004	.4465	1224.5	1340.2	1.4944	760
.4937	1234.4	1353.2	1.5121	.4725	1232.2	1350.2	1.5061	.4528	1230.0	1347.3	1.5002	770
.5003	1239.6	1359.9	1.5175	.4789	1237.5	1357.1	1.5116	.4591	1235.3	1354.3	1.5058	780
.5067	1244.7	1366.6	1.5229	.4852	1242.7	1363.9	1.5171	.4652	1240.6	1361.1	1.5114	790
.5130	1249.8	1373.2	1.5282	.4914	1247.8	1370.6	1.5224	.4713	1245.8	1367.9	1.5168	800
.5255	1259.8	1386.3	1.5385	.5036	1258.0	1383.8	1.5328	.4832	1256.1	1381.3	1.5273	820
.5378	1269.7	1399.1	1.5484	.5155	1267.9	1396.7	1.5428	.4949	1266.2	1394.4	1.5374	840
.5498	1279.4	1411.7	1.5580	.5273	1277.8	1409.5	1.5526	.5063	1276.1	1407.3	1.5473	860
.5617	1289.0	1424.1	1.5674	.5388	1287.5	1422.1	1.5620	.5175	1285.9	1420.0	1.5568	880
.5733	1298.5	1436.5	1.5765	.5501	1297.1	1434.5	1.5712	.5285	1295.6	1432.5	1.5661	900
.5848	1308.0	1448.7	1.5854	.5613	1306.6	1446.8	1.5802	.5394	1305.1	1444.9	1.5752	920
.5961	1317.4	1460.8	1.5941	.5723	1316.0	1459.0	1.5890	.5501	1314.6	1457.2	1.5840	940
.6074	1326.7	1472.8	1.6027	.5832	1325.4	1471.1	1.5976	.5607	1324.1	1469.4	1.5926	960
.6185	1336.0	1484.7	1.6110	.5940	1334.7	1483.1	1.6060	.5712	1333.5	1481.5	1.6011	980
.6295	1345.2	1496.6	1.6192	.6046	1344.0	1495.1	1.6142	.5815	1342.8	1493.5	1.6094	1000
.6403	1354.4	1508.5	1.6273	.6152	1353.3	1507.0	1.6223	.5918	1352.1	1505.5	1.6175	1020
.6511	1363.6	1520.2	1.6352	.6256	1362.5	1518.8	1.6303	.6019	1361.4	1517.4	1.6255	1040
.6618	1372.8	1532.0	1.6430	.6360	1371.7	1530.6	1.6381	.6120	1370.7	1529.2	1.6334	1060
.6725	1381.9	1543.7	1.6506	.6463	1380.9	1542.4	1.6458	.6220	1379.9	1541.0	1.6411	1080
.6830	1391.1	1555.4	1.6582	.6565	1390.1	1554.1	1.6534	.6319	1389.1	1552.8	1.6487	1100
.6935	1400.2	1567.0	1.6656	.6667	1399.3	1565.8	1.6608	.6417	1398.3	1564.6	1.6562	1120
.7039	1409.3	1578.7	1.6729	.6767	1408.4	1577.5	1.6682	.6515	1407.5	1576.3	1.6636	1140
.7143	1418.4	1590.3	1.6801	.6868	1417.6	1589.1	1.6754	.6612	1416.7	1588.0	1.6709	1160
.7246	1427.6	1601.9	1.6872	.6967	1426.7	1600.8	1.6826	.6709	1425.9	1599.7	1.6780	1180
.7348	1436.7	1613.5	1.6943	.7066	1435.9	1612.4	1.6896	.6805	1435.1	1611.4	1.6851	1200
.7450	1445.8	1625.0	1.7012	.7165	1445.0	1624.0	1.6966	.6900	1444.3	1623.0	1.6921	1220
.7551	1455.0	1636.6	1.7080	.7263	1454.2	1635.6	1.7034	.6995	1453.5	1634.7	1.6990	1240
.7652	1464.1	1648.2	1.7148	.7360	1463.4	1647.2	1.7102	.7089	1462.6	1646.3	1.7058	1260
.7753	1473.3	1659.8	1.7215	.7458	1472.5	1658.9	1.7169	.7183	1471.8	1657.9	1.7125	1280
.7853	1482.4	1671.3	1.7281	.7554	1481.7	1670.5	1.7236	.7277	1481.1	1669.6	1.7192	1300
.8102	1505.4	1700.3	1.7443	.7795	1504.7	1699.5	1.7398	.7510	1504.1	1698.7	1.7355	1350
.8349	1528.4	1729.3	1.7601	.8033	1527.8	1728.5	1.7557	.7740	1527.2	1727.8	1.7513	1400
.8594	1551.6	1758.3	1.7755	.8270	1551.0	1757.6	1.7711	.7969	1550.5	1756.9	1.7668	1450
.8838	1574.8	1787.4	1.7906	.8505	1574.3	1786.8	1.7862	.8196	1573.8	1786.1	1.7819	1500
.9321	1621.7	1845.9	1.8197	.8972	1621.3	1845.4	1.8153	.8647	1620.8	1844.8	1.8111	1600
1.0279	1717.2	1964.5	1.8747	.9895	1716.9	1964.1	1.8703	.9539	1716.5	1963.6	1.8662	1800
1.1228	1815.3	2085.4	1.9259	1.0810	1815.0	2085.1	1.9216	1.0423	1814.7	2084.7	1.9175	2000
1.2172	1916.1	2208.9	1.9742	1.1721	1915.7	2208.5	1.9699	1.1302	1915.4	2208.2	1.9658	2200
1.3115	2019.2	2334.7	2.0198	1.2630	2018.9	2334.4	2.0155	1.2179	2018.6	2334.1	2.0114	2400

Table 3. Vapor

p (t Sat.)	1450 (591.88)				1500 (596.39)				1550 (600.78)			
t	v	u	h	s	v	u	h	s	v	u	h	s
Sat.	.2888	1093.9	1171.4	1.3409	.2769	1091.8	1168.7	1.3359	.2657	1089.6	1165.9	1.3308
590	*.2863*	*1091.4*	*1168.2*	*1.3379*	*.2680*	*1082.6*	*1157.0*	*1.3248*				
600	.2992	1104.0	1184.3	1.3531	.2816	1096.6	1174.8	1.3416	*.2647*	*1088.5*	*1164.5*	*1.3295*
610	.3108	1115.2	1198.6	1.3665	.2937	1108.7	1190.2	1.3561	.2773	1101.8	1181.4	1.3454
620	.3214	1125.3	1211.5	1.3786	.3046	1119.5	1204.1	1.3690	.2886	1113.5	1196.3	1.3592
630	.3312	1134.5	1223.4	1.3896	.3146	1129.4	1216.7	1.3807	.2989	1124.0	1209.7	1.3716
640	.3405	1143.2	1234.5	1.3997	.3240	1138.5	1228.4	1.3913	.3084	1133.6	1222.0	1.3829
650	.3493	1151.3	1245.0	1.4092	.3329	1147.0	1239.4	1.4012	.3173	1142.5	1233.5	1.3933
660	.3577	1159.0	1254.9	1.4181	.3413	1155.0	1249.7	1.4105	.3258	1150.8	1244.3	1.4029
670	.3657	1166.3	1264.4	1.4265	.3493	1162.5	1259.5	1.4192	.3338	1158.7	1254.4	1.4120
680	.3735	1173.3	1273.5	1.4345	.3570	1169.8	1268.9	1.4275	.3415	1166.2	1264.2	1.4205
690	.3809	1180.1	1282.3	1.4422	.3644	1176.7	1277.9	1.4354	.3489	1173.4	1273.4	1.4287
700	.3882	1186.6	1290.7	1.4495	.3716	1183.4	1286.6	1.4429	.3561	1180.3	1282.4	1.4364
710	.3952	1192.9	1298.9	1.4566	.3786	1189.9	1295.0	1.4502	.3630	1186.9	1291.0	1.4438
720	.4021	1199.0	1306.9	1.4634	.3854	1196.2	1303.2	1.4571	.3698	1193.3	1299.4	1.4509
730	.4088	1205.0	1314.7	1.4699	.3921	1202.3	1311.2	1.4638	.3763	1199.6	1307.5	1.4578
740	.4154	1210.9	1322.3	1.4763	.3985	1208.3	1318.9	1.4703	.3827	1205.7	1315.5	1.4644
750	.4219	1216.6	1329.8	1.4825	.4049	1214.1	1326.5	1.4767	.3890	1211.6	1323.2	1.4709
760	.4282	1222.2	1337.1	1.4885	.4111	1219.9	1334.0	1.4828	.3951	1217.5	1330.8	1.4771
770	.4344	1227.7	1344.3	1.4944	.4172	1225.5	1341.3	1.4888	.4011	1223.2	1338.2	1.4832
780	.4405	1233.2	1351.4	1.5002	.4233	1231.0	1348.5	1.4946	.4071	1228.8	1345.5	1.4891
790	.4466	1238.5	1358.4	1.5058	.4292	1236.4	1355.6	1.5003	.4129	1234.3	1352.7	1.4949
800	.4525	1243.8	1365.3	1.5112	.4350	1241.8	1362.5	1.5058	.4186	1239.7	1359.8	1.5005
820	.4642	1254.2	1378.8	1.5219	.4464	1252.3	1376.2	1.5166	.4298	1250.4	1373.7	1.5115
840	.4756	1264.4	1392.0	1.5322	.4576	1262.6	1389.6	1.5270	.4407	1260.8	1387.2	1.5220
860	.4867	1274.4	1405.0	1.5421	.4685	1272.7	1402.8	1.5370	.4514	1271.0	1400.5	1.5321
880	.4977	1284.3	1417.8	1.5517	.4792	1282.7	1415.7	1.5468	.4619	1281.1	1413.6	1.5419
900	.5085	1294.1	1430.5	1.5611	.4897	1292.5	1428.5	1.5562	.4722	1291.0	1426.4	1.5515
920	.5191	1303.7	1443.0	1.5702	.5000	1302.3	1441.1	1.5654	.4823	1300.8	1439.1	1.5607
940	.5295	1313.3	1455.4	1.5791	.5102	1311.9	1453.5	1.5744	.4922	1310.5	1451.7	1.5698
960	.5398	1322.8	1467.6	1.5878	.5203	1321.5	1465.9	1.5832	.5020	1320.2	1464.1	1.5786
980	.5500	1332.2	1479.8	1.5964	.5302	1331.0	1478.1	1.5917	.5117	1329.7	1476.5	1.5872
1000	.5601	1341.6	1491.9	1.6047	.5400	1340.4	1490.3	1.6001	.5212	1339.2	1488.7	1.5957
1020	.5700	1351.0	1503.9	1.6129	.5497	1349.8	1502.4	1.6084	.5307	1348.7	1500.9	1.6040
1040	.5799	1360.3	1515.9	1.6209	.5593	1359.2	1514.5	1.6164	.5400	1358.1	1513.0	1.6121
1060	.5897	1369.6	1527.8	1.6288	.5688	1368.6	1526.4	1.6244	.5493	1367.5	1525.0	1.6201
1080	.5994	1378.9	1539.7	1.6366	.5782	1377.9	1538.4	1.6322	.5585	1376.8	1537.0	1.6279
1100	.6090	1388.1	1551.5	1.6442	.5876	1387.2	1550.3	1.6399	.5676	1386.2	1549.0	1.6356
1120	.6185	1397.4	1563.3	1.6517	.5969	1396.4	1562.1	1.6474	.5766	1395.5	1560.9	1.6432
1140	.6280	1406.6	1575.1	1.6592	.6061	1405.7	1573.9	1.6548	.5856	1404.8	1572.8	1.6507
1160	.6374	1415.8	1586.9	1.6665	.6152	1415.0	1585.7	1.6622	.5945	1414.1	1584.6	1.6580
1180	.6468	1425.1	1598.6	1.6737	.6243	1424.2	1597.5	1.6694	.6033	1423.4	1596.4	1.6653
1200	.6561	1434.3	1610.3	1.6808	.6334	1433.5	1609.3	1.6765	.6121	1432.6	1608.2	1.6724
1220	.6653	1443.5	1622.0	1.6878	.6423	1442.7	1621.0	1.6836	.6208	1441.9	1620.0	1.6795
1240	.6745	1452.7	1633.7	1.6947	.6513	1451.9	1632.7	1.6905	.6295	1451.2	1631.7	1.6864
1260	.6837	1461.9	1645.4	1.7015	.6602	1461.2	1644.4	1.6973	.6381	1460.4	1643.5	1.6933
1280	.6928	1471.1	1657.0	1.7082	.6690	1470.4	1656.1	1.7041	.6467	1469.7	1655.2	1.7001
1300	.7019	1480.4	1668.7	1.7149	.6778	1479.7	1667.8	1.7108	.6553	1479.0	1666.9	1.7068
1350	.7244	1503.5	1697.9	1.7313	.6996	1502.8	1697.1	1.7272	.6765	1502.2	1696.2	1.7232
1400	.7467	1526.7	1727.0	1.7472	.7213	1526.1	1726.3	1.7431	.6975	1525.5	1725.5	1.7392
1450	.7689	1549.9	1756.2	1.7627	.7428	1549.4	1755.5	1.7586	.7183	1548.8	1754.9	1.7547
1500	.7909	1573.3	1785.5	1.7778	.7641	1572.8	1784.9	1.7738	.7390	1572.3	1784.2	1.7699
1600	.8345	1620.4	1844.3	1.8070	.8064	1619.9	1843.7	1.8031	.7800	1619.5	1843.2	1.7992
1800	.9208	1716.1	1963.2	1.8621	.8899	1715.7	1962.7	1.8582	.8609	1715.4	1962.3	1.8544
2000	1.0062	1814.3	2084.3	1.9135	.9725	1814.0	2083.9	1.9096	.9410	1813.6	2083.5	1.9058
2200	1.0912	1915.1	2207.9	1.9618	1.0548	1914.7	2207.5	1.9579	1.0207	1914.4	2207.2	1.9542
2400	1.1760	2018.2	2333.8	2.0074	1.1368	2017.9	2333.4	2.0035	1.1002	2017.5	2333.1	1.9998

	1600 (605.06)				**1650** (609.24)				**1700** (613.32)			**p** (t Sat.)
v	u	h	s	v	u	h	s	v	u	h	s	t
.2552	1087.4	1162.9	1.3258	.2452	1085.1	1159.9	1.3208	.2358	1082.7	1156.9	1.3159	Sat.
												590
												600
.2481	*1079.6*	*1153.1*	*1.3165*	.2462	1086.2	1161.4	1.3222	*.23112*	*1077.2*	*1149.9*	*1.3094*	610
.2615	1094.4	1171.8	1.3341	.2586	1100.1	1179.1	1.3386	.24432	1092.7	1169.5	1.3276	620
.2733	1107.0	1187.9	1.3491	.2696	1112.2	1194.5	1.3529	.25577	1105.8	1186.3	1.3431	630
.2839	1118.3	1202.3	1.3624	.2795	1123.0	1208.4	1.3655	.26603	1117.4	1201.1	1.3566	640
.2936	1128.4	1215.4	1.3743									
.3027	1137.8	1227.4	1.3852	.2887	1132.9	1221.1	1.3770	.2754	1127.9	1214.5	1.3687	650
.3112	1146.5	1238.7	1.3953	.2973	1142.1	1232.9	1.3876	.2842	1137.5	1226.8	1.3798	660
.3193	1154.7	1249.2	1.4047	.3054	1150.6	1243.9	1.3974	.2924	1146.4	1238.4	1.3901	670
.3270	1162.5	1259.3	1.4136	.3132	1158.7	1254.3	1.4066	.3001	1154.8	1249.2	1.3996	680
.3343	1169.9	1268.9	1.4219	.3206	1166.3	1264.2	1.4152	.3075	1162.7	1259.4	1.4085	690
.3415	1177.0	1278.1	1.4299	.3277	1173.7	1273.7	1.4234	.3146	1170.2	1269.2	1.4170	700
.3483	1183.8	1287.0	1.4375	.3345	1180.7	1282.8	1.4313	.3214	1177.5	1278.6	1.4251	710
.3550	1190.4	1295.5	1.4448	.3411	1187.4	1291.6	1.4388	.3280	1184.4	1287.6	1.4327	720
.3615	1196.8	1303.9	1.4518	.3476	1194.0	1300.1	1.4459	.3344	1191.1	1296.3	1.4401	730
.3678	1203.0	1312.0	1.4586	.3538	1200.3	1308.4	1.4529	.3406	1197.6	1304.8	1.4472	740
.3740	1209.1	1319.9	1.4652	.3600	1206.5	1316.4	1.4596	.3467	1203.9	1313.0	1.4540	750
.3801	1215.0	1327.6	1.4715	.3659	1212.6	1324.3	1.4660	.3526	1210.1	1321.0	1.4606	760
.3860	1220.8	1335.1	1.4777	.3718	1218.5	1332.0	1.4723	.3584	1216.1	1328.8	1.4670	770
.3918	1226.5	1342.6	1.4837	.3775	1224.3	1339.5	1.4784	.3640	1222.0	1336.5	1.4732	780
.3976	1232.1	1349.9	1.4896	.3832	1230.0	1347.0	1.4844	.3696	1227.8	1344.0	1.4792	790
.4032	1237.7	1357.0	1.4953	.3887	1235.6	1354.2	1.4902	.3751	1233.4	1351.4	1.4851	800
.4142	1248.5	1371.1	1.5064	.3995	1246.5	1368.5	1.5014	.3857	1244.5	1365.8	1.4965	820
.4249	1259.0	1384.8	1.5170	.4101	1257.1	1382.4	1.5122	.3961	1255.3	1379.9	1.5074	840
.4354	1269.3	1398.2	1.5273	.4203	1267.6	1395.9	1.5225	.4061	1265.8	1393.6	1.5179	860
.4456	1279.5	1411.4	1.5372	.4304	1277.8	1409.2	1.5325	.4160	1276.2	1407.1	1.5280	880
.4557	1289.5	1424.4	1.5468	.4402	1287.9	1422.3	1.5422	.4257	1286.4	1420.3	1.5378	900
.4656	1299.4	1437.2	1.5562	.4499	1297.9	1435.3	1.5517	.4351	1296.4	1433.3	1.5473	920
.4753	1309.1	1449.9	1.5653	.4594	1307.7	1448.0	1.5608	.4444	1306.3	1446.1	1.5565	940
.4849	1318.8	1462.4	1.5741	.4688	1317.5	1460.6	1.5698	.4536	1316.2	1458.9	1.5655	960
.4943	1328.5	1474.8	1.5828	.4780	1327.2	1473.1	1.5785	.4626	1325.9	1471.4	1.5743	980
.5036	1338.0	1487.1	1.5913	.4871	1336.8	1485.5	1.5871	.4715	1335.6	1483.9	1.5830	1000
.5129	1347.5	1499.4	1.5997	.4961	1346.4	1497.8	1.5955	.4803	1345.2	1496.3	1.5914	1020
.5220	1357.0	1511.5	1.6078	.5050	1355.9	1510.1	1.6037	.4890	1354.7	1508.6	1.5996	1040
.5310	1366.4	1523.6	1.6159	.5138	1365.3	1522.2	1.6117	.4977	1364.3	1520.8	1.6077	1060
.5400	1375.8	1535.7	1.6237	.5226	1374.8	1534.3	1.6197	.5062	1373.8	1533.0	1.6157	1080
.5488	1385.2	1547.7	1.6315	.5312	1384.2	1546.4	1.6274	.5146	1383.2	1545.1	1.6235	1100
.5576	1394.5	1559.6	1.6391	.5398	1393.6	1558.4	1.6351	.5230	1392.6	1557.2	1.6312	1120
.5663	1403.9	1571.6	1.6466	.5483	1403.0	1570.4	1.6426	.5313	1402.0	1569.2	1.6387	1140
.5750	1413.2	1583.5	1.6540	.5567	1412.3	1582.3	1.6500	.5395	1411.4	1581.2	1.6462	1160
.5836	1422.5	1595.3	1.6612	.5651	1421.7	1594.2	1.6573	.5477	1420.8	1593.1	1.6535	1180
.5921	1431.8	1607.1	1.6684	.5734	1431.0	1606.1	1.6645	.5558	1430.2	1605.0	1.6607	1200
.6006	1441.1	1619.0	1.6755	.5817	1440.3	1617.9	1.6716	.5638	1439.5	1616.9	1.6679	1220
.6091	1450.4	1630.8	1.6825	.5899	1449.7	1629.8	1.6786	.5718	1448.9	1628.8	1.6749	1240
.6175	1459.7	1642.5	1.6894	.5981	1459.0	1641.6	1.6855	.5798	1458.2	1640.6	1.6818	1260
.6258	1469.0	1654.3	1.6962	.6062	1468.3	1653.4	1.6924	.5877	1467.6	1652.5	1.6887	1280
.6341	1478.3	1666.1	1.7029	.6143	1477.6	1665.2	1.6991	.5956	1476.9	1664.3	1.6954	1300
.6547	1501.6	1695.4	1.7194	.6343	1500.9	1694.6	1.7156	.6151	1500.3	1693.8	1.7120	1350
.6752	1524.9	1724.8	1.7354	.6542	1524.3	1724.1	1.7317	.6344	1523.7	1723.3	1.7280	1400
.6954	1548.3	1754.2	1.7509	.6739	1547.7	1753.5	1.7473	.6536	1547.2	1752.8	1.7437	1450
.7155	1571.8	1783.6	1.7662	.6934	1571.2	1783.0	1.7625	.6726	1570.7	1782.3	1.7590	1500
.7553	1619.0	1842.6	1.7955	.7320	1618.5	1842.1	1.7919	.7102	1618.1	1841.5	1.7884	1600
.8338	1715.0	1961.9	1.8508	.8083	1714.6	1961.4	1.8472	.7843	1714.2	1961.0	1.8438	1800
.9115	1813.3	2083.2	1.9022	.8837	1812.9	2082.8	1.8987	.8576	1812.6	2082.4	1.8952	2000
.9888	1914.1	2206.8	1.9505	.9588	1913.7	2206.5	1.9470	.9305	1913.4	2206.1	1.9436	2200
1.0658	2017.2	2332.8	1.9962	1.0336	2016.9	2332.4	1.9927	1.0032	2016.5	2332.1	1.9893	2400

Table 3. Vapor

p (t Sat.)	1750 (617.31)				1800 (621.21)				1850 (625.02)			
t	v	u	h	s	v	u	h	s	v	u	h	s
Sat.	.2268	1080.2	1153.7	1.3109	.2183	1077.7	1150.4	1.3060	.2102	1075.1	1147.0	1.3010
620	.2304	1084.6	1159.2	1.3160	*.2166*	*1075.6*	*1147.7*	*1.3035*	*.2028*	*1065.4*	*1134.9*	*1.2898*
630	.2425	1098.9	1177.5	1.3329	.2295	1091.6	1168.0	1.3222	.2168	1083.5	1157.7	1.3109
640	.2531	1111.4	1193.4	1.3474	.2406	1105.1	1185.2	1.3379	.2285	1098.3	1176.5	1.3280
650	.2627	1122.5	1207.6	1.3603	.2505	1117.0	1200.4	1.3517	.2388	1111.1	1192.8	1.3428
660	.2716	1132.7	1220.6	1.3720	.2596	1127.6	1214.1	1.3640	.2481	1122.4	1207.4	1.3558
670	.2799	1142.0	1232.6	1.3827	.2680	1137.5	1226.7	1.3752	.2567	1132.7	1220.6	1.3676
680	.2877	1150.7	1243.9	1.3926	.2759	1146.5	1238.4	1.3855	.2646	1142.2	1232.8	1.3784
690	.2951	1158.9	1254.5	1.4018	.2834	1155.1	1249.5	1.3951	.2722	1151.1	1244.3	1.3884
700	.3022	1166.7	1264.6	1.4106	.2905	1163.1	1259.9	1.4042	.2793	1159.4	1255.1	1.3977
710	.3090	1174.2	1274.2	1.4189	.2973	1170.8	1269.8	1.4127	.2861	1167.4	1265.3	1.4065
720	.3156	1181.3	1283.5	1.4267	.3038	1178.1	1279.3	1.4208	.2927	1174.9	1275.1	1.4148
730	.3220	1188.2	1292.4	1.4343	.3102	1185.2	1288.5	1.4285	.2990	1182.1	1284.5	1.4228
740	.3281	1194.8	1301.1	1.4415	.3163	1192.0	1297.3	1.4359	.3051	1189.1	1293.5	1.4304
750	.3341	1201.3	1309.5	1.4485	.3222	1198.6	1305.9	1.4430	.3110	1195.8	1302.3	1.4376
760	.3400	1207.5	1317.6	1.4552	.3280	1205.0	1314.2	1.4499	.3167	1202.4	1310.8	1.4446
770	.3457	1213.7	1325.6	1.4617	.3337	1211.2	1322.4	1.4565	.3223	1208.7	1319.1	1.4514
780	.3513	1219.7	1333.4	1.4680	.3392	1217.3	1330.3	1.4629	.3278	1214.9	1327.1	1.4579
790	.3568	1225.5	1341.1	1.4742	.3447	1223.3	1338.1	1.4692	.3332	1221.0	1335.0	1.4643
800	.3622	1231.3	1348.6	1.4802	.3500	1229.1	1345.7	1.4753	.3384	1226.9	1342.8	1.4704
810	.3675	1236.9	1355.9	1.4860	.3552	1234.8	1353.1	1.4812	.3436	1232.7	1350.3	1.4764
820	.3727	1242.5	1363.2	1.4917	.3603	1240.5	1360.5	1.4869	.3486	1238.4	1357.8	1.4823
830	.3778	1248.0	1370.3	1.4973	.3654	1246.0	1367.8	1.4926	.3536	1244.1	1365.1	1.4880
840	.3828	1253.4	1377.4	1.5027	.3704	1251.5	1374.9	1.4981	.3585	1249.6	1372.4	1.4936
850	.3878	1258.8	1384.4	1.5081	.3753	1256.9	1381.9	1.5035	.3634	1255.1	1379.5	1.4990
860	.3928	1264.1	1391.3	1.5133	.3801	1262.3	1388.9	1.5088	.3681	1260.5	1386.5	1.5044
870	.3976	1269.3	1398.1	1.5184	.3849	1267.6	1395.8	1.5140	.3728	1265.9	1393.5	1.5096
880	.4024	1274.5	1404.9	1.5235	.3896	1272.9	1402.6	1.5191	.3775	1271.2	1400.4	1.5148
890	.4072	1279.7	1411.5	1.5285	.3943	1278.1	1409.4	1.5241	.3821	1276.4	1407.2	1.5199
900	.4119	1284.8	1418.2	1.5334	.3989	1283.2	1416.1	1.5291	.3866	1281.6	1414.0	1.5249
920	.4212	1294.9	1431.3	1.5430	.4081	1293.4	1429.3	1.5388	.3956	1291.9	1427.3	1.5346
940	.4303	1304.9	1444.3	1.5523	.4170	1303.5	1442.4	1.5482	.4044	1302.1	1440.5	1.5441
960	.4393	1314.8	1457.1	1.5614	.4258	1313.4	1455.3	1.5573	.4130	1312.1	1453.5	1.5533
980	.4482	1324.6	1469.7	1.5702	.4345	1323.3	1468.0	1.5662	.4215	1322.0	1466.3	1.5623
1000	.4569	1334.3	1482.3	1.5789	.4430	1333.1	1480.7	1.5749	.4299	1331.9	1479.0	1.5711
1020	.4655	1344.0	1494.8	1.5874	.4514	1342.8	1493.2	1.5835	.4381	1341.6	1491.6	1.5796
1040	.4740	1353.6	1507.1	1.5957	.4598	1352.5	1505.6	1.5918	.4463	1351.4	1504.1	1.5880
1060	.4824	1363.2	1519.4	1.6038	.4680	1362.1	1518.0	1.6000	.4544	1361.0	1516.6	1.5962
1080	.4907	1372.7	1531.6	1.6118	.4761	1371.7	1530.3	1.6080	.4623	1370.6	1528.9	1.6043
1100	.4990	1382.2	1543.8	1.6197	.4842	1381.2	1542.5	1.6159	.4702	1380.2	1541.2	1.6122
1120	.5071	1391.7	1555.9	1.6274	.4922	1390.7	1554.7	1.6237	.4780	1389.8	1553.4	1.6200
1140	.5152	1401.1	1568.0	1.6350	.5001	1400.2	1566.8	1.6313	.4858	1399.3	1565.6	1.6277
1160	.5233	1410.6	1580.0	1.6424	.5079	1409.7	1578.8	1.6388	.4934	1408.8	1577.7	1.6352
1180	.5312	1420.0	1592.0	1.6498	.5157	1419.1	1590.9	1.6462	.5010	1418.2	1589.8	1.6426
1200	.5392	1429.4	1604.0	1.6571	.5235	1428.5	1602.9	1.6534	.5086	1427.7	1601.8	1.6499
1220	.5470	1438.7	1615.9	1.6642	.5311	1437.9	1614.9	1.6606	.5161	1437.1	1613.8	1.6571
1240	.5548	1448.1	1627.8	1.6712	.5388	1447.4	1626.8	1.6677	.5235	1446.6	1625.8	1.6642
1260	.5626	1457.5	1639.7	1.6782	.5463	1456.7	1638.7	1.6747	.5309	1456.0	1637.8	1.6712
1280	.5703	1466.9	1651.6	1.6851	.5539	1466.1	1650.6	1.6815	.5383	1465.4	1649.7	1.6781
1300	.5780	1476.2	1663.4	1.6918	.5614	1475.5	1662.5	1.6883	.5456	1474.8	1661.6	1.6849
1350	.5970	1499.7	1693.0	1.7084	.5799	1499.0	1692.2	1.7050	.5637	1498.4	1691.4	1.7016
1400	.6158	1523.1	1722.6	1.7245	.5983	1522.5	1721.8	1.7211	.5816	1521.9	1721.1	1.7178
1450	.6345	1546.6	1752.1	1.7402	.6164	1546.1	1751.4	1.7368	.5994	1545.5	1750.7	1.7335
1500	.6530	1570.2	1781.7	1.7555	.6345	1569.7	1781.0	1.7521	.6169	1569.2	1780.4	1.7489
1600	.6896	1617.6	1841.0	1.7850	.6701	1617.2	1840.4	1.7817	.6517	1616.7	1839.8	1.7784
1800	.7617	1713.9	1960.5	1.8404	.7404	1713.5	1960.1	1.8371	.7202	1713.1	1959.6	1.8339
2000	.8330	1812.3	2082.0	1.8919	.8098	1811.9	2081.7	1.8887	.7878	1811.6	2081.3	1.8855
2200	.9039	1913.1	2205.8	1.9403	.8788	1912.7	2205.4	1.9371	.8550	1912.4	2205.1	1.9339
2400	.9746	2016.2	2331.8	1.9860	.9476	2015.8	2331.5	1.9827	.9220	2015.5	2331.1	1.9796

	1900 (628.76)				**1950** (632.41)				**2000** (636.00)			**p** (t Sat.)
v	u	h	s	v	u	h	s	v	u	h	s	t
.2025	1072.3	1143.5	1.2961	.19517	1069.5	1140.0	1.2911	.18813	1066.6	1136.3	1.2861	**Sat.**
												620
.20425	1074.6	1146.4	1.2987	*.19163*	*1064.6*	*1133.8*	*1.2854*					**630**
.21673	1091.0	1167.2	1.3177	.20514	1083.1	1157.1	1.3068	.19365	1074.4	1146.1	1.2950	**640**
.2275	1104.8	1184.8	1.3336	.2164	1098.2	1176.3	1.3241	.2057	1091.1	1167.2	1.3141	**650**
.2370	1116.9	1200.3	1.3475	.2264	1111.2	1192.8	1.3389	.2160	1105.1	1185.0	1.3301	**660**
.2458	1127.8	1214.2	1.3599	.2353	1122.7	1207.6	1.3521	.2252	1117.3	1200.7	1.3440	**670**
.2539	1137.8	1227.0	1.3712	.2436	1133.1	1221.0	1.3639	.2336	1128.3	1214.8	1.3565	**680**
.2615	1147.0	1238.9	1.3816	.2512	1142.8	1233.4	1.3747	.2414	1138.4	1227.8	1.3678	**690**
.2686	1155.7	1250.1	1.3913	.2585	1151.8	1245.0	1.3848	.2487	1147.7	1239.8	1.3782	**700**
.2755	1163.8	1260.7	1.4003	.2653	1160.2	1255.9	1.3942	.2556	1156.5	1251.1	1.3879	**710**
.2820	1171.6	1270.7	1.4089	.2719	1168.2	1266.3	1.4030	.2622	1164.8	1261.8	1.3970	**720**
.2883	1179.0	1280.4	1.4171	.2782	1175.8	1276.2	1.4113	.2685	1172.6	1272.0	1.4056	**730**
.2944	1186.2	1289.7	1.4248	.2842	1183.2	1285.7	1.4193	.2745	1180.1	1281.7	1.4138	**740**
.3003	1193.0	1298.6	1.4322	.2901	1190.2	1294.9	1.4269	.2803	1187.3	1291.1	1.4216	**750**
.3060	1199.7	1307.3	1.4394	.2957	1197.0	1303.7	1.4342	.2860	1194.3	1300.1	1.4290	**760**
.3115	1206.2	1315.7	1.4463	.3013	1203.6	1312.3	1.4412	.2915	1201.0	1308.9	1.4362	**770**
.3170	1212.5	1323.9	1.4529	.3066	1210.0	1320.7	1.4480	.2968	1207.6	1317.4	1.4431	**780**
.3223	1218.6	1332.0	1.4594	.3119	1216.3	1328.8	1.4545	.3020	1213.9	1325.7	1.4498	**790**
.3275	1224.7	1339.8	1.4656	.3170	1222.4	1336.8	1.4609	.3071	1220.1	1333.8	1.4562	**800**
.3326	1230.6	1347.5	1.4717	.3221	1228.4	1344.6	1.4671	.3121	1226.2	1341.7	1.4625	**810**
.3376	1236.4	1355.1	1.4776	.3270	1234.3	1352.3	1.4731	.3170	1232.2	1349.5	1.4686	**820**
.3425	1242.1	1362.5	1.4834	.3319	1240.1	1359.8	1.4789	.3218	1238.0	1357.1	1.4745	**830**
.3473	1247.7	1369.8	1.4891	.3367	1245.8	1367.2	1.4847	.3265	1243.8	1364.6	1.4803	**840**
.3521	1253.2	1377.0	1.4946	.3414	1251.4	1374.5	1.4903	.3312	1249.5	1372.0	1.4860	**850**
.3568	1258.7	1384.2	1.5000	.3460	1256.9	1381.7	1.4957	.3357	1255.1	1379.3	1.4915	**860**
.3614	1264.1	1391.2	1.5053	.3506	1262.4	1388.9	1.5011	.3403	1260.6	1386.5	1.4969	**870**
.3660	1269.5	1398.2	1.5106	.3551	1267.8	1395.9	1.5064	.3447	1266.0	1393.6	1.5023	**880**
.3705	1274.8	1405.0	1.5157	.3595	1273.1	1402.8	1.5116	.3491	1271.4	1400.6	1.5075	**890**
.3750	1280.0	1411.9	1.5207	.3639	1278.4	1409.7	1.5166	.3534	1276.8	1407.6	1.5126	**900**
.3838	1290.4	1425.3	1.5306	.3726	1288.9	1423.3	1.5266	.3620	1287.3	1421.3	1.5226	**920**
.3924	1300.6	1438.6	1.5401	.3811	1299.2	1436.7	1.5362	.3703	1297.7	1434.8	1.5323	**940**
.4009	1310.7	1451.7	1.5494	.3894	1309.3	1449.9	1.5455	.3785	1308.0	1448.0	1.5417	**960**
.4093	1320.7	1464.6	1.5584	.3976	1319.4	1462.9	1.5546	.3865	1318.1	1461.1	1.5509	**980**
.4175	1330.6	1477.4	1.5672	.4057	1329.4	1475.7	1.5635	.3945	1328.1	1474.1	1.5598	**1000**
.4256	1340.4	1490.1	1.5759	.4136	1339.2	1488.5	1.5722	.4023	1338.0	1486.9	1.5686	**1020**
.4335	1350.2	1502.6	1.5843	.4214	1349.1	1501.1	1.5807	.4100	1347.9	1499.6	1.5771	**1040**
.4414	1359.9	1515.1	1.5926	.4292	1358.8	1513.7	1.5890	.4175	1357.7	1512.3	1.5855	**1060**
.4492	1369.6	1527.5	1.6007	.4368	1368.5	1526.2	1.5971	.4250	1367.5	1524.8	1.5937	**1080**
.4570	1379.2	1539.9	1.6087	.4444	1378.2	1538.6	1.6051	.4325	1377.2	1537.2	1.6017	**1100**
.4646	1388.8	1552.1	1.6165	.4519	1387.8	1550.9	1.6130	.4398	1386.8	1549.6	1.6096	**1120**
.4722	1398.3	1564.4	1.6242	.4593	1397.4	1563.1	1.6207	.4471	1396.5	1561.9	1.6173	**1140**
.4797	1407.9	1576.5	1.6317	.4667	1407.0	1575.4	1.6283	.4543	1406.1	1574.2	1.6249	**1160**
.4871	1417.4	1588.7	1.6392	.4739	1416.5	1587.5	1.6358	.4614	1415.6	1586.4	1.6324	**1180**
.4945	1426.9	1600.7	1.6465	.4812	1426.0	1599.7	1.6431	.4685	1425.2	1598.6	1.6398	**1200**
.5019	1436.3	1612.8	1.6537	.4884	1435.5	1611.8	1.6504	.4755	1434.7	1610.7	1.6471	**1220**
.5091	1445.8	1624.8	1.6608	.4955	1445.0	1623.8	1.6575	.4825	1444.3	1622.8	1.6542	**1240**
.5164	1455.3	1636.8	1.6678	.5026	1454.5	1635.9	1.6645	.4894	1453.8	1634.9	1.6613	**1260**
.5236	1464.7	1648.8	1.6748	.5096	1464.0	1647.9	1.6715	.4963	1463.3	1646.9	1.6683	**1280**
.5307	1474.1	1660.7	1.6816	.5166	1473.4	1659.9	1.6783	.5031	1472.7	1659.0	1.6751	**1300**
.5484	1497.7	1690.6	1.6983	.5339	1497.1	1689.7	1.6951	.5201	1496.5	1688.9	1.6919	**1350**
.5659	1521.4	1720.3	1.7145	.5510	1520.8	1719.6	1.7113	.5368	1520.2	1718.8	1.7082	**1400**
.5832	1545.0	1750.0	1.7303	.5679	1544.4	1749.3	1.7271	.5533	1543.9	1748.7	1.7241	**1450**
.6004	1568.7	1779.8	1.7457	.5846	1568.2	1779.1	1.7425	.5697	1567.7	1778.5	1.7395	**1500**
.6343	1616.3	1839.3	1.7753	.6177	1615.8	1838.7	1.7722	.6020	1615.4	1838.2	1.7692	**1600**
.7010	1712.7	1959.2	1.8308	.6829	1712.4	1958.8	1.8278	.6656	1712.0	1958.3	1.8249	**1800**
.7670	1811.2	2080.9	1.8824	.7472	1810.9	2080.5	1.8794	.7284	1810.6	2080.2	1.8765	**2000**
.8324	1912.1	2204.8	1.9308	.8111	1911.7	2204.4	1.9279	.7908	1911.4	2204.1	1.9249	**2200**
.8977	2015.2	2330.8	1.9765	.8748	2014.8	2330.5	1.9736	.8529	2014.5	2330.2	1.9707	**2400**

Table 3. Vapor

p (t Sat.)	2100 (642.95)				2200 (649.64)				2300 (656.09)			
t	v	u	h	s	v	u	h	s	v	u	h	s
Sat.	.17491	1060.6	1128.5	1.2760	.16270	1054.0	1120.3	1.2657	.15133	1047.0	1111.4	1.2551
640	*.17027*	*1053.5*	*1119.7*	*1.2680*								
650	.18457	1074.9	1146.6	1.2924	.16325	1054.9	1121.4	1.2667				
660	.19608	1091.6	1167.8	1.3114	.17676	1076.1	1148.0	1.2906	.15735	1057.1	1124.1	1.2664
670	.20599	1105.7	1185.8	1.3273	.18772	1092.7	1169.2	1.3094	.17003	1077.8	1150.2	1.2897
680	.21483	1118.1	1201.6	1.3413	.19719	1106.8	1187.1	1.3252	.18043	1094.3	1171.1	1.3081
690	.22291	1129.2	1215.8	1.3537	.20567	1119.2	1202.9	1.3391	.18946	1108.3	1189.0	1.3237
700	.2304	1139.3	1228.9	1.3650	.2134	1130.4	1217.2	1.3514	.19756	1120.7	1204.8	1.3374
710	.2374	1148.8	1241.0	1.3754	.2206	1140.6	1230.4	1.3627	.20498	1131.8	1219.1	1.3497
720	.2440	1157.6	1252.4	1.3851	.2273	1150.0	1242.6	1.3731	.21187	1142.0	1232.2	1.3609
730	.2504	1165.9	1263.2	1.3942	.2337	1158.9	1254.0	1.3828	.21835	1151.5	1244.5	1.3712
740	.2564	1173.8	1273.5	1.4028	.2398	1167.3	1264.9	1.3919	.22447	1160.4	1256.0	1.3808
750	.2622	1181.4	1283.3	1.4110	.2456	1175.2	1275.2	1.4004	.2303	1168.8	1266.9	1.3899
760	.2678	1188.7	1292.7	1.4188	.2512	1182.9	1285.1	1.4086	.2359	1176.8	1277.3	1.3984
770	.2733	1195.7	1301.9	1.4262	.2566	1190.2	1294.6	1.4164	.2413	1184.5	1287.2	1.4066
780	.2785	1202.5	1310.7	1.4334	.2618	1197.2	1303.8	1.4238	.2465	1191.9	1296.8	1.4143
790	.2837	1209.1	1319.3	1.4403	.2669	1204.1	1312.7	1.4310	.2515	1199.0	1306.0	1.4218
800	.2887	1215.5	1327.7	1.4470	.2718	1210.7	1321.4	1.4379	.2564	1205.9	1315.0	1.4289
810	.2936	1221.8	1335.9	1.4534	.2767	1217.2	1329.8	1.4445	.2612	1212.5	1323.7	1.4358
820	.2984	1227.9	1343.9	1.4597	.2814	1223.5	1338.1	1.4510	.2658	1219.1	1332.2	1.4424
830	.3031	1233.9	1351.7	1.4658	.2860	1229.7	1346.1	1.4573	.2704	1225.4	1340.5	1.4489
840	.3077	1239.8	1359.4	1.4718	.2905	1235.8	1354.0	1.4634	.2748	1231.6	1348.6	1.4552
850	.3122	1245.6	1367.0	1.4776	.2950	1241.7	1361.8	1.4693	.2792	1237.7	1356.5	1.4613
860	.3167	1251.3	1374.4	1.4832	.2993	1247.6	1369.4	1.4751	.2834	1243.7	1364.4	1.4672
870	.3211	1257.0	1381.8	1.4888	.3036	1253.3	1376.9	1.4808	.2876	1249.6	1372.0	1.4730
880	.3254	1262.6	1389.0	1.4942	.3078	1259.0	1384.3	1.4863	.2918	1255.4	1379.6	1.4787
890	.3297	1268.1	1396.2	1.4995	.3120	1264.6	1391.6	1.4918	.2958	1261.1	1387.1	1.4842
900	.3339	1273.5	1403.3	1.5048	.3161	1270.2	1398.9	1.4971	.2998	1266.8	1394.4	1.4896
920	.3422	1284.2	1417.2	1.5149	.3242	1281.1	1413.1	1.5075	.3077	1277.9	1408.9	1.5002
940	.3503	1294.8	1430.9	1.5248	.3320	1291.8	1427.0	1.5175	.3154	1288.8	1423.0	1.5104
960	.3582	1305.2	1444.4	1.5343	.3397	1302.3	1440.6	1.5272	.3229	1299.5	1436.9	1.5202
980	.3660	1315.4	1457.6	1.5436	.3473	1312.7	1454.1	1.5366	.3302	1310.0	1450.6	1.5298
1000	.3736	1325.6	1470.8	1.5527	.3547	1323.0	1467.4	1.5458	.3374	1320.4	1464.0	1.5391
1020	.3812	1335.6	1483.8	1.5615	.3620	1333.2	1480.6	1.5547	.3445	1330.7	1477.3	1.5481
1040	.3886	1345.6	1496.6	1.5702	.3692	1343.3	1493.6	1.5635	.3514	1340.9	1490.5	1.5570
1060	.3959	1355.5	1509.4	1.5786	.3763	1353.3	1506.4	1.5720	.3583	1351.0	1503.5	1.5656
1080	.4032	1365.3	1522.0	1.5869	.3832	1363.2	1519.2	1.5803	.3651	1361.0	1516.4	1.5740
1100	.4103	1375.1	1534.6	1.5950	.3901	1373.1	1531.9	1.5885	.3717	1371.0	1529.2	1.5823
1120	.4174	1384.9	1547.1	1.6029	.3970	1382.9	1544.5	1.5966	.3783	1380.9	1541.9	1.5904
1140	.4244	1394.6	1559.5	1.6108	.4037	1392.7	1557.0	1.6044	.3848	1390.8	1554.6	1.5983
1160	.4313	1404.3	1571.9	1.6184	.4104	1402.4	1569.5	1.6122	.3913	1400.6	1567.1	1.6061
1180	.4381	1413.9	1584.2	1.6260	.4170	1412.1	1581.9	1.6198	.3977	1410.4	1579.6	1.6138
1200	.4449	1423.5	1596.4	1.6334	.4235	1421.8	1594.3	1.6273	.4040	1420.1	1592.1	1.6214
1220	.4517	1433.1	1608.6	1.6407	.4300	1431.5	1606.6	1.6346	.4102	1429.9	1604.5	1.6288
1240	.4584	1442.7	1620.8	1.6480	.4365	1441.1	1618.8	1.6419	.4165	1439.5	1616.8	1.6361
1260	.4650	1452.3	1633.0	1.6551	.4429	1450.7	1631.0	1.6490	.4226	1449.2	1629.1	1.6433
1280	.4716	1461.8	1645.1	1.6621	.4492	1460.3	1643.2	1.6561	.4287	1458.9	1641.3	1.6504
1300	.4782	1471.3	1657.2	1.6690	.4555	1469.9	1655.4	1.6630	.4348	1468.5	1653.6	1.6573
1350	.4944	1495.2	1687.3	1.6858	.4711	1493.9	1685.7	1.6800	.4498	1492.6	1684.0	1.6744
1400	.5104	1519.0	1717.3	1.7022	.4865	1517.8	1715.8	1.6964	.4646	1516.6	1714.3	1.6909
1450	.5262	1542.8	1747.3	1.7181	.5016	1541.7	1745.9	1.7124	.4792	1540.6	1744.5	1.7069
1500	.5419	1566.6	1777.2	1.7336	.5167	1565.6	1775.9	1.7279	.4936	1564.6	1774.6	1.7225
1600	.5728	1614.5	1837.1	1.7634	.5463	1613.6	1836.0	1.7578	.5221	1612.7	1834.9	1.7525
1800	.6336	1711.2	1957.5	1.8192	.6045	1710.5	1956.6	1.8137	.5779	1709.7	1955.7	1.8085
2000	.6935	1809.9	2079.4	1.8709	.6619	1809.2	2078.7	1.8655	.6329	1808.5	2077.9	1.8603
2200	.7531	1910.8	2203.4	1.9193	.7188	1910.1	2202.7	1.9140	.6875	1909.5	2202.1	1.9088
2400	.8124	2013.8	2329.5	1.9650	.7755	2013.2	2328.9	1.9597	.7418	2012.5	2328.2	1.9545

v	u	h	s	v	u	h	s	v	u	h	s	t
.14067	1039.3	1101.8	1.2441	.13059	1031.0	1091.4	1.2327	.12099	1021.8	1080.0	1.2205	**Sat.**
												640
												650
.13631	*1031.5*	*1092.0*	*1.2354*									**660**
.15236	1060.0	1127.6	1.2671	.13368	1036.9	1098.7	1.2391	*.11012*	*999.3*	*1052.3*	*1.1960*	**670**
.16420	1080.1	1153.0	1.2894	.14811	1063.4	1131.9	1.2684	.13145	1042.5	1105.8	1.2432	**680**
.17403	1096.3	1173.6	1.3074	.15913	1082.8	1156.4	1.2898	.14445	1067.2	1136.7	1.2702	**690**
.18261	1110.2	1191.3	1.3228	.16839	1098.7	1176.6	1.3073	.15469	1085.9	1160.3	1.2907	**700**
.19035	1122.5	1207.0	1.3363	.17655	1112.4	1194.1	1.3223	.16341	1101.4	1180.0	1.3076	**710**
.19745	1133.6	1221.3	1.3484	.18392	1124.6	1209.6	1.3355	.17114	1114.9	1197.2	1.3222	**720**
.20406	1143.8	1234.4	1.3595	.19072	1135.6	1223.8	1.3475	.17817	1126.9	1212.6	1.3352	**730**
.21028	1153.3	1246.7	1.3697	.19705	1145.7	1236.9	1.3585	.18466	1137.8	1226.7	1.3470	**740**
.2162	1162.2	1258.2	1.3793	.2030	1155.2	1249.1	1.3686	.19073	1147.9	1239.7	1.3578	**750**
.2218	1170.6	1269.1	1.3883	.2087	1164.1	1260.7	1.3781	.19646	1157.4	1251.9	1.3678	**760**
.2272	1178.6	1279.5	1.3968	.2141	1172.5	1271.6	1.3870	.20189	1166.2	1263.4	1.3772	**770**
.2323	1186.3	1289.5	1.4049	.2193	1180.6	1282.0	1.3955	.20709	1174.7	1274.3	1.3861	**780**
.2374	1193.7	1299.1	1.4126	.2242	1188.3	1292.0	1.4035	.21208	1182.7	1284.7	1.3945	**790**
.2422	1200.8	1308.4	1.4200	.2291	1195.7	1301.7	1.4112	.2169	1190.4	1294.8	1.4024	**800**
.2469	1207.8	1317.4	1.4271	.2337	1202.9	1311.0	1.4186	.2215	1197.8	1304.4	1.4101	**810**
.2515	1214.5	1326.2	1.4340	.2383	1209.8	1320.0	1.4257	.2260	1205.0	1313.8	1.4174	**820**
.2560	1221.0	1334.7	1.4407	.2427	1216.5	1328.8	1.4325	.2304	1212.0	1322.8	1.4245	**830**
.2604	1227.4	1343.1	1.4471	.2470	1223.1	1337.4	1.4391	.2347	1218.7	1331.7	1.4313	**840**
.2646	1233.7	1351.2	1.4534	.2513	1229.5	1345.8	1.4456	.2389	1225.3	1340.3	1.4379	**850**
.2688	1239.8	1359.2	1.4594	.2554	1235.8	1354.0	1.4518	.2429	1231.8	1348.7	1.4443	**860**
.2730	1245.8	1367.1	1.4654	.2594	1242.0	1362.0	1.4579	.2469	1238.1	1356.9	1.4505	**870**
.2770	1251.8	1374.8	1.4712	.2634	1248.1	1369.9	1.4638	.2509	1244.3	1365.0	1.4566	**880**
.2810	1257.6	1382.4	1.4768	.2673	1254.0	1377.7	1.4696	.2547	1250.4	1372.9	1.4625	**890**
.2849	1263.4	1389.9	1.4824	.2712	1259.9	1385.4	1.4752	.2585	1256.4	1380.7	1.4682	**900**
.2926	1274.7	1404.6	1.4931	.2787	1271.4	1400.4	1.4862	.2658	1268.1	1396.0	1.4794	**920**
.3001	1285.7	1419.0	1.5035	.2860	1282.7	1415.0	1.4967	.2730	1279.6	1410.9	1.4901	**940**
.3074	1296.6	1433.1	1.5135	.2931	1293.7	1429.3	1.5069	.2800	1290.7	1425.5	1.5004	**960**
.3145	1307.3	1447.0	1.5232	.3001	1304.5	1443.4	1.5167	.2868	1301.7	1439.7	1.5104	**980**
.3215	1317.8	1460.6	1.5326	.3069	1315.2	1457.2	1.5262	.2935	1312.5	1453.7	1.5201	**1000**
.3284	1328.2	1474.1	1.5417	.3136	1325.7	1470.8	1.5355	.3000	1323.2	1467.5	1.5295	**1020**
.3352	1338.5	1487.4	1.5507	.3202	1336.1	1484.3	1.5445	.3064	1333.7	1481.2	1.5386	**1040**
.3418	1348.7	1500.6	1.5594	.3267	1346.5	1497.6	1.5534	.3127	1344.2	1494.6	1.5475	**1060**
.3484	1358.9	1513.6	1.5679	.3331	1356.7	1510.8	1.5620	.3189	1354.5	1507.9	1.5562	**1080**
.3549	1368.9	1526.5	1.5763	.3393	1366.8	1523.8	1.5704	.3250	1364.7	1521.1	1.5647	**1100**
.3612	1378.9	1539.4	1.5844	.3455	1376.9	1536.8	1.5786	.3310	1374.9	1534.2	1.5730	**1120**
.3675	1388.9	1552.1	1.5924	.3516	1386.9	1549.6	1.5867	.3370	1385.0	1547.1	1.5812	**1140**
.3738	1398.8	1564.8	1.6003	.3577	1396.9	1562.4	1.5947	.3428	1395.0	1560.0	1.5892	**1160**
.3800	1408.6	1577.4	1.6080	.3637	1406.8	1575.1	1.6024	.3486	1405.0	1572.8	1.5970	**1180**
.3861	1418.4	1589.9	1.6156	.3696	1416.7	1587.7	1.6101	.3544	1415.0	1585.5	1.6047	**1200**
.3921	1428.2	1602.4	1.6231	.3754	1426.6	1600.2	1.6176	.3600	1424.9	1598.1	1.6123	**1220**
.3981	1438.0	1614.8	1.6305	.3812	1436.4	1612.7	1.6250	.3657	1434.8	1610.7	1.6198	**1240**
.4041	1447.7	1627.1	1.6377	.3870	1446.2	1625.2	1.6323	.3712	1444.6	1623.2	1.6271	**1260**
.4100	1457.4	1639.5	1.6448	.3927	1455.9	1637.6	1.6395	.3768	1454.4	1635.7	1.6343	**1280**
.4158	1467.1	1651.8	1.6518	.3984	1465.7	1650.0	1.6465	.3822	1464.2	1648.1	1.6414	**1300**
.4303	1491.3	1682.4	1.6690	.4123	1490.0	1680.7	1.6638	.3958	1488.6	1679.1	1.6587	**1350**
.4445	1515.4	1712.8	1.6856	.4261	1514.2	1711.3	1.6804	.4091	1513.0	1709.8	1.6755	**1400**
.4586	1539.4	1743.1	1.7017	.4397	1538.3	1741.7	1.6966	.4222	1537.2	1740.3	1.6917	**1450**
.4725	1563.5	1773.4	1.7173	.4531	1562.5	1772.1	1.7123	.4351	1561.4	1770.8	1.7074	**1500**
.4999	1611.7	1833.8	1.7474	.4795	1610.8	1832.6	1.7424	.4606	1609.9	1831.5	1.7377	**1600**
.5536	1709.0	1954.8	1.8035	.5312	1708.2	1954.0	1.7986	.5105	1707.5	1953.1	1.7940	**1800**
.6064	1807.9	2077.2	1.8553	.5820	1807.2	2076.4	1.8506	.5595	1806.5	2075.7	1.8460	**2000**
.6588	1908.8	2201.4	1.9039	.6324	1908.1	2200.7	1.8991	.6080	1907.5	2200.0	1.8946	**2200**
.7110	2011.8	2327.6	1.9496	.6826	2011.2	2326.9	1.9449	.6564	2010.5	2326.3	1.9403	**2400**

Table 3. Vapor

p (t Sat.)	2700 (679.73)				2800 (685.16)				2900 (690.42)			
t	v	u	h	s	v	u	h	s	v	u	h	s
Sat.	.11172	1011.5	1067.4	1.2075	.10264	999.8	1053.0	1.1932	.09353	986.0	1036.2	1.1769
680	.11237	1012.9	1069.1	1.2090								
690	.12951	1048.4	1113.1	1.2475	.11334	1023.7	1082.5	1.2189	*.09196*	*982.1*	*1031.4*	*1.1728*
700	.14126	1071.4	1141.9	1.2725	.12780	1054.3	1120.5	1.2518	.11374	1033.2	1094.2	1.2272
710	.15078	1089.3	1164.7	1.2920	.13847	1075.8	1147.5	1.2750	.12626	1060.2	1128.0	1.2562
720	.15898	1104.4	1183.8	1.3083	.14731	1093.0	1169.3	1.2936	.13599	1080.4	1153.3	1.2778
730	.16632	1117.6	1200.7	1.3225	.15503	1107.7	1188.0	1.3093	.14423	1096.9	1174.3	1.2955
740	.17301	1129.5	1215.9	1.3353	.16198	1120.6	1204.5	1.3232	.15150	1111.1	1192.4	1.3107
750	.17921	1140.3	1229.8	1.3468	.16836	1132.2	1219.5	1.3356	.15808	1123.8	1208.6	1.3241
760	.18502	1150.3	1242.8	1.3575	.17428	1142.9	1233.2	1.3469	.16415	1135.2	1223.3	1.3362
770	.19052	1159.7	1254.9	1.3674	.17985	1152.9	1246.1	1.3574	.16981	1145.8	1236.9	1.3473
780	.19574	1168.5	1266.3	1.3766	.18512	1162.2	1258.1	1.3672	.17515	1155.6	1249.6	1.3576
790	.20074	1176.9	1277.2	1.3854	.19014	1171.0	1269.5	1.3763	.18021	1164.8	1261.6	1.3672
800	.2055	1185.0	1287.7	1.3937	.19495	1179.4	1280.4	1.3850	.18504	1173.6	i272.9	1.3763
810	.2102	1192.7	1297.7	1.4016	.19958	1187.4	1290.8	1.3932	.18967	1181.9	1283.7	1.3848
820	.2147	1200.1	1307.4	1.4092	.20405	1195.1	1300.8	1.4011	.19413	1189.9	1294.1	1.3930
830	.2190	1207.3	1316.7	1.4165	.20838	1202.5	1310.5	1.4086	.19843	1197.6	1304.1	1.4007
840	.2232	1214.3	1325.8	1.4235	.21257	1209.7	1319.8	1.4158	.20261	1205.0	1313.7	1.4082
850	.2274	1221.0	1334.6	1.4303	.2167	1216.7	1328.9	1.4228	.2067	1212.2	1323.1	1.4154
860	.2314	1227.7	1343.3	1.4369	.2206	1223.5	1337.8	1.4295	.2106	1219.2	1332.2	1.4223
870	.2353	1234.1	1351.7	1.4432	.2245	1230.1	1346.4	1.4361	.2145	1226.0	1341.1	1.4290
880	.2392	1240.5	1360.0	1.4494	.2283	1236.6	1354.9	1.4424	.2182	1232.6	1349.8	1.4355
890	.2430	1246.7	1368.1	1.4555	.2321	1243.0	1363.2	1.4486	.2219	1239.1	1358.2	1.4418
900	.2467	1252.8	1376.1	1.4614	.2357	1249.2	1371.3	1.4546	.2255	1245.5	1366.6	1.4479
910	.2503	1258.8	1383.9	1.4671	.2393	1255.3	1379.3	1.4605	.2291	1251.8	1374.7	1.4539
920	.2539	1264.8	1391.6	1.4727	.2429	1261.4	1387.2	1.4662	.2325	1257.9	1382.7	1.4598
930	.2575	1270.6	1399.3	1.4782	.2463	1267.3	1395.0	1.4718	.2360	1264.0	1390.6	1.4655
940	.2610	1276.4	1406.8	1.4836	.2498	1273.2	1402.6	1.4773	.2393	1270.0	1398.4	1.4710
950	.2644	1282.1	1414.2	1.4889	.2531	1279.0	1410.2	1.4827	.2426	1275.9	1406.1	1.4765
960	.2678	1287.8	1421.6	1.4941	.2565	1284.8	1417.6	1.4879	.2459	1281.7	1413.7	1.4819
970	.2711	1293.4	1428.8	1.4992	.2598	1290.4	1425.0	1.4931	.2491	1287.5	1421.2	1.4871
980	.2745	1298.9	1436.0	1.5042	.2630	1296.1	1432.3	1.4982	.2523	1293.2	1428.6	1.4923
990	.2777	1304.4	1443.2	1.5092	.2662	1301.6	1439.6	1.5032	.2555	1298.8	1435.9	1.4974
1000	.2810	1309.9	1450.2	1.5140	.2694	1307.2	1446.7	1.5081	.2586	1304.4	1443.2	1.5024
1020	.2873	1320.7	1464.2	1.5235	.2756	1318.1	1460.9	1.5178	.2647	1315.5	1457.5	1.5121
1040	.2936	1331.3	1478.0	1.5328	.2817	1328.9	1474.8	1.5271	.2706	1326.4	1471.6	1.5216
1060	.2997	1341.8	1491.6	1.5418	.2877	1339.5	1488.6	1.5362	.2765	1337.2	1485.5	1.5308
1080	.3058	1352.3	1505.0	1.5506	.2936	1350.0	1502.2	1.5451	.2823	1347.8	1499.3	1.5398
1100	.3117	1362.6	1518.4	1.5592	.2994	1360.5	1515.6	1.5538	.2879	1358.3	1512.8	1.5485
1120	.3176	1372.9	1531.5	1.5676	.3051	1370.8	1528.9	1.5623	.2935	1368.8	1526.3	1.5571
1140	.3234	1383.0	1544.6	1.5758	.3107	1381.1	1542.1	1.5706	.2990	1379.1	1539.6	1.5655
1160	.3291	1393.2	1557.6	1.5839	.3163	1391.3	1555.2	1.5787	.3044	1389.4	1552.7	1.5736
1180	.3347	1403.2	1570.5	1.5918	.3218	1401.4	1568.2	1.5867	.3097	1399.6	1565.8	1.5817
1200	.3403	1413.3	1583.3	1.5995	.3272	1411.5	1581.1	1.5945	.3150	1409.8	1578.8	1.5896
1220	.3458	1423.2	1596.0	1.6072	.3326	1421.6	1593.9	1.6022	.3202	1419.9	1591.7	1.5973
1240	.3512	1433.2	1608.7	1.6147	.3379	1431.6	1606.6	1.6097	.3254	1430.0	1604.6	1.6049
1260	.3567	1443.1	1621.3	1.6220	.3431	1441.5	1619.3	1.6171	.3305	1440.0	1617.3	1.6123
1280	.3620	1453.0	1633.8	1.6293	.3483	1451.5	1631.9	1.6244	.3356	1450.0	1630.0	1.6197
1300	.3673	1462.8	1646.3	1.6364	.3535	1461.4	1644.5	1.6316	.3406	1459.9	1642.7	1.6269
1350	.3804	1487.3	1677.4	1.6538	.3662	1486.0	1675.7	1.6491	.3529	1484.7	1674.1	1.6445
1400	.3933	1511.7	1708.3	1.6707	.3787	1510.5	1706.7	1.6660	.3651	1509.3	1705.2	1.6615
1450	.4060	1536.1	1738.9	1.6869	.3910	1535.0	1737.5	1.6823	.3770	1533.8	1736.2	1.6779
1500	.4185	1560.4	1769.5	1.7027	.4031	1559.4	1768.2	1.6982	.3888	1558.3	1766.9	1.6938
1600	.4432	1609.0	1830.4	1.7331	.4270	1608.1	1829.3	1.7286	.4119	1607.2	1828.2	1.7243
1800	.4913	1706.7	1952.2	1.7895	.4736	1706.0	1951.4	1.7852	.4570	1705.3	1950.5	1.7810
2000	.5386	1805.9	2075.0	1.8415	.5193	1805.2	2074.2	1.8373	.5012	1804.5	2073.5	1.8331
2200	.5855	1906.9	2199.4	1.8902	.5645	1906.2	2198.7	1.8859	.5450	1905.6	2198.0	1.8818
2400	.6321	2009.8	2325.7	1.9359	.6096	2009.2	2325.0	1.9317	.5886	2008.5	2324.4	1.9276

	3000 (695.52)				**3100** (700.47)				**3200** (705.27)			**p** (t Sat.)
v	u	h	s	v	u	h	s	v	u	h	s	t
.08404	968.8	1015.5	1.1575	.07322	944.8	986.8	1.1316	.05444	887.7	919.9	1.0730	Sat.
												680
												690
.09771	1003.9	1058.1	1.1944									**700**
.11382	1041.8	1105.0	1.2347	.10054	1018.5	1076.1	1.2083	.08476	984.5	1034.7	1.1714	**710**
.12486	1066.2	1135.5	1.2606	.11372	1049.8	1115.1	1.2414	.10226	1030.3	1090.8	1.2192	**720**
.13378	1085.1	1159.4	1.2808	.12359	1072.1	1143.0	1.2650	.11351	1057.4	1124.6	1.2478	**730**
.14146	1100.9	1179.5	1.2976	.13179	1089.9	1165.5	1.2839	.12241	1077.9	1150.4	1.2693	**740**
.14831	1114.7	1197.1	1.3122	.13897	1105.1	1184.9	1.2999	.13000	1094.8	1171.8	1.2871	**750**
.15456	1127.1	1212.9	1.3253	.14543	1118.5	1202.0	1.3140	.13672	1109.5	1190.4	1.3024	**760**
.16034	1138.4	1227.4	1.3371	.15136	1130.6	1217.4	1.3266	.14282	1122.4	1207.0	1.3160	**770**
.16575	1148.8	1240.8	1.3479	.15687	1141.6	1231.6	1.3382	.14844	1134.2	1222.1	1.3282	**780**
.17086	1158.5	1253.3	1.3580	.16204	1151.9	1244.9	1.3488	.15370	1145.1	1236.1	1.3394	**790**
.17572	1167.6	1265.2	1.3675	.16694	1161.5	1257.3	1.3587	.15865	1155.2	1249.1	1.3498	**800**
.18036	1176.3	1276.5	1.3764	.17161	1170.6	1269.0	1.3680	.16335	1164.6	1261.4	1.3595	**810**
.18482	1184.6	1287.2	1.3849	.17607	1179.2	1280.2	1.3768	.16783	1173.6	1273.0	1.3686	**820**
.18912	1192.6	1297.6	1.3929	.18037	1187.4	1290.9	1.3851	.17213	1182.2	1284.1	1.3773	**830**
.19328	1200.2	1307.5	1.4006	.18451	1195.4	1301.2	1.3930	.17627	1190.4	1294.8	1.3855	**840**
.19731	1207.7	1317.2	1.4080	.18853	1203.0	1311.2	1.4007	.18027	1198.3	1305.0	1.3934	**850**
.20123	1214.8	1326.5	1.4151	.19242	1210.4	1320.8	1.4080	.18414	1205.9	1314.9	1.4009	**860**
.20504	1221.8	1335.6	1.4220	.19621	1217.6	1330.1	1.4150	.18791	1213.2	1324.5	1.4081	**870**
.20877	1228.6	1344.5	1.4286	.19990	1224.6	1339.2	1.4219	.19157	1220.4	1333.8	1.4151	**880**
.21241	1235.3	1353.2	1.4351	.20350	1231.4	1348.1	1.4285	.19514	1227.4	1342.9	1.4219	**890**
.2160	1241.8	1361.7	1.4414	.2070	1238.0	1356.8	1.4349	.19863	1234.2	1351.8	1.4285	**900**
.2195	1248.2	1370.0	1.4475	.2105	1244.5	1365.3	1.4411	.20204	1240.9	1360.5	1.4348	**910**
.2229	1254.5	1378.2	1.4534	.2139	1251.0	1373.6	1.4472	.20539	1247.4	1369.0	1.4410	**920**
.2263	1260.6	1386.3	1.4592	.2172	1257.2	1381.8	1.4531	.20867	1253.8	1377.4	1.4470	**930**
.2296	1266.7	1394.2	1.4649	.2205	1263.4	1389.9	1.4589	.21189	1260.1	1385.6	1.4529	**940**
.2328	1272.7	1402.0	1.4705	.2237	1269.5	1397.8	1.4645	.2151	1266.3	1393.6	1.4587	**950**
.2361	1278.7	1409.7	1.4759	.2268	1275.6	1405.7	1.4701	.2182	1272.4	1401.6	1.4643	**960**
.2392	1284.5	1417.3	1.4813	.2299	1281.5	1413.4	1.4755	.2212	1278.5	1409.5	1.4698	**970**
.2423	1290.3	1424.8	1.4865	.2330	1287.4	1421.0	1.4808	.2243	1284.4	1417.2	1.4752	**980**
.2454	1296.0	1432.3	1.4917	.2360	1293.2	1428.6	1.4860	.2272	1290.3	1424.9	1.4805	**990**
.2485	1301.7	1439.6	1.4967	.2390	1298.9	1436.0	1.4912	.2302	1296.1	1432.4	1.4857	**1000**
.2545	1312.9	1454.2	1.5066	.2449	1310.3	1450.8	1.5012	.2359	1307.6	1447.3	1.4959	**1020**
.2603	1323.9	1468.4	1.5162	.2506	1321.4	1465.2	1.5109	.2416	1318.9	1461.9	1.5057	**1040**
.2660	1334.8	1482.5	1.5255	.2563	1332.4	1479.4	1.5203	.2471	1330.0	1476.3	1.5152	**1060**
.2717	1345.5	1496.4	1.5346	.2618	1343.3	1493.4	1.5295	.2525	1341.0	1490.5	1.5244	**1080**
.2772	1356.2	1510.1	1.5434	.2672	1354.0	1507.3	1.5384	.2578	1351.8	1504.5	1.5335	**1100**
.2826	1366.7	1523.6	1.5520	.2725	1364.6	1520.9	1.5471	.2630	1362.5	1518.3	1.5423	**1120**
.2880	1377.1	1537.0	1.5605	.2777	1375.1	1534.5	1.5556	.2681	1373.1	1531.9	1.5508	**1140**
.2933	1387.5	1550.3	1.5687	.2829	1385.6	1547.9	1.5639	.2732	1383.7	1545.4	1.5592	**1160**
.2985	1397.8	1563.5	1.5768	.2880	1396.0	1561.2	1.5721	.2781	1394.1	1558.8	1.5675	**1180**
.3036	1408.0	1576.6	1.5848	.2930	1406.3	1574.4	1.5801	.2830	1404.5	1572.1	1.5755	**1200**
.3087	1418.2	1589.6	1.5925	.2980	1416.5	1587.4	1.5879	.2879	1414.8	1585.3	1.5834	**1220**
.3138	1428.3	1602.5	1.6002	.3029	1426.7	1600.5	1.5956	.2927	1425.1	1598.4	1.5912	**1240**
.3187	1438.4	1615.4	1.6077	.3077	1436.8	1613.4	1.6032	.2974	1435.3	1611.4	1.5988	**1260**
.3237	1448.5	1628.1	1.6151	.3125	1446.9	1626.2	1.6106	.3021	1445.4	1624.3	1.6062	**1280**
.3285	1458.5	1640.9	1.6224	.3173	1457.0	1639.0	1.6179	.3067	1455.5	1637.2	1.6136	**1300**
.3406	1483.3	1672.4	1.6400	.3290	1482.0	1670.7	1.6357	.3181	1480.7	1669.1	1.6315	**1350**
.3524	1508.1	1703.7	1.6571	.3405	1506.9	1702.2	1.6528	.3293	1505.6	1700.6	1.6487	**1400**
.3639	1532.7	1734.8	1.6736	.3517	1531.6	1733.4	1.6694	.3403	1530.5	1732.0	1.6653	**1450**
.3754	1557.3	1765.7	1.6896	.3628	1556.2	1764.4	1.6854	.3511	1555.2	1763.1	1.6814	**1500**
.3978	1606.3	1827.1	1.7201	.3846	1605.4	1826.0	1.7161	.3723	1604.5	1824.9	1.7122	**1600**
.4416	1704.5	1949.6	1.7769	.4271	1703.8	1948.8	1.7730	.4136	1703.0	1947.9	1.7692	**1800**
.4844	1803.9	2072.8	1.8291	.4687	1803.2	2072.1	1.8253	.4539	1802.5	2071.3	1.8215	**2000**
.5268	1904.9	2197.4	1.8778	.5098	1904.3	2196.7	1.8740	.4938	1903.6	2196.1	1.8702	**2200**
.5691	2007.8	2323.8	1.9236	.5508	2007.2	2323.1	1.9198	.5336	2006.5	2322.5	1.9161	**2400**

Table 3. Vapor

p		3300				3400				3500		
t	v	u	h	s	v	u	h	s	v	u	h	s
650	.02511	666.1	681.4	.8654	.02500	664.8	680.5	.8642	.02491	663.5	679.7	.8630
660	.02585	682.4	698.2	.8805	.02572	680.9	697.1	.8790	.02560	679.4	696.0	.8776
670	.02674	700.0	716.3	.8966	.02657	698.1	714.8	.8948	.02641	696.3	713.4	.8931
680	.02787	719.4	736.4	.9142	.02763	716.9	734.3	.9119	.02741	714.6	732.4	.9098
690	.02943	741.9	759.8	.9347	.02904	738.3	756.6	.9314	.02871	735.1	753.7	.9284
700	.03204	771.2	790.7	.9615	.03121	764.7	784.3	.9554	.03058	759.5	779.3	.9506
710	.05175	882.0	913.6	1.0667	.03638	808.7	831.5	.9959	.03408	793.9	816.0	.9821
720	.08989	1005.4	1060.2	1.1918	.07518	969.2	1016.5	1.1534	.05449	902.3	937.6	1.0855
730	.10337	1040.5	1103.7	1.2284	.09292	1020.5	1078.9	1.2061	.08173	995.2	1048.1	1.1789
740	.11322	1064.7	1133.8	1.2537	.10411	1049.8	1115.3	1.2366	.09495	1032.8	1094.3	1.2176
750	.12132	1083.7	1157.8	1.2736	.11288	1071.7	1142.7	1.2593	.10460	1058.4	1126.1	1.2440
760	.12836	1099.8	1178.2	1.2904	.12031	1089.5	1165.2	1.2778	.11251	1078.4	1151.3	1.2647
770	.13467	1113.9	1196.1	1.3050	.12686	1104.8	1184.6	1.2937	.11936	1095.2	1172.5	1.2820
780	.14044	1126.5	1212.2	1.3181	.13280	1118.3	1201.9	1.3077	.12549	1109.8	1191.1	1.2970
790	.14578	1138.0	1227.0	1.3299	.13826	1130.6	1217.5	1.3203	.13108	1122.9	1207.7	1.3104
800	.15080	1148.6	1240.7	1.3409	.14335	1141.8	1232.0	1.3318	.13626	1134.7	1223.0	1.3226
810	.15554	1158.5	1253.5	1.3510	.14814	1152.2	1245.4	1.3424	.14111	1145.7	1237.1	1.3338
820	.16004	1167.9	1265.6	1.3605	.15268	1162.0	1258.1	1.3523	.14569	1156.0	1250.3	1.3441
830	.16436	1176.8	1277.2	1.3695	.15700	1171.3	1270.1	1.3617	.15004	1165.6	1262.8	1.3538
840	.16850	1185.3	1288.2	1.3780	.16115	1180.1	1281.5	1.3705	.15419	1174.7	1274.6	1.3629
850	.17249	1193.4	1298.8	1.3861	.16514	1188.5	1292.4	1.3788	.15818	1183.4	1285.9	1.3716
860	.17635	1201.3	1309.0	1.3939	.16899	1196.6	1302.9	1.3868	.16203	1191.8	1296.7	1.3798
870	.18009	1208.8	1318.8	1.4013	.17271	1204.4	1313.0	1.3945	.16574	1199.8	1307.1	1.3877
880	.18373	1216.2	1328.4	1.4085	.17633	1211.9	1322.8	1.4019	.16934	1207.6	1317.2	1.3953
890	.18727	1223.3	1337.7	1.4154	.17985	1219.2	1332.4	1.4089	.17284	1215.1	1327.0	1.4025
900	.19073	1230.3	1346.8	1.4221	.18328	1226.4	1341.7	1.4158	.17625	1222.4	1336.5	1.4096
910	.19411	1237.1	1355.6	1.4286	.18663	1233.3	1350.7	1.4224	.17957	1229.5	1345.8	1.4163
920	.19741	1243.8	1364.3	1.4349	.18990	1240.1	1359.6	1.4289	.18281	1236.4	1354.8	1.4229
930	.20066	1250.3	1372.8	1.4411	.19311	1246.8	1368.3	1.4352	.18598	1243.2	1363.7	1.4293
940	.20384	1256.7	1381.2	1.4471	.19625	1253.3	1376.8	1.4413	.18909	1249.9	1372.3	1.4355
950	.2070	1263.0	1389.4	1.4529	.19934	1259.7	1385.2	1.4472	.19214	1256.4	1380.8	1.4416
960	.2100	1269.3	1397.5	1.4586	.20237	1266.1	1393.4	1.4530	.19514	1262.8	1389.2	1.4475
970	.2131	1275.4	1405.5	1.4642	.20535	1272.3	1401.5	1.4587	.19808	1269.2	1397.4	1.4533
980	.2160	1281.4	1413.4	1.4697	.20828	1278.4	1409.5	1.4643	.20097	1275.4	1405.6	1.4589
990	.2190	1287.4	1421.1	1.4751	.21117	1284.5	1417.3	1.4697	.20382	1281.5	1413.5	1.4645
1000	.2219	1293.3	1428.8	1.4804	.2140	1290.5	1425.1	1.4751	.2066	1287.6	1421.4	1.4699
1020	.2275	1304.9	1443.9	1.4906	.2196	1302.3	1440.4	1.4855	.2121	1299.5	1436.9	1.4804
1040	.2331	1316.4	1458.7	1.5006	.2250	1313.8	1455.4	1.4955	.2175	1311.2	1452.1	1.4906
1060	.2385	1327.6	1473.2	1.5102	.2304	1325.2	1470.1	1.5053	.2227	1322.7	1467.0	1.5005
1080	.2438	1338.7	1487.5	1.5195	.2356	1336.4	1484.6	1.5147	.2278	1334.0	1481.6	1.5100
1100	.2490	1349.6	1501.6	1.5287	.2407	1347.4	1498.8	1.5239	.2328	1345.2	1496.0	1.5193
1120	.2541	1360.4	1515.6	1.5375	.2457	1358.3	1512.9	1.5329	.2377	1356.2	1510.2	1.5283
1140	.2591	1371.1	1529.3	1.5462	.2506	1369.1	1526.8	1.5416	.2426	1367.1	1524.2	1.5372
1160	.2640	1381.7	1543.0	1.5547	.2554	1379.8	1540.5	1.5502	.2473	1377.9	1538.0	1.5458
1180	.2689	1392.3	1556.5	1.5629	.2602	1390.4	1554.1	1.5585	.2520	1388.6	1551.7	1.5542
1200	.2737	1402.7	1569.9	1.5711	.2649	1400.9	1567.6	1.5667	.2566	1399.2	1565.3	1.5624
1220	.2784	1413.1	1583.1	1.5790	.2695	1411.4	1581.0	1.5747	.2611	1409.7	1578.8	1.5705
1240	.2831	1423.4	1596.3	1.5868	.2741	1421.8	1594.2	1.5826	.2656	1420.1	1592.1	1.5784
1260	.2877	1433.7	1609.4	1.5945	.2786	1432.1	1607.4	1.5903	.2700	1430.5	1605.4	1.5861
1280	.2923	1443.9	1622.4	1.6020	.2831	1442.4	1620.5	1.5978	.2744	1440.9	1618.6	1.5938
1300	.2968	1454.1	1635.3	1.6094	.2875	1452.6	1633.5	1.6053	.2787	1451.1	1631.7	1.6012
1350	.3080	1479.3	1667.4	1.6273	.2984	1478.0	1665.7	1.6233	.2893	1476.7	1664.0	1.6194
1400	.3189	1504.4	1699.1	1.6446	.3090	1503.2	1697.6	1.6407	.2997	1501.9	1696.1	1.6368
1450	.3296	1529.3	1730.6	1.6613	.3194	1528.2	1729.2	1.6574	.3099	1527.0	1727.8	1.6537
1500	.3401	1554.1	1761.8	1.6775	.3297	1553.1	1760.5	1.6736	.3199	1552.0	1759.2	1.6699
1600	.3607	1603.5	1823.8	1.7083	.3498	1602.6	1822.7	1.7046	.3395	1601.7	1821.6	1.7010
1800	.4008	1702.3	1947.1	1.7654	.3889	1701.5	1946.2	1.7618	.3776	1700.8	1945.4	1.7583
2000	.4401	1801.9	2070.6	1.8178	.4270	1801.2	2069.9	1.8143	.4147	1800.6	2069.2	1.8108
2200	.4789	1903.0	2195.4	1.8666	.4648	1902.3	2194.8	1.8631	.4515	1901.7	2194.1	1.8596
2400	.5175	2005.9	2321.9	1.9124	.5023	2005.2	2321.2	1.9089	.4880	2004.5	2320.6	1.9055

	3600				**3700**				**3800**			**p**
v	u	h	s	v	u	h	s	v	u	h	s	t
.02481	662.3	678.8	.8618	.02472	661.1	678.0	.8607	.02463	659.9	677.3	.8596	650
.02548	678.0	694.9	.8762	.02537	676.6	693.9	.8749	.02526	675.2	693.0	.8737	660
.02626	694.6	712.1	.8915	.02612	692.9	710.8	.8899	.02599	691.3	709.6	.8884	670
.02721	712.4	730.6	.9078	.02702	710.4	728.9	.9059	.02685	708.5	727.4	.9041	680
.02841	732.2	751.2	.9258	.02815	729.6	748.8	.9233	.02791	727.1	746.7	.9210	690
.03008	755.2	775.2	.9466	.02965	751.4	771.7	.9431	.02929	748.0	768.6	.9399	700
.03281	784.9	806.7	.9736	.03194	778.2	800.1	.9674	.03126	772.8	794.8	.9624	710
.04074	841.7	868.8	1.0265	.03669	818.9	844.0	1.0048	.03475	806.4	830.9	.9931	720
.06909	961.0	1007.0	1.1432	.05512	914.0	951.8	1.0957	.04477	870.5	902.0	1.0531	730
.08559	1012.9	1069.9	1.1958	.07580	988.8	1040.7	1.1702	.06550	959.0	1005.0	1.1394	740
.09641	1043.6	1107.8	1.2273	.08823	1027.0	1087.4	1.2090	.07997	1007.9	1064.2	1.1885	750
.10492	1066.4	1136.3	1.2507	.09748	1053.3	1120.1	1.2359	.09016	1039.0	1102.4	1.2200	760
.11212	1084.9	1159.6	1.2698	.10511	1074.0	1146.0	1.2571	.09829	1062.3	1131.4	1.2437	770
.11847	1100.8	1179.7	1.2861	.11173	1091.3	1167.8	1.2748	.10521	1081.3	1155.3	1.2630	780
.12422	1114.8	1197.6	1.3004	.11764	1106.4	1186.9	1.2901	.11132	1097.5	1175.8	1.2795	790
.12950	1127.4	1213.7	1.3133	.12304	1119.8	1204.1	1.3038	.11686	1111.9	1194.1	1.2941	800
.13442	1139.0	1228.5	1.3250	.12804	1132.0	1219.7	1.3161	.12195	1124.8	1210.6	1.3071	810
.13904	1149.7	1242.4	1.3358	.13272	1143.3	1234.2	1.3275	.12669	1136.7	1225.8	1.3190	820
.14342	1159.8	1255.3	1.3459	.13714	1153.8	1247.7	1.3380	.13115	1147.7	1239.9	1.3300	830
.14760	1169.3	1267.6	1.3554	.14133	1163.7	1260.4	1.3478	.13537	1157.9	1253.1	1.3402	840
.15159	1178.3	1279.3	1.3644	.14534	1173.0	1272.5	1.3571	.13939	1167.6	1265.6	1.3498	850
.15544	1186.9	1290.4	1.3729	.14918	1181.9	1284.0	1.3659	.14324	1176.8	1277.5	1.3589	860
.15914	1195.2	1301.2	1.3810	.15288	1190.4	1295.1	1.3742	.14694	1185.6	1288.9	1.3675	870
.16273	1203.1	1311.5	1.3887	.15646	1198.6	1305.7	1.3822	.15051	1194.0	1299.9	1.3757	880
.16621	1210.8	1321.6	1.3962	.15992	1206.5	1316.0	1.3898	.15396	1202.2	1310.4	1.3835	890
.16959	1218.3	1331.3	1.4034	.16329	1214.2	1326.0	1.3972	.15731	1210.0	1320.6	1.3911	900
.17289	1225.6	1340.8	1.4103	.16656	1221.6	1335.7	1.4043	.16056	1217.6	1330.5	1.3983	910
.17611	1232.7	1350.0	1.4170	.16976	1228.9	1345.1	1.4112	.16374	1225.0	1340.2	1.4053	920
.17925	1239.6	1359.0	1.4235	.17287	1235.9	1354.3	1.4178	.16683	1232.2	1349.6	1.4121	930
.18233	1246.4	1367.8	1.4299	.17592	1242.9	1363.3	1.4243	.16985	1239.3	1358.7	1.4187	940
.18534	1253.0	1376.5	1.4360	.17891	1249.6	1372.1	1.4305	.17281	1246.2	1367.7	1.4251	950
.18830	1259.6	1385.0	1.4420	.18184	1256.3	1380.8	1.4367	.17571	1253.0	1376.5	1.4313	960
.19121	1266.0	1393.4	1.4479	.18471	1262.8	1389.3	1.4426	.17855	1259.6	1385.1	1.4374	970
.19407	1272.3	1401.6	1.4537	.18753	1269.2	1397.6	1.4484	.18134	1266.1	1393.6	1.4433	980
.19688	1278.6	1409.7	1.4593	.19031	1275.6	1405.9	1.4541	.18408	1272.5	1402.0	1.4491	990
.19964	1284.7	1417.7	1.4648	.19304	1281.8	1414.0	1.4597	.18678	1278.9	1410.2	1.4547	1000
.20506	1296.8	1433.4	1.4755	.19838	1294.1	1429.9	1.4705	.19205	1291.3	1426.3	1.4657	1020
.21033	1308.7	1448.8	1.4858	.20358	1306.0	1445.4	1.4810	.19718	1303.4	1442.1	1.4763	1040
.21547	1320.3	1463.8	1.4957	.20864	1317.8	1460.6	1.4911	.20216	1315.3	1457.5	1.4865	1060
.22050	1331.7	1478.6	1.5054	.21359	1329.3	1475.6	1.5008	.20703	1327.0	1472.6	1.4963	1080
.2254	1342.9	1493.1	1.5148	.2184	1340.7	1490.3	1.5103	.2118	1338.4	1487.4	1.5059	1100
.2302	1354.1	1507.4	1.5239	.2232	1351.9	1504.7	1.5195	.2165	1349.8	1502.0	1.5152	1120
.2350	1365.0	1521.6	1.5328	.2278	1363.0	1519.0	1.5285	.2210	1360.9	1516.4	1.5242	1140
.2396	1375.9	1535.6	1.5415	.2324	1374.0	1533.1	1.5372	.2255	1372.0	1530.6	1.5331	1160
.2442	1386.7	1549.4	1.5499	.2369	1384.8	1547.0	1.5458	.2299	1382.9	1544.6	1.5417	1180
.2487	1397.4	1563.1	1.5582	.2413	1395.6	1560.8	1.5541	.2343	1393.8	1558.5	1.5501	1200
.2532	1408.0	1576.6	1.5664	.2457	1406.2	1574.4	1.5623	.2386	1404.5	1572.2	1.5583	1220
.2576	1418.5	1590.1	1.5743	.2500	1416.8	1588.0	1.5703	.2428	1415.1	1585.9	1.5664	1240
.2619	1428.9	1603.4	1.5821	.2542	1427.3	1601.4	1.5782	.2469	1425.7	1599.4	1.5743	1260
.2662	1439.3	1616.6	1.5898	.2584	1437.8	1614.7	1.5859	.2511	1436.2	1612.8	1.5821	1280
.2704	1449.7	1629.8	1.5973	.2626	1448.2	1628.0	1.5934	.2551	1446.7	1626.1	1.5897	1300
.2808	1475.3	1662.4	1.6155	.2727	1474.0	1660.7	1.6118	.2651	1472.6	1659.0	1.6081	1350
.2910	1500.7	1694.5	1.6331	.2827	1499.5	1693.0	1.6294	.2748	1498.2	1691.5	1.6258	1400
.3009	1525.9	1726.4	1.6500	.2924	1524.8	1725.0	1.6464	.2843	1523.6	1723.6	1.6428	1450
.3107	1551.0	1757.9	1.6663	.3019	1549.9	1756.6	1.6627	.2937	1548.9	1755.4	1.6593	1500
.3298	1600.8	1820.5	1.6974	.3206	1599.9	1819.4	1.6940	.3119	1599.0	1818.3	1.6906	1600
.3669	1700.1	1944.5	1.7549	.3568	1699.3	1943.7	1.7516	.3473	1698.6	1942.8	1.7483	1800
.4031	1799.9	2068.5	1.8075	.3921	1799.2	2067.7	1.8042	.3817	1798.6	2067.0	1.8010	2000
.4389	1901.1	2193.4	1.8563	.4270	1900.4	2192.8	1.8530	.4158	1899.8	2192.2	1.8499	2200
.4745	2003.9	2320.0	1.9022	.4617	2003.2	2319.4	1.8989	.4496	2002.6	2318.7	1.8958	2400

Table 3. Vapor

p	3900				4000				4200			
t	v	u	h	s	v	u	h	s	v	u	h	s
650	.02455	658.8	676.5	.8585	.02447	657.7	675.8	.8574	.02431	655.6	674.5	.8554
660	.02516	673.9	692.1	.8724	.02506	672.7	691.2	.8712	.02487	670.3	689.6	.8690
670	.02586	689.8	708.5	.8870	.02574	688.3	707.4	.8856	.02551	685.6	705.4	.8830
680	.02668	706.6	725.9	.9024	.02653	704.9	724.5	.9007	.02625	701.6	722.0	.8977
690	.02769	724.8	744.8	.9189	.02748	722.7	743.0	.9169	.02712	718.7	739.8	.9132
700	.02896	744.9	765.8	.9371	.02867	742.1	763.4	.9345	.02817	737.1	759.0	.9298
710	.03072	768.2	790.4	.9582	.03026	764.3	786.7	.9545	.02951	757.5	780.4	.9482
720	.03351	797.8	822.0	.9851	.03262	791.2	815.3	.9789	.03134	781.0	805.3	.9694
730	.03966	844.4	873.0	1.0281	.03698	828.8	856.2	1.0134	.03414	810.1	836.6	.9958
740	.05543	924.2	964.2	1.1044	.04751	891.8	927.0	1.0726	.03930	850.7	881.2	1.0332
750	.07162	985.9	1037.6	1.1654	.06331	960.7	1007.5	1.1395	.04933	908.3	946.7	1.0875
760	.08291	1023.1	1082.9	1.2027	.07574	1005.4	1061.4	1.1839	.06191	964.5	1012.6	1.1417
770	.09164	1049.6	1115.7	1.2295	.08512	1035.9	1098.9	1.2145	.07254	1004.8	1061.2	1.1815
780	.09891	1070.6	1142.0	1.2508	.09279	1059.3	1128.0	1.2380	.08108	1034.2	1097.3	1.2107
790	.10524	1088.3	1164.2	1.2686	.09938	1078.5	1152.1	1.2574	.08823	1057.4	1126.0	1.2337
800	.11092	1103.6	1183.7	1.2842	.10522	1095.0	1172.9	1.2740	.09445	1076.6	1150.0	1.2529
810	.11612	1117.4	1201.2	1.2980	.11053	1109.6	1191.4	1.2887	.10001	1093.2	1171.0	1.2695
820	.12093	1129.8	1217.1	1.3105	.11542	1122.8	1208.2	1.3018	.10507	1108.0	1189.6	1.2841
830	.12543	1141.3	1231.9	1.3220	.11998	1134.8	1223.7	1.3138	.10975	1121.3	1206.6	1.2973
840	.12969	1152.0	1245.6	1.3326	.12426	1146.0	1238.0	1.3249	.11413	1133.5	1222.2	1.3094
850	.13373	1162.1	1258.6	1.3426	.12833	1156.5	1251.5	1.3352	.11824	1144.8	1236.7	1.3205
860	.13758	1171.6	1270.9	1.3519	.13220	1166.3	1264.2	1.3449	.12214	1155.4	1250.4	1.3309
870	.14129	1180.7	1282.7	1.3608	.13590	1175.7	1276.3	1.3541	.12587	1165.4	1263.3	1.3406
880	.14485	1189.4	1293.9	1.3692	.13946	1184.6	1287.9	1.3627	.12943	1174.9	1275.5	1.3498
890	.14829	1197.7	1304.7	1.3773	.14290	1193.2	1299.0	1.3710	.13286	1184.0	1287.3	1.3585
900	.15163	1205.8	1315.2	1.3850	.14622	1201.5	1309.7	1.3789	.13617	1192.7	1298.5	1.3669
910	.15487	1213.6	1325.3	1.3924	.14945	1209.5	1320.1	1.3865	.13936	1201.1	1309.4	1.3748
920	.15802	1221.1	1335.2	1.3996	.15258	1217.2	1330.1	1.3938	.14247	1209.2	1319.9	1.3825
930	.16109	1228.5	1344.8	1.4065	.15563	1224.7	1339.9	1.4009	.14548	1217.0	1330.1	1.3898
940	.16409	1235.7	1354.1	1.4132	.15861	1232.0	1349.4	1.4077	.14842	1224.6	1340.0	1.3969
950	.16702	1242.7	1363.3	1.4197	.16151	1239.2	1358.8	1.4144	.15129	1232.1	1349.7	1.4038
960	.16989	1249.6	1372.2	1.4260	.16436	1246.2	1367.9	1.4208	.15409	1239.3	1359.1	1.4105
970	.17270	1256.3	1381.0	1.4322	.16715	1253.1	1376.8	1.4271	.15682	1246.4	1368.3	1.4169
980	.17546	1263.0	1389.6	1.4382	.16988	1259.8	1385.6	1.4332	.15951	1253.4	1377.4	1.4232
990	.17818	1269.5	1398.1	1.4441	.17256	1266.4	1394.2	1.4391	.16214	1260.2	1386.2	1.4294
1000	.18084	1275.9	1406.4	1.4498	.17520	1272.9	1402.6	1.4449	.16472	1266.9	1394.9	1.4354
1020	.18605	1288.5	1422.8	1.4609	.18034	1285.7	1419.2	1.4562	.16975	1280.0	1412.0	1.4470
1040	.19110	1300.8	1438.7	1.4716	.18533	1298.1	1435.3	1.4670	.17462	1292.8	1428.5	1.4581
1060	.19602	1312.8	1454.3	1.4819	.19019	1310.3	1451.1	1.4775	.17935	1305.2	1444.6	1.4687
1080	.20082	1324.6	1469.5	1.4919	.19492	1322.2	1466.5	1.4876	.18396	1317.4	1460.4	1.4790
1100	.2055	1336.2	1484.5	1.5016	.19954	1333.9	1481.6	1.4973	.18845	1329.3	1475.8	1.4890
1120	.2101	1347.6	1499.2	1.5110	.20406	1345.4	1496.5	1.5068	.19284	1341.1	1490.9	1.4986
1140	.2146	1358.9	1513.8	1.5201	.20848	1356.8	1511.1	1.5160	.19714	1352.6	1505.8	1.5080
1160	.2190	1370.0	1528.1	1.5290	.21283	1368.0	1525.6	1.5250	.20135	1364.0	1520.5	1.5171
1180	.2234	1381.0	1542.2	1.5377	.21709	1379.1	1539.8	1.5337	.20548	1375.3	1535.0	1.5260
1200	.2276	1391.9	1556.2	1.5462	.2213	1390.1	1553.9	1.5423	.2095	1386.5	1549.3	1.5347
1220	.2318	1402.8	1570.1	1.5544	.2254	1401.0	1567.9	1.5506	.2135	1397.5	1563.5	1.5432
1240	.2360	1413.5	1583.8	1.5626	.2295	1411.8	1581.7	1.5588	.2175	1408.4	1577.5	1.5514
1260	.2400	1424.1	1597.4	1.5705	.2335	1422.5	1595.3	1.5668	.2213	1419.3	1591.3	1.5595
1280	.2441	1434.7	1610.9	1.5783	.2375	1433.2	1608.9	1.5746	.2252	1430.0	1605.0	1.5675
1300	.2481	1445.2	1624.2	1.5860	.2414	1443.7	1622.4	1.5823	.2289	1440.7	1618.7	1.5753
1350	.2578	1471.2	1657.3	1.6045	.2509	1469.9	1655.6	1.6010	.2382	1467.2	1652.3	1.5941
1400	.2674	1497.0	1689.9	1.6223	.2603	1495.7	1688.4	1.6188	.2471	1493.2	1685.3	1.6121
1450	.2767	1522.5	1722.2	1.6394	.2694	1521.3	1720.8	1.6360	.2559	1519.0	1717.9	1.6294
1500	.2858	1547.8	1754.1	1.6559	.2784	1546.7	1752.8	1.6526	.2645	1544.6	1750.2	1.6461
1600	.3037	1598.0	1817.2	1.6873	.2959	1597.1	1816.1	1.6841	.2813	1595.3	1813.9	1.6778
1800	.3382	1697.9	1942.0	1.7451	.3296	1697.1	1941.1	1.7420	.3137	1695.6	1939.4	1.7360
2000	.3719	1797.9	2066.1	1.7978	.3625	1797.3	2065.6	1.7948	.3451	1796.0	2064.2	1.7889
2200	.4051	1899.2	2191.5	1.8468	.3949	1898.5	2190.9	1.8437	.3761	1897.3	2189.6	1.8379
2400	.4381	2001.9	2318.1	1.8927	.4272	2001.3	2317.5	1.8896	.4069	2000.0	2316.2	1.8838

	4400				4600				4800			p
v	u	h	s	v	u	h	s	v	u	h	s	t
.02416	653.6	673.3	.8535	.02402	651.7	672.1	.8517	.02389	649.8	671.0	.8499	650
.02470	668.0	688.1	.8668	.02454	665.8	686.7	.8648	.02439	663.7	685.4	.8628	660
.02531	682.9	703.5	.8806	.02511	680.5	701.9	.8782	.02494	678.2	700.3	.8760	670
.02600	698.6	719.8	.8948	.02576	695.8	717.7	.8922	.02555	693.1	715.8	.8897	680
.02680	715.1	736.9	.9098	.02651	711.8	734.3	.9067	.02625	708.7	732.0	.9039	690
.02775	732.7	755.3	.9257	.02738	728.7	752.0	.9221	.02705	725.1	749.1	.9187	700
.02892	751.8	775.4	.9430	.02843	746.9	771.1	.9384	.02800	742.5	767.4	.9344	710
.03043	773.2	798.0	.9622	.02973	766.8	792.1	.9563	.02915	761.3	787.2	.9512	720
.03253	798.1	824.6	.9847	.03143	789.1	815.9	.9764	.03060	782.0	809.1	.9697	730
.03579	829.1	858.3	1.0129	.03382	815.4	844.2	1.0000	.03250	805.3	834.1	.9907	740
.04147	870.8	904.6	1.0513	.03750	847.9	879.8	1.0296	.03518	832.6	863.9	1.0154	750
.05058	922.1	963.2	1.0996	.04336	888.7	925.6	1.0673	.03917	865.7	900.5	1.0455	760
.06094	969.4	1019.1	1.1452	.05149	934.0	977.9	1.1100	.04494	904.5	944.4	1.0814	770
.07011	1005.9	1063.0	1.1807	.06025	975.1	1026.3	1.1493	.05218	944.8	991.2	1.1192	780
.07782	1033.9	1097.3	1.2083	.06822	1008.1	1066.2	1.1813	.05973	981.0	1034.1	1.1537	790
.08444	1056.5	1125.3	1.2306	.07518	1034.6	1098.6	1.2071	.06674	1011.2	1070.5	1.1827	800
.09028	1075.6	1149.1	1.2495	.08129	1056.6	1125.8	1.2286	.07302	1036.2	1101.1	1.2069	810
.09554	1092.2	1170.0	1.2658	.08675	1075.3	1149.2	1.2469	.07865	1057.4	1127.3	1.2274	820
.10036	1106.9	1188.7	1.2804	.09171	1091.8	1169.8	1.2630	.08374	1075.7	1150.1	1.2452	830
.10483	1120.3	1205.7	1.2936	.09628	1106.5	1188.4	1.2774	.08841	1092.0	1170.5	1.2610	840
.10901	1132.6	1221.4	1.3056	.10054	1119.9	1205.5	1.2905	.09274	1106.6	1188.9	1.2751	850
.11296	1144.1	1236.0	1.3167	.10453	1132.2	1221.2	1.3024	.09678	1119.9	1205.9	1.2880	860
.11670	1154.8	1249.8	1.3271	.10831	1143.7	1235.9	1.3135	.10059	1132.3	1221.6	1.2999	870
.12028	1164.9	1262.8	1.3369	.11190	1154.5	1249.7	1.3239	.10420	1143.8	1236.3	1.3109	880
.12370	1174.5	1275.2	1.3461	.11533	1164.7	1262.9	1.3337	.10763	1154.6	1250.2	1.3212	890
.12700	1183.7	1287.1	1.3548	.11862	1174.4	1275.4	1.3429	.11092	1164.8	1263.4	1.3310	900
.13018	1192.5	1298.5	1.3632	.12179	1183.6	1287.3	1.3516	.11408	1174.6	1275.9	1.3402	910
.13326	1200.9	1309.5	1.3712	.12485	1192.5	1298.8	1.3600	.11713	1183.9	1288.0	1.3489	920
.13625	1209.1	1320.1	1.3789	.12781	1201.1	1309.9	1.3680	.12007	1192.9	1299.5	1.3573	930
.13915	1217.1	1330.4	1.3863	.13069	1209.4	1320.6	1.3757	.12292	1201.5	1310.7	1.3653	940
.14198	1224.8	1340.4	1.3934	.13348	1217.4	1331.0	1.3831	.12569	1209.9	1321.5	1.3730	950
.14474	1232.3	1350.2	1.4003	.13621	1225.2	1341.2	1.3903	.12839	1218.0	1332.0	1.3804	960
.14744	1239.7	1359.7	1.4070	.13887	1232.8	1351.0	1.3972	.13101	1225.8	1342.2	1.3875	970
.15007	1246.9	1369.1	1.4135	.14146	1240.2	1360.7	1.4039	.13357	1233.5	1352.2	1.3945	980
.15266	1253.9	1378.2	1.4198	.14400	1247.5	1370.1	1.4104	.13608	1241.0	1361.9	1.4012	990
.15519	1260.8	1387.2	1.4260	.14649	1254.6	1379.3	1.4168	.13853	1248.3	1371.4	1.4078	1000
.16012	1274.3	1404.7	1.4379	.15133	1268.5	1397.3	1.4290	.14328	1262.6	1389.8	1.4203	1020
.16489	1287.4	1421.6	1.4493	.15600	1281.9	1414.7	1.4407	.14786	1276.3	1407.7	1.4323	1040
.16951	1300.1	1438.1	1.4602	.16052	1294.9	1431.5	1.4519	.15229	1289.7	1424.9	1.4437	1060
.17400	1312.5	1454.2	1.4707	.16491	1307.6	1448.0	1.4626	.15659	1302.6	1441.7	1.4547	1080
.17838	1324.7	1469.9	1.4809	.16918	1320.0	1464.0	1.4730	.16076	1315.3	1458.1	1.4653	1100
.18265	1336.7	1485.4	1.4907	.17335	1332.2	1479.8	1.4830	.16483	1327.7	1474.1	1.4755	1120
.18682	1348.4	1500.5	1.5003	.17742	1344.2	1495.2	1.4927	.16880	1339.9	1489.9	1.4854	1140
.19091	1360.0	1515.5	1.5095	.18139	1356.0	1510.4	1.5021	.17268	1351.9	1505.3	1.4950	1160
.19492	1371.5	1530.2	1.5186	.18529	1367.6	1525.3	1.5113	.17647	1363.7	1520.5	1.5043	1180
.19886	1382.8	1544.7	1.5274	.18912	1379.1	1540.1	1.5202	.18020	1375.4	1535.4	1.5133	1200
.20273	1394.0	1559.0	1.5360	.19288	1390.4	1554.6	1.5290	.18385	1386.9	1550.2	1.5222	1220
.20654	1405.1	1573.2	1.5443	.19657	1401.6	1569.0	1.5375	.18744	1398.2	1564.7	1.5308	1240
.21029	1416.0	1587.3	1.5525	.20021	1412.8	1583.2	1.5458	.19097	1409.5	1579.1	1.5392	1260
.21398	1426.9	1601.1	1.5606	.20378	1423.8	1597.2	1.5539	.19444	1420.6	1593.3	1.5474	1280
.2176	1437.7	1614.9	1.5685	.2073	1434.7	1611.2	1.5619	.19787	1431.7	1607.4	1.5555	1300
.2265	1464.4	1648.9	1.5875	.2159	1461.7	1645.5	1.5811	.20623	1458.9	1642.1	1.5749	1350
.2352	1490.7	1682.3	1.6057	.2243	1488.2	1679.2	1.5994	.21434	1485.7	1676.1	1.5934	1400
.2437	1516.7	1715.1	1.6231	.2325	1514.4	1712.3	1.6170	.22224	1512.1	1709.5	1.6112	1450
.2520	1542.5	1747.6	1.6399	.2405	1540.4	1745.1	1.6340	.22996	1538.2	1742.5	1.6282	1500
.2681	1593.4	1811.7	1.6718	.2560	1591.6	1809.5	1.6661	.2450	1589.8	1807.4	1.6605	1600
.2992	1694.2	1937.7	1.7302	.2859	1692.7	1936.1	1.7247	.2738	1691.2	1934.4	1.7194	1800
.3293	1794.7	2062.8	1.7832	.3149	1793.4	2061.4	1.7778	.3016	1792.1	2060.0	1.7726	2000
.3590	1896.0	2188.3	1.8323	.3434	1894.7	2187.0	1.8269	.3291	1893.5	2185.8	1.8218	2200
.3885	1998.7	2315.0	1.8782	.3717	1997.4	2313.8	1.8729	.3563	1996.1	2312.6	1.8678	2400

Table 3. Vapor

p		5000				5200				5400		
t	v	u	h	s	v	u	h	s	v	u	h	s
650	.02377	648.0	670.0	.8482	.02365	646.3	669.1	.8465	.02354	644.6	668.2	.8449
660	.02424	661.8	684.2	.8609	.02411	659.9	683.1	.8591	.02398	658.1	682.0	.8574
670	.02477	675.9	698.9	.8739	.02461	673.8	697.5	.8719	.02446	671.8	696.3	.8700
680	.02535	690.6	714.1	.8873	.02517	688.2	712.5	.8851	.02500	686.0	711.0	.8830
690	.02601	705.9	729.9	.9012	.02579	703.2	728.0	.8987	.02559	700.6	726.2	.8963
700	.02676	721.8	746.6	.9156	.02650	718.7	744.2	.9127	.02625	715.8	742.0	.9100
710	.02763	738.6	764.2	.9307	.02730	735.0	761.3	.9274	.02701	731.7	758.6	.9243
720	.02867	756.5	783.0	.9468	.02825	752.2	779.4	.9428	.02788	748.3	776.1	.9392
730	.02993	775.9	803.6	.9641	.02937	770.6	798.9	.9592	.02890	765.9	794.8	.9549
740	.03152	797.3	826.4	.9832	.03075	790.6	820.2	.9771	.03012	784.8	814.9	.9717
750	.03364	821.4	852.6	1.0049	.03251	812.6	843.9	.9968	.03162	805.3	836.9	.9900
760	.03657	849.6	883.4	1.0303	.03482	837.5	871.0	1.0191	.03353	827.9	861.4	1.0102
770	.04072	882.4	920.0	1.0602	.03795	865.9	902.4	1.0447	.03601	853.2	889.2	1.0328
780	.04625	918.7	961.5	1.0938	.04213	897.7	938.3	1.0738	.03926	881.4	920.6	1.0583
790	.05272	954.8	1003.5	1.1276	.04734	931.4	977.0	1.1049	.04338	911.9	955.3	1.0862
800	.05932	987.2	1042.1	1.1583	.05314	964.1	1015.2	1.1353	.04824	943.1	991.3	1.1149
810	.06555	1015.0	1075.6	1.1849	.05900	993.5	1050.3	1.1631	.05348	972.9	1026.3	1.1426
820	.07124	1038.6	1104.5	1.2075	.06458	1019.2	1081.3	1.1874	.05874	999.9	1058.6	1.1679
830	.07642	1058.9	1129.6	1.2271	.06976	1041.5	1108.6	1.2087	.06378	1023.8	1087.5	1.1904
840	.08117	1076.8	1151.9	1.2442	.07453	1061.0	1132.7	1.2273	.06851	1044.8	1113.3	1.2103
850	.08556	1092.7	1171.9	1.2596	.07896	1078.3	1154.3	1.2438	.07292	1063.6	1136.4	1.2281
860	.08964	1107.2	1190.1	1.2734	.08308	1094.0	1173.9	1.2587	.07705	1080.4	1157.4	1.2440
870	.09348	1120.4	1206.9	1.2861	.08694	1108.2	1191.9	1.2723	.08092	1095.7	1176.5	1.2584
880	.09711	1132.7	1222.6	1.2979	.09058	1121.3	1208.5	1.2848	.08456	1109.7	1194.2	1.2717
890	.10056	1144.2	1237.3	1.3088	.09403	1133.6	1224.0	1.2963	.08802	1122.6	1210.6	1.2839
900	.10385	1155.1	1251.1	1.3190	.09732	1145.0	1238.7	1.3071	.09130	1134.8	1226.0	1.2952
910	.10700	1165.3	1264.3	1.3287	.10047	1155.8	1252.5	1.3173	.09444	1146.2	1240.5	1.3059
920	.11003	1175.1	1276.9	1.3379	.10349	1166.1	1265.7	1.3269	.09745	1157.0	1254.3	1.3159
930	.11296	1184.5	1289.0	1.3466	.10640	1175.9	1278.3	1.3360	.10034	1167.2	1267.5	1.3254
940	.11579	1193.5	1300.6	1.3549	.10921	1185.3	1290.4	1.3447	.10313	1177.0	1280.1	1.3345
950	.11853	1202.2	1311.9	1.3629	.11193	1194.4	1302.1	1.3530	.10583	1186.4	1292.2	1.3431
960	.12120	1210.6	1322.7	1.3706	.11457	1203.1	1313.3	1.3609	.10844	1195.5	1303.9	1.3513
970	.12379	1218.8	1333.3	1.3780	.11713	1211.6	1324.3	1.3686	.11098	1204.3	1315.2	1.3593
980	.12632	1226.7	1343.6	1.3852	.11963	1219.8	1334.9	1.3760	.11345	1212.8	1326.1	1.3669
990	.12879	1234.4	1353.6	1.3921	.12207	1227.8	1345.2	1.3831	.11586	1221.0	1336.8	1.3743
1000	.13120	1242.0	1363.4	1.3988	.12445	1235.5	1355.3	1.3901	.11821	1229.0	1347.2	1.3814
1020	.13588	1256.6	1382.3	1.4118	.12906	1250.6	1374.8	1.4033	.12275	1244.5	1367.2	1.3950
1040	.14038	1270.7	1400.6	1.4240	.13349	1265.1	1393.5	1.4159	.12711	1259.4	1386.4	1.4079
1060	.14473	1284.4	1418.3	1.4357	.13775	1279.1	1411.6	1.4279	.13131	1273.7	1404.9	1.4202
1080	.14894	1297.6	1435.5	1.4469	.14188	1292.6	1429.1	1.4393	.13536	1287.5	1422.8	1.4319
1100	.15302	1310.6	1452.2	1.4577	.14589	1305.8	1446.2	1.4504	.13929	1301.0	1440.2	1.4431
1120	.15700	1323.2	1468.5	1.4681	.14978	1318.7	1462.8	1.4610	.14310	1314.1	1457.1	1.4539
1140	.16087	1335.6	1484.5	1.4782	.15357	1331.3	1479.1	1.4712	.14681	1327.0	1473.7	1.4643
1160	.16466	1347.8	1500.2	1.4879	.15727	1343.7	1495.0	1.4811	.15043	1339.6	1489.9	1.4744
1180	.16836	1359.8	1515.6	1.4974	.16089	1355.9	1510.7	1.4907	.15397	1351.9	1505.8	1.4842
1200	.17199	1371.6	1530.8	1.5066	.16443	1367.9	1526.1	1.5000	.15743	1364.1	1521.4	1.4936
1220	.17555	1383.3	1545.7	1.5155	.16789	1379.7	1541.2	1.5091	.16081	1376.1	1536.8	1.5028
1240	.17904	1394.8	1560.5	1.5243	.17130	1391.4	1556.2	1.5180	.16414	1387.9	1551.9	1.5118
1260	.18248	1406.2	1575.0	1.5328	.17465	1402.9	1570.9	1.5266	.16740	1399.6	1566.8	1.5205
1280	.18586	1417.5	1589.4	1.5411	.17793	1414.3	1585.5	1.5350	.17061	1411.1	1581.6	1.5290
1300	.18918	1428.6	1603.7	1.5493	.18117	1425.6	1599.9	1.5432	.17376	1422.5	1596.2	1.5374
1350	.19730	1456.2	1638.7	1.5689	.18906	1453.4	1635.3	1.5631	.18144	1450.6	1631.9	1.5574
1400	.20517	1483.2	1673.0	1.5876	.19670	1480.7	1669.9	1.5819	.18887	1478.1	1666.9	1.5765
1450	.21282	1509.8	1706.7	1.6055	.20413	1507.5	1703.9	1.6000	.19608	1505.2	1701.1	1.5946
1500	.22029	1536.1	1739.9	1.6227	.21137	1534.0	1737.4	1.6173	.20312	1531.8	1734.8	1.6120
1600	.2348	1587.9	1805.2	1.6551	.2254	1586.1	1803.0	1.6499	.2167	1584.2	1800.8	1.6449
1800	.2626	1689.8	1932.7	1.7142	.2523	1688.3	1931.1	1.7093	.2428	1686.8	1929.4	1.7045
2000	.2895	1790.8	2058.6	1.7676	.2782	1789.5	2057.2	1.7628	.2679	1788.2	2055.8	1.7581
2200	.3159	1892.2	2184.5	1.8168	.3037	1891.0	2183.2	1.8120	.2925	1889.7	2182.0	1.8074
2400	.3422	1994.8	2311.3	1.8628	.3291	1993.5	2310.1	1.8580	.3170	1992.2	2308.9	1.8534

v	u	h	s	v	u	h	s	v	u	h	s	t
.02343	643.0	667.3	.8434	.02332	641.5	666.5	.8419	.02322	640.0	665.8	.8405	**650**
.02386	656.3	681.0	.8557	.02374	654.6	680.1	.8541	.02363	653.0	679.2	.8525	**660**
.02432	669.9	695.1	.8682	.02419	668.0	694.0	.8664	.02406	666.2	692.9	.8647	**670**
.02483	683.8	709.6	.8810	.02468	681.8	708.3	.8790	.02454	679.8	707.0	.8771	**680**
.02540	698.2	724.5	.8940	.02522	695.9	723.0	.8919	.02506	693.7	721.6	.8898	**690**
.02603	713.1	740.1	.9075	.02582	710.5	738.2	.9051	.02563	708.1	736.5	.9028	**700**
.02674	728.6	756.3	.9214	.02649	725.6	754.1	.9187	.02626	722.9	752.1	.9161	**710**
.02755	744.7	773.2	.9358	.02725	741.4	770.6	.9327	.02698	738.3	768.2	.9299	**720**
.02848	761.7	791.2	.9510	.02811	757.8	788.0	.9474	.02778	754.2	785.1	.9441	**730**
.02958	779.7	810.4	.9670	.02912	775.1	806.4	.9628	.02871	771.0	802.8	.9590	**740**
.03090	799.0	831.0	.9842	.03030	793.5	826.0	.9791	.02978	788.6	821.7	.9746	**750**
.03253	820.0	853.7	1.0029	.03172	813.2	847.3	.9966	.03105	807.3	841.8	.9912	**760**
.03458	843.0	878.8	1.0234	.03347	834.6	870.5	1.0156	.03257	827.4	863.5	1.0089	**770**
.03719	868.4	906.9	1.0461	.03564	857.8	896.0	1.0363	.03442	848.9	887.2	1.0280	**780**
.04049	896.0	938.0	1.0711	.03833	883.0	924.1	1.0588	.03668	872.2	912.9	1.0487	**790**
.04448	925.0	971.1	1.0975	.04162	909.7	954.4	1.0830	.03942	896.9	940.7	1.0708	**800**
.04900	953.9	1004.7	1.1240	.04545	937.1	985.9	1.1079	.04265	922.6	970.0	1.0940	**810**
.05377	981.3	1037.0	1.1494	.04964	964.0	1017.3	1.1325	.04629	948.5	999.9	1.1175	**820**
.05852	1006.2	1066.9	1.1726	.05400	989.3	1047.3	1.1558	.05019	973.5	1029.2	1.1404	**830**
.06311	1028.6	1094.0	1.1936	.05834	1012.5	1075.1	1.1774	.05420	997.1	1057.3	1.1620	**840**
.06745	1048.6	1118.5	1.2123	.06253	1033.6	1100.7	1.1969	.05818	1018.8	1083.4	1.1820	**850**
.07154	1066.5	1140.7	1.2292	.06654	1052.6	1124.0	1.2147	.06204	1038.7	1107.6	1.2004	**860**
.07539	1082.8	1161.0	1.2446	.07033	1069.9	1145.4	1.2308	.06575	1056.8	1129.8	1.2172	**870**
.07902	1097.7	1179.6	1.2585	.07394	1085.6	1165.0	1.2455	.06928	1073.5	1150.4	1.2326	**880**
.08247	1111.5	1197.0	1.2714	.07735	1100.2	1183.2	1.2591	.07266	1088.8	1169.4	1.2468	**890**
.08574	1124.3	1213.2	1.2834	.08061	1113.7	1200.2	1.2716	.07588	1102.9	1187.2	1.2599	**900**
.08886	1136.3	1228.4	1.2946	.08371	1126.3	1216.1	1.2833	.07895	1116.2	1203.8	1.2721	**910**
.09186	1147.6	1242.8	1.3050	.08669	1138.2	1231.2	1.2942	.08190	1128.6	1219.5	1.2835	**920**
.09473	1158.4	1256.5	1.3150	.08954	1149.4	1245.5	1.3045	.08473	1140.3	1234.4	1.2942	**930**
.09750	1168.6	1269.6	1.3243	.09229	1160.1	1259.1	1.3143	.08746	1151.4	1248.5	1.3044	**940**
.10018	1178.4	1282.2	1.3333	.09494	1170.2	1272.1	1.3236	.09008	1162.0	1262.0	1.3140	**950**
.10277	1187.8	1294.3	1.3419	.09751	1180.0	1284.7	1.3325	.09263	1172.1	1275.0	1.3232	**960**
.10529	1196.9	1306.0	1.3501	.10000	1189.4	1296.7	1.3409	.09509	1181.9	1287.5	1.3319	**970**
.10773	1205.7	1317.3	1.3579	.10242	1198.5	1308.4	1.3491	.09748	1191.3	1299.5	1.3403	**980**
.11011	1214.2	1328.3	1.3655	.10477	1207.3	1319.7	1.3569	.09980	1200.3	1311.1	1.3484	**990**
.11243	1222.5	1339.0	1.3729	.10706	1215.8	1330.7	1.3645	.10207	1209.1	1322.4	1.3561	**1000**
.11691	1238.4	1359.5	1.3869	.11148	1232.2	1351.9	1.3788	.10643	1226.0	1344.1	1.3709	**1020**
.12120	1253.6	1379.2	1.4001	.11571	1247.8	1372.0	1.3924	.11060	1242.0	1364.8	1.3848	**1040**
.12533	1268.3	1398.1	1.4126	.11977	1262.8	1391.4	1.4052	.11460	1257.3	1384.6	1.3979	**1060**
.12931	1282.4	1416.4	1.4246	.12369	1277.3	1410.0	1.4174	.11846	1272.1	1403.6	1.4103	**1080**
.13317	1296.2	1434.2	1.4360	.12748	1291.3	1428.1	1.4291	.12218	1286.4	1422.1	1.4222	**1100**
.13691	1309.6	1451.4	1.4470	.13115	1304.9	1445.7	1.4403	.12579	1300.3	1440.0	1.4336	**1120**
.14055	1322.6	1468.3	1.4576	.13472	1318.2	1462.8	1.4510	.12930	1313.8	1457.4	1.4446	**1140**
.14409	1335.4	1484.7	1.4679	.13820	1331.2	1479.6	1.4614	.13271	1327.1	1474.4	1.4552	**1160**
.14755	1348.0	1500.9	1.4778	.14159	1344.0	1496.0	1.4715	.13603	1340.0	1491.0	1.4654	**1180**
.15094	1360.3	1516.7	1.4874	.14490	1356.5	1512.0	1.4812	.13927	1352.7	1507.3	1.4752	**1200**
.15425	1372.5	1532.3	1.4967	.14814	1368.8	1527.8	1.4907	.14245	1365.2	1523.3	1.4848	**1220**
.15749	1384.4	1547.6	1.5058	.15131	1380.9	1543.4	1.4999	.14555	1377.5	1539.1	1.4941	**1240**
.16068	1396.2	1562.8	1.5146	.15443	1392.9	1558.7	1.5088	.14860	1389.6	1554.6	1.5032	**1260**
.16381	1407.9	1577.7	1.5232	.15748	1404.7	1573.7	1.5176	.15159	1401.5	1569.8	1.5120	**1280**
.16688	1419.5	1592.4	1.5317	.16049	1416.4	1588.6	1.5261	.15453	1413.3	1584.9	1.5206	**1300**
.17437	1447.8	1628.5	1.5519	.16779	1445.0	1625.1	1.5465	.16166	1442.2	1621.7	1.5413	**1350**
.18160	1475.6	1663.8	1.5711	.17484	1473.0	1660.7	1.5659	.16854	1470.5	1657.6	1.5608	**1400**
.18862	1502.8	1698.3	1.5894	.18168	1500.5	1695.5	1.5844	.17520	1498.2	1692.7	1.5795	**1450**
.19546	1529.7	1732.2	1.6070	.18833	1527.5	1729.7	1.6020	.18168	1525.4	1727.1	1.5972	**1500**
.2087	1582.4	1798.6	1.6400	.2012	1580.5	1796.5	1.6353	.19420	1578.7	1794.3	1.6307	**1600**
.2339	1685.3	1927.8	1.6999	.2257	1683.9	1926.1	1.6954	.21801	1682.4	1924.5	1.6910	**1800**
.2582	1786.9	2054.5	1.7536	.2492	1785.6	2053.1	1.7492	.24087	1784.3	2051.7	1.7450	**2000**
.2821	1888.5	2180.7	1.8030	.2723	1887.2	2179.5	1.7986	.26328	1886.0	2178.3	1.7945	**2200**
.3057	1990.9	2307.7	1.8490	.2953	1989.6	2306.5	1.8447	.28553	1988.3	2305.3	1.8405	**2400**

Table 3.   Vapor

p		6200				6400				6600		
t	v	u	h	s	v	u	h	s	v	u	h	s
650	.02313	638.5	665.1	.8390	.02303	637.1	664.4	.8377	.02294	635.7	663.8	.8363
660	.02352	651.4	678.4	.8510	.02341	649.8	677.6	.8495	.02331	648.4	676.8	.8480
670	.02394	664.5	692.0	.8631	.02382	662.8	691.0	.8615	.02371	661.2	690.2	.8599
680	.02440	677.9	705.9	.8753	.02427	676.1	704.8	.8736	.02414	674.3	703.8	.8719
690	.02490	691.7	720.2	.8878	.02475	689.6	719.0	.8859	.02461	687.7	717.8	.8841
700	.02545	705.8	735.0	.9006	.02527	703.6	733.5	.8985	.02511	701.4	732.1	.8965
710	.02605	720.3	750.2	.9137	.02585	717.8	748.5	.9114	.02567	715.5	746.8	.9092
720	.02673	735.3	766.0	.9271	.02649	732.6	764.0	.9246	.02628	730.0	762.1	.9221
730	.02748	750.9	782.4	.9410	.02721	747.8	780.0	.9381	.02695	744.9	777.8	.9354
740	.02834	767.2	799.7	.9554	.02801	763.6	796.8	.9522	.02771	760.3	794.2	.9491
750	.02932	784.2	817.8	.9705	.02892	780.1	814.3	.9667	.02856	776.3	811.2	.9633
760	.03047	802.1	837.0	.9863	.02997	797.3	832.8	.9820	.02952	793.0	829.1	.9780
770	.03182	821.1	857.6	1.0031	.03118	815.5	852.4	.9980	.03063	810.5	847.9	.9934
780	.03343	841.4	879.7	1.0210	.03261	834.7	873.3	1.0149	.03192	828.8	867.8	1.0095
790	.03537	863.0	903.6	1.0402	.03430	855.1	895.7	1.0329	.03342	848.2	889.0	1.0265
800	.03769	886.0	929.2	1.0607	.03631	876.7	919.7	1.0520	.03517	868.5	911.5	1.0444
810	.04044	910.1	956.5	1.0822	.03867	899.3	945.1	1.0721	.03723	889.9	935.4	1.0633
820	.04358	934.7	984.7	1.1043	.04138	922.6	971.6	1.0929	.03959	912.0	960.3	1.0829
830	.04703	959.1	1013.0	1.1264	.04441	946.1	998.7	1.1140	.04225	934.5	986.1	1.1029
840	.05066	982.5	1040.7	1.1477	.04768	969.1	1025.6	1.1347	.04516	956.8	1012.0	1.1230
850	.05437	1004.6	1067.0	1.1679	.05107	991.2	1051.7	1.1547	.04825	978.6	1037.6	1.1426
860	.05803	1025.1	1091.7	1.1867	.05451	1011.9	1076.5	1.1736	.05143	999.5	1062.3	1.1613
870	.06161	1043.9	1114.6	1.2040	.05791	1031.3	1099.9	1.1913	.05463	1019.1	1085.8	1.1791
880	.06505	1061.3	1135.9	1.2200	.06123	1049.2	1121.8	1.2077	.05780	1037.5	1108.1	1.1958
890	.06836	1077.3	1155.7	1.2347	.06444	1065.9	1142.2	1.2229	.06089	1054.6	1129.0	1.2113
900	.07153	1092.1	1174.2	1.2483	.06754	1081.3	1161.3	1.2370	.06390	1070.6	1148.6	1.2259
910	.07457	1105.9	1191.5	1.2610	.07052	1095.7	1179.2	1.2501	.06681	1085.5	1167.1	1.2394
920	.07748	1118.9	1207.8	1.2729	.07339	1109.2	1196.1	1.2624	.06963	1099.5	1184.5	1.2521
930	.08028	1131.1	1223.2	1.2840	.07615	1121.9	1212.1	1.2739	.07234	1112.7	1201.0	1.2640
940	.08297	1142.7	1237.9	1.2945	.07881	1133.9	1227.2	1.2848	.07496	1125.1	1216.6	1.2752
950	.08557	1153.7	1251.9	1.3045	.08138	1145.3	1241.7	1.2951	.07749	1136.9	1231.5	1.2858
960	.08809	1164.2	1265.3	1.3139	.08387	1156.2	1255.5	1.3048	.07994	1148.2	1245.8	1.2959
970	.09052	1174.3	1278.1	1.3230	.08627	1166.6	1268.8	1.3142	.08231	1158.9	1259.4	1.3054
980	.09289	1184.0	1290.5	1.3316	.08861	1176.6	1281.5	1.3231	.08462	1169.2	1272.6	1.3146
990	.09518	1193.3	1302.5	1.3399	.09088	1186.3	1293.9	1.3316	.08686	1179.2	1285.2	1.3234
1000	.09742	1202.4	1314.1	1.3479	.09308	1195.6	1305.8	1.3398	.08904	1188.7	1297.5	1.3318
1020	.10173	1219.7	1336.4	1.3631	.09733	1213.4	1328.6	1.3553	.09323	1207.0	1320.9	1.3477
1040	.10584	1236.1	1357.5	1.3773	.10139	1230.2	1350.3	1.3699	.09723	1224.3	1343.0	1.3626
1060	.10978	1251.8	1377.8	1.3907	.10528	1246.3	1371.0	1.3836	.10106	1240.7	1364.2	1.3766
1080	.11358	1266.9	1397.2	1.4034	.10901	1261.7	1390.8	1.3965	.10474	1256.5	1384.4	1.3898
1100	.11724	1281.5	1416.0	1.4155	.11262	1276.6	1410.0	1.4089	.10829	1271.6	1403.9	1.4024
1120	.12079	1295.7	1434.2	1.4271	.11611	1291.0	1428.5	1.4207	.11172	1286.3	1422.8	1.4144
1140	.12423	1309.4	1452.0	1.4383	.11949	1305.0	1446.5	1.4320	.11505	1300.5	1441.1	1.4259
1160	.12758	1322.8	1469.2	1.4490	.12278	1318.6	1464.0	1.4429	.11828	1314.4	1458.9	1.4370
1180	.13084	1336.0	1486.1	1.4593	.12598	1332.0	1481.2	1.4534	.12143	1327.9	1476.2	1.4476
1200	.13402	1348.9	1502.6	1.4694	.12910	1345.0	1497.9	1.4636	.12449	1341.2	1493.2	1.4579
1220	.13713	1361.5	1518.8	1.4791	.13215	1357.9	1514.4	1.4734	.12749	1354.2	1509.9	1.4679
1240	.14017	1374.0	1534.8	1.4885	.13514	1370.5	1530.5	1.4830	.13041	1366.9	1526.2	1.4776
1260	.14316	1386.2	1550.5	1.4977	.13806	1382.9	1546.4	1.4923	.13328	1379.5	1542.3	1.4870
1280	.14608	1398.3	1565.9	1.5066	.14092	1395.1	1562.0	1.5013	.13609	1391.8	1558.1	1.4961
1300	.14895	1410.2	1581.1	1.5153	.14374	1407.1	1577.4	1.5101	.13884	1404.0	1573.6	1.5050
1350	.15593	1439.4	1618.3	1.5362	.15056	1436.6	1615.0	1.5312	.14553	1433.8	1611.6	1.5263
1400	.16265	1467.9	1654.5	1.5559	.15713	1465.4	1651.5	1.5511	.15195	1462.8	1648.4	1.5463
1450	.16915	1495.8	1689.9	1.5747	.16348	1493.5	1687.1	1.5700	.15816	1491.1	1684.3	1.5654
1500	.17547	1523.2	1724.5	1.5926	.16965	1521.1	1722.0	1.5880	.16419	1518.9	1719.4	1.5836
1600	.18767	1576.8	1792.2	1.6262	.18154	1575.0	1790.0	1.6219	.17580	1573.1	1787.8	1.6176
1800	.21083	1680.9	1922.8	1.6868	.20410	1679.5	1921.2	1.6827	.19778	1678.0	1919.6	1.6786
2000	.23304	1783.0	2050.4	1.7409	.22570	1781.7	2049.0	1.7369	.21882	1780.4	2047.7	1.7330
2200	.25481	1884.7	2177.0	1.7904	.24686	1883.5	2175.8	1.7864	.23940	1882.2	2174.6	1.7826
2400	.27640	1987.0	2304.1	1.8365	.26785	1985.7	2303.0	1.8325	.25982	1984.5	2301.8	1.8287

	6800				7000				7200			p
v	u	h	s	v	u	h	s	v	u	h	s	t
.02286	634.4	663.2	.8350	.02277	633.1	662.6	.8337	.02269	631.8	662.1	.8325	**650**
.02322	646.9	676.1	.8466	.02313	645.5	675.5	.8453	.02304	644.1	674.8	.8440	**660**
.02361	659.6	689.3	.8584	.02350	658.1	688.6	.8569	.02340	656.6	687.8	.8555	**670**
.02402	672.6	702.8	.8703	.02391	671.0	701.9	.8687	.02380	669.4	701.1	.8672	**680**
.02447	685.9	716.6	.8824	.02434	684.1	715.6	.8807	.02422	682.3	714.6	.8790	**690**
.02496	699.4	730.8	.8946	.02481	697.4	729.6	.8928	.02468	695.5	728.4	.8910	**700**
.02549	713.3	745.3	.9071	.02533	711.1	743.9	.9051	.02517	709.0	742.6	.9031	**710**
.02607	727.5	760.3	.9198	.02589	725.1	758.7	.9176	.02571	722.9	757.1	.9155	**720**
.02672	742.1	775.7	.9329	.02650	739.5	773.8	.9304	.02629	737.0	772.0	.9281	**730**
.02743	757.2	791.7	.9463	.02717	754.3	789.5	.9435	.02694	751.5	787.4	.9410	**740**
.02823	772.8	808.3	.9600	.02792	769.6	805.7	.9570	.02765	766.5	803.3	.9542	**750**
.02912	789.0	825.7	.9743	.02876	785.3	822.6	.9709	.02844	781.9	819.8	.9677	**760**
.03014	805.9	843.8	.9891	.02971	801.7	840.2	.9853	.02932	797.8	836.9	.9817	**770**
.03131	823.5	862.9	1.0046	.03078	818.7	858.6	1.0002	.03031	814.3	854.7	.9961	**780**
.03266	842.0	883.1	1.0208	.03201	836.5	877.9	1.0157	.03143	831.4	873.3	1.0111	**790**
.03422	861.4	904.5	1.0378	.03341	855.0	898.3	1.0319	.03271	849.3	892.8	1.0266	**800**
.03603	881.6	927.0	1.0556	.03503	874.3	919.7	1.0489	.03416	867.8	913.3	1.0428	**810**
.03811	902.6	950.6	1.0742	.03687	894.3	942.1	1.0664	.03582	886.9	934.7	1.0596	**820**
.04046	924.1	975.0	1.0932	.03895	914.9	965.4	1.0845	.03768	906.6	956.8	1.0768	**830**
.04305	945.7	999.9	1.1124	.04127	935.7	989.2	1.1029	.03975	926.7	979.6	1.0945	**840**
.04584	967.1	1024.7	1.1314	.04378	956.5	1013.2	1.1213	.04202	946.8	1002.8	1.1122	**850**
.04875	987.7	1049.1	1.1499	.04644	976.8	1037.0	1.1394	.04444	966.7	1025.9	1.1298	**860**
.05174	1007.4	1072.6	1.1677	.04920	996.4	1060.2	1.1570	.04699	986.1	1048.7	1.1470	**870**
.05473	1026.1	1095.0	1.1845	.05201	1015.2	1082.6	1.1737	.04960	1004.8	1070.9	1.1636	**880**
.05769	1043.6	1116.2	1.2003	.05482	1032.9	1104.0	1.1896	.05225	1022.7	1092.3	1.1796	**890**
.06060	1060.0	1136.3	1.2151	.05760	1049.7	1124.3	1.2047	.05489	1039.7	1112.8	1.1947	**900**
.06342	1075.4	1155.2	1.2289	.06032	1065.5	1143.6	1.2188	.05751	1055.7	1132.4	1.2090	**910**
.06616	1089.8	1173.1	1.2420	.06299	1080.3	1161.9	1.2321	.06008	1070.9	1151.0	1.2225	**920**
.06882	1103.4	1190.0	1.2542	.06557	1094.3	1179.2	1.2446	.06259	1085.3	1168.7	1.2353	**930**
.07139	1116.3	1206.1	1.2657	.06809	1107.5	1195.7	1.2565	.06504	1098.9	1185.5	1.2474	**940**
.07388	1128.5	1221.5	1.2766	.07053	1120.1	1211.5	1.2677	.06743	1111.8	1201.6	1.2589	**950**
.07629	1140.1	1236.1	1.2870	.07290	1132.1	1226.5	1.2783	.06974	1124.1	1217.0	1.2697	**960**
.07863	1151.2	1250.1	1.2968	.07520	1143.5	1240.9	1.2884	.07200	1135.8	1231.7	1.2801	**970**
.08090	1161.8	1263.6	1.3063	.07743	1154.4	1254.7	1.2980	.07420	1147.0	1245.9	1.2899	**980**
.08311	1172.0	1276.6	1.3152	.07961	1164.9	1268.0	1.3072	.07634	1157.8	1259.5	1.2994	**990**
.08526	1181.9	1289.2	1.3239	.08172	1175.0	1280.9	1.3161	.07842	1168.2	1272.7	1.3084	**1000**
.08939	1200.6	1313.1	1.3402	.08580	1194.3	1305.4	1.3328	.08243	1187.9	1297.7	1.3255	**1020**
.09333	1218.3	1335.8	1.3554	.08968	1212.4	1328.5	1.3483	.08626	1206.4	1321.3	1.3413	**1040**
.09711	1235.1	1357.3	1.3697	.09340	1229.6	1350.5	1.3629	.08992	1224.0	1343.8	1.3562	**1060**
.10074	1251.2	1378.0	1.3832	.09698	1246.0	1371.6	1.3766	.09344	1240.7	1365.2	1.3702	**1080**
.10423	1266.7	1397.8	1.3960	.10042	1261.7	1391.8	1.3897	.09683	1256.7	1385.8	1.3835	**1100**
.10761	1281.6	1417.0	1.4082	.10374	1276.9	1411.3	1.4021	.10011	1272.2	1405.6	1.3961	**1120**
.11088	1296.1	1435.6	1.4199	.10697	1291.6	1430.2	1.4140	.10328	1287.2	1424.8	1.4082	**1140**
.11406	1310.2	1453.7	1.4311	.11009	1305.9	1448.5	1.4254	.10635	1301.7	1443.4	1.4197	**1160**
.11715	1323.9	1471.3	1.4419	.11313	1319.9	1466.4	1.4364	.10934	1315.8	1461.5	1.4308	**1180**
.12016	1337.3	1488.5	1.4524	.11609	1333.5	1483.9	1.4469	.11225	1329.6	1479.2	1.4416	**1200**
.12310	1350.5	1505.4	1.4625	.11898	1346.8	1500.9	1.4572	.11509	1343.1	1496.5	1.4519	**1220**
.12597	1363.4	1521.9	1.4723	.12180	1359.9	1517.7	1.4671	.11786	1356.4	1513.4	1.4619	**1240**
.12879	1376.1	1538.2	1.4818	.12456	1372.7	1534.1	1.4767	.12057	1369.4	1530.0	1.4717	**1260**
.13154	1388.6	1554.1	1.4910	.12726	1385.4	1550.2	1.4860	.12323	1382.1	1546.3	1.4811	**1280**
.13424	1400.9	1569.9	1.5000	.12991	1397.8	1566.1	1.4951	.12583	1394.7	1562.4	1.4903	**1300**
.14079	1431.0	1608.2	1.5215	.13633	1428.2	1604.8	1.5168	.13213	1425.4	1601.4	1.5121	**1350**
.14708	1460.3	1645.3	1.5417	.14250	1457.7	1642.3	1.5372	.13817	1455.1	1639.2	1.5328	**1400**
.15316	1488.8	1681.5	1.5609	.14845	1486.5	1678.7	1.5565	.14400	1484.1	1676.0	1.5523	**1450**
.15905	1516.8	1716.9	1.5792	.15421	1514.6	1714.4	1.5749	.14965	1512.4	1711.8	1.5708	**1500**
.17039	1571.3	1785.7	1.6134	.16530	1569.4	1783.5	1.6094	.16050	1567.6	1781.4	1.6054	**1600**
.19184	1676.5	1917.9	1.6747	.18625	1675.1	1916.3	1.6709	.18096	1673.6	1914.7	1.6672	**1800**
.21234	1779.1	2046.3	1.7292	.20623	1777.8	2045.0	1.7255	.20047	1776.5	2043.6	1.7219	**2000**
.23239	1881.0	2173.4	1.7789	.22578	1879.7	2172.2	1.7752	.21953	1878.5	2171.0	1.7717	**2200**
.25227	1983.2	2300.6	1.8250	.24515	1981.9	2299.4	1.8214	.23843	1980.6	2298.3	1.8178	**2400**

Table 3. Vapor

p		7400				7600				7800		
t	v	u	h	s	v	u	h	s	v	u	h	s
650	.02261	630.6	661.5	.8313	.02254	629.4	661.1	.8301	.02246	628.2	660.6	.8289
660	.02295	642.8	674.2	.8427	.02287	641.5	673.6	.8414	.02278	640.2	673.1	.8402
670	.02331	655.2	687.1	.8541	.02321	653.8	686.4	.8528	.02312	652.4	685.8	.8515
680	.02369	667.8	700.2	.8657	.02359	666.3	699.5	.8643	.02349	664.8	698.7	.8628
690	.02410	680.6	713.6	.8774	.02399	679.0	712.8	.8759	.02388	677.4	711.9	.8743
700	.02454	693.7	727.3	.8893	.02442	692.0	726.3	.8876	.02429	690.3	725.3	.8860
710	.02502	707.1	741.3	.9013	.02488	705.2	740.1	.8995	.02474	703.3	739.0	.8977
720	.02554	720.7	755.7	.9135	.02538	718.6	754.3	.9115	.02522	716.6	753.0	.9096
730	.02610	734.6	770.4	.9259	.02592	732.3	768.8	.9238	.02574	730.2	767.3	.9217
740	.02671	748.9	785.5	.9385	.02650	746.4	783.7	.9362	.02631	744.0	782.0	.9340
750	.02739	763.6	801.1	.9515	.02715	760.8	799.0	.9489	.02692	758.2	797.0	.9465
760	.02813	778.6	817.2	.9647	.02786	775.6	814.8	.9619	.02760	772.7	812.5	.9592
770	.02896	794.2	833.8	.9783	.02864	790.8	831.0	.9752	.02834	787.6	828.5	.9723
780	.02989	810.2	851.2	.9924	.02950	806.4	847.9	.9889	.02915	802.9	844.9	.9856
790	.03092	826.8	869.2	1.0068	.03047	822.6	865.4	1.0029	.03005	818.6	862.0	.9993
800	.03209	844.0	888.0	1.0218	.03155	839.2	883.6	1.0174	.03106	834.8	879.6	1.0133
810	.03342	861.8	907.6	1.0374	.03276	856.4	902.5	1.0324	.03218	851.5	897.9	1.0278
820	.03491	880.3	928.1	1.0534	.03412	874.2	922.2	1.0478	.03343	868.7	916.9	1.0427
830	.03659	899.2	949.3	1.0699	.03565	892.4	942.6	1.0637	.03482	886.3	936.5	1.0580
840	.03846	918.5	971.2	1.0868	.03734	911.1	963.6	1.0799	.03636	904.3	956.8	1.0736
850	.04051	938.0	993.4	1.1039	.03920	929.9	985.1	1.0964	.03806	922.6	977.5	1.0895
860	.04272	957.4	1015.9	1.1210	.04121	948.8	1006.8	1.1129	.03990	941.0	998.5	1.1055
870	.04505	976.5	1038.2	1.1378	.04336	967.6	1028.5	1.1293	.04187	959.3	1019.7	1.1215
880	.04748	995.1	1060.1	1.1542	.04561	985.9	1050.0	1.1454	.04395	977.3	1040.8	1.1373
890	.04996	1012.9	1081.4	1.1700	.04792	1003.7	1071.1	1.1611	.04611	994.9	1061.5	1.1527
900	.05246	1030.0	1101.9	1.1852	.05028	1020.8	1091.5	1.1762	.04833	1012.0	1081.8	1.1676
910	.05496	1046.3	1121.6	1.1996	.05265	1037.2	1111.3	1.1906	.05057	1028.5	1101.5	1.1821
920	.05743	1061.8	1140.4	1.2133	.05501	1052.8	1130.2	1.2044	.05282	1044.2	1120.5	1.1959
930	.05985	1076.4	1158.4	1.2263	.05735	1067.7	1148.4	1.2175	.05506	1059.3	1138.8	1.2091
940	.06223	1090.3	1175.5	1.2386	.05965	1081.9	1165.8	1.2300	.05727	1073.7	1156.3	1.2217
950	.06456	1103.5	1191.9	1.2503	.06190	1095.4	1182.5	1.2419	.05946	1087.4	1173.2	1.2337
960	.06682	1116.1	1207.6	1.2614	.06411	1108.3	1198.4	1.2532	.06160	1100.5	1189.4	1.2452
970	.06903	1128.1	1222.7	1.2719	.06627	1120.6	1213.8	1.2639	.06370	1113.1	1205.0	1.2561
980	.07118	1139.7	1237.1	1.2820	.06838	1132.3	1228.5	1.2742	.06576	1125.1	1220.0	1.2666
990	.07328	1150.7	1251.1	1.2916	.07043	1143.6	1242.7	1.2840	.06777	1136.6	1234.5	1.2766
1000	.07533	1161.3	1264.5	1.3009	.07244	1154.5	1256.4	1.2935	.06974	1147.7	1248.4	1.2862
1020	.07928	1181.5	1290.1	1.3183	.07632	1175.1	1282.5	1.3112	.07355	1168.8	1275.0	1.3043
1040	.08305	1200.4	1314.2	1.3345	.08003	1194.5	1307.0	1.3277	.07719	1188.5	1300.0	1.3210
1060	.08665	1218.4	1337.0	1.3496	.08358	1212.8	1330.3	1.3431	.08068	1207.2	1323.6	1.3367
1080	.09012	1235.4	1358.8	1.3638	.08699	1230.1	1352.5	1.3576	.08404	1224.9	1346.2	1.3515
1100	.09346	1251.8	1379.7	1.3773	.09028	1246.8	1373.7	1.3713	.08728	1241.8	1367.8	1.3654
1120	.09668	1267.5	1399.9	1.3902	.09345	1262.8	1394.2	1.3844	.09040	1258.1	1388.5	1.3786
1140	.09980	1282.7	1419.4	1.4024	.09652	1278.2	1414.0	1.3968	.09343	1273.7	1408.6	1.3912
1160	.10283	1297.4	1438.2	1.4142	.09950	1293.2	1433.1	1.4087	.09636	1288.9	1428.0	1.4033
1180	.10577	1311.8	1456.6	1.4254	.10240	1307.7	1451.7	1.4201	.09921	1303.7	1446.9	1.4149
1200	.10863	1325.8	1474.5	1.4363	.10521	1321.9	1469.9	1.4311	.10198	1318.0	1465.2	1.4260
1220	.11142	1339.4	1492.0	1.4468	.10796	1335.8	1487.6	1.4417	.10468	1332.1	1483.2	1.4367
1240	.11415	1352.8	1509.2	1.4569	.11064	1349.3	1504.9	1.4520	.10732	1345.8	1500.7	1.4471
1260	.11681	1366.0	1525.9	1.4667	.11326	1362.6	1521.9	1.4619	.10989	1359.2	1517.8	1.4571
1280	.11942	1378.9	1542.4	1.4763	.11582	1375.7	1538.5	1.4715	.11241	1372.4	1534.7	1.4669
1300	.12198	1391.6	1558.6	1.4855	.11833	1388.5	1554.9	1.4809	.11488	1385.4	1551.2	1.4763
1350	.12816	1422.6	1598.1	1.5076	.12440	1419.8	1594.7	1.5032	.12085	1416.9	1591.4	1.4988
1400	.13409	1452.6	1636.2	1.5284	.13023	1450.0	1633.1	1.5241	.12656	1447.4	1630.1	1.5200
1450	.13980	1481.8	1673.2	1.5481	.13583	1479.4	1670.4	1.5439	.13207	1477.1	1667.7	1.5399
1500	.14534	1510.3	1709.3	1.5667	.14125	1508.1	1706.8	1.5627	.13739	1505.9	1704.3	1.5588
1600	.15596	1565.7	1779.3	1.6015	.15166	1563.8	1777.1	1.5977	.14759	1562.0	1775.0	1.5940
1800	.17597	1672.1	1913.1	1.6636	.17124	1670.6	1911.5	1.6600	.16676	1669.2	1909.9	1.6565
2000	.19503	1775.3	2042.3	1.7184	.18987	1774.0	2041.0	1.7149	.18498	1772.7	2039.7	1.7116
2200	.21363	1877.2	2169.8	1.7682	.20805	1876.0	2168.6	1.7648	.20275	1874.8	2167.4	1.7615
2400	.23208	1979.3	2297.1	1.8144	.22606	1978.0	2296.0	1.8110	.22035	1976.8	2294.8	1.8077

	8000				**8200**				**8400**			**p**
v	**u**	**h**	**s**	**v**	**u**	**h**	**s**	**v**	**u**	**h**	**s**	**t**
.02239	627.0	660.2	.8278	.02232	625.9	659.7	.8267	.02225	624.8	659.4	.8256	650
.02270	639.0	672.6	.8389	.02263	637.8	672.1	.8378	.02255	636.6	671.6	.8366	660
.02304	651.1	685.2	.8502	.02295	649.8	684.6	.8489	.02287	648.5	684.1	.8477	670
.02339	663.4	698.0	.8615	.02330	662.0	697.4	.8601	.02321	660.7	696.7	.8588	680
.02377	675.9	711.1	.8729	.02367	674.4	710.3	.8714	.02357	673.0	709.6	.8701	690
.02418	688.6	724.4	.8844	.02406	687.0	723.5	.8829	.02396	685.4	722.7	.8814	700
.02461	701.5	737.9	.8960	.02449	699.8	736.9	.8944	.02437	698.1	736.0	.8928	710
.02508	714.7	751.8	.9078	.02494	712.8	750.6	.9060	.02480	711.0	749.5	.9043	720
.02558	728.1	765.9	.9197	.02542	726.0	764.6	.9178	.02527	724.1	763.4	.9160	730
.02612	741.7	780.4	.9319	.02595	739.5	778.9	.9298	.02578	737.4	777.5	.9278	740
.02671	755.7	795.2	.9442	.02651	753.3	793.5	.9419	.02633	751.0	791.9	.9398	750
.02735	769.9	810.4	.9567	.02713	767.3	808.5	.9543	.02691	764.8	806.7	.9519	760
.02806	784.5	826.1	.9695	.02780	781.7	823.9	.9668	.02755	778.9	821.8	.9643	770
.02883	799.5	842.2	.9825	.02853	796.4	839.7	.9796	.02825	793.4	837.3	.9768	780
.02968	814.9	858.8	.9959	.02933	811.4	855.9	.9927	.02901	808.1	853.2	.9896	790
.03061	830.7	876.0	1.0096	.03021	826.8	872.7	1.0060	.02984	823.2	869.6	1.0027	800
.03166	846.9	893.8	1.0236	.03118	842.6	890.0	1.0197	.03075	838.7	886.5	1.0160	810
.03281	863.6	912.1	1.0380	.03226	858.8	907.8	1.0337	.03176	854.5	903.8	1.0297	820
.03409	880.6	931.1	1.0528	.03344	875.4	926.2	1.0480	.03286	870.6	921.7	1.0436	830
.03551	898.1	950.7	1.0679	.03475	892.4	945.1	1.0626	.03407	887.1	940.1	1.0577	840
.03706	915.8	970.7	1.0832	.03618	909.6	964.5	1.0775	.03539	903.8	958.9	1.0722	850
.03875	933.7	991.1	1.0987	.03773	927.0	984.3	1.0925	.03683	920.8	978.0	1.0868	860
.04056	951.6	1011.6	1.1143	.03941	944.5	1004.3	1.1076	.03838	937.8	997.5	1.1014	870
.04249	969.3	1032.2	1.1297	.04119	961.8	1024.3	1.1226	.04003	954.9	1017.1	1.1161	880
.04450	986.7	1052.6	1.1448	.04306	979.0	1044.3	1.1375	.04177	971.7	1036.7	1.1307	890
.04657	1003.7	1072.6	1.1596	.04500	995.8	1064.1	1.1521	.04359	988.3	1056.1	1.1450	900
.04869	1020.1	1092.2	1.1740	.04700	1012.1	1083.5	1.1663	.04547	1004.6	1075.3	1.1591	910
.05083	1035.9	1111.2	1.1878	.04902	1028.0	1102.4	1.1800	.04738	1020.4	1094.0	1.1727	920
.05297	1051.1	1129.5	1.2010	.05106	1043.2	1120.7	1.1933	.04932	1035.6	1112.3	1.1859	930
.05510	1065.7	1147.2	1.2137	.05310	1057.9	1138.5	1.2060	.05127	1050.4	1130.1	1.1987	940
.05720	1079.6	1164.3	1.2259	.05513	1072.0	1155.6	1.2183	.05322	1064.6	1147.3	1.2109	950
.05928	1092.9	1180.7	1.2375	.05714	1085.5	1172.2	1.2300	.05516	1078.2	1164.0	1.2227	960
.06132	1105.7	1196.5	1.2485	.05912	1098.4	1188.2	1.2412	.05707	1091.4	1180.1	1.2340	970
.06333	1117.9	1211.7	1.2592	.06107	1110.9	1203.6	1.2519	.05896	1104.0	1195.6	1.2449	980
.06529	1129.7	1226.4	1.2693	.06298	1122.9	1218.4	1.2622	.06083	1116.1	1210.7	1.2553	990
.06722	1141.0	1240.5	1.2791	.06486	1134.4	1232.8	1.2721	.06266	1127.8	1225.2	1.2653	1000
.07095	1162.5	1267.5	1.2974	.06851	1156.3	1260.2	1.2907	.06623	1150.1	1253.0	1.2842	1020
.07452	1182.6	1293.0	1.3145	.07202	1176.7	1286.0	1.3081	.06966	1170.9	1279.2	1.3018	1040
.07796	1201.6	1317.0	1.3304	.07539	1196.1	1310.5	1.3243	.07297	1190.6	1304.0	1.3182	1060
.08126	1219.6	1339.9	1.3454	.07864	1214.4	1333.7	1.3395	.07617	1209.2	1327.6	1.3336	1080
.08445	1236.8	1361.9	1.3596	.08177	1231.9	1356.0	1.3538	.07925	1227.0	1350.1	1.3482	1100
.08752	1253.4	1382.9	1.3730	.08480	1248.7	1377.3	1.3674	.08223	1244.0	1371.8	1.3620	1120
.09050	1269.3	1403.3	1.3858	.08773	1264.8	1397.9	1.3804	.08511	1260.4	1392.7	1.3751	1140
.09339	1284.7	1422.9	1.3980	.09057	1280.4	1417.9	1.3928	.08791	1276.2	1412.9	1.3876	1160
.09619	1299.6	1442.0	1.4097	.09333	1295.6	1437.2	1.4046	.09062	1291.6	1432.4	1.3996	1180
.09892	1314.2	1460.6	1.4210	.09602	1310.3	1456.0	1.4160	.09327	1306.5	1451.5	1.4112	1200
.10158	1328.4	1478.8	1.4318	.09864	1324.7	1474.4	1.4270	.09584	1321.0	1470.0	1.4223	1220
.10417	1342.3	1496.5	1.4423	.10119	1338.7	1492.3	1.4376	.09836	1335.2	1488.1	1.4330	1240
.10670	1355.8	1513.8	1.4525	.10368	1352.5	1509.8	1.4479	.10081	1349.1	1505.8	1.4433	1260
.10918	1369.2	1530.8	1.4623	.10612	1365.9	1527.0	1.4578	.10321	1362.7	1523.1	1.4534	1280
.11161	1382.3	1547.5	1.4718	.10851	1379.2	1543.8	1.4674	.10556	1376.1	1540.1	1.4631	1300
.11748	1414.1	1588.0	1.4946	.11428	1411.3	1584.7	1.4903	.11123	1408.5	1581.4	1.4862	1350
.12309	1444.9	1627.1	1.5158	.11980	1442.3	1624.1	1.5118	.11666	1439.7	1621.1	1.5078	1400
.12849	1474.7	1664.9	1.5359	.12510	1472.3	1662.2	1.5320	.12188	1470.0	1659.4	1.5282	1450
.13372	1503.8	1701.7	1.5550	.13023	1501.6	1699.2	1.5512	.12692	1499.5	1696.7	1.5475	1500
.14372	1560.1	1772.9	1.5904	.14005	1558.3	1770.8	1.5868	.13656	1556.4	1768.7	1.5833	1600
.16251	1667.7	1908.3	1.6531	.15847	1666.2	1906.7	1.6498	.15462	1664.7	1905.1	1.6465	1800
.18034	1771.4	2038.4	1.7083	.17593	1770.1	2037.1	1.7051	.17173	1768.8	2035.8	1.7019	2000
.19772	1873.5	2166.2	1.7583	.19294	1872.3	2165.1	1.7551	.18839	1871.0	2163.9	1.7520	2200
.21494	1975.5	2293.7	1.8045	.20979	1974.2	2292.5	1.8013	.20488	1972.9	2291.4	1.7983	2400

Table 3. Vapor

t	8600 v	u	h	s	8800 v	u	h	s	9000 v	u	h	s
650	.02218	623.7	659.0	.8245	.02212	622.6	658.6	.8234	.02206	621.6	658.3	.8224
660	.02248	635.4	671.2	.8355	.02241	634.3	670.8	.8343	.02234	633.2	670.4	.8332
670	.02279	647.3	683.6	.8465	.02272	646.1	683.1	.8453	.02264	644.9	682.6	.8441
680	.02312	659.3	696.1	.8575	.02304	658.1	695.6	.8563	.02296	656.8	695.0	.8551
690	.02348	671.6	708.9	.8687	.02338	670.2	708.3	.8674	.02330	668.8	707.6	.8661
700	.02385	683.9	721.9	.8799	.02375	682.5	721.1	.8785	.02365	681.0	720.4	.8771
710	.02425	696.5	735.1	.8913	.02414	694.9	734.2	.8897	.02403	693.4	733.4	.8883
720	.02468	709.2	748.5	.9027	.02455	707.5	747.5	.9011	.02443	705.9	746.6	.8995
730	.02513	722.2	762.2	.9142	.02500	720.4	761.1	.9125	.02486	718.6	760.0	.9108
740	.02562	735.4	776.1	.9259	.02547	733.4	774.9	.9241	.02532	731.5	773.7	.9223
750	.02615	748.8	790.4	.9377	.02598	746.7	788.9	.9358	.02581	744.6	787.6	.9338
760	.02671	762.4	804.9	.9497	.02652	760.1	803.3	.9476	.02634	757.9	801.8	.9455
770	.02733	776.3	819.8	.9619	.02711	773.8	818.0	.9596	.02691	771.5	816.3	.9573
780	.02799	790.5	835.1	.9742	.02775	787.8	833.0	.9717	.02752	785.2	831.1	.9693
790	.02871	805.0	850.7	.9868	.02844	802.1	848.4	.9841	.02818	799.2	846.2	.9815
800	.02950	819.8	866.8	.9996	.02918	816.6	864.1	.9966	.02889	813.5	861.6	.9938
810	.03036	834.9	883.2	1.0126	.03000	831.4	880.2	1.0094	.02966	828.1	877.5	1.0063
820	.03130	850.4	900.2	1.0259	.03088	846.5	896.8	1.0224	.03050	842.9	893.7	1.0190
830	.03233	866.1	917.6	1.0394	.03185	861.9	913.8	1.0356	.03141	858.0	910.3	1.0320
840	.03346	882.2	935.4	1.0532	.03291	877.6	931.2	1.0490	.03240	873.3	927.3	1.0451
850	.03469	898.5	953.7	1.0672	.03405	893.5	949.0	1.0627	.03348	888.9	944.6	1.0584
860	.03602	915.0	972.4	1.0814	.03529	909.7	967.2	1.0765	.03464	904.7	962.4	1.0719
870	.03746	931.7	991.3	1.0957	.03663	926.0	985.6	1.0904	.03589	920.6	980.4	1.0855
880	.03899	948.4	1010.4	1.1100	.03806	942.3	1004.3	1.1044	.03723	936.6	998.6	1.0991
890	.04062	964.9	1029.6	1.1243	.03958	958.6	1023.0	1.1183	.03865	952.6	1016.9	1.1128
900	.04232	981.3	1048.7	1.1384	.04118	974.7	1041.8	1.1322	.04014	968.5	1035.3	1.1263
910	.04409	997.4	1067.6	1.1522	.04284	990.6	1060.4	1.1458	.04170	984.2	1053.6	1.1397
920	.04590	1013.1	1086.1	1.1658	.04455	1006.2	1078.7	1.1592	.04332	999.6	1071.8	1.1529
930	.04774	1028.4	1104.3	1.1789	.04629	1021.4	1096.8	1.1722	.04497	1014.7	1089.6	1.1658
940	.04960	1043.1	1122.1	1.1916	.04807	1036.2	1114.4	1.1849	.04666	1029.5	1107.2	1.1784
950	.05147	1057.4	1139.3	1.2039	.04985	1050.5	1131.7	1.1971	.04837	1043.8	1124.3	1.1906
960	.05333	1071.2	1156.0	1.2157	.05164	1064.3	1148.4	1.2090	.05008	1057.7	1141.1	1.2025
970	.05518	1084.4	1172.2	1.2271	.05342	1077.7	1164.7	1.2204	.05180	1071.1	1157.4	1.2139
980	.05701	1097.2	1187.9	1.2380	.05519	1090.6	1180.4	1.2314	.05351	1084.1	1173.2	1.2250
990	.05882	1109.5	1203.1	1.2485	.05695	1103.0	1195.8	1.2420	.05521	1096.7	1188.6	1.2356
1000	.06060	1121.4	1217.8	1.2586	.05868	1115.1	1210.6	1.2522	.05689	1108.8	1203.6	1.2459
1020	.06408	1144.0	1246.0	1.2778	.06207	1137.9	1239.0	1.2715	.06019	1132.0	1232.2	1.2654
1040	.06745	1165.1	1272.5	1.2956	.06536	1159.4	1265.8	1.2895	.06340	1153.7	1259.3	1.2836
1060	.07069	1185.1	1297.6	1.3122	.06855	1179.7	1291.3	1.3064	.06652	1174.3	1285.1	1.3006
1080	.07383	1204.0	1321.5	1.3279	.07162	1198.9	1315.5	1.3222	.06954	1193.8	1309.6	1.3167
1100	.07686	1222.0	1344.4	1.3426	.07460	1217.2	1338.6	1.3371	.07246	1212.3	1333.0	1.3318
1120	.07979	1239.3	1366.3	1.3566	.07748	1234.7	1360.9	1.3513	.07529	1230.1	1355.5	1.3461
1140	.08263	1255.9	1387.4	1.3699	.08027	1251.5	1382.2	1.3648	.07804	1247.1	1377.1	1.3597
1160	.08538	1272.0	1407.9	1.3826	.08298	1267.8	1402.9	1.3776	.08070	1263.6	1398.0	1.3727
1180	.08805	1287.5	1427.7	1.3947	.08561	1283.5	1422.9	1.3899	.08329	1279.5	1418.2	1.3851
1200	.09066	1302.6	1446.9	1.4064	.08818	1298.8	1442.4	1.4017	.08582	1295.0	1437.9	1.3970
1220	.09319	1317.3	1465.6	1.4176	.09067	1313.7	1461.3	1.4130	.08828	1310.0	1457.0	1.4085
1240	.09567	1331.7	1483.9	1.4284	.09311	1328.2	1479.8	1.4239	.09067	1324.7	1475.7	1.4195
1260	.09808	1345.7	1501.8	1.4389	.09549	1342.4	1497.9	1.4345	.09302	1339.0	1493.9	1.4302
1280	.10044	1359.5	1519.3	1.4490	.09781	1356.2	1515.5	1.4447	.09531	1353.0	1511.7	1.4405
1300	.10276	1373.0	1536.5	1.4588	.10009	1369.9	1532.8	1.4546	.09755	1366.8	1529.2	1.4505
1350	.10834	1405.7	1578.1	1.4821	.10559	1402.8	1574.8	1.4781	.10296	1400.0	1571.5	1.4742
1400	.11368	1437.2	1618.1	1.5039	.11084	1434.6	1615.1	1.5001	.10813	1432.0	1612.1	1.4963
1450	.11881	1467.6	1656.7	1.5244	.11588	1465.3	1654.0	1.5208	.11309	1462.9	1651.3	1.5171
1500	.12376	1497.3	1694.2	1.5438	.12075	1495.1	1691.8	1.5403	.11788	1493.0	1689.3	1.5368
1600	.13323	1554.5	1766.6	1.5798	.13006	1552.7	1764.5	1.5765	.12704	1550.8	1762.4	1.5731
1800	.15096	1663.3	1903.5	1.6433	.14746	1661.8	1901.9	1.6402	.14413	1660.3	1900.4	1.6371
2000	.16773	1767.5	2034.5	1.6989	.16392	1766.2	2033.2	1.6958	.16028	1764.9	2031.9	1.6929
2200	.18406	1869.8	2162.7	1.7490	.17992	1868.6	2161.5	1.7460	.17597	1867.3	2160.4	1.7431
2400	.20021	1971.7	2290.3	1.7953	.19576	1970.4	2289.2	1.7923	.19150	1969.1	2288.0	1.7894

	9200				**9400**				**9600**			**p**
v	u	h	s	v	u	h	s	v	u	h	s	t
.02199	620.6	658.0	.8214	.02193	619.5	657.7	.8204	.02187	618.6	657.4	.8194	650
.02227	632.1	670.0	.8322	.02221	631.0	669.7	.8311	.02214	630.0	669.3	.8301	660
.02257	643.8	682.2	.8430	.02250	642.6	681.8	.8419	.02243	641.5	681.4	.8408	670
.02288	655.6	694.5	.8539	.02280	654.4	694.0	.8527	.02273	653.2	693.6	.8515	680
.02321	667.5	707.0	.8648	.02312	666.2	706.5	.8636	.02304	665.0	705.9	.8623	690
.02356	679.6	719.7	.8758	.02347	678.3	719.1	.8745	.02338	676.9	718.4	.8732	700
.02393	691.9	732.6	.8869	.02383	690.4	731.9	.8855	.02373	689.0	731.1	.8841	710
.02432	704.3	745.7	.8980	.02421	702.7	744.8	.8965	.02410	701.2	744.0	.8951	720
.02474	716.9	759.0	.9092	.02462	715.2	758.0	.9077	.02450	713.6	757.1	.9061	730
.02518	729.7	772.5	.9206	.02505	727.9	771.4	.9189	.02492	726.1	770.4	.9172	740
.02566	742.6	786.3	.9320	.02551	740.7	785.1	.9302	.02537	738.9	783.9	.9285	750
.02617	755.8	800.3	.9435	.02600	753.7	799.0	.9416	.02585	751.7	797.7	.9398	760
.02672	769.2	814.6	.9552	.02653	766.9	813.1	.9532	.02636	764.8	811.6	.9512	770
.02730	782.7	829.2	.9670	.02710	780.4	827.5	.9648	.02690	778.1	825.9	.9627	780
.02793	796.6	844.1	.9790	.02770	794.0	842.2	.9766	.02749	791.5	840.3	.9743	790
.02861	810.6	859.3	.9911	.02836	807.8	857.2	.9886	.02811	805.2	855.1	.9861	800
.02935	824.9	874.9	1.0034	.02906	821.9	872.4	1.0006	.02878	819.0	870.2	.9980	810
.03015	839.4	890.8	1.0159	.02982	836.2	888.1	1.0129	.02951	833.1	885.5	1.0101	820
.03101	854.2	907.0	1.0285	.03063	850.7	904.0	1.0253	.03029	847.4	901.2	1.0222	830
.03194	869.3	923.7	1.0414	.03152	865.5	920.3	1.0379	.03112	861.9	917.2	1.0346	840
.03295	884.5	940.6	1.0544	.03247	880.4	936.9	1.0506	.03202	876.5	933.4	1.0471	850
.03404	900.0	957.9	1.0675	.03349	895.6	953.8	1.0635	.03299	891.4	950.0	1.0597	860
.03521	915.6	975.5	1.0808	.03460	910.8	971.0	1.0765	.03403	906.4	966.8	1.0724	870
.03647	931.2	993.3	1.0942	.03577	926.2	988.4	1.0895	.03514	921.5	983.9	1.0852	880
.03780	946.9	1011.3	1.1075	.03703	941.6	1006.0	1.1026	.03632	936.6	1001.2	1.0980	890
.03920	962.6	1029.3	1.1208	.03835	957.0	1023.7	1.1157	.03757	951.8	1018.5	1.1108	900
.04067	978.1	1047.3	1.1340	.03973	972.3	1041.4	1.1286	.03887	966.8	1035.9	1.1235	910
.04220	993.4	1065.2	1.1470	.04117	987.4	1059.0	1.1415	.04024	981.8	1053.3	1.1362	920
.04377	1008.4	1082.9	1.1598	.04266	1002.3	1076.5	1.1541	.04165	996.5	1070.5	1.1486	930
.04537	1023.0	1100.3	1.1723	.04419	1016.9	1093.8	1.1664	.04310	1011.0	1087.6	1.1609	940
.04700	1037.3	1117.4	1.1844	.04574	1031.2	1110.7	1.1785	.04458	1025.2	1104.4	1.1729	950
.04864	1051.3	1134.1	1.1962	.04732	1045.1	1127.4	1.1903	.04609	1039.1	1121.0	1.1846	960
.05029	1064.7	1150.4	1.2077	.04890	1058.6	1143.6	1.2017	.04761	1052.6	1137.2	1.1959	970
.05194	1077.8	1166.3	1.2188	.05049	1071.7	1159.5	1.2128	.04914	1065.8	1153.1	1.2070	980
.05358	1090.5	1181.7	1.2295	.05207	1084.5	1175.0	1.2235	.05066	1078.6	1168.6	1.2177	990
.05521	1102.8	1196.8	1.2398	.05365	1096.8	1190.1	1.2339	.05219	1091.0	1183.7	1.2282	1000
.05843	1126.2	1225.6	1.2594	.05677	1120.4	1219.2	1.2537	.05522	1114.8	1212.9	1.2480	1020
.06156	1148.2	1253.0	1.2778	.05983	1142.7	1246.7	1.2722	.05820	1137.3	1240.7	1.2666	1040
.06461	1169.0	1279.0	1.2950	.06280	1163.7	1273.0	1.2895	.06110	1158.5	1267.1	1.2842	1060
.06757	1188.7	1303.7	1.3112	.06570	1183.7	1298.0	1.3059	.06394	1178.7	1292.3	1.3007	1080
.07043	1207.5	1327.4	1.3265	.06851	1202.7	1321.9	1.3213	.06670	1198.0	1316.5	1.3162	1100
.07321	1225.5	1350.1	1.3410	.07125	1220.9	1344.9	1.3359	.06938	1216.4	1339.7	1.3310	1120
.07591	1242.8	1372.0	1.3547	.07390	1238.4	1367.0	1.3499	.07199	1234.1	1362.0	1.3451	1140
.07854	1259.4	1393.1	1.3679	.07648	1255.3	1388.3	1.3631	.07452	1251.1	1383.5	1.3584	1160
.08109	1275.5	1413.6	1.3804	.07899	1271.6	1409.0	1.3758	.07699	1267.6	1404.4	1.3712	1180
.08357	1291.2	1433.4	1.3925	.08144	1287.4	1429.0	1.3880	.07940	1283.6	1424.6	1.3835	1200
.08599	1306.4	1452.8	1.4040	.08382	1302.7	1448.5	1.3996	.08175	1299.1	1444.3	1.3953	1220
.08836	1321.2	1471.6	1.4152	.08615	1317.7	1467.5	1.4109	.08404	1314.2	1463.5	1.4067	1240
.09066	1335.6	1490.0	1.4259	.08842	1332.3	1486.1	1.4217	.08628	1329.0	1482.2	1.4176	1260
.09292	1349.8	1508.0	1.4363	.09064	1346.6	1504.3	1.4322	.08847	1343.4	1500.6	1.4282	1280
.09513	1363.7	1525.6	1.4464	.09281	1360.6	1522.0	1.4424	.09061	1357.5	1518.5	1.4384	1300
.10046	1397.2	1568.3	1.4703	.09806	1394.4	1565.0	1.4665	.09578	1391.6	1561.8	1.4627	1350
.10555	1429.5	1609.2	1.4926	.10308	1426.9	1606.2	1.4889	.10072	1424.4	1603.3	1.4853	1400
.11043	1460.6	1648.6	1.5135	.10789	1458.2	1645.9	1.5100	.10546	1455.9	1643.2	1.5065	1450
.11515	1490.8	1686.8	1.5333	.11253	1488.6	1684.4	1.5299	.11002	1486.5	1681.9	1.5265	1500
.12414	1548.9	1760.3	1.5699	.12138	1547.1	1758.2	1.5667	.11874	1545.2	1756.1	1.5635	1600
.14094	1658.8	1898.8	1.6341	.13789	1657.4	1897.2	1.6311	.13498	1655.9	1895.7	1.6282	1800
.15680	1763.6	2030.6	1.6900	.15347	1762.3	2029.3	1.6871	.15028	1761.1	2028.0	1.6843	2000
.17220	1866.1	2159.2	1.7403	.16859	1864.8	2158.1	1.7375	.16513	1863.6	2156.9	1.7347	2200
.18743	1967.8	2286.9	1.7866	.18354	1966.6	2285.8	1.7838	.17981	1965.3	2284.7	1.7810	2400

Table 3. Vapor

p	9800				10 000				10 500			
t	v	u	h	s	v	u	h	s	v	u	h	s
650	.02181	617.6	657.2	.8184	.02176	616.6	656.9	.8175	.02162	614.3	656.3	.8152
660	.02208	629.0	669.0	.8291	.02202	628.0	668.7	.8281	.02187	625.5	668.0	.8256
670	.02236	640.4	681.0	.8397	.02229	639.4	680.6	.8387	.02214	636.8	679.8	.8361
680	.02265	652.0	693.1	.8504	.02258	650.9	692.7	.8493	.02241	648.2	691.7	.8466
690	.02296	663.8	705.4	.8612	.02289	662.6	704.9	.8600	.02270	659.7	703.8	.8572
700	.02329	675.6	717.9	.8719	.02321	674.3	717.3	.8707	.02301	671.3	716.0	.8677
710	.02364	687.6	730.5	.8828	.02354	686.3	729.8	.8815	.02333	683.0	728.3	.8783
720	.02400	699.7	743.3	.8937	.02390	698.3	742.5	.8923	.02367	694.9	740.8	.8890
730	.02439	712.0	756.3	.9046	.02428	710.5	755.4	.9032	.02402	706.8	753.5	.8997
740	.02480	724.5	769.4	.9157	.02468	722.8	768.5	.9141	.02440	718.9	766.3	.9104
750	.02523	737.1	782.8	.9268	.02510	735.3	781.8	.9251	.02480	731.2	779.3	.9212
760	.02570	749.8	796.4	.9380	.02555	748.0	795.2	.9362	.02522	743.5	792.5	.9321
770	.02619	762.8	810.2	.9493	.02603	760.8	808.9	.9474	.02566	756.1	805.9	.9430
780	.02672	775.9	824.3	.9606	.02654	773.7	822.8	.9587	.02613	768.7	819.5	.9540
790	.02728	789.1	838.6	.9721	.02708	786.9	837.0	.9700	.02663	781.5	833.2	.9651
800	.02788	802.6	853.2	.9838	.02766	800.2	851.4	.9815	.02717	794.4	847.2	.9762
810	.02853	816.3	868.0	.9955	.02828	813.7	866.0	.9931	.02773	807.5	861.4	.9874
820	.02922	830.1	883.1	1.0073	.02895	827.3	880.9	1.0047	.02833	820.8	875.8	.9987
830	.02996	844.2	898.5	1.0193	.02966	841.2	896.1	1.0165	.02897	834.2	890.5	1.0101
840	.03076	858.4	914.2	1.0314	.03042	855.2	911.5	1.0285	.02966	847.7	905.3	1.0216
850	.03161	872.9	930.2	1.0437	.03123	869.4	927.2	1.0405	.03038	861.4	920.4	1.0331
860	.03253	887.5	946.5	1.0561	.03210	883.7	943.1	1.0526	.03116	875.2	935.7	1.0448
870	.03351	902.2	963.0	1.0685	.03303	898.2	959.3	1.0649	.03198	889.1	951.2	1.0565
880	.03456	917.0	979.7	1.0810	.03402	912.8	975.7	1.0772	.03285	903.1	967.0	1.0683
890	.03567	931.9	996.6	1.0936	.03508	927.4	992.3	1.0895	.03378	917.2	982.9	1.0801
900	.03685	946.8	1013.6	1.1062	.03619	942.1	1009.1	1.1019	.03476	931.4	998.9	1.0919
910	.03809	961.7	1030.7	1.1187	.03736	956.8	1025.9	1.1142	.03579	945.5	1015.0	1.1038
920	.03938	976.4	1047.8	1.1312	.03859	971.3	1042.7	1.1264	.03686	959.6	1031.2	1.1156
930	.04072	991.0	1064.9	1.1435	.03986	985.8	1059.5	1.1385	.03799	973.7	1047.5	1.1273
940	.04210	1005.4	1081.7	1.1556	.04117	1000.0	1076.2	1.1505	.03915	987.6	1063.7	1.1389
950	.04351	1019.5	1098.4	1.1674	.04252	1014.1	1092.7	1.1623	.04035	1001.3	1079.8	1.1504
960	.04495	1033.3	1114.9	1.1791	.04390	1027.8	1109.1	1.1738	.04158	1014.9	1095.7	1.1616
970	.04641	1046.9	1131.0	1.1904	.04530	1041.3	1125.1	1.1851	.04284	1028.3	1111.5	1.1727
980	.04788	1060.0	1146.9	1.2015	.04671	1054.5	1140.9	1.1961	.04412	1041.4	1127.1	1.1836
990	.04935	1072.9	1162.4	1.2122	.04813	1067.3	1156.4	1.2068	.04541	1054.2	1142.5	1.1942
1000	.05083	1085.4	1177.5	1.2226	.04955	1079.9	1171.6	1.2172	.04672	1066.8	1157.6	1.2046
1020	.05376	1109.3	1206.8	1.2425	.05240	1103.9	1200.9	1.2372	.04934	1091.0	1186.9	1.2246
1040	.05666	1132.0	1234.7	1.2613	.05522	1126.8	1228.9	1.2560	.05196	1114.2	1215.1	1.2435
1060	.05950	1153.4	1261.3	1.2789	.05798	1148.4	1255.7	1.2738	.05455	1136.2	1242.2	1.2615
1080	.06227	1173.8	1286.8	1.2955	.06069	1169.0	1281.3	1.2905	.05711	1157.2	1268.2	1.2785
1100	.06498	1193.3	1311.1	1.3113	.06334	1188.7	1305.9	1.3064	.05962	1177.3	1293.2	1.2946
1120	.06761	1211.9	1334.5	1.3262	.06592	1207.5	1329.5	1.3214	.06207	1196.6	1317.2	1.3099
1140	.07017	1229.8	1357.1	1.3403	.06844	1225.6	1352.2	1.3357	.06447	1215.1	1340.4	1.3245
1160	.07266	1247.0	1378.8	1.3539	.07089	1243.0	1374.2	1.3493	.06682	1232.9	1362.8	1.3384
1180	.07509	1263.7	1399.9	1.3668	.07328	1259.8	1395.4	1.3624	.06911	1250.2	1384.5	1.3517
1200	.07746	1279.8	1420.3	1.3792	.07561	1276.1	1416.0	1.3749	.07135	1266.9	1405.5	1.3644
1220	.07977	1295.5	1440.2	1.3911	.07789	1291.9	1436.1	1.3869	.07354	1283.0	1425.9	1.3767
1240	.08203	1310.8	1459.6	1.4025	.08011	1307.3	1455.6	1.3984	.07567	1298.8	1445.8	1.3884
1260	.08423	1325.7	1478.4	1.4136	.08228	1322.3	1474.6	1.4096	.07776	1314.1	1465.2	1.3998
1280	.08639	1340.2	1496.9	1.4242	.08440	1337.0	1493.2	1.4203	.07981	1329.1	1484.2	1.4108
1300	.08850	1354.4	1514.9	1.4346	.08648	1351.4	1511.4	1.4307	.08181	1343.8	1502.8	1.4214
1350	.09359	1388.8	1558.6	1.4590	.09150	1386.1	1555.4	1.4554	.08666	1379.1	1547.5	1.4464
1400	.09846	1421.8	1600.4	1.4818	.09630	1419.3	1597.5	1.4783	.09128	1412.9	1590.3	1.4698
1450	.10313	1453.6	1640.6	1.5031	.10090	1451.2	1637.9	1.4998	.09572	1445.4	1631.4	1.4916
1500	.10763	1484.3	1679.5	1.5232	.10533	1482.1	1677.1	1.5200	.09999	1476.8	1671.0	1.5121
1600	.11621	1543.3	1754.1	1.5604	.11378	1541.5	1752.0	1.5573	.10813	1536.8	1746.9	1.5499
1800	.13218	1654.4	1894.1	1.6253	.12950	1652.9	1892.6	1.6225	.12326	1649.2	1888.7	1.6156
2000	.14723	1759.8	2026.8	1.6816	.14430	1758.5	2025.5	1.6789	.13747	1755.2	2022.3	1.6723
2200	.16182	1862.4	2155.8	1.7320	.15864	1861.1	2154.7	1.7294	.15123	1858.0	2151.9	1.7229
2400	.17624	1964.0	2283.6	1.7783	.17281	1962.7	2282.5	1.7757	.16482	1959.6	2279.8	1.7693

		11 000				11 500				12 000		p
v	u	h	s	v	u	h	s	v	u	h	s	t
.02149	612.1	655.9	.8130	.02137	610.0	655.5	.8108	.02125	608.0	655.1	.8087	**650**
.02173	623.2	667.4	.8233	.02160	620.9	666.9	.8211	.02147	618.8	666.5	.8189	**660**
.02199	634.3	679.1	.8337	.02184	632.0	678.5	.8313	.02171	629.7	677.9	.8291	**670**
.02225	645.6	690.9	.8441	.02210	643.1	690.1	.8416	.02195	640.7	689.5	.8393	**680**
.02253	656.9	702.8	.8545	.02236	654.3	701.9	.8519	.02221	651.8	701.1	.8494	**690**
.02282	668.4	714.8	.8649	.02264	665.6	713.8	.8622	.02248	663.0	712.9	.8596	**700**
.02312	679.9	727.0	.8754	.02293	677.0	725.8	.8725	.02276	674.2	724.8	.8698	**710**
.02345	691.6	739.3	.8859	.02324	688.5	738.0	.8829	.02305	685.6	736.8	.8801	**720**
.02378	703.4	751.8	.8964	.02356	700.1	750.3	.8933	.02335	697.0	748.9	.8903	**730**
.02414	715.3	764.4	.9070	.02390	711.8	762.7	.9037	.02368	708.6	761.2	.9005	**740**
.02451	727.3	777.2	.9176	.02425	723.7	775.3	.9141	.02401	720.2	773.5	.9108	**750**
.02491	739.4	790.1	.9282	.02463	735.6	788.0	.9246	.02436	732.0	786.0	.9211	**760**
.02533	751.7	803.2	.9389	.02502	747.6	800.8	.9351	.02474	743.8	798.7	.9315	**770**
.02577	764.1	816.5	.9497	.02543	759.7	813.9	.9456	.02512	755.7	811.5	.9418	**780**
.02623	776.6	830.0	.9605	.02587	772.0	827.0	.9562	.02553	767.7	824.4	.9522	**790**
.02672	789.2	843.6	.9713	.02632	784.3	840.4	.9668	.02596	779.8	837.5	.9626	**800**
.02724	801.9	857.4	.9822	.02681	796.8	853.8	.9775	.02641	792.0	850.7	.9730	**810**
.02779	814.8	871.4	.9932	.02732	809.4	867.5	.9882	.02689	804.3	864.0	.9835	**820**
.02838	827.8	885.6	1.0043	.02785	822.0	881.3	.9989	.02738	816.7	877.5	.9940	**830**
.02900	840.9	900.0	1.0154	.02842	834.8	895.3	1.0097	.02791	829.2	891.2	1.0046	**840**
.02965	854.2	914.6	1.0266	.02902	847.7	909.5	1.0206	.02846	841.8	905.0	1.0151	**850**
.03035	867.5	929.3	1.0378	.02965	860.7	923.8	1.0315	.02904	854.4	918.9	1.0257	**860**
.03109	881.0	944.3	1.0491	.03032	873.7	938.3	1.0424	.02965	867.1	933.0	1.0364	**870**
.03187	894.6	959.4	1.0604	.03103	886.9	952.9	1.0534	.03029	879.9	947.2	1.0470	**880**
.03269	908.2	974.7	1.0718	.03177	900.1	967.7	1.0644	.03097	892.8	961.5	1.0577	**890**
.03356	921.8	990.2	1.0832	.03255	913.3	982.6	1.0754	.03168	905.6	976.0	1.0684	**900**
.03447	935.5	1005.7	1.0946	.03337	926.6	997.6	1.0864	.03242	918.6	990.6	1.0790	**910**
.03543	949.2	1021.3	1.1060	.03423	939.9	1012.7	1.0974	.03319	931.5	1005.2	1.0897	**920**
.03643	962.9	1037.0	1.1173	.03512	953.2	1027.9	1.1084	.03400	944.5	1020.0	1.1003	**930**
.03747	976.4	1052.7	1.1285	.03605	966.4	1043.1	1.1193	.03484	957.4	1034.7	1.1109	**940**
.03854	989.9	1068.4	1.1397	.03702	979.6	1058.4	1.1301	.03572	970.2	1049.6	1.1215	**950**
.03965	1003.2	1084.0	1.1507	.03801	992.7	1073.6	1.1409	.03662	983.0	1064.4	1.1319	**960**
.04078	1016.4	1099.4	1.1616	.03904	1005.6	1088.7	1.1515	.03755	995.8	1079.1	1.1423	**970**
.04194	1029.4	1114.8	1.1722	.04009	1018.4	1103.7	1.1619	.03850	1008.3	1093.9	1.1526	**980**
.04312	1042.1	1129.9	1.1827	.04116	1031.0	1118.6	1.1723	.03948	1020.8	1108.5	1.1627	**990**
.04431	1054.7	1144.9	1.1930	.04225	1043.5	1133.4	1.1824	.04048	1033.1	1123.0	1.1727	**1000**
.04672	1079.0	1174.1	1.2129	.04447	1067.7	1162.3	1.2021	.04252	1057.2	1151.6	1.1922	**1020**
.04915	1102.3	1202.3	1.2319	.04672	1091.1	1190.5	1.2210	.04460	1080.6	1179.6	1.2110	**1040**
.05157	1124.6	1229.6	1.2499	.04898	1113.6	1217.8	1.2391	.04671	1103.1	1206.9	1.2290	**1060**
.05398	1146.0	1255.8	1.2671	.05123	1135.2	1244.2	1.2564	.04882	1124.9	1233.3	1.2463	**1080**
.05635	1166.4	1281.1	1.2834	.05347	1155.9	1269.7	1.2728	.05093	1145.8	1258.9	1.2628	**1100**
.05868	1186.0	1305.5	1.2989	.05568	1175.8	1294.3	1.2885	.05302	1166.0	1283.8	1.2786	**1120**
.06096	1204.9	1329.0	1.3137	.05785	1195.1	1318.2	1.3035	.05508	1185.5	1307.8	1.2938	**1140**
.06320	1223.1	1351.8	1.3279	.05999	1213.6	1341.3	1.3179	.05712	1204.3	1331.2	1.3083	**1160**
.06540	1240.7	1373.9	1.3414	.06209	1231.5	1363.6	1.3316	.05912	1222.5	1353.8	1.3222	**1180**
.06755	1257.8	1395.3	1.3544	.06414	1248.9	1385.4	1.3448	.06109	1240.2	1375.9	1.3355	**1200**
.06965	1274.3	1416.1	1.3669	.06616	1265.7	1406.5	1.3574	.06303	1257.3	1397.3	1.3484	**1220**
.07170	1290.4	1436.3	1.3789	.06814	1282.1	1427.1	1.3696	.06493	1274.0	1418.2	1.3607	**1240**
.07372	1306.0	1456.1	1.3904	.07007	1298.1	1447.2	1.3814	.06679	1290.2	1438.5	1.3727	**1260**
.07569	1321.3	1475.4	1.4016	.07197	1313.6	1466.8	1.3927	.06862	1306.1	1458.4	1.3842	**1280**
.07762	1336.3	1494.3	1.4124	.07384	1328.8	1486.0	1.4037	.07042	1321.5	1477.9	1.3953	**1300**
.08229	1372.3	1539.8	1.4379	.07835	1365.4	1532.2	1.4296	.07477	1358.7	1524.7	1.4215	**1350**
.08676	1406.6	1583.2	1.4615	.08266	1400.3	1576.2	1.4536	.07894	1394.1	1569.4	1.4459	**1400**
.09104	1439.6	1624.9	1.4837	.08680	1433.8	1618.5	1.4760	.08295	1428.1	1612.3	1.4686	**1450**
.09517	1471.4	1665.1	1.5044	.09079	1466.0	1659.2	1.4971	.08681	1460.7	1653.5	1.4899	**1500**
.10303	1532.2	1741.9	1.5427	.09839	1527.5	1736.9	1.5357	.09416	1522.9	1732.0	1.5290	**1600**
.11761	1645.5	1884.9	1.6090	.11247	1641.8	1881.2	1.6026	.10778	1638.1	1877.4	1.5965	**1800**
.13129	1752.0	2019.2	1.6660	.12566	1748.8	2016.2	1.6599	.12051	1745.5	2013.1	1.6540	**2000**
.14452	1854.9	2149.1	1.7167	.13840	1851.8	2146.4	1.7108	.13281	1848.7	2143.6	1.7050	**2200**
.15758	1956.4	2277.1	1.7632	.15097	1953.2	2274.5	1.7572	.14493	1950.1	2271.9	1.7515	**2400**

Table 3. Vapor

p	12 500				13 000				13 500			
t	v	u	h	s	v	u	h	s	v	u	h	s
650	.02114	606.0	654.9	.8067	.02103	604.1	654.7	.8048	.02092	602.3	654.5	.8029
660	.02135	616.7	666.1	.8168	.02124	614.8	665.8	.8148	.02113	612.8	665.6	.8129
670	.02158	627.6	677.5	.8269	.02146	625.5	677.1	.8248	.02134	623.5	676.8	.8228
680	.02181	638.5	688.9	.8370	.02168	636.3	688.4	.8348	.02156	634.2	688.0	.8327
690	.02206	649.4	700.4	.8471	.02192	647.1	699.9	.8448	.02179	644.9	699.3	.8426
700	.02232	660.5	712.1	.8572	.02217	658.1	711.4	.8548	.02203	655.7	710.8	.8525
710	.02259	671.6	723.8	.8672	.02243	669.1	723.0	.8648	.02227	666.6	722.3	.8624
720	.02287	682.8	735.7	.8773	.02270	680.2	734.7	.8748	.02253	677.6	733.9	.8723
730	.02316	694.1	747.7	.8875	.02298	691.3	746.6	.8847	.02280	688.7	745.6	.8821
740	.02347	705.5	759.8	.8976	.02327	702.6	758.5	.8947	.02308	699.8	757.4	.8920
750	.02379	717.0	772.0	.9077	.02357	713.9	770.6	.9048	.02338	710.9	769.3	.9019
760	.02412	728.5	784.3	.9179	.02389	725.3	782.8	.9148	.02368	722.2	781.4	.9118
770	.02447	740.2	796.8	.9280	.02423	736.8	795.0	.9248	.02400	733.5	793.5	.9217
780	.02484	751.9	809.4	.9382	.02458	748.3	807.4	.9348	.02433	744.9	805.7	.9316
790	.02523	763.7	822.1	.9484	.02494	759.9	819.9	.9449	.02468	756.4	818.0	.9415
800	.02563	775.6	834.9	.9587	.02532	771.7	832.6	.9549	.02504	767.9	830.5	.9514
810	.02605	787.6	847.8	.9689	.02572	783.4	845.3	.9650	.02542	779.5	843.0	.9614
820	.02650	799.7	860.9	.9792	.02614	795.3	858.2	.9751	.02581	791.2	855.7	.9713
830	.02696	811.8	874.2	.9895	.02658	807.2	871.1	.9852	.02622	802.9	868.4	.9812
840	.02745	824.0	887.5	.9998	.02703	819.2	884.2	.9953	.02665	814.7	881.3	.9911
850	.02796	836.3	901.0	1.0101	.02751	831.2	897.4	1.0054	.02710	826.5	894.2	1.0010
860	.02850	848.7	914.6	1.0204	.02801	843.3	910.7	1.0155	.02757	838.4	907.3	1.0110
870	.02906	861.1	928.3	1.0308	.02853	855.5	924.1	1.0257	.02805	850.3	920.4	1.0209
880	.02965	873.6	942.1	1.0412	.02908	867.7	937.7	1.0358	.02856	862.3	933.7	1.0308
890	.03027	886.1	956.1	1.0516	.02965	880.0	951.3	1.0459	.02909	874.3	947.0	1.0407
900	.03092	898.7	970.2	1.0619	.03024	892.3	965.0	1.0561	.02965	886.4	960.4	1.0507
910	.03159	911.3	984.3	1.0723	.03087	904.6	978.8	1.0662	.03022	898.4	973.9	1.0606
920	.03230	923.9	998.6	1.0827	.03151	916.9	992.7	1.0763	.03082	910.5	987.5	1.0704
930	.03304	936.5	1012.9	1.0931	.03219	929.3	1006.7	1.0864	.03144	922.6	1001.2	1.0803
940	.03380	949.1	1027.3	1.1034	.03289	941.6	1020.8	1.0965	.03209	934.7	1014.9	1.0901
950	.03459	961.7	1041.8	1.1136	.03362	954.0	1034.8	1.1065	.03276	946.8	1028.6	1.0999
960	.03542	974.3	1056.2	1.1239	.03437	966.2	1048.9	1.1165	.03345	958.9	1042.4	1.1097
970	.03626	986.7	1070.6	1.1340	.03515	978.5	1063.0	1.1264	.03416	970.9	1056.2	1.1194
980	.03714	999.1	1085.0	1.1440	.03595	990.7	1077.1	1.1362	.03490	982.9	1070.1	1.1290
990	.03803	1011.4	1099.4	1.1540	.03677	1002.8	1091.2	1.1459	.03566	994.8	1083.9	1.1385
1000	.03895	1023.6	1113.7	1.1638	.03761	1014.8	1105.2	1.1556	.03643	1006.6	1097.6	1.1480
1020	.04082	1047.5	1141.9	1.1830	.03934	1038.4	1133.1	1.1745	.03804	1030.0	1125.0	1.1666
1040	.04275	1070.7	1169.6	1.2016	.04113	1061.5	1160.5	1.1929	.03970	1052.9	1152.1	1.1848
1060	.04472	1093.3	1196.7	1.2196	.04296	1084.0	1187.4	1.2107	.04140	1075.3	1178.8	1.2025
1080	.04669	1115.1	1223.1	1.2368	.04481	1105.9	1213.7	1.2279	.04314	1097.1	1204.9	1.2196
1100	.04868	1136.2	1248.8	1.2534	.04668	1127.1	1239.4	1.2445	.04489	1118.4	1230.5	1.2361
1120	.05065	1156.6	1273.8	1.2693	.04854	1147.6	1264.4	1.2604	.04665	1139.0	1255.5	1.2520
1140	.05261	1176.3	1298.0	1.2845	.05040	1167.5	1288.7	1.2757	.04842	1159.0	1279.9	1.2674
1160	.05455	1195.4	1321.6	1.2992	.05224	1186.7	1312.4	1.2905	.05017	1178.4	1303.7	1.2822
1180	.05646	1213.8	1344.4	1.3132	.05407	1205.4	1335.5	1.3046	.05191	1197.2	1326.9	1.2964
1200	.05835	1231.7	1366.7	1.3267	.05587	1223.5	1357.9	1.3182	.05363	1215.6	1349.5	1.3101
1220	.06020	1249.1	1388.4	1.3397	.05765	1241.2	1379.8	1.3314	.05534	1233.4	1371.6	1.3234
1240	.06203	1266.1	1409.5	1.3522	.05940	1258.3	1401.2	1.3440	.05702	1250.7	1393.2	1.3361
1260	.06382	1282.5	1430.2	1.3643	.06113	1275.0	1422.1	1.3562	.05868	1267.6	1414.2	1.3484
1280	.06558	1298.6	1450.3	1.3759	.06282	1291.3	1442.4	1.3680	.06031	1284.2	1434.8	1.3603
1300	.06732	1314.3	1470.0	1.3872	.06450	1307.2	1462.4	1.3794	.06192	1300.3	1455.0	1.3718
1350	.07152	1352.1	1517.5	1.4138	.06855	1345.5	1510.4	1.4063	.06584	1339.0	1503.5	1.3991
1400	.07555	1388.0	1562.7	1.4385	.07246	1381.9	1556.2	1.4313	.06962	1375.9	1549.8	1.4243
1450	.07943	1422.4	1606.1	1.4615	.07621	1416.7	1600.1	1.4545	.07326	1411.1	1594.2	1.4478
1500	.08317	1455.4	1647.8	1.4830	.07983	1450.2	1642.2	1.4763	.07677	1445.0	1636.8	1.4698
1600	.09029	1518.3	1727.1	1.5225	.08673	1513.7	1722.3	1.5162	.08347	1509.1	1717.6	1.5101
1800	.10347	1634.4	1873.7	1.5905	.09952	1630.7	1870.1	1.5847	.09587	1627.0	1866.5	1.5791
2000	.11579	1742.3	2010.1	1.6483	.11145	1739.0	2007.1	1.6429	.10744	1735.8	2004.2	1.6376
2200	.12768	1845.6	2141.0	1.6995	.12295	1842.5	2138.3	1.6942	.11859	1839.5	2135.7	1.6890
2400	.13938	1946.9	2269.3	1.7460	.13428	1943.8	2266.8	1.7407	.12956	1940.6	2264.3	1.7356

	14 000				**14 500**				**15 000**			**p**
v	**u**	**h**	**s**	**v**	**u**	**h**	**s**	**v**	**u**	**h**	**s**	**t**
.02082	600.5	654.4	.8011	.02073	598.8	654.4	.7993	.02063	597.1	654.4	.7976	**650**
.02102	611.0	665.4	.8110	.02092	609.2	665.3	.8091	.02082	607.5	665.3	.8073	**660**
.02123	621.5	676.5	.8208	.02112	619.7	676.3	.8189	.02101	617.9	676.2	.8171	**670**
.02144	632.1	687.7	.8306	.02132	630.2	687.4	.8287	.02121	628.3	687.2	.8267	**680**
.02166	642.8	698.9	.8405	.02154	640.8	698.5	.8384	.02142	638.8	698.3	.8364	**690**
.02189	653.5	710.2	.8503	.02176	651.4	709.8	.8481	.02164	649.3	709.4	.8461	**700**
.02213	664.3	721.7	.8601	.02199	662.1	721.1	.8579	.02186	659.9	720.6	.8557	**710**
.02238	675.2	733.2	.8699	.02223	672.8	732.5	.8676	.02209	670.6	731.9	.8653	**720**
.02264	686.1	744.8	.8797	.02248	683.7	744.0	.8773	.02233	681.3	743.3	.8749	**730**
.02291	697.1	756.4	.8894	.02274	694.5	755.6	.8869	.02258	692.1	754.8	.8845	**740**
.02319	708.2	768.2	.8992	.02301	705.5	767.2	.8966	.02284	702.9	766.3	.8941	**750**
.02348	719.3	780.1	.9090	.02329	716.5	779.0	.9063	.02311	713.8	777.9	.9037	**760**
.02379	730.5	792.1	.9188	.02358	727.5	790.8	.9160	.02339	724.7	789.7	.9133	**770**
.02410	741.7	804.1	.9286	.02389	738.6	802.7	.9256	.02368	735.7	801.5	.9228	**780**
.02443	753.0	816.3	.9383	.02420	749.8	814.8	.9353	.02399	746.8	813.4	.9324	**790**
.02478	764.4	828.6	.9481	.02453	761.1	826.9	.9449	.02430	757.9	825.3	.9419	**800**
.02514	775.8	840.9	.9579	.02487	772.3	839.1	.9546	.02463	769.0	837.4	.9514	**810**
.02551	787.3	853.4	.9677	.02523	783.7	851.4	.9642	.02496	780.2	849.5	.9610	**820**
.02590	798.9	866.0	.9774	.02560	795.0	863.7	.9738	.02532	791.4	861.7	.9705	**830**
.02630	810.5	878.6	.9872	.02598	806.5	876.2	.9835	.02568	802.7	874.0	.9799	**840**
.02672	822.1	891.3	.9969	.02638	817.9	888.7	.9931	.02606	814.0	886.3	.9894	**850**
.02716	833.8	904.1	1.0067	.02679	829.4	901.3	1.0027	.02645	825.4	898.8	.9989	**860**
.02762	845.5	917.1	1.0164	.02722	841.0	914.0	1.0123	.02686	836.7	911.3	1.0083	**870**
.02810	857.3	930.1	1.0262	.02767	852.6	926.8	1.0218	.02728	848.1	923.9	1.0177	**880**
.02859	869.1	943.1	1.0359	.02813	864.2	939.7	1.0314	.02772	859.6	936.5	1.0271	**890**
.02911	880.9	956.3	1.0456	.02862	875.8	952.6	1.0409	.02817	871.0	949.2	1.0365	**900**
.02964	892.7	969.5	1.0553	.02912	887.4	965.6	1.0504	.02864	882.5	962.0	1.0459	**910**
.03020	904.6	982.8	1.0650	.02964	899.1	978.6	1.0599	.02913	894.0	974.8	1.0552	**920**
.03077	916.5	996.2	1.0747	.03017	910.8	991.7	1.0694	.02963	905.5	987.7	1.0645	**930**
.03137	928.4	1009.6	1.0843	.03073	922.4	1004.9	1.0788	.03015	916.9	1000.6	1.0738	**940**
.03199	940.2	1023.1	1.0939	.03131	934.1	1018.1	1.0882	.03069	928.4	1013.6	1.0830	**950**
.03263	952.1	1036.6	1.1034	.03190	945.8	1031.4	1.0976	.03124	939.9	1026.6	1.0922	**960**
.03329	963.9	1050.1	1.1129	.03252	957.4	1044.6	1.1069	.03182	951.3	1039.6	1.1014	**970**
.03397	975.7	1063.7	1.1224	.03315	969.0	1057.9	1.1162	.03241	962.7	1052.7	1.1105	**980**
.03467	987.4	1077.2	1.1317	.03380	980.5	1071.2	1.1254	.03301	974.1	1065.8	1.1195	**990**
.03539	999.1	1090.7	1.1410	.03447	992.0	1084.5	1.1345	.03363	985.5	1078.8	1.1285	**1000**
.03688	1022.1	1117.7	1.1594	.03585	1014.8	1111.0	1.1526	.03493	1008.0	1105.0	1.1463	**1020**
.03842	1044.9	1144.4	1.1773	.03729	1037.3	1137.4	1.1703	.03627	1030.3	1130.9	1.1637	**1040**
.04002	1067.1	1170.8	1.1948	.03878	1059.4	1163.5	1.1876	.03767	1052.2	1156.7	1.1808	**1060**
.04164	1088.9	1196.8	1.2118	.04031	1081.1	1189.2	1.2044	.03910	1073.7	1182.2	1.1975	**1080**
.04330	1110.1	1222.3	1.2282	.04186	1102.2	1214.6	1.2207	.04057	1094.8	1207.4	1.2137	**1100**
.04496	1130.7	1247.2	1.2441	.04343	1122.9	1239.4	1.2366	.04205	1115.4	1232.2	1.2295	**1120**
.04663	1150.8	1271.6	1.2595	.04502	1143.0	1263.8	1.2519	.04355	1135.6	1256.5	1.2448	**1140**
.04830	1170.4	1295.5	1.2743	.04660	1162.7	1287.7	1.2668	.04506	1155.3	1280.4	1.2596	**1160**
.04996	1189.4	1318.8	1.2886	.04818	1181.8	1311.1	1.2811	.04657	1174.5	1303.7	1.2740	**1180**
.05160	1207.8	1341.5	1.3024	.04976	1200.4	1333.9	1.2949	.04808	1193.2	1326.6	1.2878	**1200**
.05324	1225.8	1363.8	1.3157	.05132	1218.5	1356.2	1.3083	.04957	1211.4	1349.1	1.3013	**1220**
.05485	1243.4	1385.5	1.3285	.05287	1236.2	1378.1	1.3213	.05106	1229.3	1371.0	1.3143	**1240**
.05645	1260.5	1406.7	1.3410	.05441	1253.5	1399.5	1.3338	.05254	1246.7	1392.5	1.3268	**1260**
.05802	1277.2	1427.5	1.3530	.05592	1270.3	1420.4	1.3459	.05400	1263.7	1413.6	1.3390	**1280**
.05957	1293.5	1447.8	1.3646	.05742	1286.8	1440.9	1.3576	.05544	1280.3	1434.2	1.3508	**1300**
.06336	1332.7	1496.8	1.3921	.06108	1326.5	1490.4	1.3853	.05898	1320.3	1484.1	1.3787	**1350**
.06701	1370.0	1543.6	1.4176	.06462	1364.2	1537.5	1.4110	.06240	1358.4	1531.6	1.4047	**1400**
.07054	1405.6	1588.4	1.4413	.06803	1400.2	1582.7	1.4350	.06572	1394.8	1577.2	1.4289	**1450**
.07395	1439.8	1631.4	1.4635	.07134	1434.7	1626.1	1.4574	.06893	1429.6	1621.0	1.4515	**1500**
.08045	1504.6	1713.0	1.5042	.07766	1500.1	1708.4	1.4984	.07507	1495.6	1704.0	1.4928	**1600**
.09250	1623.3	1862.9	1.5737	.08938	1619.6	1859.4	1.5684	.08647	1616.0	1856.0	1.5633	**1800**
.10374	1732.6	2001.3	1.6324	.10029	1729.3	1998.4	1.6274	.09709	1726.1	1995.6	1.6225	**2000**
.11455	1836.4	2133.1	1.6840	.11079	1833.3	2130.6	1.6791	.10730	1830.3	2128.1	1.6743	**2200**
.12518	1937.5	2261.8	1.7306	.12112	1934.4	2259.4	1.7258	.11734	1931.3	2257.0	1.7211	**2400**

Table 4. Liquid

p (t Sat.)	0				500 (467.13)				1000 (544.75)			
t	v	u	h	s	v	u	h	s	v	u	h	s
Sat.					.019748	447.70	449.53	.64904	.021591	538.39	542.38	.74320
32	.016022	-.01	-.01	-.00003	.015994	.00	1.49	.00000	.015967	.03	2.99	.00005
50	.016024	18.06	18.06	.03607	.015998	18.02	19.50	.03599	.015972	17.99	20.94	.03592
100	.016130	68.05	68.05	.12963	.016106	67.87	69.36	.12932	.016082	67.70	70.68	.12901
150	.016343	117.95	117.95	.21504	.016318	117.66	119.17	.21457	.016293	117.38	120.40	.21410
200	.016635	168.05	168.05	.29402	.016608	167.65	169.19	.29341	.016580	167.26	170.32	.29281
250	.017003	218.52	218.52	.36777	.016972	217.99	219.56	.36702	.016941	217.47	220.61	.36628
300	.017453	269.61	269.61	.43732	.017416	268.92	270.53	.43641	.017379	268.24	271.46	.43552
350	.018000	321.59	321.59	.50359	.017954	320.71	322.37	.50249	.017909	319.83	323.15	.50140
400	.018668	374.85	374.85	.56740	.018608	373.68	375.40	.56604	.018550	372.55	375.98	.56472
450	.019503	429.96	429.96	.62970	.019420	428.40	430.19	.62798	.019340	426.89	430.47	.62632
500	.02060	488.1	488.1	.6919	.02048	485.9	487.8	.6896	.02036	483.8	487.5	.6874
510	.02087	500.3	500.3	.7046	.02073	497.9	499.8	.7021	.02060	495.6	499.4	.6997
520	.02116	512.7	512.7	.7173	.02100	510.1	512.0	.7146	.02086	507.6	511.5	.7121
530	.02148	525.5	525.5	.7303	.02130	522.6	524.5	.7273	.02114	519.9	523.8	.7245
540	.02182	538.6	538.6	.7434	.02162	535.3	537.3	.7402	.02144	532.4	536.3	.7372
550	.02221	552.1	552.1	.7569	.02198	548.4	550.5	.7532	.02177	545.1	549.2	.7499
560	.02265	566.1	566.1	.7707	.02237	562.0	564.0	.7666	.02213	558.3	562.4	.7630
570	.02315	580.8	580.8	.7851	.02281	576.0	578.1	.7804	.02253	571.8	576.0	.7763
580					.02332	590.8	592.9	.7946	.02298	585.9	590.1	.7899
590					.02392	606.4	608.6	.8096	.02349	600.6	604.9	.8041
600									.02409	616.2	620.6	.8189
610									.02482	632.9	637.5	.8348

p (t Sat.)	1500 (596.39)				2000 (636.00)				2500 (668.31)			
t	v	u	h	s	v	u	h	s	v	u	h	s
Sat.	.023461	604.97	611.48	.80824	.025649	662.40	671.89	.86227	.028605	717.66	730.89	.91306
32	.015939	.05	4.47	.00007	.015912	.06	5.95	.00008	.015885	.08	7.43	.00009
50	.015946	17.95	22.38	.03584	.015920	17.91	23.81	.03575	.015895	17.88	25.23	.03566
100	.016058	67.53	71.99	.12870	.016034	67.37	73.30	.12839	.016010	67.20	74.61	.12808
150	.016268	117.10	121.62	.21364	.016244	116.83	122.84	.21318	.016220	116.56	124.07	.21272
200	.016554	166.87	171.46	.29221	.016527	166.49	172.60	.29162	.016501	166.11	173.75	.29104
250	.016910	216.96	221.65	.36554	.016880	216.46	222.70	.36482	.016851	215.96	223.75	.36410
300	.017343	267.58	272.39	.43463	.017308	266.93	273.33	.43376	.017274	266.29	274.28	.43290
350	.017865	318.98	323.94	.50034	.017822	318.15	324.74	.49929	.017780	317.33	325.56	.49826
400	.018493	371.45	376.59	.56343	.018439	370.38	377.21	.56216	.018386	369.34	377.84	.56092
450	.019264	425.44	430.79	.62470	.019191	424.04	431.14	.62313	.019120	422.68	431.52	.62160
500	.02024	481.8	487.4	.6853	.02014	479.8	487.3	.6832	.02004	478.0	487.3	.6813
510	.02048	493.4	499.1	.6974	.02036	491.4	498.9	.6953	.02025	489.4	498.8	.6932
520	.02072	505.3	511.0	.7096	.02060	503.1	510.7	.7073	.02048	501.0	510.4	.7051
530	.02099	517.3	523.1	.7219	.02085	514.9	522.6	.7195	.02072	512.6	522.2	.7171
540	.02127	529.6	535.5	.7343	.02112	527.0	534.8	.7317	.02098	524.5	534.2	.7292
550	.02158	542.1	548.1	.7469	.02141	539.2	547.2	.7440	.02125	536.6	546.4	.7413
560	.02191	554.9	561.0	.7596	.02172	551.8	559.8	.7565	.02154	548.9	558.8	.7536
570	.02228	568.0	574.2	.7725	.02206	564.6	572.8	.7691	.02186	561.4	571.5	.7659
580	.02269	581.6	587.9	.7857	.02243	577.8	586.1	.7820	.02221	574.3	584.5	.7785
590	.02314	595.7	602.1	.7993	.02284	591.3	599.8	.7951	.02258	587.4	597.9	.7913
600	.02366	610.4	616.9	.8134	.02330	605.4	614.0	.8086	.02300	601.0	611.6	.8043
610	.02426	625.8	632.6	.8281	.02382	620.0	628.8	.8225	.02346	615.0	625.9	.8177
620	.02498	642.5	649.4	.8437	.02443	635.4	644.5	.8371	.02399	629.6	640.7	.8315
630	.02590	660.8	668.0	.8609	.02514	651.9	661.2	.8525	.02459	644.9	656.3	.8459
640					.02603	669.8	679.4	.8691	.02530	661.2	672.9	.8610
650					.02724	690.3	700.4	.8881	.02616	678.7	690.8	.8773
660									.02729	698.4	711.0	.8954
670									.02895	722.1	735.5	.9172

Table 4. Liquid

p (t Sat.)	3000 (695.52)				3204 (705.44)				3500			
t	v	u	h	s	v	u	h	s	v	u	h	s
Sat.	.034310	783.45	802.50	.97320	.050533	872.58	902.53	1.05803				
32	.015859	.09	8.90	.00009	.015848	.10	9.49	.00009	.015833	.10	10.36	.00009
50	.015870	17.84	26.65	.03555	.015860	17.82	27.23	.03551	.015845	17.80	28.06	.03545
100	.015987	67.04	75.91	.12777	.015978	66.97	76.45	.12764	.015964	66.88	77.22	.12746
150	.016196	116.30	125.29	.21226	.016187	116.19	125.79	.21208	.016173	116.03	126.51	.21181
200	.016476	165.74	174.89	.29046	.016465	165.59	175.36	.29022	.016450	165.38	176.03	.28988
250	.016822	215.47	224.81	.36340	.016810	215.27	225.24	.36311	.016793	214.99	225.87	.36269
300	.017240	265.66	275.23	.43205	.017226	265.41	275.62	.43170	.017206	265.04	276.19	.43121
350	.017739	316.53	326.38	.49725	.017722	316.21	326.71	.49684	.017699	315.75	327.21	.49625
400	.018334	368.32	378.50	.55970	.018313	367.91	378.76	.55921	.018284	367.32	379.16	.55851
450	.019053	421.36	431.93	.62011	.019026	420.83	432.11	.61952	.018987	420.08	432.37	.61866
500	.019944	476.2	487.3	.6794	.019906	475.5	487.3	.6786	.019853	474.5	487.4	.6775
520	.020367	498.9	510.2	.7030	.020324	498.1	510.2	.7022	.020262	497.0	510.1	.7010
540	.020842	522.2	533.8	.7268	.020791	521.3	533.6	.7258	.020717	520.0	533.4	.7245
560	.021382	546.2	558.0	.7508	.021319	545.1	557.7	.7497	.021231	543.6	557.3	.7482
580	.022004	571.0	583.2	.7753	.021926	569.8	582.8	.7740	.021818	568.0	582.1	.7723
600	.02274	597.0	609.6	.8004	.02264	595.5	608.9	.7989	.02250	593.4	607.9	.7969
620	.02362	624.5	637.7	.8266	.02349	622.6	636.6	.8248	.02331	620.0	635.1	.8223
640	.02475	654.3	668.0	.8545	.02456	651.8	666.4	.8521	.02431	648.4	664.2	.8489
660	.02629	687.6	702.2	.8853	.02598	684.0	699.4	.8819	.02560	679.4	696.0	.8776
680	.02879	728.4	744.3	.9226	.02813	722.0	738.7	.9167	.02741	714.6	732.4	.9098
700					.03323	779.8	799.5	.9695	.03058	759.5	779.3	.9506
710									.03408	793.9	816.0	.9821

p	4000				5000				6000			
t	v	u	h	s	v	u	h	s	v	u	h	s
32	.015807	.10	11.80	.00005	.015755	.11	14.70	-.00001	.015705	.11	17.55	-.00013
50	.015821	17.76	29.47	.03534	.015773	17.67	32.26	.03508	.015726	17.57	35.03	.03480
100	.015942	66.72	78.52	.12714	.015897	66.40	81.11	.12651	.015853	66.09	83.70	.12588
150	.016150	115.77	127.73	.21136	.016104	115.27	130.17	.21046	.016059	114.77	132.60	.20957
200	.016425	165.02	177.18	.28931	.016376	164.32	179.47	.28818	.016328	163.63	181.76	.28707
250	.016765	214.52	226.93	.36200	.016710	213.59	229.05	.36063	.016656	212.70	231.19	.35929
300	.017174	264.43	277.15	.43038	.017110	263.25	279.08	.42875	.017048	262.10	281.03	.42716
350	.017659	314.98	328.05	.49526	.017583	313.48	329.75	.49334	.017510	312.04	331.48	.49147
400	.018235	366.35	379.85	.55734	.018141	364.47	381.25	.55506	.018051	362.66	382.71	.55285
450	.018924	418.83	432.84	.61725	.018803	416.44	433.84	.61451	.018689	414.17	434.92	.61189
500	.019766	472.9	487.5	.6758	.019603	469.8	487.9	.6724	.019451	466.9	488.5	.6692
520	.020161	495.2	510.1	.6990	.019974	491.7	510.2	.6953	.019802	488.4	510.4	.6918
540	.020600	517.9	533.1	.7223	.020382	513.9	532.8	.7182	.020185	510.3	532.7	.7144
560	.021091	541.2	556.8	.7457	.020835	536.7	556.0	.7411	.020606	532.6	555.5	.7369
580	.021648	565.2	581.2	.7694	.021341	560.0	579.7	.7642	.021072	555.3	578.7	.7595
600	.02229	590.0	606.5	.7936	.02191	584.0	604.2	.7876	.02159	578.6	602.6	.7822
620	.02304	616.0	633.0	.8183	.02257	608.8	629.7	.8113	.02218	602.6	627.2	.8052
640	.02394	643.3	661.1	.8441	.02334	634.6	656.2	.8357	.02285	627.3	652.7	.8286
660	.02506	672.7	691.2	.8712	.02424	661.8	684.2	.8609	.02363	653.0	679.2	.8525
680	.02653	704.9	724.5	.9007	.02535	690.6	714.1	.8873	.02454	679.8	707.0	.8771
700	.02867	742.1	763.4	.9345	.02676	721.8	746.6	.9156	.02563	708.1	736.5	.9028
710	.03026	764.3	786.7	.9545	.02763	738.6	764.2	.9307	.02626	722.9	752.1	.9161

Table 4. Liquid

p	7000				8000				9000			
t	v	u	h	s	v	u	h	s	v	u	h	s
32	.015656	.10	20.38	-.00028	.015608	.06	23.17	-.00048	.015562	.03	25.95	-.00069
50	.015680	17.48	37.79	.03451	.015635	17.38	40.53	.03419	.015590	17.28	43.24	.03385
100	.015810	65.79	86.27	.12525	.015767	65.49	88.84	.12461	.015725	65.20	91.39	.12397
150	.016015	114.29	135.03	.20869	.015972	113.82	137.46	.20782	.015929	113.35	139.88	.20696
200	.016281	162.97	184.06	.28597	.016234	162.32	186.35	.28489	.016189	161.69	188.65	.28383
250	.016604	211.82	233.33	.35797	.016553	210.98	235.48	.35668	.016503	210.15	237.63	.35540
300	.016988	260.99	283.00	.42560	.016930	259.91	284.98	.42407	.016873	258.87	286.97	.42258
350	.017439	310.65	333.24	.48964	.017371	309.30	335.01	.48786	.017305	308.00	336.81	.48612
400	.017965	360.93	384.20	.55072	.017883	359.27	385.74	.54865	.017804	357.66	387.31	.54663
450	.018581	412.01	436.08	.60937	.018479	409.95	437.30	.60694	.018381	407.97	438.58	.60460
500	.019311	464.2	489.2	.6662	.019179	461.6	490.0	.6633	.019055	459.1	490.8	.6605
520	.019643	485.4	510.8	.6885	.019495	482.6	511.4	.6854	.019357	479.9	512.1	.6824
540	.020004	506.9	532.8	.7108	.019837	503.8	533.1	.7074	.019682	500.8	533.6	.7042
560	.020398	528.8	555.2	.7329	.020208	525.3	555.2	.7292	.020033	522.0	555.4	.7257
580	.020831	551.1	578.0	.7551	.020613	547.1	577.7	.7510	.020414	543.5	577.5	.7472
600	.02131	573.8	601.4	.7773	.02106	569.4	600.5	.7728	.02083	565.3	600.0	.7687
620	.02184	597.0	625.3	.7997	.02155	592.0	623.9	.7947	.02128	587.5	622.9	.7901
640	.02244	620.9	650.0	.8223	.02209	615.2	647.9	.8167	.02179	610.1	646.4	.8116
660	.02313	645.5	675.5	.8453	.02270	639.0	672.6	.8389	.02234	633.2	670.4	.8332
680	.02391	671.0	701.9	.8687	.02339	663.4	698.0	.8615	.02296	656.8	695.0	.8551
700	.02481	697.4	729.6	.8928	.02418	688.6	724.4	.8844	.02365	681.0	720.4	.8771
710	.02533	711.1	743.9	.9051	.02461	701.5	737.9	.8960	.02403	693.4	733.4	.8883

p	10 000				11 000				12 000			
t	v	u	h	s	v	u	h	s	v	u	h	s
32	.015516	-.01	28.70	-.00094	.015471	-.07	31.42	-.00124	.015428	-.12	34.13	-.00154
50	.015547	17.17	45.94	.03349	.015504	17.06	48.62	.03311	.015462	16.95	51.29	.03272
100	.015684	64.92	93.94	.12333	.015644	64.64	96.48	.12269	.015604	64.36	99.01	.12205
150	.015888	112.90	142.30	.20610	.015846	112.47	144.72	.20525	.015806	112.04	147.13	.20440
200	.016145	161.07	190.94	.28277	.016101	160.47	193.24	.28173	.016058	159.88	195.54	.28070
250	.016454	209.35	239.79	.35415	.016406	208.56	241.96	.35292	.016359	207.80	244.13	.35170
300	.016818	257.85	288.97	.42111	.016764	256.86	290.98	.41967	.016712	255.90	293.00	.41825
350	.017241	306.73	338.63	.48442	.017179	305.51	340.47	.48276	.017119	304.32	342.33	.48113
400	.017728	356.11	388.92	.54468	.017655	354.61	390.55	.54277	.017584	353.16	392.21	.54091
450	.018288	406.08	439.92	.60234	.018200	404.26	441.30	.60015	.018115	402.50	442.73	.59802
500	.018938	456.8	491.8	.6579	.018827	454.5	492.9	.6553	.018721	452.4	494.0	.6529
520	.019227	477.3	512.9	.6796	.019105	474.9	513.8	.6769	.018989	472.6	514.8	.6743
540	.019538	498.0	534.2	.7011	.019402	495.4	534.9	.6982	.019274	492.9	535.7	.6954
560	.019871	519.0	555.7	.7225	.019720	516.1	556.2	.7193	.019578	513.3	556.8	.7164
580	.020231	540.1	577.6	.7437	.020061	537.0	577.8	.7403	.019903	534.0	578.2	.7371
600	.02062	561.6	599.7	.7648	.02043	558.1	599.7	.7611	.02025	554.8	599.8	.7577
620	.02105	583.3	622.3	.7859	.02083	579.5	621.9	.7819	.02063	575.9	621.7	.7782
640	.02151	605.4	645.2	.8069	.02126	601.2	644.4	.8026	.02103	597.2	643.9	.7986
660	.02202	628.0	668.7	.8281	.02173	623.2	667.4	.8233	.02147	618.8	666.5	.8189
680	.02258	650.9	692.7	.8493	.02225	645.6	690.9	.8441	.02195	640.7	689.5	.8393
700	.02321	674.3	717.3	.8707	.02282	668.4	714.8	.8649	.02248	663.0	712.9	.8596
710	.02354	686.3	729.8	.8815	.02312	679.9	727.0	.8754	.02276	674.2	724.8	.8698

Table 4.   Liquid

p	13 000				14 000				15 000			
t	v	u	h	s	v	u	h	s	v	u	h	s
32	.015385	- .20	36.81	-.00188	.015344	- .26	39.49	-.00222	.015303	- .33	42.15	-.00259
50	.015421	16.84	53.94	.03232	.015381	16.73	56.58	.03191	.015341	16.63	59.21	.03149
100	.015564	64.09	101.53	.12140	.015525	63.83	104.05	.12075	.015486	63.57	106.56	.12011
150	.015765	111.62	149.54	.20356	.015726	111.20	151.94	.20272	.015687	110.80	154.34	.20189
200	.016016	159.31	197.83	.27969	.015974	158.75	200.13	.27868	.015933	158.20	202.42	.27768
250	.016313	207.06	246.30	.35050	.016268	206.33	248.48	.34932	.016224	205.62	250.65	.34815
300	.016660	254.96	295.04	.41686	.016610	254.04	297.07	.41549	.016562	253.15	299.11	.41414
350	.017060	303.16	344.20	.47954	.017004	302.03	346.08	.47797	.016949	300.93	347.98	.47643
400	.017516	351.76	393.90	.53909	.017450	350.40	395.61	.53732	.017387	349.08	397.34	.53558
450	.018033	400.81	444.19	.59596	.017955	399.18	445.69	.59395	.017879	397.60	447.23	.59199
500	.018621	450.4	495.2	.6505	.018526	448.4	496.4	.6482	.018434	446.5	497.7	.6460
520	.018880	470.4	515.8	.6718	.018776	468.3	516.9	.6694	.018676	466.3	518.1	.6670
540	.019154	490.5	536.6	.6928	.019040	488.2	537.5	.6902	.018931	486.0	538.6	.6877
560	.019445	510.8	557.5	.7135	.019319	508.3	558.3	.7108	.019200	505.9	559.2	.7082
580	.019755	531.2	578.7	.7341	.019616	528.5	579.3	.7312	.019485	526.0	580.0	.7284
600	.02009	551.7	600.1	.7544	.01993	548.8	600.5	.7513	.01979	546.1	601.0	.7484
620	.02044	572.5	621.7	.7747	.02027	569.4	621.9	.7713	.02011	566.4	622.2	.7682
640	.02082	593.5	643.6	.7948	.02063	590.1	643.5	.7912	.02045	586.8	643.6	.7878
660	.02124	614.8	665.8	.8148	.02102	611.0	665.4	.8110	.02082	607.5	665.3	.8073
680	.02168	636.3	688.4	.8348	.02144	632.1	687.7	.8306	.02121	628.3	687.2	.8267
700	.02217	658.1	711.4	.8548	.02189	653.5	710.2	.8503	.02164	649.3	709.4	.8461
710	.02243	669.1	723.0	.8648	.02213	664.3	721.7	.8601	.02186	659.9	720.6	.8557

p	16 000				18 000				20 000			
t	v	u	h	s	v	u	h	s	v	u	h	s
32	.015264	- .40	44.79	-.00296	.015188	- .54	50.05	-.00372	.015116	- .65	55.30	-.00446
50	.015303	16.52	61.82	.03107	.015227	16.32	67.04	.03021	.015154	16.14	72.23	.02936
100	.015448	63.32	109.05	.11946	.015372	62.83	114.03	.11817	.015298	62.37	118.99	.11688
150	.015648	110.41	156.74	.20106	.015572	109.64	161.51	.19941	.015497	108.91	166.26	.19778
200	.015893	157.66	204.71	.27669	.015813	156.62	209.29	.27474	.015736	155.62	213.86	.27281
250	.016181	204.92	252.83	.34699	.016096	203.57	257.19	.34472	.016015	202.28	261.55	.34250
300	.016514	252.27	301.17	.41281	.016422	250.58	305.28	.41021	.016333	248.96	309.41	.40766
350	.016895	299.86	349.89	.47492	.016792	297.80	353.73	.47197	.016693	295.83	357.61	.46911
400	.017325	347.79	399.08	.53388	.017207	345.32	402.63	.53057	.017096	342.97	406.24	.52738
450	.017807	396.07	448.80	.59008'	.017669	393.15	452.00	.58639	.017541	390.39	455.31	.58286
500	.018347	444.7	499.1	.6439	.018183	441.3	501.9	.6397	.018031	438.1	504.8	.6358
520	.018582	464.3	519.3	.6648	.018404	460.6	521.9	.6605	.018242	457.2	524.7	.6564
540	.018828	484.0	539.7	.6854	.018636	480.0	542.1	.6808	.018461	476.4	544.7	.6766
560	.019088	503.7	560.2	.7057	.018879	499.5	562.4	.7009	.018689	495.6	564.8	.6965
580	.019362	523.5	580.9	.7257	.019134	519.0	582.8	.7207	.018928	514.9	585.0	.7160
600	.01965	543.5	601.7	.7456	.01940	538.7	603.3	.7403	.01918	534.3	605.2	.7354
620	.01996	563.6	622.7	.7652	.01968	558.4	623.9	.7596	.01944	553.7	625.6	.7544
640	.02029	583.8	643.9	.7846	.01998	578.2	644.8	.7787	.01972	573.2	646.1	.7732
660	.02063	604.2	665.3	.8039	.02030	598.1	665.7	.7976	.02001	592.7	666.8	.7918
680	.02101	624.7	686.9	.8231	.02064	618.2	686.9	.8163	.02031	612.4	687.6	.8102
700	.02140	645.4	708.8	.8421	.02099	638.4	708.3	.8349	.02063	632.1	708.5	.8285
710	.02161	655.9	719.9	.8516	.02118	648.5	719.1	.8442	.02080	642.1	719.1	.8375

Table 5.   Critical Region

v (t Sat.)	**.026** (640.90)				**.027** (653.11)				**.028** (663.14)			
t	p	u	h	s	p	u	h	s	p	u	h	s
Sat.	2070.	670.1	680.1	.8694	2253.	690.2	701.5	.8879	2414.	707.9	720.4	.9042
650	2599.	676.9	689.4	.8755	*2090.*	*687.9*	*698.3*	*.8858*	*1783.2*	*697.7*	*707.0*	*.8950*
655	2893.	680.6	694.5	.8789	2353.	691.7	703.4	.8892	*2020.6*	*701.6*	*712.1*	*.8985*
660	3189.	684.2	699.6	.8821	2620.	695.4	708.5	.8926	*2261.2*	*705.5*	*717.2*	*.9020*
665	3488.	687.9	704.7	.8854	2889.	699.1	713.6	.8959	2504.7	709.4	722.3	.9054
670	3788.	691.5	709.7	.8886	3161.	702.9	718.6	.8992	2750.9	713.2	727.4	.9088
675	4091.	695.1	714.8	.8918	3435.	706.5	723.7	.9025	3000.	716.9	732.5	.9122
680	4395.	698.6	719.8	.8949	3711.	710.2	728.7	.9057	3251.	720.7	737.5	.9154
685	4701.	702.2	724.8	.8980	3990.	713.8	733.7	.9088	3505.	724.4	742.6	.9187
690	5008.	705.7	729.8	.9011	4270.	717.4	738.7	.9120	3760.	728.1	747.6	.9219
695	5317.	709.2	734.8	.9041	4551.	721.0	743.7	.9151	4018.	731.7	752.5	.9251
700	5627.	712.7	739.8	.9071	4835.	724.5	748.7	.9181	4277.	735.3	757.5	.9282
705	5938.	716.2	744.8	.9101	5119.	728.0	753.6	.9212	4538.	738.9	762.5	.9313
710	6250.	719.7	749.7	.9131	5405.	731.6	758.6	.9242	4801.	742.5	767.4	.9344
715	6563.	723.1	754.7	.9160	5693.	735.1	763.5	.9272	5066.	746.1	772.3	.9374
720	6877.	726.6	759.6	.9190	5981.	738.5	768.4	.9301	5331.	749.6	777.2	.9404
725	7191.	730.0	764.6	.9219	6271.	742.0	773.3	.9330	5598.	753.1	782.1	.9433
730	7506.	733.4	769.5	.9247	6561.	745.4	778.2	.9359	5866.	756.6	787.0	.9463
735	7822.	736.8	774.4	.9276	6852.	748.9	783.1	.9388	6135.	760.1	791.9	.9492
740	8138.	740.2	779.3	.9304	7144.	752.3	788.0	.9417	6405.	763.5	796.7	.9521
745	8454.	743.6	784.2	.9332	7437.	755.7	792.8	.9445	6676.	767.0	801.5	.9549
750	8771.	746.9	789.1	.9360	7730.	759.1	797.7	.9473	6948.	770.4	806.4	.9578

v (t Sat.)	**.029** (671.38)				**.030** (678.14)				**.032** (688.16)			
t	p	u	h	s	p	u	h	s	p	u	h	s
Sat.	2552.	723.6	737.3	.9185	2671.	737.6	752.5	.9313	2857.	761.4	778.3	.9529
650	*1612.6*	*706.6*	*715.3*	*.9033*	*1531.5*	*714.6*	*723.1*	*.9108*	*1518.5*	*728.6*	*737.6*	*.9239*
655	*1827.1*	*710.7*	*720.5*	*.9070*	*1726.5*	*718.8*	*728.4*	*.9146*	*1683.2*	*733.1*	*743.1*	*.9280*
660	*2045.0*	*714.7*	*725.7*	*.9106*	*1925.0*	*723.0*	*733.7*	*.9183*	*1851.5*	*737.6*	*748.5*	*.9319*
665	*2266.1*	*718.6*	*730.8*	*.9141*	*2126.8*	*727.1*	*738.9*	*.9220*	*2023.1*	*741.9*	*753.9*	*.9358*
670	*2490.1*	*722.6*	*735.9*	*.9176*	*2331.7*	*731.1*	*744.1*	*.9256*	*2197.8*	*746.2*	*759.3*	*.9397*
675	2717.	726.5	741.0	.9210	*2539.*	*735.1*	*749.2*	*.9291*	2376.	750.5	764.5	.9434
680	2946.	730.3	746.1	.9244	2750.	739.1	754.4	.9326	*2556.*	*754.7*	*769.8*	*.9471*
685	3178.	734.1	751.2	.9277	2963.	743.0	759.5	.9360	*2739.*	*758.8*	*775.0*	*.9507*
690	3412.	737.9	756.2	.9310	3179.	746.9	764.5	.9394	2925.	762.9	780.2	.9542
695	3649.	741.6	761.2	.9342	3397.	750.7	769.6	.9427	3114.	766.9	785.3	.9577
700	3887.	745.3	766.2	.9374	3617.	754.5	774.6	.9460	3304.	770.9	790.4	.9612
705	4128.	749.0	771.1	.9406	3839.	758.3	779.6	.9492	3497.	774.8	795.5	.9646
710	4370.	752.6	776.1	.9437	4063.	762.0	784.5	.9524	3691.	778.7	800.6	.9679
715	4614.	756.2	781.0	.9468	4289.	765.7	789.5	.9555	3888.	782.5	805.6	.9712
720	4859.	759.8	785.9	.9499	4517.	769.3	794.4	.9586	4086.	786.3	810.5	.9744
725	5106.	763.4	790.8	.9529	4746.	772.9	799.3	.9617	4287.	790.1	815.5	.9776
730	5354.	766.9	795.7	.9559	4976.	776.6	804.2	.9647	4488.	793.9	820.4	.9807
735	5604.	770.5	800.5	.9588	5208.	780.1	809.0	.9677	4691.	797.6	825.3	.9838
740	5854.	774.0	805.4	.9617	5442.	783.7	813.9	.9707	4896.	801.2	830.2	.9869
745	6106.	777.4	810.2	.9646	5676.	787.2	818.7	.9736	5102.	804.9	835.1	.9899
750	6359.	780.9	815.0	.9675	5912.	790.7	823.5	.9765	5309.	808.5	839.9	.9929

Table 5. Critical Region

v (t Sat.)	.035 (697.10)				.040 (703.26)				.045 (704.94)			
t	p	u	h	s	p	u	h	s	p	u	h	s
Sat.	3032.	789.2	808.8	.9785	3158.	822.9	846.3	1.0100	3193.	848.5	875.1	1.0346
650	1636.9	745.9	756.5	.9403	1840.1	769.2	782.8	.9627	1964.3	789.5	805.9	.9826
655	1772.8	750.8	762.3	.9447	1953.5	774.7	789.2	.9677	2069.2	795.5	812.7	.9880
660	1911.9	755.7	768.1	.9490	2069.4	780.2	795.5	.9726	2175.6	801.3	819.4	.9932
665	2054.2	760.4	773.7	.9533	2187.6	785.5	801.7	.9773	2283.7	807.0	826.0	.9983
670	2199.4	765.1	779.4	.9574	2307.9	790.7	807.8	.9819	2393.1	812.6	832.5	1.0032
675	2347.	769.7	784.9	.9615	2430.	795.8	813.8	.9864	2504.	818.0	838.9	1.0080
680	2498.	774.2	790.4	.9655	2555.	800.8	819.7	.9908	2616.	823.4	845.2	1.0127
685	2651.	778.7	795.9	.9694	2681.	805.7	825.5	.9951	2729.	828.6	851.3	1.0173
690	2807.	783.1	801.3	.9732	2810.	810.5	831.3	.9993	2844.	833.7	857.4	1.0218
695	2965.	787.4	806.6	.9769	2939.	815.3	837.0	1.0034	2960.	838.8	863.4	1.0262
700	3125.	791.7	811.9	.9806	3071.	819.9	842.6	1.0075	3077.	843.7	869.3	1.0304
705	3287.	795.9	817.1	.9842	3204.	824.5	848.2	1.0114	3194.	848.6	875.2	1.0346
710	3450.	800.0	822.3	.9878	3339.	829.0	853.7	1.0153	3313.	853.3	880.9	1.0387
715	3616.	804.1	827.5	.9913	3475.	833.4	859.1	1.0190	3433.	858.0	886.6	1.0427
720	3784.	808.1	832.6	.9947	3612.	837.8	864.5	1.0227	3553.	862.6	892.2	1.0466
725	3953.	812.1	837.7	.9981	3751.	842.1	869.8	1.0264	3675.	867.1	897.7	1.0504
730	4124.	816.0	842.7	1.0014	3891.	846.3	875.1	1.0299	3797.	871.6	903.2	1.0541
735	4296.	819.9	847.7	1.0047	4032.	850.5	880.3	1.0334	3920.	875.9	908.6	1.0578
740	4469.	823.8	852.7	1.0079	4174.	854.6	885.5	1.0369	4044.	880.2	913.9	1.0614
745	4644.	827.6	857.7	1.0110	4317.	858.7	890.6	1.0403	4168.	884.5	919.2	1.0649
750	4820.	831.4	862.6	1.0142	4461.	862.7	895.7	1.0436	4293.	888.7	924.4	1.0684

v (t Sat.)	.050 (705.44)				.05053 (705.44)				.055 (705.13)			
t	p	u	h	s	p	u	h	s	p	u	h	s
Sat.	3203.	870.4	900.0	1.0559	3204.	872.6	902.5	1.0580	3197.	889.7	922.2	1.0750
650	2042.	808.6	827.5	1.0015	2049.	810.6	829.7	1.0035	2102.	827.0	848.4	1.0198
655	2143.	814.8	834.6	1.0071	2150.	816.8	836.9	1.0091	2200.	833.3	855.7	1.0255
660	2244.	820.9	841.7	1.0125	2251.	822.9	844.0	1.0145	2298.	839.5	862.9	1.0311
665	2347.	826.8	848.5	1.0178	2353.	828.9	850.9	1.0198	2396.	845.6	870.0	1.0365
670	2450.	832.6	855.3	1.0230	2456.	834.7	857.7	1.0250	2495.	851.5	876.9	1.0417
675	2555.	838.3	862.0	1.0280	2559.	840.4	864.3	1.0300	2594.	857.3	883.7	1.0468
680	2659.	843.9	868.5	1.0329	2664.	846.0	870.9	1.0349	2693.	863.0	890.4	1.0518
685	2765.	849.3	874.9	1.0376	2769.	851.4	877.3	1.0397	2793.	868.5	896.9	1.0566
690	2871.	854.6	881.2	1.0423	2874.	856.8	883.6	1.0444	2893.	873.9	903.4	1.0614
695	2978.	859.8	887.4	1.0468	2980.	862.0	889.9	1.0489	2993.	879.2	909.7	1.0660
700	3086.	865.0	893.5	1.0512	3087.	867.1	896.0	1.0533	3094.	884.4	915.9	1.0705
705	3194.	870.0	899.5	1.0555	3194.	872.1	902.0	1.0576	3194.	889.5	922.0	1.0748
710	3302.	874.9	905.4	1.0597	3302.	877.1	907.9	1.0619	3295.	894.5	928.1	1.0791
715	3412.	879.7	911.3	1.0638	3410.	881.9	913.8	1.0660	3396.	899.4	934.0	1.0833
720	3521.	884.4	917.0	1.0679	3519.	886.6	919.5	1.0700	3498.	904.2	939.8	1.0874
725	3631.	889.1	922.7	1.0718	3628.	891.3	925.2	1.0740	3599.	908.9	945.6	1.0914
730	3742.	893.7	928.3	1.0757	3737.	895.9	930.8	1.0778	3701.	913.6	951.2	1.0953
735	3853.	898.2	933.8	1.0794	3847.	900.4	936.4	1.0816	3803.	918.1	956.8	1.0991
740	3964.	902.6	939.3	1.0831	3957.	904.8	941.8	1.0853	3905.	922.6	962.3	1.1028
745	4076.	906.9	944.6	1.0867	4068.	909.2	947.2	1.0889	4007.	927.0	967.8	1.1065
750	4188.	911.2	950.0	1.0903	4179.	913.5	952.5	1.0925	4109.	931.3	973.2	1.1101

Table 5. Critical Region

v (t Sat.)	.06 (704.42)				.07 (701.70)				.08 (697.51)			
t	p	u	h	s	p	u	h	s	p	u	h	s
Sat.	3182.	907.0	942.3	1.0924	3125.	936.6	977.1	1.1229	3040.	960.5	1005.5	1.1484
650	2154.	844.7	868.6	1.0376	2244.	878.0	907.1	1.0712	2312.	907.9	942.2	1.1020
655	2249.	851.1	876.1	1.0433	2331.	884.3	914.5	1.0769	2391.	914.0	949.4	1.1075
660	2343.	857.4	883.4	1.0489	2418.	890.4	921.8	1.0824	2469.	920.0	956.5	1.1128
665	2437.	863.5	890.5	1.0543	2504.	896.4	928.9	1.0877	2547.	925.8	963.5	1.1179
670	2532.	869.4	897.5	1.0596	2590.	902.3	935.8	1.0929	2624.	931.4	970.3	1.1230
675	2626.	875.2	904.4	1.0647	2675.	908.0	942.7	1.0980	2701.	937.0	976.9	1.1279
680	2721.	880.9	911.1	1.0697	2760.	913.6	949.4	1.1029	2777.	942.4	983.5	1.1326
685	2815.	886.5	917.7	1.0746	2845.	919.1	956.0	1.1077	2852.	947.7	989.9	1.1373
690	2910.	891.9	924.2	1.0794	2929.	924.5	962.4	1.1124	2928.	952.9	996.2	1.1418
695	3004.	897.3	930.6	1.0840	3013.	929.7	968.8	1.1169	3002.	958.0	1002.4	1.1462
700	3099.	902.5	936.9	1.0885	3097.	934.9	975.0	1.1214	3077.	963.0	1008.5	1.1505
705	3193.	907.6	943.0	1.0929	3180.	939.9	981.1	1.1257	3151.	967.8	1014.5	1.1547
710	3287.	912.6	949.1	1.0972	3263.	944.9	987.1	1.1300	3224.	972.6	1020.4	1.1588
715	3382.	917.5	955.1	1.1014	3346.	949.7	993.1	1.1341	3297.	977.3	1026.2	1.1628
720	3476.	922.3	960.9	1.1055	3429.	954.5	998.9	1.1381	3370.	982.0	1031.8	1.1668
725	3570.	927.1	966.7	1.1095	3511.	959.1	1004.6	1.1421	3442.	986.5	1037.4	1.1706
730	3665.	931.7	972.4	1.1134	3593.	963.7	1010.2	1.1459	3515.	990.9	1043.0	1.1743
735	3759.	936.3	978.0	1.1172	3675.	968.2	1015.8	1.1497	3586.	995.3	1048.4	1.1780
740	3853.	940.8	983.6	1.1210	3757.	972.6	1021.3	1.1534	3658.	999.6	1053.7	1.1816
745	3947.	945.2	989.0	1.1246	3838.	977.0	1026.7	1.1570	3729.	1003.8	1059.0	1.1851
750	4041.	949.5	994.4	1.1282	3919.	981.2	1032.0	1.1606	3800.	1008.0	1064.2	1.1886

v (t Sat.)	.09 (692.40)				.10 (686.72)				.11 (680.76)			
t	p	u	h	s	p	u	h	s	p	u	h	s
Sat.	2938.	980.0	1028.9	1.1701	2829.	996.0	1048.4	1.1887	2719.	1009.5	1064.8	1.2049
650	2354.	934.5	973.8	1.1299	2374.	958.0	1001.9	1.1549	2374.	978.6	1027.0	1.1775
655	2425.	940.4	980.7	1.1351	2438.	963.5	1008.6	1.1599	2431.	983.9	1033.4	1.1823
660	2496.	946.0	987.6	1.1402	2501.	969.0	1015.2	1.1648	2488.	989.1	1039.7	1.1869
665	2566.	951.6	994.3	1.1451	2564.	974.3	1021.7	1.1695	2545.	994.2	1046.0	1.1914
670	2635.	957.0	1000.9	1.1499	2626.	979.4	1028.0	1.1741	2600.	999.1	1052.0	1.1958
675	2704.	962.3	1007.4	1.1546	2687.	984.5	1034.3	1.1786	2656.	1004.0	1058.0	1.2001
680	2772.	967.5	1013.7	1.1592	2748.	989.5	1040.4	1.1830	2711.	1008.7	1063.9	1.2043
685	2839.	972.6	1019.9	1.1637	2809.	994.4	1046.4	1.1872	2765.	1013.4	1069.7	1.2084
690	2906.	977.6	1026.0	1.1680	2869.	999.2	1052.2	1.1914	2819.	1018.0	1075.4	1.2124
695	2973.	982.5	1032.0	1.1723	2928.	1003.9	1058.0	1.1955	2872.	1022.5	1080.9	1.2163
700	3039.	987.3	1037.9	1.1764	2987.	1008.5	1063.7	1.1995	2925.	1026.9	1086.4	1.2201
705	3105.	992.0	1043.7	1.1805	3046.	1013.0	1069.3	1.2033	2978.	1031.2	1091.8	1.2238
710	3170.	996.6	1049.4	1.1844	3104.	1017.4	1074.8	1.2071	3030.	1035.5	1097.2	1.2275
715	3235.	1001.2	1055.0	1.1883	3162.	1021.8	1080.3	1.2109	3082.	1039.7	1102.4	1.2310
720	3299.	1005.6	1060.5	1.1920	3219.	1026.0	1085.6	1.2145	3133.	1043.8	1107.6	1.2345
725	3363.	1010.0	1066.0	1.1957	3276.	1030.2	1090.9	1.2180	3184.	1047.9	1112.7	1.2380
730	3427.	1014.3	1071.3	1.1994	3333.	1034.4	1096.1	1.2215	3235.	1051.8	1117.7	1.2413
735	3490.	1018.5	1076.6	1.2029	3389.	1038.5	1101.2	1.2249	3285.	1055.8	1122.6	1.2446
740	3553.	1022.6	1081.8	1.2064	3445.	1042.5	1106.2	1.2283	3335.	1059.6	1127.5	1.2478
745	3616.	1026.7	1087.0	1.2098	3501.	1046.4	1111.2	1.2316	3385.	1063.5	1132.4	1.2510
750	3678.	1030.8	1092.0	1.2131	3556.	1050.3	1116.1	1.2348	3435.	1067.2	1137.1	1.2541

Table 5.   Critical Region

v (t Sat.)   .12 (674.71)   .13 (668.67)   .14 (662.70)

t	p	u	h	s	p	u	h	s	p	u	h	s
Sat.	2611.	1020.8	1078.8	1.2192	2506.	1030.5	1090.8	1.2320	2406.	1038.8	1101.2	1.2434
650	*2359.*	*996.9*	*1049.2*	*1.1979*	*2332.*	*1013.0*	*1069.1*	*1.2163*	*2298.*	*1027.3*	*1086.8*	*1.2331*
655	*2411.*	*1001.9*	*1055.4*	*1.2024*	*2380.*	*1017.8*	*1075.0*	*1.2206*	*2341.*	*1031.9*	*1092.5*	*1.2372*
660	*2462.*	*1006.8*	*1061.5*	*1.2068*	*2426.*	*1022.5*	*1080.9*	*1.2249*	*2384.*	*1036.4*	*1098.2*	*1.2413*
665	*2513.*	*1011.7*	*1067.5*	*1.2111*	*2473.*	*1027.1*	*1086.6*	*1.2290*	2426.	1040.9	1103.7	1.2452
670	*2563.*	*1016.4*	*1073.3*	*1.2153*	2518.	1031.7	1092.3	1.2330	2467.	1045.2	1109.2	1.2491
675	2613.	1021.1	1079.1	1.2194	2564.	1036.1	1097.8	1.2370	2509.	1049.5	1114.5	1.2529
680	2663.	1025.6	1084.8	1.2235	2608.	1040.5	1103.3	1.2408	2550.	1053.7	1119.8	1.2566
685	2712.	1030.1	1090.3	1.2274	2653.	1044.8	1108.6	1.2446	2590.	1057.9	1125.0	1.2602
690	2760.	1034.5	1095.8	1.2312	2697.	1049.0	1113.9	1.2483	2630.	1061.9	1130.1	1.2637
695	2809.	1038.8	1101.2	1.2350	2741.	1053.2	1119.1	1.2519	2670.	1065.9	1135.1	1.2672
700	2856.	1043.1	1106.5	1.2386	2784.	1057.3	1124.2	1.2554	2709.	1069.9	1140.1	1.2706
705	2904.	1047.2	1111.7	1.2422	2827.	1061.3	1129.3	1.2588	2749.	1073.7	1145.0	1.2740
710	2951.	1051.3	1116.8	1.2457	2869.	1065.2	1134.3	1.2622	2787.	1077.6	1149.8	1.2772
715	2997.	1055.4	1121.9	1.2492	2912.	1069.1	1139.2	1.2655	2826.	1081.3	1154.5	1.2804
720	3044.	1059.3	1126.9	1.2525	2954.	1073.0	1144.0	1.2688	2864.	1085.0	1159.2	1.2836
725	3090.	1063.2	1131.8	1.2558	2995.	1076.7	1148.8	1.2720	2902.	1088.7	1163.9	1.2867
730	3136.	1067.1	1136.7	1.2591	3037.	1080.5	1153.5	1.2751	2940.	1092.3	1168.5	1.2897
735	3181.	1070.9	1141.5	1.2623	3078.	1084.1	1158.2	1.2782	2978.	1095.8	1173.0	1.2927
740	3226.	1074.6	1146.2	1.2654	3119.	1087.8	1162.8	1.2812	3015.	1099.4	1177.5	1.2956
745	3271.	1078.3	1150.9	1.2684	3160.	1091.3	1167.3	1.2842	3052.	1102.8	1181.9	1.2985
750	3316.	1081.9	1155.6	1.2715	3200.	1094.9	1171.8	1.2871	3089.	1106.3	1186.3	1.3014

v (t Sat.)   .15 (656.86)   .16 (651.16)   .17 (645.62)

t	p	u	h	s	p	u	h	s	p	u	h	s
Sat.	2312.	1046.1	1110.3	1.2538	2223.	1052.4	1118.3	1.2633	2139.	1058.0	1125.3	1.2720
650	*2258.*	*1040.0*	*1102.7*	*1.2484*	*2215.*	*1051.5*	*1117.0*	*1.2624*	2169.	1061.7	1129.9	1.2753
655	*2298.*	*1044.5*	*1108.2*	*1.2523*	2251.	1055.7	1122.4	1.2662	2202.	1065.8	1135.1	1.2790
660	2337.	1048.8	1113.7	1.2562	2287.	1059.9	1127.6	1.2700	2235.	1069.9	1140.2	1.2826
665	2375.	1053.1	1119.0	1.2600	2322.	1064.0	1132.8	1.2736	2268.	1073.9	1145.2	1.2862
670	2413.	1057.3	1124.3	1.2638	2357.	1068.1	1137.9	1.2772	2300.	1077.8	1150.2	1.2896
675	2451.	1061.4	1129.5	1.2674	2392.	1072.1	1142.9	1.2808	2332.	1081.7	1155.0	1.2930
680	2488.	1065.5	1134.6	1.2710	2426.	1076.0	1147.8	1.2842	2364.	1085.4	1159.8	1.2964
685	2525.	1069.5	1139.6	1.2745	2460.	1079.9	1152.7	1.2876	2396.	1089.2	1164.5	1.2997
690	2562.	1073.4	1144.5	1.2779	2494.	1083.7	1157.5	1.2909	2427.	1092.9	1169.2	1.3029
695	2599.	1077.3	1149.4	1.2813	2528.	1087.4	1162.2	1.2941	2458.	1096.5	1173.8	1.3060
700	2635.	1081.1	1154.2	1.2845	2561.	1091.1	1166.9	1.2973	2489.	1100.1	1178.4	1.3091
705	2671.	1084.8	1159.0	1.2878	2594.	1094.7	1171.5	1.3005	2519.	1103.6	1182.9	1.3122
710	2706.	1088.5	1163.6	1.2909	2627.	1098.3	1176.1	1.3035	2549.	1107.1	1187.3	1.3151
715	2742.	1092.2	1168.3	1.2941	2659.	1101.8	1180.6	1.3066	2579.	1110.5	1191.7	1.3181
720	2777.	1095.8	1172.8	1.2971	2692.	1105.3	1185.0	1.3095	2609.	1113.9	1196.0	1.3210
725	2812.	1099.3	1177.3	1.3001	2724.	1108.8	1189.4	1.3124	2639.	1117.3	1200.3	1.3238
730	2846.	1102.8	1181.8	1.3031	2755.	1112.2	1193.8	1.3153	2668.	1120.6	1204.6	1.3266
735	2881.	1106.3	1186.2	1.3059	2787.	1115.6	1198.1	1.3181	2698.	1123.9	1208.8	1.3294
740	2915.	1109.7	1190.6	1.3088	2819.	1118.9	1202.3	1.3209	2727.	1127.1	1212.9	1.3321
745	2949.	1113.0	1194.9	1.3116	2850.	1122.2	1206.5	1.3236	2756.	1130.4	1217.0	1.3347
750	2982.	1116.4	1199.2	1.3144	2881.	1125.4	1210.7	1.3263	2785.	1133.5	1221.1	1.3374

Table 6.   Saturation: Solid-Vapor

		Specific Volume		Internal Energy			Enthalpy			Entropy		
Temp. Fahr. t	Press. (lbf/in^2) p	Sat. Solid v_i	Sat. Vapor $v_g \times 10^{-3}$	Sat. Solid u_i	Subl. u_{ig}	Sat. Vapor u_g	Sat. Solid h_i	Subl. h_{ig}	Sat. Vapor h_g	Sat. Solid s_i	Subl. s_{ig}	Sat. Vapor s_g
32.018	.0887	.01747	3.302	−143.34	1164.6	1021.2	−143.34	1218.7	1075.4	−.292	2.479	2.187
32	.0886	.01747	3.305	−143.35	1164.6	1021.2	−143.35	1218.7	1075.4	−.292	2.479	2.187
30	.0808	.01747	3.607	−144.35	1164.9	1020.5	−144.35	1218.9	1074.5	−.294	2.489	2.195
25	.0641	.01746	4.506	−146.84	1165.7	1018.9	−146.84	1219.1	1072.3	−.299	2.515	2.216
20	.0505	.01745	5.655	−149.31	1166.5	1017.2	−149.31	1219.4	1070.1	−.304	2.542	2.238
15	.0396	.01745	7.13	−151.75	1167.3	1015.5	−151.75	1219.7	1067.9	−.309	2.569	2.260
10	.0309	.01744	9.04	−154.17	1168.1	1013.9	−154.17	1219.9	1065.7	−.314	2.597	2.283
5	.0240	.01743	11.52	−156.56	1168.8	1012.2	−156.56	1220.1	1063.5	−.320	2.626	2.306
0	.0185	.01743	14.77	−158.93	1169.5	1010.6	−158.93	1220.2	1061.2	−.325	2.655	2.330
−5	.0142	.01742	19.03	−161.27	1170.2	1008.9	−161.27	1220.3	1059.0	−.330	2.684	2.354
−10	.0109	.01741	24.66	−163.59	1170.9	1007.3	−163.59	1220.4	1056.8	−.335	2.714	2.379
−15	.0082	.01740	32.2	−165.89	1171.5	1005.6	−165.89	1220.5	1054.6	−.340	2.745	2.405
−20	.0062	.01740	42.2	−168.16	1172.1	1003.9	−168.16	1220.6	1052.4	−.345	2.776	2.431
−25	.0046	.01739	55.7	−170.40	1172.7	1002.3	−170.40	1220.6	1050.2	−.351	2.808	2.457
−30	.0035	.01738	74.1	−172.63	1173.2	1000.6	−172.63	1220.6	1048.0	−.356	2.841	2.485
−35	.0026	.01737	99.2	−174.82	1173.8	998.9	−174.82	1220.6	1045.8	−.361	2.874	2.513
−40	.0019	.01737	133.8	−177.00	1174.3	997.3	−177.00	1220.6	1043.6	−.366	2.908	2.542

Table 7. Dynamic viscosity (micropoise)

Pressure bars	lbf/in²	Temp. °C °F	0 32	50 122	100 212	150 302	200 392	250 482	300 572	350 662	375 707
1	14.504		17500	5440	121.1	141.5	161.8	182.2	202.5	223	233
5	72.52		17500	5440	2790	1810	160.2	181.4	202.3		234
10	145.04		17500	5440	2790	1810	158.5	180.6	202.2		234
25	362.6		17500	5440	2800	1820	1340	177.8	201.6		236
50	725.2		17500	5450	2800	1820	1350	1070	200.6		240
75	1087.8		17500	5450	2800	1830	1350	1080	199.2		244
100	1450.4		17500	5450	2810	1830	1360	1080	905		249
125	1813.0		17500	5460	2810	1840	1360	1090	911		254
150	2176		17400	5460	2820	1840	1370	1100	917		262
175	2538		17400	5460	2820	1850	1380	1100	924		273
200	2901		17400	5460	2830	1860	1380	1110	930	735	291
225	3263		17400	5460	2830	1860	1390	1120	936	747	491
250	3626		17400	5470	2840	1870	1390	1120	943	760	597
275	3989		17400	5470	2840	1870	1400	1130	949	772	633
300	4351		17400	5470	2850	1880	1400	1130	955	785	657
350	5076		17300	5480	2860	1890	1420	1150	968	805	693
400	5802		17300	5480	2870	1900	1430	1160	981	825	721
450	6527		17300	5490	2880	1910	1440	1170	993	837	743
500	7252		17200	5490	2890	1920	1450	1180	1010	850	762
550	7977		17200	5500	2900	1930	1460	1200	1020	860	780
600	8702		17200	5500	2910	1940	1480	1210	1030	870	795
650	9427		17200	5510	2920	1960	1490	1220	1040	882	809
700	10153		17100	5510	2930	1970	1500	1230	1060	895	822
750	10878		17100	5520	2940	1980	1510	1240	1070	905	835
800	11603		17100	5520	2950	1990	1520	1260	1080	915	846

Pressure bars	lbf/in²	Temp. °C °F	400 752	425 797	450 842	475 887	500 932	550 1022	600 1112	650 1202	700 1292
1	14.504		243	253	264	274	284	304	325	345	365
5	72.52		244	254	264	274	284	305	325	345	366
10	145.04		244	255	265	275	285	305	326	346	366
25	362.6		246	256	266	276	287	307	327	347	367
50	725.2		250	259	269	279	289	309	329	349	369
75	1087.8		253	263	273	282	292	312	332	352	372
100	1450.4		258	267	276	286	295	315	334	354	374
125	1813.0		263	271	280	289	299	318	337	357	376
150	2176		269	276	285	294	302	321	340	359	379
175	2538		276	282	290	298	307	324	343	362	381
200	2901		286	289	296	303	311	328	346	365	384
225	3263		299	298	302	309	316	332	350	368	386
250	3626		321	309	310	315	321	336	353	371	389
275	3989		367	324	320	322	327	341	357	374	392
300	4351		458	345	331	330	334	346	361	377	395
350	5076		573	416	363	351	349	357	369	385	401
400	5802		628	503	411	379	369	369	379	392	408
450	6527		664	565	468	415	393	383	389	401	415
500	7252		693	609	521	456	421	400	401	410	423
550	7977		716	643	564	497	453	418	414	420	431
600	8702		736	670	600	534	485	439	428	430	439
650	9427		754	693	629	567	516	460	442	441	448
700	10153		770	713	654	596	545	482	458	453	458
750	10878		784	732	676	621	572	504	474	466	468
800	11603		798	748	695	644	596	526	491	478	478

Conversion Factors for Viscosity

10^6 micropoise = 0.0020885 lbf × sec/ft²   = 241.91 lb/(hr × ft)
        = 0.067197 lb/(ft × sec)   = 0.58015 × 10⁻⁶ lbf × hr/ft²   = 0.1 kg/(m × sec)

Table 8. Kinematic viscosity $\times 10^3$ (cm²/sec)

Pressure bars	lbf/in²	Temp. °C / °F	0 / 32	50 / 122	100 / 212	150 / 302	200 / 392	250 / 482	300 / 572	350 / 662	375 / 707
1	14.504		17.5	5.51	205.4	273.9	351.4	438.3	534.4	640.2	695.9
5	72.52		17.5	5.51	2.91	1.98	68.1	86.1	105.7		138.9
10	145.04		17.5	5.50	2.91	1.98	32.7	42.0	52.2		68.9
25	362.6		17.5	5.50	2.91	1.98	1.55	15.47	19.94		27.1
50	725.2		17.5	5.50	2.91	1.98	1.55	1.34	9.09		13.2
75	1087.8		17.4	5.50	2.92	1.99	1.56	1.34	5.32		8.5
100	1450.4		17.4	5.49	2.92	1.99	1.56	1.34	1.26		6.11
125	1813.0		17.4	5.49	2.92	1.99	1.56	1.35	1.26		4.64
150	2176		17.3	5.49	2.92	2.00	1.57	1.35	1.26		3.64
175	2538		17.3	5.48	2.92	2.00	1.57	1.36	1.27		2.88
200	2901		17.2	5.48	2.93	2.00	1.58	1.36	1.27	1.22	2.24
225	3263		17.2	5.48	2.93	2.00	1.58	1.36	1.27	1.22	1.23
250	3626		17.2	5.47	2.93	2.01	1.58	1.37	1.27	1.22	1.18
275	3989		17.1	5.47	2.93	2.01	1.59	1.37	1.27	1.22	1.18
300	4351		17.1	5.47	2.93	2.01	1.59	1.37	1.27	1.22	1.18
350	5076		17.0	5.46	2.94	2.02	1.60	1.38	1.28	1.22	1.18
400	5802		17.0	5.45	2.94	2.03	1.61	1.39	1.28	1.22	1.19
450	6527		16.9	5.45	2.94	2.03	1.61	1.40	1.29	1.23	1.19
500	7252		16.8	5.44	2.95	2.04	1.62	1.40	1.30	1.23	1.19

Pressure bars	lbf/in²	Temp. °C / °F	400 / 752	425 / 797	450 / 842	475 / 887	500 / 932	550 / 1022	600 / 1112	650 / 1202	700 / 1292
1	14.504		754.0	814.4	880	945	1013	1154	1309	1469	1639
5	72.52		150.6	162.8	175	188	202	231	261	293	328
10	145.04		74.8	81.2	88	94	101	115	131	147	164
25	362.6		29.5	32.0	34.6	37.3	40.2	46.0	52.1	58.6	65.4
50	725.2		14.5	15.7	17.0	18.4	19.8	22.8	25.9	29.2	32.7
75	1087.8		9.3	10.3	11.2	12.1	13.1	15.1	17.2	19.4	21.8
100	1450.4		6.82	7.51	8.21	8.95	9.67	11.23	12.82	14.52	16.30
125	1813.0		5.26	5.84	6.44	7.03	7.66	8.91	10.21	11.60	13.01
150	2176		4.21	4.73	5.26	5.78	6.28	7.36	8.47	9.62	10.85
175	2538		3.44	3.93	4.40	4.86	5.33	6.25	7.23	8.23	9.27
200	2901		2.85	3.31	3.76	4.18	4.59	5.43	6.29	7.19	8.12
225	3263		2.35	2.83	3.25	3.65	4.03	4.79	5.58	6.38	7.20
250	3626		1.93	2.44	2.84	3.22	3.58	4.28	4.99	5.73	6.48
275	3989		1.54	2.11	2.51	2.87	3.20	3.86	4.52	5.19	5.89
300	4351		1.28	1.83	2.23	2.57	2.90	3.52	4.13	4.75	5.40
350	5076		1.21	1.43	1.80	2.13	2.42	2.98	3.52	4.07	4.63
400	5802		1.20	1.28	1.52	1.81	2.08	2.58	3.07	3.55	4.06
450	6527		1.20	1.24	1.37	1.59	1.82	2.27	2.72	3.17	3.62
500	7252		1.20	1.22	1.30	1.45	1.64	2.05	2.45	2.86	3.27

Conversion Factors for Kinematic Viscosity

1 cm²/sec (Stoke) = 0.001076 ft²/sec
= 3.8750   ft²/hr
= 0.36   m²/hr

Table 9. Thermal conductivity (milliwatts/m°K)

Pressure bars	lbf/in.²	Temp. °C / °F	0 / 32	50 / 122	100 / 212	150 / 302	200 / 392	250 / 482	300 / 572	350 / 662	375 / 707
1	14.504		569	643	24.8	28.7	33.2	38.2	43.4	49.0	51.9
5	72.52		569	644	681	687	33.8	38.6	43.8	49.4	52.3
10	145.04		570	644	681	687	35.1	39.3	44.4	49.9	52.8
25	362.6		571	645	682	688	665	42.9	46.5	51.6	54.3
50	725.2		573	647	684	690	668	618	52.5	55.4	57.6
75	1087.8		575	649	686	691	670	622	63.7	60.8	62.0
100	1450.4		577	651	688	693	672	625	545	68.8	67.9
125	1813.0		579	653	689	695	674	629	552	81.3	75.9
150	2176		581	655	691	696	676	633	559	104.0	87.5
175	2538		583	657	693	698	679	636	565	442	106.0
200	2901		585	659	695	700	681	639	571	454	126
225	3263		587	661	696	701	683	642	577	465	297
250	3626		589	662	698	703	685	646	582	476	376
275	3989		591	664	699	705	687	649	588	486	402
300	4351		592	666	701	706	689	652	592	496	419
350	5076		596	669	704	710	693	657	601	514	444
400	5802		599	672	707	713	697	662	609	529	468
450	6527		603	675	710	716	701	667	616	541	486
500	7252		606	678	713	720	704	671	622	552	501

Pressure bars	lbf/in.²	Temp. °C / °F	400 / 752	425 / 797	450 / 842	475 / 887	500 / 932	550 / 1022	600 / 1112	650 / 1202	700 / 1292
1	14.504		54.9	58.0	61.1	64.2	67.4	73.9	80.6	87.4	94.3
5	72.52		55.3	58.3	61.4	64.5	67.7	74.3	80.9	87.7	94.6
10	145.04		55.7	58.8	61.8	65.0	68.2	74.7	81.4	88.2	95.0
25	362.6		57.2	60.2	63.3	66.4	69.6	76.1	82.7	89.5	96.3
50	725.2		60.2	63.0	65.9	68.9	72.0	78.4	85.0	91.7	98.6
75	1087.8		63.9	66.3	68.9	71.7	74.7	80.9	87.4	94.0	101.0
100	1450.4		68.6	70.2	72.4	74.9	77.6	83.5	89.8	96	103
125	1813.0		74.5	74.9	76.4	78.4	80.8	86.3	92.4	99	105
150	2176		82.2	80.7	81.0	82.4	84.3	89.3	95.1	101	108
175	2538		92.6	87.9	86.5	86.9	88.3	92.5	98.0	104	110
200	2901		107	97	93	92	93	96	101	107	113
225	3263		130	109	101	98	97	100	104	110	115
250	3626		157	125	111	105	103	104	107	112	118
275	3989		200	147	123	113	109	108	111	115	121
300	4351		264	171	138	122	116	112	114	118	124
350	5076		351	239	182	147	132	122	122	125	129
400	5802		390	296	220	177	153	134	130	132	135
450	6527		416	338	264	210	180	148	139	139	142
500	7252		436	370	301	246	206	163	149	147	148

Conversion Factors for Thermal Conductivity

$$1000 \text{ mW/m}°K = 0.57779 \quad \text{Btu/(hr} \times \text{ft} \times °\text{F)}$$
$$= 449.62 \quad \text{lbf/(hr} \times °\text{F)}$$
$$= 0.16933 \quad \text{W/(ft} \times °\text{F)}$$
$$= 367.10 \quad \text{kgf/(hr} \times °\text{K)}$$
$$= 0.0023885 \quad \text{cal/(sec} \times \text{cm} \times °\text{K)}$$
$$= 0.85985 \quad \text{kcal/(hr} \times \text{m} \times °\text{K)}$$

Table 10. Conversion Factors
Basic Equivalents

Length: One inch = 2.54000 cm.
Mass: One pound (lb) = 453.59237 grams (g)
Force: One kilogram (kgf) = 980665 dynes
               One pound (1bf) = 0.45359237 kilograms (kgf)
Pressure: One bar = 10^6 dynes/cm^2
               One atmosphere = 1.01325 bars
Energy: One calorie (Int. Table) = 4.1868 joules
               One joule = one watt-second = 10^7 ergs
                 = 10^7 dyne-cm

Specific Entropy and Specific Heat Capacity:
               One Btu/lb deg Rankine = 1 cal/g deg Kelvin
Temperature: On the Kelvin scale the temperature of the
               triple point of water is 273.16.
               T (Rankine) = $1.8T$ (Kelvin)
               t (Celsius) = T (Kelvin) $-$ 273.15
               t **(Fahrenheit) = 32 + 1.8** t **(Celsius)**
               The Kelvin, Rankine, Celsius, and
               Fahrenheit scales used here are all
               thermodynamic scales of temperature.

Specific Volume

1 ft^3/lb = 62.428 cm^3/g   1 cm^3/g = 0.0160185 ft^3/lb.

Pressure[1]

	atm	kgf/cm^2	lbf/in^2	bar	mm Hg	in Hg	ft H$_2$O
1 atm	1*	1.03323	14.6959	1.01325*	760.000	29.9213	33.8985
1 kgf/cm^2	0.967841	1*	14.2233	0.980665*	735.559	28.9590	32.8084
10 lbf/in^2	0.680460	0.703070	10*	0.689476	517.149	20.3602	23.0666
1 bar	0.986923	1.01972	14.5038	1*	750.062	29.5300	33.4553
10^3 mm Hg	1.31579	1.35951*	19.3368	1.33322	1000*	39.3701	44.6033
10 in Hg	0.334211	0.345316	4.91154	0.338639	254*	10*	11.32925*
10^2 ft H$_2$O	2.94998	3.048*	43.3528	2.98907	2241.98	88.2671	100*

The units mm Hg, in Hg and ft H$_2$O are conventional barometric and manometric units of pressure. The conventional "mercury" and conventional "water" involved are, by definition, fluids having exactly invariable densities of 13.5951 g/cm^3 and 1 g/cm^3. Gravity is 980.665 cm/sec^2.

Energy[1]

	joule	kgf m	ft lbf	W h	cal	Btu	(lbf/in^2) $\times$ ft^3	atm cm^3	hp h
10^4 joules	10000*	1019.72	7375.62	2.77778	2388.46	9.47817	51.2196	98692.3	0.00372506
10^4 kgf m	98066.5*	10000*	72330.1	27.2407	23422.8	92.9491	502.293	967841	0.0365304
10^4 ft lbf	13558.2	1382.55	10000*	3.76616	3238.32	12.8507	69.4444	133809	0.00505051
10^{-3} kW h	3600*	367.098	2655.22	1*	859.845	3.41214	18.4391	35529.2	0.00134102
10^4 cal	41868*	4269.35	30880.3	11.63*	10000*	39.6832	214.446	413205	0.0155961
10 Btu	10550.6	1075.86	7781.69	2.93071	2519.96	10*	54.0395	104126	0.00393015
10^2(lbf/in^2) $\times$ft^3	19523.8	1990.87	14400*	5.42327	4663.17	18.5050	100*	192685	0.00727273
10^5 atm $\times$cm^3	10132.5*	1033.23	7473.35	2.81458	2420.11	9.60376	51.8983	100000*	0.00377442
10^{-3} hp h	2684.52	273.745	1980*	0.745700	641.186	2.54443	13.75*	26494.1	0.001*
10^{23} el-volts	16020.7	1633.66	11816.26	4.45020	3826.48	15.18469	82.0574	158112.0	0.00596781

The joule is the unit formerly called the "Absolute" joule. The "International" joule was abandoned on 31st December, 1947.

The calorie used in this table is the "International Table Calorie (1956)".

Specific Energy[1]

	joule / gram	kgf m / g	ft lbf / lb	W h / g	cal / g	Btu / lb	(lbf/in²) / (lb/ft³)	atm cm³ / g	hp h / lb
$10^3 \dfrac{j}{g}$	1000*	101.972	334553	0.277778	238.846	429.923	2323.28	9869.23	0.168966
$10 \dfrac{kgf\ m}{g}$	98.0665*	10*	32808.4	0.0272407	23.4228	42.1610	227.836	967.841	0.0165699
$10^6 \dfrac{ft\ lbf}{lb}$	2989.07	304.8*	1000000*	0.830296	713.926	1285.07	6944.44	29499.8	0.505051
$10^{-5} \dfrac{kW\ h}{g}$	36*	3.67098	12043.9	0.01*	8.59845	15.4772	83.6381	355.292	0.00608277
$10 \dfrac{cal}{g}$	41.868*	4.26935	14007.0	0.01163*	10*	18*	97.2712	413.205	0.00707427
$10^2 \dfrac{Btu}{lb}$	232.6*	23.7186	77816.9	0.0646111	55.5556	100*	540.395	2295.58	0.0393015
$10^4 \dfrac{(lbf/in^2)}{(lb/ft^3)}$	4304.26	438.912	1440000*	1.19563	1028.05	1850.50	10000*	42479.7	0.727273
$10^4 \dfrac{atm\ cm^3}{g}$	1013.25*	103.323	338985	0.281458	242.011	435.619	2354.07	10000*	0.171205
$1 \dfrac{hp\ h}{lb}$	5918.35	603.504	1980000*	1.64399	1413.57	2544.43	13750*	58409.6	1*

* Exact value by definition

[1] Lefevre, E. J., "Conversion Factors for Specific Energy, Energy, and Pressure" Heat Division Paper, DSIR, Aug., 1956.

Table 10 (Cont.)

Temperature Conversion. Celcius (C) to Fahrenheit (F)

°C	°F	°C	°F	°C	°F	°C	°F	°C	°F	°C	°F	°C	°F
−50	−58	150	302	350	662	550	1022	750	1382	950	1742	1150	2102
−45	−49	155	311	355	671	555	1031	755	1391	955	1751	1155	2111
−40	−40	160	320	360	680	560	1040	760	1400	960	1760	1160	2120
−35	−31	165	329	365	689	565	1049	765	1409	965	1769	1165	2129
−30	−22	170	338	370	698	570	1058	770	1418	970	1778	1170	2138
−25	−13	175	347	375	707	575	1067	775	1427	975	1787	1175	2147
−20	−4	180	356	380	716	580	1076	780	1436	980	1796	1180	2156
−15	5	185	365	385	725	585	1085	785	1445	985	1805	1185	2165
−10	14	190	374	390	734	590	1094	790	1454	990	1814	1190	2174
−5	23	195	383	395	743	595	1103	795	1463	995	1823	1195	2183
0	32	200	392	400	752	600	1112	800	1472	1000	1832	1200	2192
5	41	205	401	405	761	605	1121	805	1481	1005	1841	1205	2201
10	50	210	410	410	770	610	1130	810	1490	1010	1850	1210	2210
15	59	215	419	415	779	615	1139	815	1499	1015	1859	1215	2219
20	68	220	428	420	788	620	1148	820	1508	1020	1868	1220	2228
25	77	225	437	425	797	625	1157	825	1517	1025	1877	1225	2237
30	86	230	446	430	806	630	1166	830	1526	1030	1886	1230	2246
35	95	235	455	435	815	635	1175	835	1535	1035	1895	1235	2255
40	104	240	464	440	824	640	1184	840	1544	1040	1904	1240	2264
45	113	245	473	445	833	645	1193	845	1553	1045	1913	1245	2273
50	122	250	482	450	842	650	1202	850	1562	1050	1922	1250	2282
55	131	255	491	455	851	655	1211	855	1571	1055	1931	1255	2291
60	140	260	500	460	860	660	1220	860	1580	1060	1940	1260	2300
65	149	265	509	465	869	665	1229	865	1589	1065	1949	1265	2309
70	158	270	518	470	878	670	1238	870	1598	1070	1958	1270	2318
75	167	275	527	475	887	675	1247	875	1607	1075	1967	1275	2327
80	176	280	536	480	896	680	1256	880	1616	1080	1976	1280	2336
85	185	285	545	485	905	685	1265	885	1625	1085	1985	1285	2345
90	194	290	554	490	914	690	1274	890	1634	1090	1994	1290	2354
95	203	295	563	495	923	695	1283	895	1643	1095	2003	1295	2363
100	212	300	572	500	932	700	1292	900	1652	1100	2012	1300	2372
105	221	305	581	505	941	705	1301	905	1661	1105	2021	1305	2381
110	230	310	590	510	950	710	1310	910	1670	1110	2030	1310	2390
115	239	315	599	515	959	715	1319	915	1679	1115	2039	1315	2399
120	248	320	608	520	968	720	1328	920	1688	1120	2048	1320	2408
125	257	325	617	525	977	725	1337	925	1697	1125	2057	1325	2417
130	266	330	626	530	986	730	1346	930	1706	1130	2066	1330	2426
135	275	335	635	535	995	735	1355	935	1715	1135	2075	1335	2435
140	284	340	644	540	1004	740	1364	940	1724	1140	2084	1340	2444
145	293	345	653	545	1013	745	1373	945	1733	1145	2093	1345	2453

TABLE OF VALUES FOR INTERPOLATION IN ABOVE

1°C = 1.8°F	4°C = 7.2°F	7°C = 12.6°F	1°F = 0.55°C	4°F = 2.22°C	7°F = 3.88°C
2 = 3.6	5 = 9.0	8 = 14.4	2 = 1.11	5 = 2.77	8 = 4.44
3 = 5.4	6 = 10.8	9 = 16.2	3 = 1.66	6 = 3.33	9 = 5.00

All decimals are exact.　　　　　　　　　　　　　All decimals are repeating decimals.

Temperature Conversion. International Practical Scale* (int) to Thermodynamic (th) Fahrenheit Scales

t_{int}	$t_{th}-t_{int}$	t_{int}	$t_{th}-t_{int}$	t_{int}	$t_{th}-t_{int}$	t_{int}	$t_{th}-t_{int}$	t_{int}	$t_{th}-t_{int}$	t_{int}	$t_{th}-t_{int}$	t_{int}	$t_{th}-t_{int}$
		200	−0.014	400	0.078	600	0.184	800	0.20	1100	0.17	1500	0.62
		210	−0.011	410	0.084	610	0.188	810	0.19	1120	0.18	1600	0.83
		220	−0.008	420	0.090	620	0.191	820	0.19	1140	0.19	1700	1.06
32	−0.000	230	−0.005	430	0.096	630	0.195	830	0.18	1160	0.20	1800	1.34
40	−0.004	240	−0.002	440	0.101	640	0.198	840	0.18	1180	0.21	1900	1.65
50	−0.008	250	0.002	450	0.107	650	0.200	850	0.17	1200	0.23	2000	1.99
60	−0.011	260	0.006	460	0.113	660	0.203	860	0.17	1220	0.24	2100	2.37
70	−0.014	270	0.010	470	0.119	670	0.205	870	0.16	1240	0.26	2200	2.79
80	−0.017	280	0.014	480	0.125	680	0.206	880	0.16	1260	0.28	2300	3.24
90	−0.019	290	0.019	490	0.130	690	0.208	890	0.16	1280	0.30	2400	3.73
100	−0.020	300	0.023	500	0.136	700	0.209	900	0.15	1300	0.32		
110	−0.021	310	0.028	510	0.142	710	0.209	920	0.15	1320	0.35		
120	−0.022	320	0.033	520	0.147	720	0.210	940	0.14	1340	0.37		
130	−0.022	330	0.039	530	0.152	730	0.209	960	0.14	1360	0.40		
140	−0.022	340	0.044	540	0.157	740	0.209	980	0.14	1380	0.43		
150	−0.022	350	0.049	550	0.162	750	0.208	1000	0.14	1400	0.46		
160	−0.021	360	0.055	560	0.167	760	0.206	1020	0.14	1420	0.49		
170	−0.020	370	0.061	570	0.172	770	0.204	1040	0.15	1440	0.52		
180	−0.018	380	0.066	580	0.176	780	0.202	1060	0.15	1460	0.55		
190	−0.016	390	0.072	590	0.180	790	0.199	1080	0.16	1480	0.59		

*1948

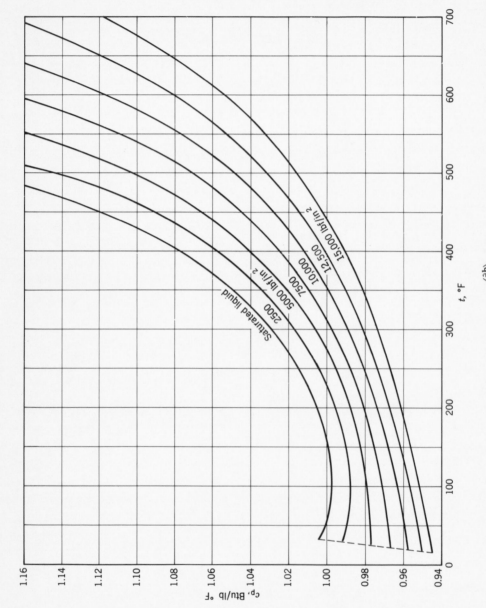

Fig. 2. Specific heat capacity of the liquid at constant pressure $\left(\frac{\partial h}{\partial t}\right)_p$. Below 150°F the curves have been smoothed, as compared with calculated values, by less than one part in 250.

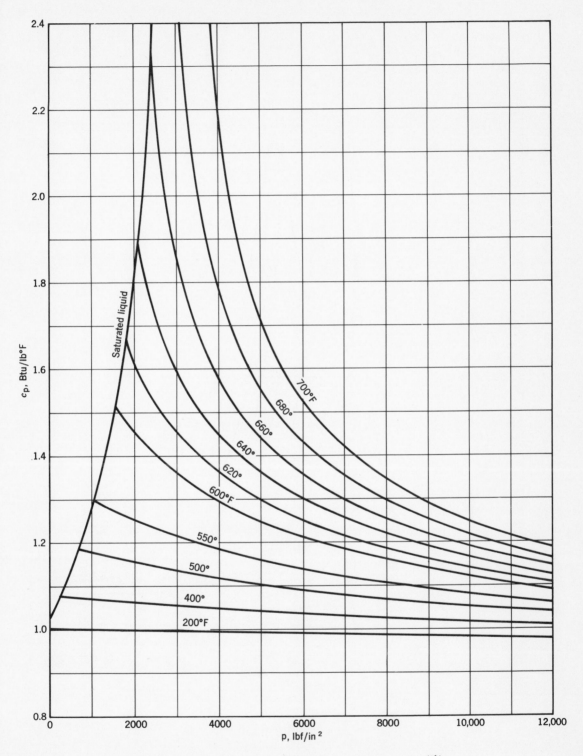

Fig. 3. Specific heat capacity of the liquid at constant pressure $\left(\dfrac{\partial h}{\partial t}\right)_p$;

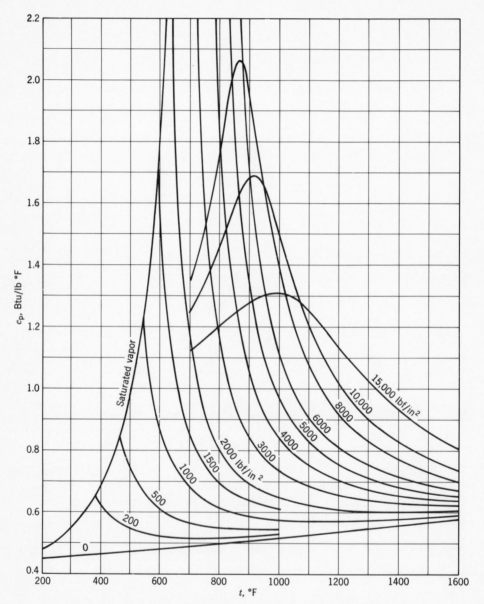

Fig. 4. Specific heat capacity of the vapor at constant pressure $\left(\dfrac{\partial h}{\partial t}\right)_p$;

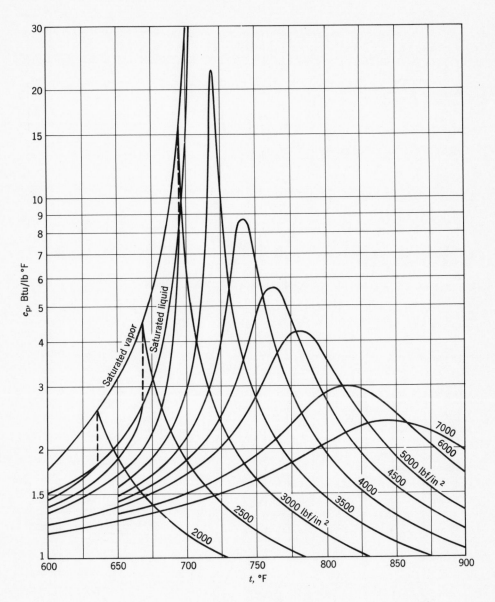

Fig. 5.   Specific heat capacity at constant pressure near the critical point $\left(\dfrac{\partial h}{\partial t}\right)_p$.

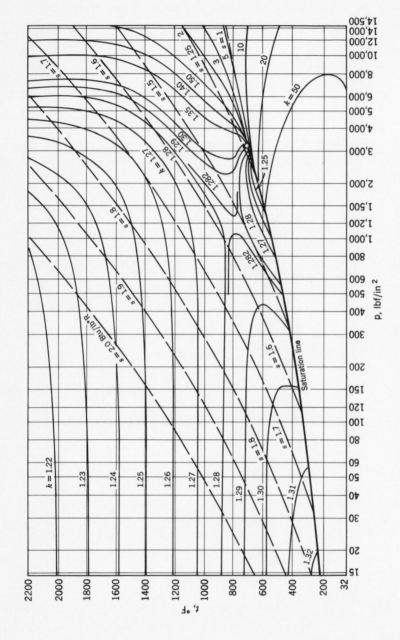

Fig. 6. Isentropic expansion exponents

$$k = -\left[\frac{\partial(\log p)}{\partial(\log v)}\right]_s = -\frac{v}{p}\left(\frac{\partial p}{\partial v}\right)_s = -\frac{v}{p}\left(\frac{\partial p}{\partial v}\right)_T \frac{c_p}{c_v}.$$

For small changes in pressure (or volume) along an isentrope, pv^k = constant.

(124)

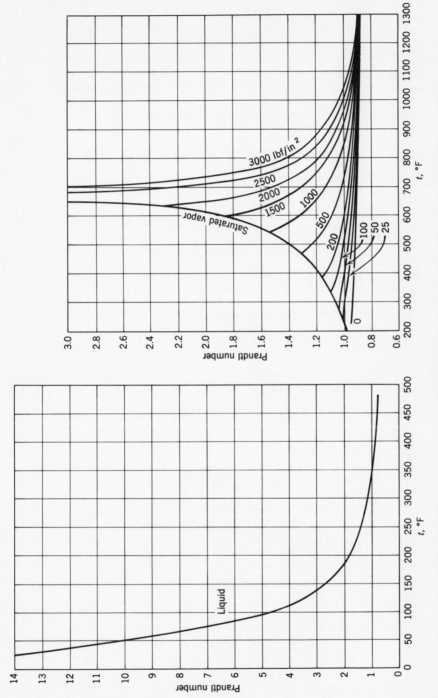

Fig. 7. (a) Prandtl number of liquid water below 1000 lbf/in.²; (b) Prandtl number of superheated water vapor.

Table 11. Logarithms to the Base 10

	0	1	2	3	4	5	6	7	8	9	10
1.00	0.0000	0004	0009	0013	0017	0022	0026	0030	0035	0039	0043
1.01	0043	0048	0052	0056	0060	0065	0069	0073	0077	0082	0086
1.02	0086	0090	0095	0099	0103	0107	0111	0116	0120	0124	0128
1.03	0128	0133	0137	0141	0145	0149	0154	0158	0162	0166	0170
1.04	0170	0175	0179	0183	0187	0191	0195	0199	0204	0208	0212
1.05	0212	0216	0220	0224	0228	0233	0237	0241	0245	0249	0253
1.06	0253	0257	0261	0265	0269	0273	0278	0282	0286	0290	0294
1.07	0294	0298	0302	0306	0310	0314	0318	0322	0326	0330	0334
1.08	0334	0338	0342	0346	0350	0354	0358	0362	0366	0370	0374
1.09	0374	0378	0382	0386	0390	0394	0398	0402	0406	0410	0414
1.10	0.0414	0418	0422	0426	0430	0434	0438	0441	0445	0449	0453
1.11	0453	0457	0461	0465	0469	0473	0477	0481	0484	0488	0492
1.12	0492	0496	0500	0504	0508	0512	0515	0519	0523	0527	0531
1.13	0531	0535	0538	0542	0546	0550	0554	0558	0561	0565	0569
1.14	0569	0573	0577	0580	0584	0588	0592	0596	0599	0603	0607
1.15	0607	0611	0615	0618	0622	0626	0630	0633	0637	0641	0645
1.16	0645	0648	0652	0656	0660	0663	0667	0671	0674	0678	0682
1.17	0682	0686	0689	0693	0697	0700	0704	0708	0711	0715	0719
1.18	0719	0722	0726	0730	0734	0737	0741	0745	0748	0752	0755
1.19	0755	0759	0763	0766	0770	0774	0777	0781	0785	0788	0792
1.20	0.0792	0795	0799	0803	0806	0810	0813	0817	0821	0824	0828
1.21	0828	0831	0835	0839	0842	0846	0849	0853	0856	0860	0864
1.22	0864	0867	0871	0874	0878	0881	0885	0888	0892	0896	0899
1.23	0899	0903	0906	0910	0913	0917	0920	0924	0927	0931	0934
1.24	0934	0938	0941	0945	0948	0952	0955	0959	0962	0966	0969
1.25	0969	0973	0976	0980	0983	0986	0990	0993	0997	1000	1004
1.26	1004	1007	1011	1014	1017	1021	1024	1028	1031	1035	1038
1.27	1038	1041	1045	1048	1052	1055	1059	1062	1065	1069	1072
1.28	1072	1075	1079	1082	1086	1089	1092	1096	1099	1103	1106
1.29	1106	1109	1113	1116	1119	1123	1126	1129	1133	1136	1139
1.30	0.1139	1143	1146	1149	1153	1156	1159	1163	1166	1169	1173
1.31	1173	1176	1179	1183	1186	1189	1193	1196	1199	1202	1206
1.32	1206	1209	1212	1216	1219	1222	1225	1229	1232	1235	1239
1.33	1239	1242	1245	1248	1252	1255	1258	1261	1265	1268	1271
1.34	1271	1274	1278	1281	1284	1287	1290	1294	1297	1300	1303
1.35	1303	1307	1310	1313	1316	1319	1323	1326	1329	1332	1335
1.36	1335	1339	1342	1345	1348	1351	1355	1358	1361	1364	1367
1.37	1367	1370	1374	1377	1380	1383	1386	1389	1392	1396	1399
1.38	1399	1402	1405	1408	1411	1414	1418	1421	1424	1427	1430
1.39	1430	1433	1436	1440	1443	1446	1449	1452	1455	1458	1461
1.40	0.1461	1464	1467	1471	1474	1477	1480	1483	1486	1489	1492
1.41	1492	1495	1498	1501	1504	1508	1511	1514	1517	1520	1523
1.42	1523	1526	1529	1532	1535	1538	1541	1544	1547	1550	1553
1.43	1553	1556	1559	1562	1565	1569	1572	1575	1578	1581	1584
1.44	1584	1587	1590	1593	1596	1599	1602	1605	1608	1611	1614
1.45	1614	1617	1620	1623	1626	1629	1632	1635	1638	1641	1644
1.46	1644	1647	1649	1652	1655	1658	1661	1664	1667	1670	1673
1.47	1673	1676	1679	1682	1685	1688	1691	1694	1697	1700	1703
1.48	1703	1706	1708	1711	1714	1717	1720	1723	1726	1729	1732
1.49	1732	1735	1738	1741	1744	1746	1749	1752	1755	1758	1761

Table 11.   Logarithms to the Base 10

	0	1	2	3	4	5	6	7	8	9	10
1.50	0.1761	1764	1767	1770	1772	1775	1778	1781	1784	1787	1790
1.51	1790	1793	1796	1798	1801	1804	1807	1810	1813	1816	1818
1.52	1818	1821	1824	1827	1830	1833	1836	1838	1841	1844	1847
1.53	1847	1850	1853	1855	1858	1861	1864	1867	1870	1872	1875
1.54	1875	1878	1881	1884	1886	1889	1892	1895	1898	1901	1903
1.55	1903	1906	1909	1912	1915	1917	1920	1923	1926	1928	1931
1.56	1931	1934	1937	1940	1942	1945	1948	1951	1953	1956	1959
1.57	1959	1962	1965	1967	1970	1973	1976	1978	1981	1984	1987
1.58	1987	1989	1992	1995	1998	2000	2003	2006	2009	2011	2014
1.59	2014	2017	2019	2022	2025	2028	2030	2033	2036	2038	2041
1.60	0.2041	2044	2047	2049	2052	2055	2057	2060	2063	2066	2068
1.61	2068	2071	2074	2076	2079	2082	2084	2087	2090	2092	2095
1.62	2095	2098	2101	2103	2106	2109	2111	2114	2117	2119	2122
1.63	2122	2125	2127	2130	2133	2135	2138	2140	2143	2146	2148
1.64	2148	2151	2154	2156	2159	2162	2164	2167	2170	2172	2175
1.65	2175	2177	2180	2183	2185	2188	2191	2193	2196	2198	2201
1.66	2201	2204	2206	2209	2212	2214	2217	2219	2222	2225	2227
1.67	2227	2230	2232	2235	2238	2240	2243	2245	2248	2251	2253
1.68	2253	2256	2258	2261	2263	2266	2269	2271	2274	2276	2279
1.69	2279	2281	2284	2287	2289	2292	2294	2297	2299	2302	2304
1.70	0.2304	2307	2310	2312	2315	2317	2320	2322	2325	2327	2330
1.71	2330	2333	2335	2338	2340	2343	2345	2348	2350	2353	2355
1.72	2355	2358	2360	2363	2365	2368	2370	2373	2375	2378	2380
1.73	2380	2383	2385	2388	2390	2393	2395	2398	2400	2403	2405
1.74	2405	2408	2410	2413	2415	2418	2420	2423	2425	2428	2430
1.75	2430	2433	2435	2438	2440	2443	2445	2448	2450	2453	2455
1.76	2455	2458	2460	2463	2465	2467	2470	2472	2475	2477	2480
1.77	2480	2482	2485	2487	2490	2492	2494	2497	2499	2502	2504
1.78	2504	2507	2509	2512	2514	2516	2519	2521	2524	2526	2529
1.79	2529	2531	2533	2536	2538	2541	2543	2545	2548	2550	2553
1.80	0.2553	2555	2558	2560	2562	2565	2567	2570	2572	2574	2577
1.81	2577	2579	2582	2584	2586	2589	2591	2594	2596	2598	2601
1.82	2601	2603	2605	2608	2610	2613	2615	2617	2620	2622	2625
1.83	2625	2627	2629	2632	2634	2636	2639	2641	2643	2646	2648
1.84	2648	2651	2653	2655	2658	2660	2662	2665	2667	2669	2672
1.85	2672	2674	2676	2679	2681	2683	2686	2688	2690	2693	2695
1.86	2695	2697	2700	2702	2704	2707	2709	2711	2714	2716	2718
1.87	2718	2721	2723	2725	2728	2730	2732	2735	2737	2739	2742
1.88	2742	2744	2746	2749	2751	2753	2755	2758	2760	2762	2765
1.89	2765	2767	2769	2772	2774	2776	2778	2781	2783	2785	2788
1.90	0.2788	2790	2792	2794	2797	2799	2801	2804	2806	2808	2810
1.91	2810	2813	2815	2817	2819	2822	2824	2826	2828	2831	2833
1.92	2833	2835	2838	2840	2842	2844	2847	2849	2851	2853	2856
1.93	2856	2858	2860	2862	2865	2867	2869	2871	2874	2876	2878
1.94	2878	2880	2882	2885	2887	2889	2891	2894	2896	2898	2900
1.95	2900	2903	2905	2907	2909	2911	2914	2916	2918	2920	2923
1.96	2923	2925	2927	2929	2931	2934	2936	2938	2940	2942	2945
1.97	2945	2947	2949	2951	2953	2956	2958	2960	2962	2964	2967
1.98	2967	2969	2971	2973	2975	2978	2980	2982	2984	2986	2989
1.99	2989	2991	2993	2995	2997	2999	3002	3004	3006	3008	3010

Table 11. Logarithms to the Base 10

These two pages give the common logarithms of numbers between 1 and 10, correct to four places. Moving the decimal point n places to the right (or left) in the number is equivalent to adding n (or $-n$) to the logarithm. Thus, log $0.017453 = 0.2419 - 2$ [$= \overline{2}.2419$].

To facilitate interpolation, the tenths of the tabular differences are given at the end of each line, so that the differences themselves need not be considered. In using these aids, first find the nearest tabular entry, and then add (to move to the right) or subtract (to move to the left), as the case may require.

	0	1	2	3	4	5	6	7	8	9	10	1	2	3	4	5
												Tenths of the Tabular Difference				
1.0	0.0000	0043	0086	0128	0170	0212	0253	0294	0334	0374	0414					
1.1	0414	0453	0492	0531	0569	0607	0645	0682	0719	0755	0792					
1.2	0792	0828	0864	0899	0934	0969	1004	1038	1072	1106	1139					
1.3	1139	1173	1206	1239	1271	1303	1335	1367	1399	1430	1461					
1.4	1461	1492	1523	1553	1584	1614	1644	1673	1703	1732	1761					
1.5	1761	1790	1818	1847	1875	1903	1931	1959	1987	2014	2041					
1.6	2041	2068	2095	2122	2148	2175	2201	2227	2253	2279	2304					
1.7	2304	2330	2355	2380	2405	2430	2455	2480	2504	2529	2553					
1.8	2553	2577	2601	2625	2648	2672	2695	2718	2742	2765	2788					
1.9	2788	2810	2833	2856	2878	2900	2923	2945	2967	2989	3010					
2.0	0.3010	3032	3054	3075	3096	3118	3139	3160	3181	3201	3222	2	4	6	8	11
2.1	3222	3243	3263	3284	3304	3324	3345	3365	3385	3404	3424	2	4	6	8	10
2.2	3424	3444	3464	3483	3502	3522	3541	3560	3579	3598	3617	2	4	6	8	10
2.3	3617	3636	3655	3674	3692	3711	3729	3747	3766	3784	3802	2	4	5	7	9
2.4	3802	3820	3838	3856	3874	3892	3909	3927	3945	3962	3979	2	4	5	7	9
2.5	3979	3997	4014	4031	4048	4065	4082	4099	4116	4133	4150	2	3	5	7	9
2.6	4150	4166	4183	4200	4216	4232	4249	4265	4281	4298	4314	2	3	5	7	8
2.7	4314	4330	4346	4362	4378	4393	4409	4425	4440	4456	4472	2	3	5	6	8
2.8	4472	4487	4502	4518	4533	4548	4564	4579	4594	4609	4624	2	3	5	6	8
2.9	4624	4639	4654	4669	4683	4698	4713	4728	4742	4757	4771	1	3	4	6	7
3.0	0.4771	4786	4800	4814	4829	4843	4857	4871	4886	4900	4914	1	3	4	6	7
3.1	4914	4928	4942	4955	4969	4983	4997	5011	5024	5038	5051	1	3	4	6	7
3.2	5051	5065	5079	5092	5105	5119	5132	5145	5159	5172	5185	1	3	4	5	7
3.3	5185	5198	5211	5224	5237	5250	5263	5276	5289	5302	5315	1	3	4	5	6
3.4	5315	5328	5340	5353	5366	5378	5391	5403	5416	5428	5441	1	3	4	5	6
3.5	5441	5453	5465	5478	5490	5502	5514	5527	5539	5551	5563	1	2	4	5	6
3.6	5563	5575	5587	5599	5611	5623	5635	5647	5658	5670	5682	1	2	4	5	6
3.7	5682	5694	5705	5717	5729	5740	5752	5763	5775	5786	5798	1	2	3	5	6
3.8	5798	5809	5821	5832	5843	5855	5866	5877	5888	5899	5911	1	2	3	5	6
3.9	5911	5922	5933	5944	5955	5966	5977	5988	5999	6010	6021	1	2	3	4	6
4.0	0.6021	6031	6042	6053	6064	6075	6085	6096	6107	6117	6128	1	2	3	4	5
4.1	6128	6138	6149	6160	6170	6180	6191	6201	6212	6222	6232	1	2	3	4	5
4.2	6232	6243	6253	6263	6274	6284	6294	6304	6314	6325	6335	1	2	3	4	5
4.3	6335	6345	6355	6365	6375	6385	6395	6405	6415	6425	6435	1	2	3	4	5
4.4	6435	6444	6454	6464	6474	6484	6493	6503	6513	6522	6532	1	2	3	4	5
4.5	6532	6542	6551	6561	6571	6580	6590	6599	6609	6618	6628	1	2	3	4	5
4.6	6628	6637	6646	6656	6665	6675	6684	6693	6702	6712	6721	1	2	3	4	5
4.7	6721	6730	6739	6749	6758	6767	6776	6785	6794	6803	6812	1	2	3	4	5
4.8	6812	6821	6830	6839	6848	6857	6866	6875	6884	6893	6902	1	2	3	4	4
4.9	6902	6911	6920	6928	6937	6946	6955	6964	6972	6981	6990	1	2	3	4	4

To avoid Interpolation in the first ten lines, use the special table on the preceding page.

Table 11. Logarithms to the Base 10

	0	1	2	3	4	5	6	7	8	9	10	1	2	3	4	5
5.0	0.6990	6998	7007	7016	7024	7033	7042	7050	7059	7067	7076	1	2	3	3	4
5.1	7076	7084	7093	7101	7110	7118	7126	7135	7143	7152	7160	1	2	3	3	4
5.2	7160	7168	7177	7185	7193	7202	7210	7218	7226	7235	7243	1	2	2	3	4
5.3	7243	7251	7259	7267	7275	7284	7292	7300	7308	7316	7324	1	2	2	3	4
5.4	7324	7332	7340	7348	7356	7364	7372	7380	7388	7396	7404	1	2	2	3	4
5.5	7404	7412	7419	7427	7435	7443	7451	7459	7466	7474	7482	1	2	2	3	4
5.6	7482	7490	7497	7505	7513	7520	7528	7536	7543	7551	7559	1	2	2	3	4
5.7	7559	7566	7574	7582	7589	7597	7604	7612	7619	7627	7634	1	2	2	3	4
5.8	7634	7642	7649	7657	7664	7672	7679	7686	7694	7701	7709	1	1	2	3	4
5.9	7709	7716	7723	7731	7738	7745	7752	7760	7767	7774	7782	1	1	2	3	4
6.0	0.7782	7789	7796	7803	7810	7818	7825	7832	7839	7846	7853	1	1	2	3	4
6.1	7853	7860	7868	7875	7882	7889	7896	7903	7910	7917	7924	1	1	2	3	4
6.2	7924	7931	7938	7945	7952	7959	7966	7973	7980	7987	7993	1	1	2	3	3
6.3	7993	8000	8007	8014	8021	8028	8035	8041	8048	8055	8062	1	1	2	3	3
6.4	8062	8069	8075	8082	8089	8096	8102	8109	8116	8122	8129	1	1	2	3	3
6.5	8129	8136	8142	8149	8156	8162	8169	8176	8182	8189	8195	1	1	2	3	3
6.6	8195	8202	8209	8215	8222	8228	8235	8241	8248	8254	8261	1	1	2	3	3
6.7	8261	8267	8274	8280	8287	8293	8299	8306	8312	8319	8325	1	1	2	3	3
6.8	8325	8331	8338	8344	8351	8357	8363	8370	8376	8382	8388	1	1	2	3	3
6.9	8388	8395	8401	8407	8414	8420	8426	8432	8439	8445	8451	1	1	2	3	3
7.0	0.8451	8457	8463	8470	8476	8482	8488	8494	8500	8506	8513	1	1	2	2	3
7.1	8513	8519	8525	8531	8537	8543	8549	8555	8561	8567	8573	1	1	2	2	3
7.2	8573	8579	8585	8591	8597	8603	8609	8615	8621	8627	8633	1	1	2	2	3
7.3	8633	8639	8645	8651	8657	8663	8669	8675	8681	8686	8692	1	1	2	2	3
7.4	8692	8698	8704	8710	8716	8722	8727	8733	8739	8745	8751	1	1	2	2	3
7.5	8751	8756	8762	8768	8774	8779	8785	8791	8797	8802	8808	1	1	2	2	3
7.6	8808	8814	8820	8825	8831	8837	8842	8848	8854	8859	8865	1	1	2	2	3
7.7	8865	8871	8876	8882	8887	8893	8899	8904	8910	8915	8921	1	1	2	2	3
7.8	8921	8927	8932	8938	8943	8949	8954	8960	8965	8971	8976	1	1	2	2	3
7.9	8976	8982	8987	8993	8998	9004	9009	9015	9020	9025	9031	1	1	2	2	3
8.0	0.9031	9036	9042	9047	9053	9058	9063	9069	9074	9079	9085	1	1	2	2	3
8.1	9085	9090	9096	9101	9106	9112	9117	9122	9128	9133	9138	1	1	2	2	3
8.2	9138	9143	9149	9154	9159	9165	9170	9175	9180	9186	9191	1	1	2	2	3
8.3	9191	9196	9201	9206	9212	9217	9222	9227	9232	9238·	9243	1	1	2	2	3
8.4	9243	9248	9253	9258	9263	9269	9274	9279	9284	9289	9294	1	1	2	2	3
8.5	9294	9299	9304	9309	9315	9320	9325	9330	9335	9340	9345	1	1	2	2	3
8.6	9345	9350	9355	9360	9365	9370	9375	9380	9385	9390	9395	1	1	2	2	3
8.7	9395	9400	9405	9410	9415	9420	9425	9430	9435	9440	9445	0	1	1	2	2
8.8	9445	9450	9455	9460	9465	9469	9474	9479	9484	9489	9494	0	1	1	2	2
8.9	9494	9499	9504	9509	9513	9518	9523	9528	9533	9538	9542	0	1	1	2	2
9.0	0.9542	9547	9552	9557	9562	9566	9571	9576	9581	9586	9590	0	1	1	2	2
9.1	9590	9595	9600	9605	9609·	9614	9619	9624	9628	9633	9638	0	1	1	2	2
9.2	9638	9643	9647	9652	9657	9661	9666	9671	9675	9680	9685	0	1	1	2	2
9.3	9685	9689	9694	9699	9703	9708	9713	9717	9722	9727	9731	0	1	1	2	2
9.4	9731	9736	9741	9745	9750	9754	9759	9763	9768	9773	9777	0	1	1	2	2
9.5	9777	9782	9786	9791	9795	9800	9805	9809	9814	9818	9823	0	1	1	2	2
9.6	9823	9827	9832	9836	9841	9845	9850	9854	9859	9863	9868	0	1	1	2	2
9.7	9868	9872	9877	9881	9886	9890	9894	9899	9903	9908	9912	0	1	1	2	2
9.8	9912	9917	9921	9926	9930	9934	9939	9943	9948	9952	9956	0	1	1	2	2
9.9	9956	9961	9965	9969	9974	9978	9983	9987	9991	9996		0	1	1	2	2

The columns 1 2 3 4 5 at the right are headed: Tenths of the Tabular Difference

Table 12. Logarithms to the Base e

These two pages give the natural (hyperbolic, or Napierian) logarithms of numbers between 1 and 10, correct to four places. Moving the decimal point n places to the right (or left) in the number is equivalent to adding n times 2.3026 (or n times $\overline{3}$.6974) to the logarithm.

1	**2.3026**	1	**0.6974–3**
2	4.6052	2	0.3948–5
3	6.9078	3	0.0922–7
4	9.2103	4	0.7897–10
5	11.5129	5	0.4871–12
6	13.8155	6	0.1845–14
7	16.1181	7	0.8819–17
8	18.4207	8	0.5793–19
9	20.7233	9	0.2767–21

	0	1	2	3	4	5	6	7	8	9	10	Tenths of the Tabular Difference 1 2 3 4 5
1.0	0.0000	0100	0198	0296	0392	0488	0583	0677	0770	0862	0.0953	10 19 29 38 48
1.1	0953	1044	1133	1222	1310	1398	1484	1570	1655	1740	1823	9 17 26 35 44
1.2	1823	1906	1989	2070	2151	2231	2311	2390	2469	2546	2624	8 16 24 32 40
1.3	2624	2700	2776	2852	2927	3001	3075	3148	3221	3293	3365	7 15 22 30 37
1.4	3365	3436	3507	3577	3646	3716	3784	3853	3920	3988	4055	7 14 21 28 34
1.5	4055	4121	4187	4253	4318	4383	4447	4511	4574	4637	4700	6 13 19 26 32
1.6	4700	4762	4824	4886	4947	5008	5068	5128	5188	5247	5306	6 12 18 24 30
1.7	5306	5365	5423	5481	5539	5596	5653	5710	5766	5822	5878	6 11 17 23 29
1.8	5878	5933	5988	6043	6098	6152	6206	6259	6313	6366	6419	5 11 16 22 27
1.9	6419	6471	6523	6575	6627	6678	6729	6780	6831	6881	0.6931	5 10 15 21 26
2.0	0.6931	6981	7031	7080	7129	7178	7227	7275	7324	7372	7419	5 10 15 20 24
2.1	7419	7467	7514	7561	7608	7655	7701	7747	7793	7839	7885	5 9 14 19 23
2.2	7885	7930	7975	8020	8065	8109	8154	8198	8242	8286	8329	4 9 13 18 22
2.3	8329	8372	8416	8459	8502	8544	8587	8629	8671	8713	8755	4 9 13 17 21
2.4	8755	8796	8838	8879	8920	8961	9002	9042	9083	9123	9163	4 8 12 16 20
2.5	9163	9203	9243	9282	9322	9361	9400	9439	9478	9517	9555	4 8 12 16 20
2.6	9555	9594	9632	9670	9708	9746	9783	9821	9858	9895	0.9933	4 8 11 15 19
2.7	0.9933	9969	0006	0043	0080	0116	0152	0188	0225	0260	1.0296	4 7 11 15 18
2.8	1.0296	0332	0367	0403	0438	0473	0508	0543	0578	0613	0647	4 7 11 14 18
2.9	0647	0682	0716	0750	0784	0818	0852	0886	0919	0953	1.0986	3 7 10 14 17
3.0	1.0986	1019	1053	1086	1119	1151	1184	1217	1249	1282	1314	3 7 10 13 16
3.1	1314	1346	1378	1410	1442	1474	1506	1537	1569	1600	1632	3 6 10 13 16
3.2	1632	1663	1694	1725	1756	1787	1817	1848	1878	1909	1939	3 6 9 12 15
3.3	1939	1969	2000	2030	2060	2090	2119	2149	2179	2208	2238	3 6 9 12 15
3.4	2238	2267	2296	2326	2355	2384	2413	2442	2470	2499	2528	3 6 9 12 14
3.5	2528	2556	2585	2613	2641	2669	2698	2726	2754	2782	2809	3 6 8 11 14
3.6	2809	2837	2865	2892	2920	2947	2975	3002	3029	3056	3083	3 5 8 11 14
3.7	3083	3110	3137	3164	3191	3218	3244	3271	3297	3324	3350	3 5 8 11 13
3.8	3350	3376	3403	3429	3455	3481	3507	3533	3558	3584	3610	3 5 8 10 13
3.9	3610	3635	3661	3686	3712	3737	3762	3788	3813	3838	1.3863	3 5 8 10 13
4.0	1.3863	3888	3913	3938	3962	3987	4012	4036	4061	4085	4110	2 5 7 10 12
4.1	4110	4134	4159	4183	4207	4231	4255	4279	4303	4327	4351	2 5 7 10 12
4.2	4351	4375	4398	4422	4446	4469	4493	4516	4540	4563	4586	2 5 7 9 12
4.3	4586	4609	4633	4656	4679	4702	4725	4748	4770	4793	4816	2 5 7 9 11
4.4	4816	4839	4861	4884	4907	4929	4951	4974	4996	5019	5041	2 4 7 9 11
4.5	5041	5063	5085	5107	5129	5151	5173	5195	5217	5239	5261	2 4 7 9 11
4.6	5261	5282	5304	5326	5347	5369	5390	5412	5433	5454	5476	2 4 6 9 11
4.7	5476	5497	5518	5539	5560	5581	5602	5623	5644	5665	5686	2 4 6 8 11
4.8	5686	5707	5728	5748	5769	5790	5810	5831	5851	5872	5892	2 4 6 8 10
4.9	5892	5913	5933	5953	5974	5994	6014	6034	6054	6074	1.6094	2 4 6 8 10

Table 12. Log$_e$ (Base e = 2.71828+)

	0	1	2	3	4	5	6	7	8	9	10	Tenths of the Tabular Difference 1	2	3	4	5
5.0	1.6094	6114	6134	6154	6174	6194	6214	6233	6253	6273	6292	2	4	6	8	10
5.1	6292	6312	6332	6351	6371	6390	6409	6429	6448	6467	6487	2	4	6	8	10
5.2	6487	6506	6525	6544	6563	6582	6601	6620	6639	6658	6677	2	4	6	8	10
5.3	6677	6696	6715	6734	6752	6771	6790	6808	6827	6845	6864	2	4	6	7	9
5.4	6864	6882	6901	6919	6938	6956	6974	6993	7011	7029	7047	2	4	6	7	9
5.5	7047	7066	7084	7102	7120	7138	7156	7174	7192	7210	7228	2	4	5	7	9
5.6	7228	7246	7263	7281	7299	7317	7334	7352	7370	7387	7405	2	4	5	7	9
5.7	7405	7422	7440	7457	7475	7492	7509	7527	7544	7561	7579	2	3	5	7	9
5.8	7579	7596	7613	7630	7647	7664	7681	7699	7716	7733	7750	2	3	5	7	9
5.9	7750	7766	7783	7800	7817	7834	7851	7867	7884	7901	1.7918	2	3	5	7	8
6.0	1.7918	7934	7951	7967	7984	8001	8017	8034	8050	8066	8083	2	3	5	7	8
6.1	8083	8099	8116	8132	8148	8165	8181	8197	8213	8229	8245	2	3	5	7	8
6.2	8245	8262	8278	8294	8310	8326	8342	8358	8374	8390	8405	2	3	5	6	8
6.3	8405	8421	8437	8453	8469	8485	8500	8516	8532	8547	8563	2	3	5	6	8
6.4	8563	8579	8594	8610	8625	8641	8656	8672	8687	8703	8718	2	3	5	6	8
6.5	8718	8733	8749	8764	8779	8795	8810	8825	8840	8856	8871	2	3	5	6	8
6.6	8871	8886	8901	8916	8931	8946	8961	8976	8991	9006	9021	2	3	5	6	8
6.7	9021	9036	9051	9066	9081	9095	9110	9125	9140	9155	9169	1	3	4	6	7
6.8	9169	9184	9199	9213	9228	9242	9257	9272	9286	9301	9315	1	3	4	6	7
6.9	9315	9330	9344	9359	9373	9387	9402	9416	9430	9445	1.9459	1	3	4	6	7
7.0	1.9459	9473	9488	9502	9516	9530	9544	9559	9573	9587	9601	1	3	4	6	7
7.1	9601	9615	9629	9643	9657	9671	9685	9699	9713	9727	9741	1	3	4	6	7
7.2	9741	9755	9769	9782	9796	9810	9824	9838	9851	9865	1.9879	1	3	4	6	7
7.3	1.9879	9892	9906	9920	9933	9947	9961	9974	9988	0001	2.0015	1	3	4	5	7
7.4	2.0015	0028	0042	0055	0069	0082	0096	0109	0122	0136	0149	1	3	4	5	7
7.5	0149	0162	0176	0189	0202	0215	0229	0242	0255	0268	0281	1	3	4	5	7
7.6	0281	0295	0308	0321	0334	0347	0360	0373	0386	0399	0412	1	3	4	5	7
7.7	0412	0425	0438	0451	0464	0477	0490	0503	0516	0528	0541	1	3	4	5	6
7.8	0541	0554	0567	0580	0592	0605	0618	0631	0643	0656	0669	1	3	4	5	6
7.9	0669	0681	0694	0707	0719	0732	0744	0757	0769	0782	2.0794	1	3	4	5	6
8.0	2.0794	0807	0819	0832	0844	0857	0869	0882	0894	0906	0919	1	2	4	5	6
8.1	0919	0931	0943	0956	0968	0980	0992	1005	1017	1029	1041	1	2	4	5	6
8.2	1041	1054	1066	1078	1090	1102	1114	1126	1138	1150	1163	1	2	4	5	6
8.3	1163	1175	1187	1199	1211	1223	1235	1247	1258	1270	1282	1	2	4	5	6
8.4	1282	1294	1306	1318	1330	1342	1353	1365	1377	1389	1401	1	2	4	5	6
8.5	1401	1412	1424	1436	1448	1459	1471	1483	1494	1506	1518	1	2	4	5	6
8.6	1518	1529	1541	1552	1564	1576	1587	1599	1610	1622	1633	1	2	3	5	6
8.7	1633	1645	1656	1668	1679	1691	1702	1713	1725	1736	1748	1	2	3	5	6
8.8	1748	1759	1770	1782	1793	1804	1815	1827	1838	1849	1861	1	2	3	5	6
8.9	1861	1872	1883	1894	1905	1917	1928	1939	1950	1961	2.1972	1	2	3	4	6
9.0	2.1972	1983	1994	2006	2017	2028	2039	2050	2061	2072	2083	1	2	3	4	6
9.1	2083	2094	2105	2116	2127	2138	2148	2159	2170	2181	2192	1	2	3	4	5
9.2	2192	2203	2214	2225	2235	2246	2257	2268	2279	2289	2300	1	2	3	4	5
9.3	2300	2311	2322	2332	2343	2354	2364	2375	2386	2396	2407	1	2	3	4	5
9.4	2407	2418	2428	2439	2450	2460	2471	2481	2492	2502	2513	1	2	3	4	5
9.5	2513	2523	2534	2544	2555	2565	2576	2586	2597	2607	2618	1	2	3	4	5
9.6	2618	2628	2638	2649	2659	2670	2680	2690	2701	2711	2721	1	2	3	4	5
9.7	2721	2732	2742	2752	2762	2773	2783	2793	2803	2814	2824	1	2	3	4	5
9.8	2824	2834	2844	2854	2865	2875	2885	2895	2905	2915	2925	1	2	3	4	5
9.9	2925	2935	2946	2956	2966	2976	2986	2996	3006	3016	2.3026	1	2	3	4	5

Appendix

Appendix

THE FUNDAMENTAL EQUATION

All values that make up the present tables of thermodynamic properties of liquid and vapor water (i.e., exclusive of values for the solid and for transport properties) were obtained from the following fundamental equation which expresses the characteristic function ψ, called the Helmholtz free energy, in terms of the independent variables density ρ and temperature on the Kelvin scale T:

$$\psi = \psi_0(T) + RT\,[\ln\rho + \rho Q(\rho,\tau)], \tag{1}$$

where

$$\psi_0 = \sum_{i=1}^{6} C_i/\tau^{i-1} + C_7 \ln T + C_8\,(\ln T)/\tau \tag{2}$$

and

$$Q = (\tau - \tau_c)\sum_{j=1}^{7} (\tau - \tau_{aj})^{j-2}\left[\sum_{i=1}^{8} A_{ij}\,(\rho - \rho_{aj})^{i-1} + e^{-E\rho}\sum_{i=9}^{10} A_{ij}\,\rho^{i-9}\right]. \tag{3}$$

In (1), (2), and (3) T denotes temperature on the Kelvin scale, τ denotes $1000/T$, ρ denotes density in g/cm^3, $R = 4.6151$ bar $cm^3/g°K$ or 0.46151 $j/g°K$, $\tau_c \equiv 1000/T_{crit} = 1.544912$, $E = 4.8$, and

$$\tau_{aj} = \tau_c(j=1) \qquad \rho_{aj} = 0.634\,(j=1)$$
$$= 2.5(j>1), \qquad\quad = 1.0\,(j>1).$$

The coefficients for ψ_0 in joules per gram are given as follows:

$C_1 = 1857.065$	$C_4 = 36.6649$	$C_7 = 46.$
$C_2 = 3229.12$	$C_5 = -20.5516$	$C_8 = -1011.249$
$C_3 = -419.465$	$C_6 = 4.85233$	

the coefficients A_{ij} are listed in Table I.

THERMODYNAMIC PROPERTIES

The advantage of using a fundamental equation is that all thermodynamic properties can be obtained through derivatives of the characteristic function. Because differentiation, unlike integration, results in no undetermined functions

Table I. The Coefficients A_{ij} in Equation 3

i/j	1	2	3	4	5	6	7
1	29.492937	−5.1985860	6.8335354	−0.1564104	−6.3972405	−3.9661401	−0.69048554
2	−132.13917	7.7779182	−26.149751	−0.72546108	26.409282	15.453061	2.7407416
3	274.64632	−33.301902	65.326396	−9.2734289	−47.740374	−29.142470	−5.1028070
4	−360.93828	−16.254622	−26.181978	4.3125840	56.323130	29.568796	3.9636085
5	342.18431	−177.31074	0	0	0	0	0
6	−244.50042	127.48742	0	0	0	0	0
7	155.18535	137.46153	0	0	0	0	0
8	5.9728487	155.97836	0	0	0	0	0
9	−410.30848	337.31180	−137.46618	6.7874983	136.87317	79.847970	13.041253
10	−416.05860	−209.88866	−733.96848	10.401717	645.81880	399.17570	71.531353

or constants, the information yielded is complete and unambiguous.

The basic relations [1] for determining values of pressure, specific internal energy, and specific entropy are as follows:

$$p = \rho^2 \left(\frac{\partial \psi}{\partial \rho}\right)_T, \tag{4}$$

$$u = \left[\frac{\partial(\psi\tau)}{\partial\tau}\right]_\rho, \tag{5}$$

$$s = -\left(\frac{\partial\psi}{\partial T}\right)_\rho. \tag{6}$$

Values for specific enthalpy and specific Gibbs free energy are found in turn from their definitions in these forms:

$$h \equiv u + pv, \tag{7}$$

and

$$\zeta \equiv \psi + pv. \tag{8}$$

Specific heat capacities at constant volume and constant pressure are found by further differentiation:

$$c_v = \left(\frac{\partial u}{\partial T}\right)_\rho \tag{9}$$

and

$$c_p = \left(\frac{\partial h}{\partial T}\right)_p = \left(\frac{\partial h}{\partial T}\right)_\rho - \frac{(\partial h/\partial\rho)_T\,(\partial p/\partial T)_\rho}{(\partial p/\partial\rho)_T}. \tag{10}$$

Substitution from (1) into (4) to (6) yields

$$p = \rho RT \left[1 + \rho Q + \rho^2 \left(\frac{\partial Q}{\partial \rho}\right)_\tau\right], \tag{11}$$

$$u = RT\,\rho\tau \left(\frac{\partial Q}{\partial\tau}\right)_\rho + \frac{d(\psi_0\tau)}{d\tau}, \tag{12}$$

and

$$s = -R\left[\ln\rho + \rho Q - \rho\tau\left(\frac{\partial Q}{\partial\tau}\right)_\rho\right] - \frac{d\psi_0}{dT}. \tag{13}$$

It follows from (7), (11), and (12) that

$$h = RT\left[\rho\tau\left(\frac{\partial Q}{\partial\tau}\right)_\rho + 1 + \rho Q + \rho^2\left(\frac{\partial Q}{\partial\rho}\right)_\tau\right] + \frac{d(\psi_0\tau)}{d\tau}. \tag{14}$$

Other properties of thermodynamic interest may likewise be expressed in terms of the function Q; for example, the second virial coefficient B is given by

$$B \equiv \left\{\left[\frac{\partial(p/\rho RT)}{\partial\rho}\right]_T\right\}_{\rho=0} = Q_{\rho=0}. \tag{15}$$

(136)

TEMPERATURE SCALES

Experimental observations on the properties of water are almost invariably reported in terms of the International Practical Temperature Scale (the I. P. scale) [2]. In order, however, to relate one kind of observation to another, it is necessary to associate these observed quantities with a thermodynamic scale of temperature. For this purpose the curve of differences between temperatures on the I. P. scale and the Celsius thermodynamic scale shown in Fig. A-1 was devised. Between 0°C and the sulfur boiling point of 444.6°C the differences in Fig. A-1 conform to the observations of James A. Beattie [3] and his colleagues. At the gold point the Celsius thermodynamic temperature was taken to be in excess of the International Practical temperature by one degree. A curve continuing the curve of Beattie and passing through +1 at the gold-point temperature was drawn as shown in Figure A-1. At the temperatures of the silver and gold points this curve falls within the range of spread of experimental determinations [4] of thermodynamic temperature.

Thermodynamic temperatures for all experimental observations used in determining the coefficients A_{ij} or in comparisons with values calculated from (1) were obtained by adding to the reported observed temperatures the corrections given by Figure A-1. Thus a measured temperature of 600°C (I.P.) is taken to be 600.096°C (thermodynamic) or 873.246°K.

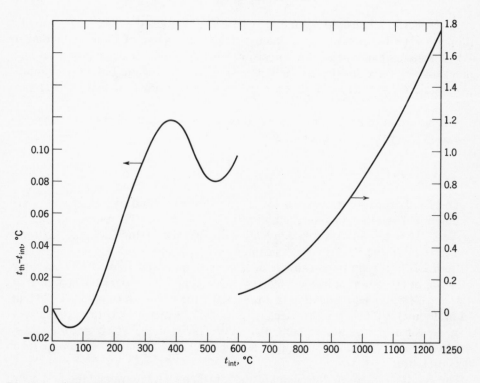

Fig. A-1. Difference between thermodynamic Celsius scale and International Practical Scale of Temperature.

Because the I.P. scale will doubtless be revised to bring it into better accord with the thermodynamic Celsius scale, it was decided to use thermodynamic Celsius temperatures. For the present, therefore, any temperature measured on an instrument calibrated in the usual way by a national calibrating laboratory should be corrected in accordance with Figure A-1 before it is used for calculating properties from (1) or for finding values from the tables. Whether or not this correction is made will in most engineering applications be a matter of indifference. Moreover, on any subsequent revision of the I.P. scale, the correspondingly revised correction should shrink to a quantity with an effect of a smaller order of magnitude than many other uncertainties in the available experimental data on water vapor.

DETERMINING COEFFICIENTS FROM EXPERIMENTAL DATA

The coefficients of the fundamental equation (1) appear in the expressions for pressure (4), energy (5), entropy (6), enthalpy (7), second virial coefficient (15), and heat capacity, (9) and (10). To determine these coefficients a number of fixed conditions smaller than the number of coefficients were established, and a large number of selected observed values were weighted and introduced into a least-squares procedure. The procedure was to minimize a quantity of the form

$$\Sigma \left\{ \frac{P_{obs} - P_{calc}}{\Delta} \right\}^2 + \lambda_1 G_1 + \lambda_2 G_2 + \cdots \tag{16}$$

with respect to all desired A_{ij} and λ_n, where P_{obs} denotes the observed value of a property P (either pressure or enthalpy), P_{calc} is the value of P corresponding to (1), Δ is an assigned measure of precision which is small for observations of high precision and large for those of low precision, λ_1, λ_2, $\cdots$ are LaGrangian multipliers, and $G_1 = 0$, $G_2 = 0$, $\cdots$ are the constraints corresponding to fixed conditions.

The selection of fixed constraints and observed values used in minimizing (16) is explained in the following paragraphs.

FIXED CONSTRAINTS

The fixed constraints satisfied by (1), which has a total of 59 constants, are outlined in Table II and shown graphically in Figure A-2. The purpose of item 1 in Table II is to conform to a convention established by the Fifth International Conference on the Properties of Steam.

Item 2 is in recognition of the detailed observations and study of the modes of behavior of the water molecule which have been carried out over the last three or four decades. The equation of item 2 represents the values of Friedman and Haar faithfully to 1300°C (2400°F). Because at 1300°C water dissociates appreciably at a pressure of one atmosphere, the values given here, which are for undissociated water vapor at this high temperature, are of limited usefulness near zero pressure.

The conditions imposed by items 1 and 2 in Table II determined the first eight coefficients in (1) as they appear in (2). These conditions therefore did not determine LaGrangian multipliers and were omitted from the minimization procedure.

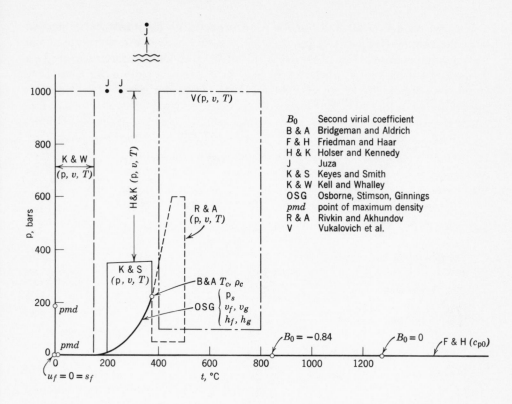

Fig. A-2. Fixed constraints and observed values selected for determining coefficients of equation (1).

A detailed and painstaking re-examination of the measurements of saturation properties by the U.S. National Bureau of Standards was made by O. C. Bridgeman and E. W. Aldrich [6]. Among their conclusions were values for temperature, density, and pressure at the critical state. Item 3 adopts these values for temperature and density. The corresponding calculated value for the pressure is 220.88 bars, compared with 220.91 bars from Bridgeman and Aldrich.

Item 4 was introduced to control the characteristics of the fundamental equation at temperatures above the highest at which observed values of pressure and density are available. It is known from kinetic theory [7] that the second virial coefficient changes sign at a temperature (the Boyle temperature) well in excess of the critical temperature. The value of that temperature chosen here is consistent with an extrapolation of second virial coefficients found from the data of Vukalovich, discussed below, and, in terms of the critical temperature, is consistent with that found by F. G. Keyes for other vapors [8].

The observed pressure-volume-temperature data for the liquid region, by their influence on the coefficients in the least-squares procedure, resulted in a line of maximum density such as liquid water is known to exhibit. The location of this line, however, was not at first in accurate accord with the observed course of the line [9, 10], in pressure and temperature. By means of item 5 the accord was made as good as the available data seem to warrant.

Table II.   Fixed Values for Determining Coefficients

1.   For the saturated liquid at the triple point, $t_{tp} = 0.01°C$, $p_{tp} = 0.006113$ bar, $u_f = 0 = s_f$.

2.   Heat capacity at zero pressure and density is given by the following equation which represents values obtained from spectroscopic observations:

$$c_{po} = 0.046\tau + 1.47276 + 0.83893/\tau$$
$$-0.219989/\tau^2 + 0.246619/\tau^3$$
$$-0.0970466/\tau^4, \text{ in joules/gram °K.}$$

3.   The critical point is fixed at the values of temperature (thermodynamic) and density given by O. C. Bridgeman and E. W. Aldrich [11]: for $t_c = 374.136°C$ and $\rho_c = 0.317$ g/cm³

$$\left(\frac{\partial p}{\partial \rho}\right)_T = 0 = \left(\frac{\partial^2 p}{\partial \rho^2}\right)_T.$$

The value of the critical pressure was not fixed.

4.   Two values of the second virial coefficient B_0 are fixed as follows: for $\tau = 0.9$ ($t \cong 840°C$), $B_0 = -0.84$ cm³/g; for $\tau = 0.65$ ($t \cong 1265.5°C$), $B_0 = 0$.

5.   Two points on the line of maximum density in the liquid region are fixed for $\rho = 1$ g/cm³, $t = 4°C$ ($p \cong 2.5$ bars); for $\rho = 1.0089$, $t = 0°C$ ($p \cong 182$ bars)

$$\left(\frac{\partial \rho}{\partial T}\right)_p = 0.$$

(The equivalent condition $(\partial p/\partial T)_\rho = 0$ was used instead as a matter of convenience.)

OBSERVED VALUES FOR LEAST-SQUARES METHOD

The observed values selected for use in the least-squares procedure are indicated in Figure A-2. The total number of observed values of all kinds was about 600.

Along the line of saturated states the specific volumes and enthalpies were taken from Osborne, Stimson, and Ginnings (O.S.G.) [11], except at temperatures below 50°C, where the liquid specific volumes were based on the data of Kell and Whalley [12]. The O.S.G. data were corrected for the temperature derivative of the difference between the I.P. temperature scale, as used by O.S.G., and the thermodynamic temperature scale.*

For the compressed liquid region the precise measurements of Kell and Whalley were used from 0 to 150°C. From 200 to 370°C to a maximum pressure of 360 bars the coefficients were controlled by the older measurements of Keyes and Smith [13]. The measurements of Holser and Kennedy [14] at 300°C to 1000 bars and those of Juza [15] at 200 and 250°C at 1000 bars and at 350°C at 1500 bars provided control at higher pressures.

In the critical region the excellent measurements of S. L. Rivkin and T. C. Akhundov [16] provided an important control which was modified, as explained

---

*At 0°C, however, the uncorrected value for vapor volume is in better accord with the calculated one. In view of the slight departure here from ideality, either the measured value of O.S.G., the effect of the temperature correction, or both must be in doubt by one part in one or two thousand.

below, by the critical point of Bridgeman and Aldrich. Extension of knowledge of properties to higher pressures and temperatures than before was represented by the extensive and high quality observations of M. P. Vukalovich and his colleagues [17]. The importance of these measurements is reflected in the large area covered by them in Figure A-2.

It will be observed that no measured values except those along the saturation line were used in the vapor region at temperatures below the critical temperature. Built into the fundamental equation (1), of course, is the condition at zero pressure that pv/RT is unity. This along with saturation values and the Rivkin data at the critical temperature provided triangular boundary conditions that by all subsequent tests proved to be sufficient to define the subcritical vapor region.

SATURATION STATES

It is characteristic of a fluid consisting of a single molecular species that two states can coexist in equilibrium over a certain range of temperature. The lower limit of this range is the triple-point temperature below which no stable equilibrium high-density fluid state exists in equilibrium with a low-density fluid state. The upper limit is the critical-point temperature above which no reversible constant-pressure, constant-temperature transition between high- and low-density states can occur. The states within this range that can coexist are called saturated-liquid and saturated-vapor states.

The conjugate saturation states could be found from (1) by identifying at each temperature those two states for which the Gibbs free energy and the pressure are identical. Alternatively, pairs of states may be found for which the pressure is the observed vapor pressure, that is, the pressure of equilibrium of coexisting states. Because, in determining the coefficients of the fundamental equation, the observed values of properties of saturation states used were consistent with equality of the Gibbs free energy at the corresponding vapor pressures, the two methods should yield substantially the same result.

A pair of saturation states at any temperature was taken to be the states of highest and lowest density corresponding to the vapor pressure given by Bridgeman and Aldrich [18] for that temperature. As a matter of convenience and for exact consistency with the critical point of (1) their values of vapor pressure were represented by the following equation:

$$\frac{p_s}{p_c} = \exp\left[\tau \, 10^{-5} (t_c - t) \sum_{i=1}^{8} F_i \, (0.65 - 0.01t)^{i-1}\right], \tag{17}$$

where p_s denotes vapor pressure, t saturation temperature (°C$_{th}$), p_c critical pressure (220.88 bars), t_c critical temperature (374.136°C$_{th}$), $\tau = 1000/T°K$, and F_i is given by the following table:

F_1	F_2	F_3	F_4	F_5	F_6	F_7	F_8
−741.9242	−29.72100	−11.55286	−0.8685635	0.1094098	0.439993	0.2520658	0.05218684

The standard deviation of the values of Bridgeman and Aldrich from (17) is about one part in 12,000; the maximum deviation is about one part in 6000.

METASTABLE STATES

The fundamental equation (1) is a representation of a continuum of equilibrium states which includes both liquid and vapor regions. Liquid and vapor states that can coexist in equilibrium at any one temperature may be identified, as explained above, by finding a pair of states for which pressure and Gibbs free energy are identical. Such a pair is made up of a saturated liquid state and a saturated vapor state. Single-phase states corresponding to (1) which are intermediate between saturated liquid and vapor states are, as in the Van der Waals equation, either metastable or unstable.

At temperatures of 300°C to the critical the curve of nonstable states between saturation states has, like the Van der Waals equation, a single maximum and a single minimum. At lower temperatures two maxima and two minima may be found. No significance is attached to this number of extrema. It is doubtless a natural consequence of having a large region of states, such as that between saturation states at low temperatures, without observed values to control the characteristics of an equation as complicated as (1). Because no measurements of properties of nonstable states are available, or likely to become available in the foreseeable future, no control in this region was considered necessary.

Nevertheless, metastable states are of sufficient engineering and scientific interest to justify offering corresponding calculated values of properties as the best values available. Properties of superheated liquid and supersaturated vapor are therefore given in italics as extensions of the tables for compressed liquid and superheated vapor. These extensions are stopped short of the first extremum at each temperature.

COMPARISON WITH OBSERVED VOLUMES

Between 0 and 220°C the calculated specific volume of the saturated liquid is generally within one part in 10,000 of the observed values of Osborne, Stimson, and Ginnings [11], as shown in Figure A-3. To a temperature of 360°C the agreement is within one part in 2000 on volume; near the critical point the agreement on pressure is within one part in 2000.

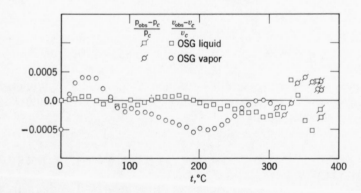

Fig. A-3. Specific volumes of saturated liquid and vapor. Comparison of observed and calculated values.

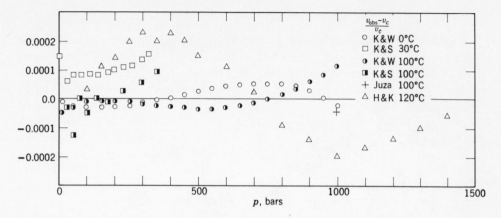

Fig. A-4. Specific volumes of compressed liquid, 0° to 120°C. Comparison of observed and calculated values.

Because the precision of measurement of properties of the vapor is less at low temperatures than for properties of the liquid, the difference between O.S.G. observations of saturated vapor volumes and calculated values is somewhat greater than for liquid. Below 320°C the volume differences are generally less than one part in 2000. Above that temperature the differences in pressure are within the same limit (Figure A-3).

In the compressed-liquid region for temperatures of 150°C and below Kell and Whalley [12] have made extremely precise measurements of pressure, volume, and temperature. Equation 1 represents these measurements within one part in 10,000 on volume (Figure A-4). Although better agreement could doubtless have been attained, further refinement of the coefficients of (1) seemed unwarranted for the present purpose.

The Keyes and Smith p-v-T measurements of nearly 40 years ago [13], which extend to a temperature of 374°C and a pressure of 350 bars, still meet high standards of quality. The agreement between these measurements and the calculated values is with few exceptions within one part in 1500 (Figure A-4, A-5, A-6, A-8).

Juza [15] made measurements at rather wide intervals at 1000 bars and higher and at 100 to 350°C by 50° intervals which were invaluable in providing some control of the coefficients of (1), particularly at high pressures above 150°C. Between 350 and 1000 bars and between 150 and 400°C the measurements by Holser and Kennedy [14] were the sole recourse. Their measurements at 300°C were used in determining the coefficients. Equation 1 represents the measurements of Juza and of Holser and Kennedy in the liquid region to within one part in 1000 to 1000 bars (Figures A-4, A-5, A-6). Although (1) was not intended for use above 1000 bars, it appears that to 350°C it may be useful to 2000 bars.

SUPERHEATED VAPOR

The p-v-T measurements of Keyes, Smith, and Gerry (K.S.G.) [19] for subcritical pressures and temperatures were not used in determining the coefficients

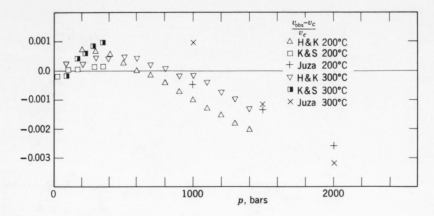

Fig. A-5. Specific volumes of compressed liquid at 200° and 300°C. Comparison of observed and calculated values.

in (1). At the lower temperatures and at specific volumes greater than about 25 cm³/g these measurements are doubtless affected by unknown amounts of adsorption of water on the inner wall of the container. The agreement between measured and calculated values is shown in Figure A-7.

NEAR THE CRITICAL TEMPERATURE

The careful measurements by Rivkin and Akhundov [16] have for the first time yielded detailed information concerning isotherms near the critical point. The differences from calculated values are shown in Figures A-6, A-8, and A-9 to be generally within one part in 1000 except for the immediate neighborhood of the critical point. There a definite N-shaped pattern appears in several isotherms. That this pattern is a result of a basic incompatibility between the Rivkin data and the Bridgeman and Aldrich critical point is indicated in Figure A-10. The ordinate of this figure is the slope of an isotherm on a pressure-density plot, that

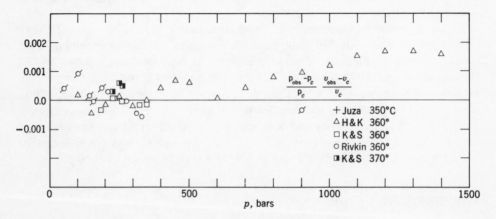

Fig. A-6. Specific volumes of compressed liquid, 350° to 370°C. Comparison of observed and calculated values.

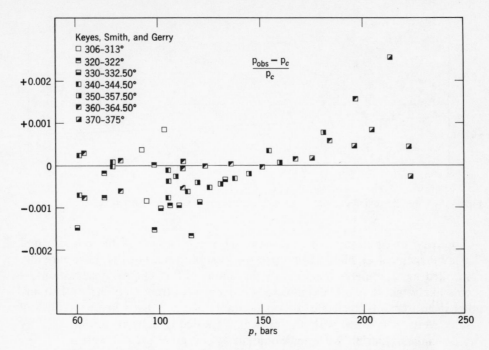

Fig. A-7. Keyes, Smith, and Gerry observations of specific volumes of vapor, 306° to 375°C. Comparison of observed and calculated values.

is, $(\partial p/\partial \rho)_T$, at the critical density. The line represents values calculated from (1). It passes through zero value of the slope at the critical temperature, 374.14°C thermodynamic (374.02°C I.P.), given by Bridgeman and Aldrich. The circles represent values obtained graphically from the Rivkin measurements at temperatures of 374.15, 375, and 380°C (I.P.). It is clear from Figure A-10 that the Rivkin isotherm at 374.15° is incompatible with the Bridgeman and Aldrich critical state and with any probable value of the critical temperature.

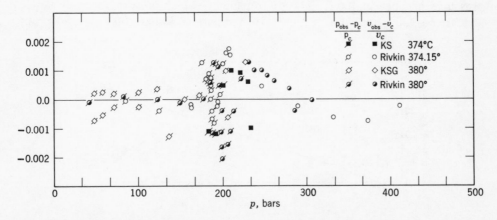

Fig. A-8. Specific volumes near the critical temperature. Comparison of observed and calculated values.

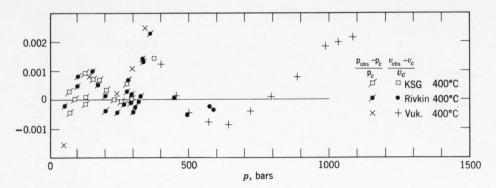

FIG. A-9.   Specific volumes at 400°C. Comparison of observed and calculated values.

Approach to equilibrium is so slow in the neighborhood of the critical point that attainment of equilibrium is open to question in all but the most carefully controlled measurements. In view of the nature of the O.S.G. observations, it seems likely that they are less affected by departures from equilibrium than any other measurements available. It was therefore decided that when experimental results were in conflict with the critical state deduced from O.S.G. data by Bridgeman and Aldrich the latter would always be given greater weight.

SUPERCRITICAL TEMPERATURES

At temperatures above the critical temperature the excellent and extensive series of measurements by M. P. Vukalovich and his colleagues [17] are the principal resource. They are supplemented by those of Keyes, Smith, and Gerry [19] to 460°C, by those of Rivkin and Akhundov [16] to 500°C, and by the very extensive but somewhat less precise measurements of Holser and Kennedy [14].

For all temperatures of 400 to 800°C the agreement with Vukalovich is generally within one part in 1000 (Figure A-9, A-12). A slightly larger discrepancy occurs at 400°C and 1000 bars, probably because of a paucity of data at high pressures just below 400°C. The discrepancies, which are progressively larger at 850 and 900°C, attain a maximum of one part in 300 at 900°C and 800 bars

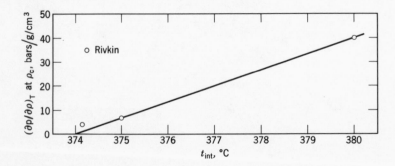

Fig. A-10.   Values of $\left(\dfrac{\partial p}{\partial \rho}\right)_T$ at the critical density. Comparison of observed and calculated values.

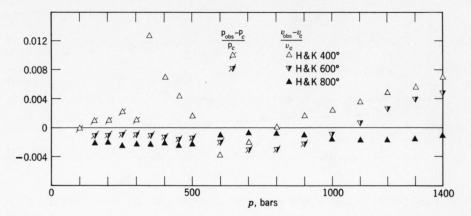

Fig. A-11. Holser and Kennedy observations of specific volumes at 400°, 600°, and 800°C. Comparison of observed and calculated values.

(Figure A-13). As explained above, the observed values at 900°C were not used in determining coefficients. At this temperature, according to the experimenters [17], the container experiences plastic deformation so that the magnitude of the enclosed volume is in doubt.

Comparisons with the data of Holser and Kennedy between 400 and 1000°C are shown in Figures A-11 and A-13.

SECOND VIRIAL COEFFICIENT

The second virial coefficient is B in the virial expansion

$$\frac{pv}{RT} = 1 + B\rho + C\rho^2 + \cdots.$$

For equations of state in general, all of which may be expanded into the virial form,

$$B \equiv \left\{ \left[\frac{\partial(pv/RT)}{\partial \rho} \right]_T \right\}_{\rho=0}$$

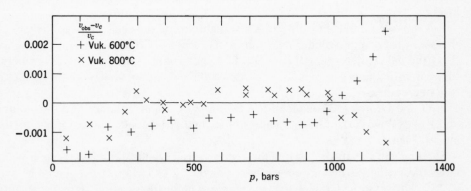

Fig. A-12. Vukalovich observations of specific volumes at 600° and 800°C. Comparison of observed and calculated values.

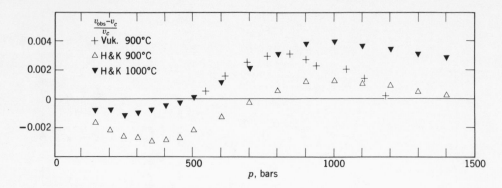

Fig. A-13. Specific volumes at 900° and 1000°C. Comparison of observed and calculated values.

Second virial coefficients are usually deduced from p-v-T measurements at low pressure. Because of adsorption effects on measurements of specific volume, this method is open to question at temperatures below the critical temperature.

In Figure A-14 calculated values of second virial coefficients are compared with those deduced from p-v-T measurements by Keyes, Smith, and Gerry [19] and by Kell, McLaurin, and Whalley [20]. The calculated values are in closer accord with the former. In 1949 Keyes [21] published an equation that gave second virial coefficients, but later high-temperature data prompted a revision that led to the equation

$$B = 2.0624 - 2.61204\tau \; 10^{(0.1008\tau^2/(1 + 0.0349\tau^2))}.$$

The values from this revised equation, shown in Figure A-14, are in close accord with the constant-temperature coefficient $(\partial h/\partial p)_T$ measured by Collins and Keyes [22] from 38 to 125°C.

The Collins and Keyes measurements, being flow measurements, should be free from the effects of adsorption. For a different reason the saturated vapor volumes, determined by O.S.G., are also free of these effects. The observations were calorimetric under circumstances in which adsorption could not be significant. Moreover, at low temperatures the saturated vapor volume must closely determine the second virial coefficient because of the small deviation from ideality.

If it is assumed that an isotherm is a straight line on the chart of pv/RT versus ρ, the saturated vapor volume determines the second virial coefficient. In Figure A-14 the values determined from O.S.G. vapor volumes are plotted for temperatures of 150°C and less. They converge on the calculated values as the temperature decreases.

Goff and Gratch [23] deduced from the O.S.G. vapor volumes and the Collins and Keyes flow measurements values of second virial coefficients at low temperatures (Figure A-14).

Stockmayer's analysis [24] of the second virial coefficient of water as a polar molecule yielded excellent agreement with the K.S.G. values between 127 and 477°C with assumed molecular constants which are in good agreement with other evidence. An equally good agreement could doubtless be found with the present calculated values.

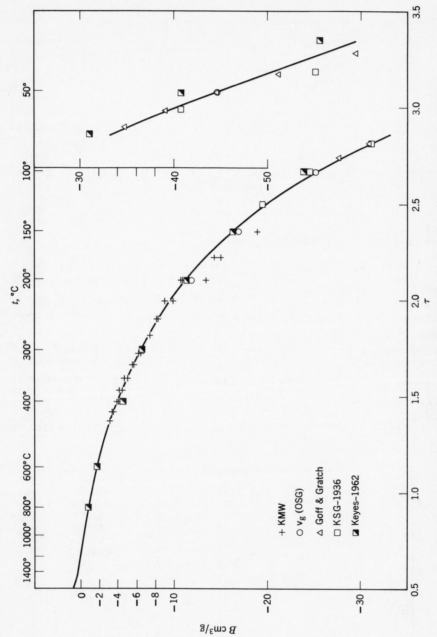

Fig. A-14. Second virial coefficient. Comparison of values deduced from observation with calculated values.

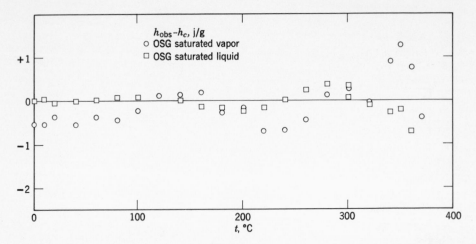

Fig. A-15. Osborne, Stimson, and Ginnings observations of enthalpy of saturated liquid and vapor. Comparison of observed and calculated values.

COMPARISONS WITH OBSERVED ENTHALPIES

The O.S.G. values of enthalpy of saturated liquid and saturated vapor are compared with calculated values in Figure A-15. Between 0 and 150°C the saturated-liquid discrepancies are less than 0.1 j/g. From 0 to 370°C for both saturation states they are in general less than 1 j/g, which for the vapor is approximately one part in 2500.

In the major experimental program on the properties of steam in the 1920's and 1930's enthalpy of superheated vapor was measured by Havlicek and Miskowsky [25] and by Egerton and Callendar [26]. The methods, which consisted of measuring heat on condensation to a liquid state, were similar.

In Figure A-16 these observations are compared with calculated values for 1 to 25 bars. The two sets of observations are in accord with each other, but they show

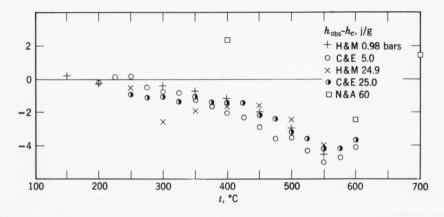

Fig. A-16. Enthalpy of vapor, 0.98 to 60 bars. Comparison of observed and calculated values.

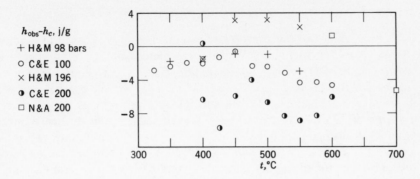

$h_{obs}-h_c$, j/g
+ H&M 98 bars
○ C&E 100
× H&M 196
◐ C&E 200
□ N&A 200

Fig. A-17. Enthalpy of vapor, 98 to 200 bars. Comparison of observed and calculated values.

a difference from calculated values that increases with increasing temperature to a maximum of about 5 j/g at 550°C. It is clear from the constancy of this difference as the pressure decreases toward zero that the observed values are discordant with the calculated values of enthalpy at zero pressure. This discord, which is an incompatibility between the enthalpy observations and the spectroscopically observed heat capacity at zero pressure, was pointed out by F. G. Keyes in 1949 [21].

The comparisons are extended to higher pressures in Figures A-17 and A-18. Also shown are comparisons with the recent excellent measurements by Newitt and Angus (N and A) [27]. At the lowest N and A observed pressure of 60 bars the differences indicate no conflict with the spectroscopic observations (Figure A-16). At 200 to 800 bars the N and A differences are of both signs and of reasonable magnitude. The observed difference in enthalpy between 800 and 1000 bars at 400°C is inordinately large in view of the relatively moderate values of derivatives in this region. This difference is in conflict with the observed p-v-t values of Vukalovich.

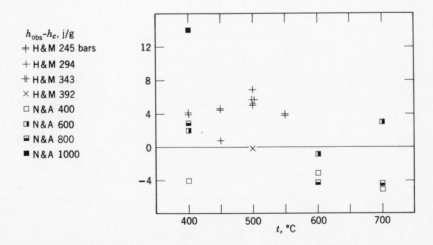

$h_{obs}-h_c$, j/g
+ H&M 245 bars
+ H&M 294
⊞ H&M 343
× H&M 392
□ N&A 400
◩ N&A 600
◪ N&A 800
■ N&A 1000

Fig. A-18. Enthalpy of vapor, 245 to 1000 bars. Comparison of observed and calculated values.

COMPARISONS WITH THROTTLING EXPERIMENTS

No observed values of the Joule-Thomson coefficient were used to determine the coefficients of (1). In Figure A-19 the calculated values are compared with the measurements of Davis and Kleinschmidt (Davis and Keenan [28]) of more than 40 years ago. The agreement is excellent except for low pressures at which the observed values decline with pressure as zero pressure is approached. At temperatures of 225°C and above the observed values indicate nearly straight isotherms at all pressures above 3 bars. Below this pressure a downward trend develops as if some new influence were coming into play with decreasing density. Because the behavior of vapors generally simplifies as the intermolecular dis-

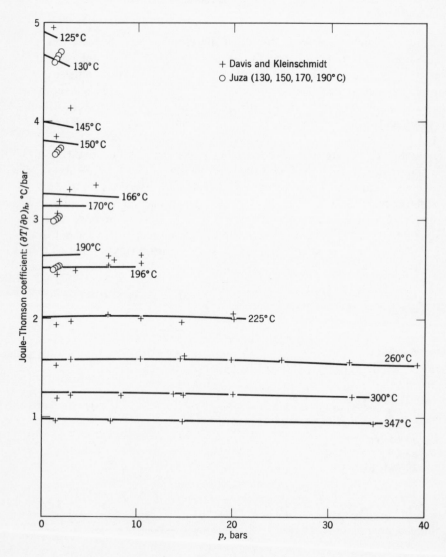

Fig. A-19. Joule-Thomson coefficients, 125° to 347°C. Comparison of observed and calculated values.

tances increase, this trend is open to a suspicion of some kind of experimental error.

The suspicion is strengthened by the exaggeration of this trend in observed values at lower temperatures at which Juza et al. have recently made some measurements [29]. These measurements are in good accord with those of Davis and Kleinschmidt. Although the calculated values for each isotherm are on the average compatible with the observations, the trend of the calculated isotherm with pressure is slightly less than zero, whereas the observed is substantially greater. It can be shown that adoption of an approximately linear isotherm passing through the observed values would not correspond to the O.S.G. enthalpies of the saturated vapor. Only by adopting a strongly curved inverted-U-shaped isotherm could observed values of the Joule-Thomson coefficient and O.S.G. saturated-vapor enthalpies be brought into accord. Taken all together, the evidence seems to cast doubt on the observed slopes of all isotherms in Figure A-19 at low pressures.

For temperatures of 38 to 125°C Collins and Keyes [22] measured the isothermal coefficient $(\partial h/\partial p)_T$ in a throttling experiment and extrapolated to zero pressure. From 60 to 125°C the observed values confirm the calculated ones (Figure A-20). Because the Joule-Thomson coefficient is simply the negative of the quotient of the isothermal coefficient and the heat capacity, this confirmation extends to the calculated Joule-Thomson coefficient. The small discrepancy between observed and calculated values at 38°C is of little significance in view of the small range of pressure for vapor states at that temperature.

COMPARISONS WITH OBSERVED HEAT CAPACITY

Over several decades measurements of heat capacity initiated by Professor Oscar Knoblauch [30, 31] were continued in Germany and were extended to 200 atm by Koch [32]. More recently an extensive series of measurements has been carried out by Sirota and Maltsev [33]. None of these were used in determining the coefficients of (1).

At low pressures the calculated values represent only a slight modification of the zero-pressure spectroscopic measurements and should therefore be highly reliable. The measurements of Knoblauch and Winkhaus [30] at pressures of less than 2 bars are in good accord.

At pressures approaching 200 atm the calculated values rise a little above the the Koch observations near saturation (Figure A-21) but they represent accurately the more recent values of Sirota and Maltsev at 200 kg/cm² (Figure A-22). In fact the accord with the Sirota and Maltsev values at high pressures is extremely good, as shown in Figure A-22. Even near the critical point where the heat capacity rises to infinity the agreement is good. As shown in Figure A-23, a displacement of a fraction of a degree in temperature is all that is needed to bring perfect agreement.

It has been pointed out how difficult it is to attain an equilibrium state in measurements near the critical state, even in static measurements such as p-v-T observations. Because heat-capacity measurements are observations on a flowing fluid, it seems unlikely that equilibrium will prevail for states near the critical

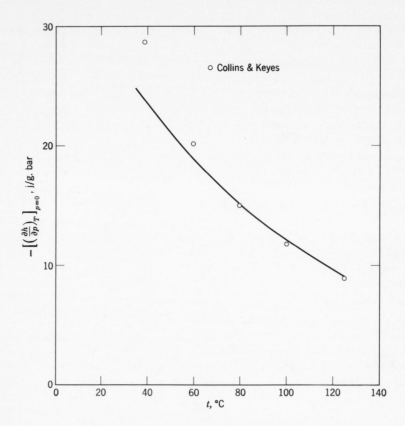

Fig. A-20. Collins and Keyes observations of $\left(\frac{\partial h}{\partial p}\right)_T$ at zero pressure. Comparison of observed and calculated values.

point. It is quite possible that a formulation based heavily in this region on the nonflow O.S.G. measurements and reflecting the p-v-T measurements beyond a short radius from the critical point would be at least as reliable as any flow measurement in this difficult region.

COMPARISON WITH SKELETON TABLES OF 1963

The present tables are generally in good accord with the Skeleton Tables of the Sixth International Conference on the Properties of Steam of 1963 [34], but because the fundamental equation (1) is the product of an entirely independent study of the available experimental data, which was completed subsequent to the Sixth Conference, some areas of small disagreement should be expected to appear.

All states for which the calculated specific volume differs from the corresponding value in the Skeleton Tables by more than the Skeleton Table tolerance are given in Table III. Complete tables with deviations appear in [35]. Here the calculated value of specific volume is given for each tabulated state; immediately below that is the Skeleton Table value minus the calculated value, followed by

Table III. Comparisons with International Skeleton Tables of 1963

Liquid and vapor saturation points outside the tolerance

Temp deg C	Pressure Bars	Enthalpy j/g liquid
30.		125.75 −.09 dev .08 tol −.03 osg
40	0.073791 −.000041 dev .000038 tol −.000013 b,a	
50	0.123419 −.00007 dev .00006 tol −.00002 b,a	
374.02 +.13 dev .10 tol 0.0 b,a	220.89 −.3 dev .1 tol +.03 b,a	

Specific volume cm³/g points outside the tolerance

T deg C / P bars	300	350	375	400	425
500	1.28620 +.002 dev 001 tol +.0005 h, k				
600	1.26828 +.002 dev .001 tol +.0005 h, k				
700	1.25237 +.0007 dev .001 tol +.0002 h, k	1.37541 +.002 dev .003 tol +.0007 h, k*			
750		1.36224 +.005 dev .003 tol +.001 h, k*	1.44112 +.005 dev .004 tol −.001 h, k*		
800		1.35003 +.005 dev .003 tol +.001 h, k*	1.42423 +.006 dev .004 tol −.001 h, k*		
850		1.33864 +.006 dev .004 tol +.001 h, k*	1.40874 +.006 dev .004 tol +.0003 hk*	1.49427 +.004 dev .003 tol .0006 v*	1.57222 +.004 dev .003 tol +.0005 v*
900		1.32798 +.006 dev .004 tol +.001 h, k*	1.39444 +.007 dev .004 tol +.001 h, k*	1.47468 +.005 dev .003 tol +.001 v*	1.54769 +.004 dev .003 tol +.001 v*
950		1.31797 +.006 dev .004 tol +.001 h, k*	1.38116 +.007 dev .004 tol +.002 h, k*	1.45677 +.006 dev .003 tol +.002 v*	1.52549 +.005 dev .003 tol +.001 v*
1000		1.30853 +.005 dev .004 tol +.001 h, k*	1.36879 +.007 dev .004 tol +.002 h, k*	1.44029 +.007 dev .003 tol +.003 v*	

Specific volume cm³/g points outside the tolerance

T deg C / P bars	25	425	450	475	500	550	600
400	125.17 −.2 dev .1 tol −.03 ksg*						
450	130.16 −.2 dev .1 tol −.03 ksg*	2.53609 +.01 dev .009 tol +.004 r*					
500				3.82495 −.011 dev .010 tol +.002 r*	3.89456 −.011 dev .008 tol +.0003 v*		
550					3.34982 −.008 dev .007 tol +.0002 v*		
600					2.95734 −.007 dev .006 tol −.0003 v*		
650						3.55104 −.008 dev .007 tol −.002 v*	
700						3.22824 −.007 dev .006 tol −.003 v*	
850							3.16178 −.007 dev .006 tol −.002 v*

Specific volume cm³/g points outside the tolerance

T deg C / P bars	150	200	250
50	1.08744 +.0004 dev .0003 tol +.00002 k, w	1.08852 −.0006 dev .0005 tol −.0008 h, k	
100	1.08414 +.0005 dev .0004 tol +.00005 k, w	1.08561 −.0008 dev .0005 tol −.0010 h, k*	
125	1.08252 +.0005 dev .0004 tol +.00007 k, w	1.08276 −.0010 dev .0005 tol −.0012 h, k	
300			1.21054 +.0006 dev .0005 tol +.0007 k, s*

Values are from Equation 1

dev Sixth International Conference value minus Equation 1
tol Sixth International Conference tolerance

Observed deviations given are observed minus calculated

b,a Bridgeman and Aldrich
h,k Holser and Kennedy
j   Juza
k,s Keyes and Smith
ksg Keyes, Smith and Gerry
k,w Kell and Whalley
osg Osborne, Stimson and Ginnings
r   Rivkin
v   Vukalovich
*   Interpolated

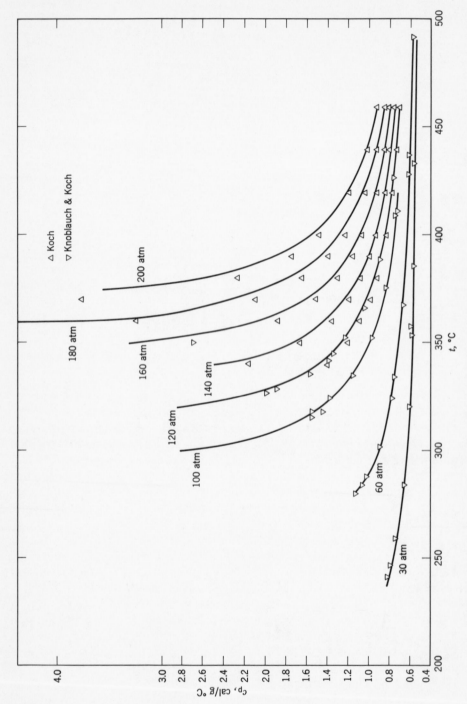

Fig. A-21. Specific heat capacity, 30 to 200 bars. Comparison of observed and calculated values.

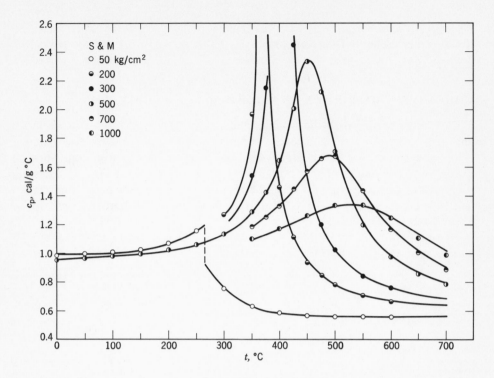

Fig. A-22. Sirota and Maltsev observations of specific heat capacity, 50 to 1000 kg/cm². Comparison of observed and calculated values.

the value of the tolerance in the Skeleton Tables. Finally there is an observed value for that state minus the calculated value.

From Table III it appears that the observed values are in substantially better agreement with (1) than with the Skeleton Tables. At 200°C and 900 to 1000 bars, since Table III shows that the spread of observed values from different experimenters exceeds the stated tolerances, an increase in tolerance seems called for. Because the volume at 250°C and 300 bars lies in a region in which the observations scatter by one part in 1000 on either side of the calculated values, an increase in tolerance again seems called for.

All calculated values of enthalpy but one saturation value differ from the Skeleton Table values by an amount usually less than but occasionally equal to the Skeleton Table tolerance. The largest actual differences appear at the highest pressures and temperatures, where they are less than one part in 300. Two calculated values of vapor pressure and one of saturated liquid enthalpy fall outside the tolerances. These values are given in Table III which shows that in each instance the observed value is in better agreement with the calculated value than with that from the Skeleton Table. The calculated critical temperature, which here is that of Bridgeman and Aldrich, is also outside the tolerance.

Equation 1 provides a critical test of the values of the International Skeleton Tables, as would any independent correlation of the available experimental data. In general it confirms the values of the Skeleton Tables of 1963, although it suggests that small modifications of Skeleton Table specific-volume values should be made. It also suggests that improvements could be made by small modifications of the Skeleton Table enthalpy values at high pressures and temperatures. In some parts of these tables an increase in tolerances seems to be required.

COMPARISON WITH KEENAN AND KEYES TABLES OF 1936

The amounts by which the Keenan and Keyes tables [36] differ from (1) are shown in [35]. Volume differences are generally small fractions of 1 percent. They become as large as 1 percent only above 4000 psi and are at a maximum at about 1300°F, being smaller at both lower and higher temperatures. A somewhat similar characteristic appears in enthalpy differences. They attain a maximum above 4000 psi in excess of 10 Btu/lb at about 1000°F, where the enthalpy is about 1400 Btu/lb.

The upper limit of observed specific volumes available to Keenan and Keyes in 1936 was at 460°C, or 860°F. The extrapolation above this temperature by means

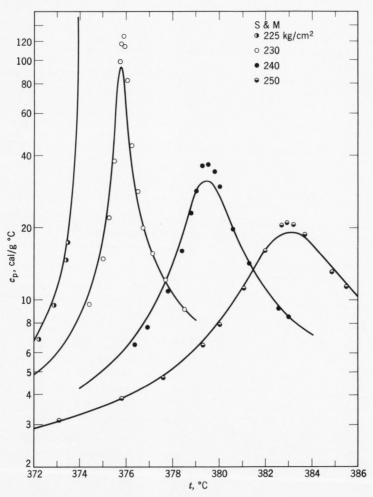

Fig. A-23. Sirota and Maltsev observations of specific heat capacity near the critical point. Comparison of observed and calculated values.

of (13) and (15) in [36] resulted in an increasing error in volume to about 1200°F and a decreasing error from 1200 to 1600°F. At the highest temperature of extra-polation the error is less than ½ percent for all pressures below 3500 psi.

Absence of the "stretching correction" in the Gordon values of heat capacity at zero pressure resulted in slightly lower values of enthalpy at zero pressure for the Keenan and Keyes tables. Errors in the p-v-T relation had the fortunate effect of compensating for the zero-pressure errors over most of the vapor region. The line of zero error in enthalpy stretches diagonally across the vapor region and results in excellent agreement between the old and the new tables except for the immediate neighborhood of the upper bound of pressure in the older tables. Agreement at the critical point in both enthalpy and volume is remarkably good.

TRANSPORT PROPERTIES

For reasons given in the Preface, values for viscosity and thermal conductivity are taken essentially from the Skeleton Tables prepared by a committee of the Sixth International Conference on the Properties of Steam [37]. The thermal con-ductivity of steam requires further investigation and correlation along with detailed exploration of the critical region at least to 50 Celsius degrees above and below the critical point.

References

[1] G. N. Hatsopoulos and J. H. Keenan, *Principles of General Thermodynamics*, Wiley, New York, 1965, p. 256.

[2] H. F. Stimson, *J. Res. Nat. Bur. Stand.*—A. Physics and Chemistry, **65A**, No. 3, 139—145 (May-June 1961).

[3] J. A. Beattie, *Proc. Am. Acad. Arts Sci.*, **77**, 255—335 (1949).

[4] H. F. Stimson, *J. Nat. Bur. Stand.*, 131 (May-June 1961).

[5] A. S. Friedman and L. Haar, *J. Chem. Phys.*, **22**, 2051—2058 (1954).

[6] O. C. Bridgeman and E. W. Aldrich, *Trans. ASME*, 266—274 (May 1965).

[7] J. O. Hirschfelder, C. F. Curtiss, and R. B. Bird, *Molecular Theory of Gases and Liquids*, Wiley, New York, 1954, p. 164.

[8] F. G. Keyes, *Temperature, Its Measurement and Control in Science and Industry*, Reinhold, New York, 1941, p. 59.

[9] N. E. Dorsey, *Properties of Ordinary Water Substance*, Reinhold, New York, 1940, pp. 275—277.

[10] R. J. Zaworski and J. H. Keenan, *Trans. ASME; J. Appl. Mech.*, **34**, Series E, 478—483 (1967).

[11] N. S. Osborne, H. F. Stimson and D. C. Ginnings, Thermal Properties of Saturated Water and Steam, *Nat. Bur. Stand., Res. Paper RP1229*, 1939.

[12] G. S. Kell and E. Whalley, *Phil. Trans. Roy. Soc. (London)*, **258**, 565—617 (1965).

[13] F. G. Keyes and L. B. Smith, *Proc. Am. Acad. Arts Sci.*, **69**, 285—314 (1934).

[14] W. T. Holser and G. C. Kennedy, *Am. J. Sci.*, **256**, 744—753 (1958), and **257**, 71—77 (1959).

[15] J. Juza, V. Kmonicek and O. Sifner, Appendix to J. Juza. An Equation of State for Water and Steam, *Acad., Naklad. Cesk. Acad. Ved*, Prague, 131—142 (1966).

[16] S. L. Rivkin and T. C. Akhundov, *Teploenerg.*, No. 1, 57—65 (1962), and No. 9, 66—68 (1963).

[17] M. P. Vukalovich, V. N. Zubarev, and A. A. Alexandrov, *Teploenerg.*, No. 10, 79—85 (1961), and No. 1, 49—51 (1962).

[18] O. C. Bridgeman and E. W. Aldrich, *Trans. ASME, J. Heat Transfer Series C-D*, **86**, 279—286 (1964).

[19] F. G. Keyes, L. B. Smith and H. T. Gerry, *Proc. Am. Acad. Arts Sci.*, **70**, 319—364 (1936).

[20] G. S. Kell, G. E. McLaurin and E. Whalley, Report No. 8316, Division of Applied Chemistry, National Research Council, Canada, 1966.

[21] F. G. Keyes, *J. Chem. Phys.*, **17**, 923—934 (1949).

[22] S. C. Collins and F. G. Keyes, *Proc. Am. Acad. Arts Sci.*, **72**, 283—299 (1938).

[23] J. Goff and S. Gratch, *Trans. A.S.H.V.E.*, **51**, 125—158 (1945).

[24] W. H. Stockmayer, *J. Chem. Phys.*, **9**, 398—402 (1941).

[25] J. Havlicek and L. Miskowsky, *Hel. Phys. Acta.*, **9**, 161—207 (1936).

[26] A. Egerton and G. S. Callendar, *Phil. Trans. Roy. Soc. (London)*, **252**, 133—164 (1960).

[27] S. Angus and D. M. Newitt, *Phil. Trans. Roy. Soc. (London)*, **259**, 107—132 (1966).

[28] Reported in H. N. Davis and J. H. Keenan, *World. Eng. Cong. Rept.* No. 455, Tokyo, 1929.

[29] J. Juza, V. Kmonicek and K. Schovanec, document of Mechanical Engineering Research Institute of the Czechoslovak Academy of Sciences, Prague, 1963.

[30] O. Knoblauch and A. Winkhaus, *Mitt. Forsch. Gebiete. Ingenieurw.*, **195**, 1–20 (1917), and *Z. Ver. deutsch. Ingenieure*, **59**, 1915, pp. 376–379, 400–405.

[31] O. Knoblauch and W. Koch, *Z. Ver. deutsch. Ingenieure*, **72**, 1733–1739 (1928).

[32] W. Koch. *Forsch. Gebiete Ingenieurw*, **3**, 1–10, 189 (1932).

[33] A. M. Sirota and B. K. Maltsev. Teploenerg., No. 1, 52–57 (1962), No. 5, 64 (1963) and No. 8, 61 (1966).

[34] Proceedings of the Sixth International Conference on the Properties of Steam, October 1963, issued by ASME, New York.

[35] F. G. Keyes, J. H. Keenan, P. G. Hill and J. G. Moore, "A Fundamental Equation for Liquid and Vapor Water." Paper presented at the Seventh International Conference on the Properties of Steam, Tokyo, Japan, 1968.

[36] J. H. Keenan and F. G. Keyes, *Thermodynamic Properties of Steam*, Wiley, New York, 1936.

[37] Proceedings of the Sixth International Conference on the Properties of Steam, Supplementary Release on Transport Properties, November 1964.